The Moon

Resources, Future Development, and Settlement

David G. Schrunk, Burton L. Sharpe,
Bonnie L. Cooper and Madhu Thangavelu

The Moon

Resources, Future Development, and Settlement

Second Edition

 Springer

Published in association with
Praxis Publishing
Chichester, UK

Dr David G. Schrunk MD
Poway
California
USA

Dr Bonnie L. Cooper
Oceaneering Space Systems
Houston
Texas
USA

Mr Burton L. Sharpe
Lebanon
Illinois
USA

Mr Madhu Thangavelu
Conductor
Space Exploration Architectures
Concept Synthesis Studio
Department of Astronautics and Space Technology
Viterbi School of Engineering and
    The School of Architecture
University of Southern California
California
USA

SPRINGER–PRAXIS BOOKS IN SPACE EXPLORATION
SUBJECT *ADVISORY EDITOR*: John Mason, M.Sc., B.Sc., Ph.D.

ISBN 978-0-387-36055-3 Springer Berlin Heidelberg New York

Springer is part of Springer-Science + Business Media (springer.com)

Library of Congress Control Number: 2007921270

Cover design: Jim Wilkie
Typesetting: Originator Publishing Services Ltd, Gt Yarmouth, Norfolk, UK

Printed on acid-free paper

Contents

This book is dedicated to the next generation of space adventurers:

Owen James, age 4, Tucson, Arizona,

> *"I would like to see the stars there."*
> *"I would like to bring my boat there."*

Olivia James, age 7, Tucson, Arizona,

> *"I would like to see the flag and how it looks."*
> *"I would like to float around."*

Brigitte Schrunk, age 9, Poway, California,

> *"I would like to do ballet on the Moon and leap clear across the stage."*

Erik Schrunk, age 10, Poway, California,

> *"I would like to play basketball on the Moon and make a slam dunk every time."*
> *"I would like to fly on the Moon."*

Chloe Saras Thangavelu, age 12, Palos Verdes, California,

> *"I would like to be on the US Volleyball team for the inaugural Lunar Olympics in 2032. Lunar gravity and genetic engineering will make me a fine competitor at 44 years old."*

Chelsea Manon Lakshmi Thangavelu, age 9, Palos Verdes, California,

> *"I want to set up the first spare parts factory for the train and the rover. I want to be the first lunar millionaire."*

O'Paul Roy Thangavelu, age 3, Palos Verdes, California,

... on seeing the illustration of the 345 Maglev train gliding down the lunar highlands into the mares:

> *"Can I ride in it?"*

Preface

In the first edition of this book published in 1999, we outlined the process by which the Moon could be transformed into an inhabited sister planet of the Earth during the twenty-first century (the "Planet Moon Project"). The rationale for our proposal was that the Moon is the logical place for the next stage of human activity in space: it is our closest celestial neighbor and it has an abundance of resources that can be used to support space exploration and development projects. In the eight years since the publication of the first edition, much has happened to bolster our view that Planet Moon will become a reality in the coming decades:

- Valuable experience has been gained with human operations in space, particularly with the International Space Station.
- An increasing number of nations have developed space launch capabilities and have expressed interest in lunar exploration programs, including manned programs.
- By White House directive, the U.S. space effort is now re-aligned to return humans to the Moon by 2020 as the first step towards the exploration of Mars and beyond.
- Commercial enterprises are becoming increasingly involved in the design and operation of space systems, including propulsion systems and habitats.
- Advances in information technology, particularly the Internet, have enabled knowledge to be widely available to people everywhere.
- Robotics technology, which will play an increasingly important role in all phases of human exploration in space, continues to make advances.
- Upcoming unmanned lunar missions by space agencies and commercial enterprises promise to yield increasing knowledge that will prepare the way for the establishment of permanent lunar settlements.

As a result of continuing advances in the space sciences, technologies, and related disciplines, we decided to write this updated edition of *The Moon*. We reorganized the middle chapters of the book (Chapters 5 through 8) to describe four phases of lunar development from the advent of the initial lunar landers to the realization of a globally-inhabited "Planet Moon". We also substituted "settlement" for "colonization" in the title of the book.

We did not attempt to cost the Planet Moon Project because we believe that such an effort would be futile. The concepts and projects presented in these chapters do not have precisely calculable precedents and it is not possible to provide credible estimates of the cost of establishing a new civilization in space. While the investments in the initial phases of exploration and development of the Moon will be high, we strongly believe that the long term benefits, for virtually every aspect of the human condition, will far outweigh the costs.

The polar regions of the Moon will likely play a strategic role in the first phases of lunar exploration and development. The summit of Mons Malapert, according to our calculations, always has the Earth and Shackleton Crater (at the lunar south pole) in view for direct and continuous high-bandwidth communications. Moreover, it may receive sunlight for solar power generation for as much as 90 percent of the lunar year. These favorable characteristics will have to be verified by analysis of imaging data from satellites (such as SMART-1 and the planned Lunar Reconnaissance Orbiter), and by lander missions that obtain "ground truth" information. If our calculations are confirmed, Mons Malapert would offer great advantages for establishing a permanent foothold on the Moon.

In the first edition, our choice for the location of the first permanent lunar base was adjacent to "Newton Crater" in the south polar region – hence the name, "Newton Base". However, our determination of the advantages of Mons Malapert led us to select the summit of that mountain as the site of the first base. Perhaps we should have named it "Malapert Base", but we decided to retain the original name from the first edition. Thus, "Newton Base" continues to be the designated name for the first base, but its location has been moved, in this second edition, to the summit of Mons Malapert.

In the final chapter we offer a speculative vision of how the industrialized Moon might be used to benefit the people of the Earth and to explore space. Given the rapid pace of knowledge accumulation and technological advancements, the realities of space exploits will undoubtedly be much different than our speculations – but we suggest that our estimates may be too conservative. The advent of television and the development of vaccines for polio, the unleashing the power of the atom, the breaking the sound barrier by a manned aircraft, the launch of the first Earth satellite, the first landing of humans on the Moon, and the advent of the personal computer, lasers, genetic engineering, and the Internet all occurred within the lifetimes of the authors. What advances will occur in the lifetime of the next set of explorers on the space frontier? We anticipate that change will probably occur more rapidly than anyone now expects and that the benefits of that change will be greater than any of us now predict.

Acknowledgements

We give special thanks to Steve Durst of Space Age Publishing for his advice and generous support in the preparation of this publication, and to Paul DiMare, our illustrator, whose artistry is truly unique and inspirational. We sincerely appreciate the advice, assistance, and vision provided by the following: (at NASA) Chris Culbert, Tom Simon, Sue Wentworth, David McKay, Carl Allen, Kriss Kennedy, Wendell Mendell, Doug Cooke, Larry Kuznetz, Simon "Pete" Worden, Larry Kellogg, and Paul Lowman, and (at Oceaneering) Mark Gittleman, Kent Copeland, Andrew Curtis, James Galbraith, and Frank Sager. Many thanks also to Manny Pimenta, David Livingstone, Jim Benson, Peter Kokh, Boris Fritz, Tom Matula, Dennis Laurie, and Phil Harris.

Many leaders and visionaries in the field of space research have contributed to our book through personal communications or through their work with scientific organizations, conferences and publications. At the university level we wish to recognize the following leaders in space research and innovation: Professors John Logsdon of George Washington University, Paul Spudis of Johns Hopkins University, Haym Benaroya of Rutgers, N.M. Komerath of Georgia Institute of Technology, David Akin of the University of Maryland, Larry Bell, Alex Ignatiev, and David Criswell of the University of Houston, John Lewis, Ron Greeley and Phil Sadler of the University of Arizona, Mike Duke of Colorado School of Mines, William "Red" Whittaker of Carnegie Mellon, Gerald Kulsinski, University of Wisconsin, Eligar Sadeh, University of North Dakota, Graham Dorrington, Queen Mary's College, London, Bruce Lusignan of Stanford, Ed Crawley and Dava Neuman of MIT, Harvey Wichman of Claremont Graduate School, Tom Shih of Iowa State University, George Morganthaler at the University of Colorado at Boulder, Sally Ride and Michael Wiskerchen of the University of California at San Diego, and Michael Gruntman, Firdaus Udwadia, Robert Brodsky, Darryl Judge, Doug Noble, and Marc Schiler at the University of Southern California (USC).

At the policy level, President Bush presented a bold plan for the United States to establish settlements on the Moon ("Moon, Mars and Beyond") and NASA administrator Michael Griffin is aligning NASA's resources and efforts to fulfill that vision. John Marburger, the White House director of the Office of Science and Technology Policy (OSTP), has stated the goal of bringing the solar system within humankind's economic sphere of influence. Our work has been influenced by the United States' presidential commissions, including the Stafford Synthesis Group, which proposed concepts for lunar return, the more recent Gehman commission which studied the safety of the national space transportation system, and the Aldridge commission, which recommended bold new approaches for NASA's return to the Moon.

We wish to note the works of the statesman and scholar Prof. A.P.J Abdul Kalam, the president of India, who has articulated the positive role of space activities in the advancement of humankind. We also recognize Prof. U.R. Rao, Narendra Bhandari and Hemant Dave of the Indian space program, Bernard Foing of ESA, Ouyang Ziyuan of the Chinese space program, Kohtaro Matsumoto and Hitoshi Mizutani of the Japanese space program, Viacheslav "Slava" Ivashkin of Russia, and Jesus Raygoza of the emerging Mexican space program. Sir Arthur Clarke continues to offer creative visions and fresh concepts, and Apollo astronauts Buzz Aldrin, Harrison Schmitt, and John Young are active proponents of lunar exploration and development.

Corporations such as Boeing, Lockheed Martin, Northrop Grumman,Loral and Oceaneering Space Systems are engineering leaders in the space systems arena and Alenia Spazio, Deutche Aerospace, Daimler, EADS, BAE, and Marconi should also be mentioned. Corporations such as Bechtel, Fluor, Jacobs, Parsons, Halliburton, Caterpillar, and Shimizu, with experience in large infrastructure development projects, have long had an interest in extraterrestrial economic activity. In the USA, the Aerospace Corp. plays a unique role in evaluating competing concepts and certifying complex space projects. Equally important, space entrepreneurs who are creating new companies, such as SpaceDev, Bigelow Aerospace, Rocketplane Kistler, SpaceX, Virgin Intergalactic, and many others, are blazing a trail for the future through their pioneering efforts in commercializing space transportation.

We recognize the following professional organizations, societies, and foundations, which have been a source of knowledge and ideas for our book: American Institute of Aeronautics and Astronautics, American Astronautical Society, International Aerospace Federation, International Space University, British Interplanetary Society, International Lunar Exploration Working Group, Space Resources Roundtable, National Space Society, Planetary Society, Mars Society, Moon Society, Space Tourism Society, Space Studies Institute, SETI Institute, Sasakawa International Center for Space Architecture, National Space Foundation, Ansari-X Prize Foundation, Sir Arthur Clarke Foundation, and Heinlein Foundation.

Finally, we extend sincere thanks to Clive Horwood, Mike Shayler, and Neil Shuttlewood for their expertise and patience in the preparation of the manuscript, and to Sijia Schrunk and Catherine Girardey Thangavelu.

Foreword

*by Buzz Aldrin**

With the United States setting the goal of returning humans to the Moon by 2020, we are at long last on track for the rebirth of crewed space exploration. Space exploration is not an endless circling of Earth; it is going to other worlds. But why return to the Moon? Haven't we been there, done that?

Hardly. There is still much to learn from the Moon, resources to utilize, and unlimited economies to launch. An observatory on the back side of the Moon would be a giant leap for astronomy, but there is a far more important reason to return. As when youths leave home for college, the Moon can become a schooling place, a stepping-stone to the boundless horizons of human destiny.

We return to the Moon to practice living off the extraterrestrial land and to test not only engineering systems but also political and social prerequisites. With the experience gained from research stations on the Moon, people from Earth will one day walk the ancient river valleys of Mars, dive the ice seas of Europa, climb the Great Wall of Miranda, and cross the far edge of the solar system.

The continued exploration of the solar system is a challenge that can unite nations, inspire youth, advance science, and ultimately end our confinement to one fragile planet.

The familiar photo that Neil Armstrong snapped of me as I stood on the surface of the Moon has become a popular icon, not because the Moon itself was some kind of culmination but because it suggests the open-ended future that awaits humanity,

---

* A version of this foreword was originally published in the January/February 2007 special "Back to the Moon" issue of *The Planetary Report*, the periodical of The Planetary Society. The Planetary Society is calling for the inauguration of the "International Lunar Decade" (ILD) this year. The ILD emphasizes the coordination of space exploration plans by the spacefaring nations and the creation of opportunities for participation by the not-yet-spacefaring nations. For more information visit *http://planetary.org/programs/projects/lunar_decade/*.

No other Apollo photograph has been reproduced as often as this portrait of Buzz Aldrin taken by Neil Armstrong during the first Moon landing. Neil is, of course, visible in reflection on Buzz's visor. Image courtesy NASA.

poised on the threshold of space. Beyond robotics and Earth-serving space stations lies the infinite journey.

We covered the globe in the old millennium and will inhabit the solar system in the new. Escaping dependence on one vulnerable world, we will found new cultures and new species of awareness, spreading consciousness into the cosmos.

But a lasting human presence in space won't result from sudden leaps like Apollo; it must move outward on a broad base of permanent support. This will require the cooperation not only of nations but also of the public and private sectors. Any permanent presence on the Moon should extend beyond government and NASA. Private industry and quasi-private consortia can help pay the costs of expansion and share in the benefits.

It is hard to know which industries will initially find such investments worthwhile. One showing immediate promise is space tourism—the one industry in which private investments in space are now being made.

I have championed access to space for tourists, not only in the hope that more people can share the adventure of which I was privileged to be a part but also in the belief that public and private interests, working side by side, will benefit from a more affordable space infrastructure.

I am delighted to introduce this visionary book on lunar exploration and settlement by David Schrunk, Burton Sharpe, Bonnie Cooper, and Madhu Thangavelu. The authors portray a broad range of ideas—each promising an optimistic and exciting future for all of humanity, based on logical and technically feasible extensions of human space activity today.

I have pondered many of these visions myself and have seen some of them develop over the years in Madhu's graduate studio at the University of Southern California. I am confident that many of these ideas will become a reality through the coordinated efforts of the global space community.

Four decades ago, the Cold War was the catalyst for a two-nation race to the Moon. Today, with the Vision for Space Exploration in America, Aurora in Europe, GLOBE and Soyuz in Russia, *SELENE* in Japan, *Chang'E* and Shenzhou in China, *Chandrayaan* in India, and the efforts of private entrepreneurs everywhere, we will fulfill the promise of the plaque Neil and I left on the Moon 38 years ago.

In the 21st century, we will truly go "in peace for all mankind."

Buzz Aldrin

On July 20, 1969, Apollo 11 astronaut Buzz Aldrin became one of the first humans to walk on the Moon. Since that day, Aldrin has remained at the forefront of efforts to ensure a continued leading role for the United States in human space exploration. He founded a rocket design company, Starcraft Boosters, Inc., and the ShareSpace Foundation, a nonprofit organization devoted to opening the doors to space tourism for all people.

Figures

Tables

Author biographies

David Schrunk is an aerospace engineer and medical doctor. His work experience includes periods at the US Army Aeromedical Research Laboratory, NASA Manned Spacecraft Center, and McDonnell-Douglas Astronautics, and a medical career in nuclear medicine and diagnostic radiology. He is the founder of the Quality of Laws Institute in Poway, California.

Bonnie L. Cooper received her Ph.D. in lunar geology and geophysics from the University of Texas in 1992 and joined Oceaneering Space Systems in 1997. She has researched planetary resources and remote-sensing topics with the goal of identifying optimum sites for a lunar outpost. Currently, Dr. Cooper is supporting NASA Johnson Space Center's Astromaterials Research and Exploration Science Group.

Burton L. Sharpe participated in mission operations planning and execution of the Gemini (Agena Target Vehicle), Apollo (ALSEP), and Viking NASA space programs. He looks forward to the day when teams of tele-operated devices on the Moon, controlled from multiple locations worldwide, will build moonbases and assist human crews with their operation.

Madhu Thangavelu conducts the Space Exploration Architectures Concept Synthesis Studio in the Astronautics and Space Technology Department of the Viterbi School of Engineering and the School of Architecture at USC. He is an Alumnus and frequent lecturer of the International Space University, and has published original concepts of space system architectures. He is former Vice Chairman for Education Programs in the Los Angeles Section of the American Institute of Aeronautics and Astronautics.

Introduction

I.1 THE SPACEFARING AGE

The promise of the Spacefaring Age – the era of human activities in space – is that it will link human expertise to the unlimited resources of space. When that linkage is established, the people of Earth will be supplied with an abundance of energy and material resources from space; the solar system will be explored in depth; human settlements will be established on multiple planets and moons; and voyages to the stars will be undertaken. The present "closed Earth" mindset – of limits to human potential related to the depletion of Earth's resources and the vital need to preserve Earth's environment – will be replaced by a much grander "open space" vision of broad-scale human advancements based upon access to unlimited resources and the opening of endless frontiers.

Within the coming decades, unmanned spacecraft will land on the Moon and begin experiments with *in-situ* resource utilization (ISRU). One of those robotic devices, after maneuvering to a suitable location with guidance from Earth-based controllers, will use a mechanical arm to scoop up a sample of lunar soil. Through a series of manufacturing steps, the onboard materials-processing unit will use the soil sample to fabricate a functioning solar cell. This "proof of concept" ISRU experiment – even if it produces only one milliwatt of power – will mark a critical milestone in the advancement of humankind on the space frontier. The use of tools and the energy and material resources of space to generate electricity and create new tools, anable the development of self-sustaining human activities in space independent from Earth resources will evolve. A significant expansion of space exploration and development activities will then occur, and the full promise of the Spacefaring Age for the benefit of all humankind will be realized.

I.2 ADVANTAGES AND OPPORTUNITIES

The Moon offers many significant advantages over other candidate locations in space for the near-term realization of the potential of the Spacefaring Age and the evolution of humankind into a multi-world species. Among these advantages are proximity to Earth, resources, gravity, protection from space hazards, and new opportunities for science, commerce, education, international cooperation, survival of the human species, and space exploration.

I.2.1 Proximity to Earth

The Moon is in orbit around the Earth at an approximate distance of 377,000 km (234,000 miles). This relatively small distance means that existing propulsion systems can be adapted to place the first elements of a permanent base on the Moon. The Earth–Moon separation is small enough (a round trip speed-of-light time of less than three seconds) to permit operators on Earth to direct near-real-time command and control of tele-operated and semi-autonomous robotic devices on the lunar surface. Since the Moon always presents the same face to Earth as the result of "tidal locking", direct communication links with devices on the lunar surface can be maintained continuously. These advantages, and the use of robotic devices, will allow for a rapid build-up of lunar base facilities as well as world-wide coordination of exploration and development efforts. For follow-on human operations, the proximity of the Moon to the Earth will allow astronauts to return to the Earth in three days in the event of emergencies.

I.2.2 Availability of energy and material resources

The Moon, which lacks an atmosphere, receives abundant, energy from the Sun. Sunlight can be converted into electricity with lunar-made solar panels to supply the Moon with all of the power needed for global exploration and development. Sunlight could also be used for operating solar ovens, heat engines, and thermal management systems.

The Moon has a wealth of raw materials that can be applied to the construction and operation of lunar bases. The lunar soil (regolith) contains iron, aluminum, calcium, silicon, titanium, and oxygen, as well as trace amounts of lighter elements such as carbon, sulfur, and nitrogen. Increased concentrations of hydrogen, detected by the Lunar Prospector Satellite, suggest the presence of water ice in the polar regions. With due precautions to preserve important geologic information, the lunar regolith will become the feedstock for lunar base industrial processes that manufacture wires, lenses, solar cells, and construction materials.

I.2.3 Gravity

For the development of the first off-world industrial base, a location that provides gravity is preferred to a location in free space. The world's industrial expertise was developed in Earth's gravity, and it will be much easier to adapt those same industrial processes to lunar gravity ($\frac{1}{6}$th that of the Earth) than to the microgravity conditions of Earth orbit or another location in free space. For example, the use of wheeled vehicles to transport goods and the separation of materials by gravity-gradient processes can be modified to be used in lunar conditions.

The first lunar settlement activities will be unmanned, but humans will follow soon afterwards because of their ability to solve complex problems and to adapt to unforseen situations. The microgravity conditions of orbiting space stations are known to have adverse effects on human physiology, including cardiac decompensation and loss of bone mass. The long-term physiologic effects of $\frac{1}{6}$G are as yet unknown. However, the Moon, whose gravity will permit humans to move about with comparable bipedal posture and locomotion that are used on Earth, will be more Earth-like and "user-friendly" than the microgravity conditions of orbiting space stations or the minimal gravity of near-Earth objects.

I.2.4   Protection from space hazards

The free space environment is hazardous to both biological systems and hardware. Background cosmic radiation levels in free space are ten times greater than the maximum acceptable levels for individuals who work in radiation environments on Earth. In addition, solar flare radiation will be fatal, in a matter of hours to days, to anyone in space who is not protected by adequate shielding. Similarly, cosmic and solar flare radiations degrade electro-mechanical systems. Micrometeorites are a direct threat to humans and equipment, and solar-thermal stress is a significant problem for any structure in space. The solution to these problems is to simply "go underground". By creating underground compartments on the Moon, humans and equipment can be completely shielded from the hazards of space in Earth-like conditions, and the only serious risk to astronauts will occur during the three-day transit between the Earth and the Moon.

I.2.5   Science opportunities

A lunar base will be a superb platform for scientific activities of, on, and from the Moon. For example, the Moon is a much more stable platform for the operation of space telescopes than Earth orbit or free space, and the lunar regolith can be used to shield instruments from ionizing radiation, micrometeorites, and temperature extremes. Interferometry, the high-precision telescopic technique that yields images of very high resolution in optical and longer wavelengths, can be fully exploited on the Moon. The far side of the Moon is also free from all radio interference from the Earth, and is therefore the ideal site in the solar system for the operation of radio telescopes, including the search for extraterrestrial intelligence (SETI). The Moon will eventually become a coordinated astronomical observatory that will greatly expand humankind's knowledge of the universe. The Moon is also a treasure trove of information on the geologic history of the solar system, and a global program of geologic exploration of mountains, rilles, maria, and lava tubes will commence with the permanent return of humanity to the Moon.

The scientific/industrial base of the Moon will be an excellent site for research and development of "spacefaring" technologies such as *in-situ* resource utilization, electromagnetic propulsion, power beaming, materials science, agronomy, pharmacology, and life support systems. These advances will benefit the people of the Earth and will be applied to the exploration of more distant sites in the solar system such as Mars. Few of the processes or tools for lunar base operations now exist in a mature form – they will have to be developed incrementally from existing technologies. As the technology is developed, additional copies of working systems can be made on the Moon, essentially "bootstrapping" into full deployment, starting from small caches of Earth-manufactured machine tools, communications devices, and other portions of payloads yet to be defined. Once these "off-Earth" technologies and innovations have been developed for lunar purposes, however, they will have ubiquitous applications in space, on asteroids, planets, and moons throughout the solar system.

I.2.6 Education

The direct, real-time communication link between the Earth and the Moon will provide educational opportunities for students everywhere on Earth. Through connections with the Internet, it will be possible for students to observe all phases of lunar exploration and development. In some cases, students will be able to participate directly in lunar science, exploration, and other activities.

I.2.7 Commerce

The development plan of a properly designed lunar base will involve all interested parties, including governments, universities, commercial enterprises, and non-profit organizations. Of these institutions, commercial enterprises may have the most to offer. Investor-funded commercial operations in space (e.g., communication satellites in geostationary Earth orbit) have grown steadily, so that they now exceed (taxpayer-funded) governmental space activities by a substantial margin. The same trend of economic activity may be expected for the development of lunar infrastructure projects.

Using the present economic systems of the Earth as a model, governments will coordinate scientific projects on the Moon, and investor-based commercial (for-profit) enterprises will undertake the task of building infrastructure projects such as the lunar railroad (discussed later in this book). One such commercial project will be to build solar panels on the Moon from lunar regolith materials. The electric energy from the "Lunar Power Company" will supply the energy markets on the Moon and, with sufficient growth, will deliver energy profitably to Earth markets. The growth of energy, mining, manufacturing, telecommunications and transportation infrastructures on the Moon will give rise to a self-supporting lunar economy that will be financed primarily by commercial investors. When permanent human habitation becomes possible, tourism and related commercial industries will add to the economic activity between the Earth and the Moon, and a robust and growing "bi-planetary" Earth–Moon economy will come into being.

I.2.8 Earth benefits

The construction of the first lunar base will provide business opportunities and jobs for people on the Earth in many diverse fields such as aerospace, robotics, and environmental sciences. In addition to their economic benefits, these efforts will yield advances in virtually every science and engineering discipline, thus directly benefiting the quality of life of people on the Earth. The establishment of a Moon-based solar power system will potentially deliver abundant, low-cost, clean solar power to every region of the Earth, thus improving the living standards of all people, especially those in developing nations.

Living standards will also increase with the delivery to Earth of high-value elements and materials that are mined from near-Earth objects. Quality of life standards will benefit from the use of excess electric power from space to clean up pollution, desalinate ocean water, and pump the desalinated potable water to arid regions of the world (discussed in Chapter 10). The delivery of energy and materials to the Earth from space has the promise of dramatically reducing the need for mining operations on Earth and the consumption of fossil fuels (that release greenhouse gases) and nuclear fission fuels.

I.2.9 International cooperation

The development of a lunar base will lend itself to cooperative, resource-pooling international efforts. Non-cooperative lunar development efforts by nations on a "go-it-alone" approach, as done in the past, would be unnecessarily redundant and limited in scope. From a practical standpoint, a cooperative effort between nations will be much more expedient in establishing the first productive lunar base. Also, it is the intention of outer space treaties that space development should be a peaceful effort for the benefit of all people on Earth. To maximize the potential benefits for the people of the Earth, therefore, it is important that as many nations as possible participate in the lunar base effort. The bonus of international cooperation in lunar base planning, construction, and operations is that it has the potential of reinforcing peaceful relations among nations.

I.2.10 Human survival

The motivation for survival is inherent within every biological species. The problem is that a natural catastrophic event, such as the collision of a large asteroid or comet with the Earth, could destroy all human life. Although steps are being taken to minimize the threat to our survival from human-made and natural disasters, the best assurance for the indefinite continuance of the human species is to establish permanent, self-sustaining settlements on other worlds. Beyond survival, there is a second consideration for the advancement of humans into the space frontier: the concept that humans are, on balance, a positive force in nature. Humans obviously have shortcomings, but the trend of human cultural development is, arguably, in a positive direction. For example, the global record on human rights is gradually improving and there is increasing international attention being paid to environmental protection.

There is the possibility that other intelligent life exists in the universe but, until contact is made with intelligent beings, we should assume that our civilization is unique. Based upon these considerations, the migration of humankind to the Moon and other destinations in the solar system will not only assure our survival as a species, but will extend the positive force of human life to the farthese reaches of space.

I.2.11   Exploration of space

The lunar industrial base will eventually be able to produce all of the hardware that is needed for a greatly expanded program of space exploration from the Moon. Mass production techniques will be used to create large numbers of miniature, specific-function satellites, and all of the components of larger manned and unmanned spacecraft, including computers, cameras, sensors, and aerobraking heat shields. These spacecraft can be launched by electromagnetic mass drivers from the Moon to the planets, their satellites, and to the asteroid belt. In this manner, the Moon will replace the Earth as the principal base for the exploration of space.

Near-Earth objects will be explored and mined for their resources. In particular, those objects that pose a threat of collision with the Earth or the Moon will be maneuvered out of harm's way and mined for their water, hydrocarbon, and metal contents – which will then be delivered to the Earth and the Moon. The space exploration capabilities of the "Planet Moon" will enable humankind to undertake the global exploration and human settlement of Mars and conduct in-depth explorations of other planets and their moons. With continuing advances, it may be possible to launch robotic missions to nearby star systems within two or three decades after the commissioning of the first lunar base.

I.3   THE FUTURE MOON

The prediction of future activities on the Moon is an inaccurate forecast, at best. Yet predictions are worthwhile because they force planners to estimate the consequences of alternative courses of action, including no action, and then to calculate the benefits and costs of those alternatives. What, then, will happen during the coming century of human activities on the Moon? Given the growth of scientific knowledge and technology, and the application of that technology to lunar exploration and development, what will Planet Moon be like in 2100 AD? Here is our prediction:

- An autonomous, self-governing society of more than 100,000 people will be living on the Moon in 2100 AD. A global utility infrastructure will be in place, providing electric power, communication, and surface transportation for the entire Moon.
- Millions of megawatts (terawatts) of electric power will be generated on the Moon by solar arrays that have been constructed from lunar regolith materials. Lunar-generated electric power will supply all of the needs of the Moon through a global lunar electric grid network, and substantial amounts of energy will be beamed to the Earth and to other sites in the solar system.
- The sophisticated manufacturing facilities that have been constructed on the Moon will use lunar regolith, other space-derived materials, and an abundance of solar electric power to produce all of the needs of the lunar economy, such as computers, robots, construction materials, and communication equipment.
- With perfect observing conditions, the Moon will be the principal astronomical observation platform in the solar system. Thousands of lunar-made telescopes will make high-resolution, long-duration observations of objects of interest in the universe from the Moon at every wavelength of the electromagnetic spectrum. Very large aperture optical interferometry telescope arrays on the Moon will not only detect planets around other stars, but also analyze their atmospheres and characterize their habitability. These achievements will lay the foundation for the human interstellar migration that is often referred to as the "Great Diaspora".
- Human and robotic explorers will have mapped every important geologic feature of the Moon. Humans will have circumnavigated the Moon at the equator and scaled its highest peaks. Every mountain range, rille, and lava tube will have been explored. Large cities will be present on the near and far sides of the Moon, and a substantial tourism industry will exist. Transportation from any point on the surface of the Moon to any other point will be accomplished in less than 24 hours by means of a global rail network.
- The Moon will replace the Earth as the primary site for the construction and launch of spacecraft. Satellites, probes, and autonomous mobile robots equipped with television cameras and scientific instruments will be manufactured on and launched from the Moon to every area of interest in the solar system. As a result, robots from the Moon will have been sent to explore the geography of every accessible solar system body, perhaps including submarine robots that will explore the (now suspected) oceans of Jupiter's moon, Europa. A fleet of solar-sail cargo ships will ply the reaches of the solar system on commercial and scientific missions.
- Thousands of near-Earth objects (asteroids and comets) will have been analyzed by telescopes on the Moon and by spacecraft that were launched from the Moon. Those objects that once presented a threat of collision with the Earth or the Moon will have had their orbits altered by Moon-constructed spacecraft. Several of the asteroids and comets that approach the Earth–Moon system will have been mined. They will provide both planets with virtually unlimited raw materials, of hydrocarbons, water, and metals such as cobalt, palladium, platinum, nickel, and chromium.

- Solar power satellites that have been manufactured on the Moon will have been placed in orbit around a number of planets and moons, providing nearby planetary outposts with continuous electric power. Lunar-made scientific and industrial equipment will have been delivered from the Moon to other planets, such as Mercury and Mars, where they will be used for global exploration and development projects. An autonomous and self-governing Martian population, based upon the model of the Moon experience, will have been established, and an active inter-planetary economy will exist between the Earth, Moon, and Mars.
- Telescopes made on the Moon will enable observatories to be established at distant points in the solar system to provide high-resolution images of objects throughout the universe. Spacecraft launched from the Moon will have explored the Kuiper Belt and Oort Cloud regions, and robotic missions to the nearest stars will have been initiated.

The Earth itself will be the clear beneficiary of the development of the Moon, its technologies, and social institutions. Earth will no longer be dependent on fossil fuels. Social and technological advances that would otherwise have been impossible will provide across-the-board improvements for the lives of all dwellers on the Earth.

I.4   THE MOON BECKONS

The desire to explore and settle new lands is a defining characteristic of the human species; to remain in a state of ignorance of any aspect of the physical universe, when the means to end that ignorance are available, is completely contrary to human nature. It is inevitable, therefore, that, in the coming decades, we will undertake the global exploration and human settlement of the Moon (the "Planet Moon Project") and become a multi-world species. The Spacefaring Age will thus come to fruition, with the promise of significant benefits for all humankind.

1

Lunar origins and physical features

1.1 THE ORIGIN OF THE MOON

People have speculated about the origin of the Moon for centuries. There are, however, only three major scientific theories that have been proposed. The first theory was that at some time in the distant past, the Earth had somehow spawned the Moon. Perhaps the Earth was not as round then as it is today, and that imbalance caused it to split in two. This is the "fission hypothesis", first proposed by George Darwin (son of Charles Darwin) in 1878.[1] George Darwin knew that the length of time of the Moon's revolution around the Earth is slowly increasing. According to Kepler's Third Law, this means that the Moon must be gradually moving farther away from the Earth. If that is the case, then there must have been a time in the distant past when the Moon was much closer to the Earth than it is now. Understanding this, G.H. Darwin proposed that the Moon must have once been a part of the Earth. Another early scientist, Osmond Fisher, suggested that perhaps the Pacific Ocean was the scar left from the separation of the two bodies.

However, there is a physical limit to how close the Moon could have been to the Earth. This is the "Roche Limit", which is approximately two planetary radii (in the case of the Earth, that would be about 12,000 km). Inside this distance, the Moon could not have been a solid body; it could only have existed as a ring of debris much like the rings of Saturn or Jupiter, which are close to the planet, while the satellites are

---

[1] Discussion of the origin of the Moon is largely derived from the following sources: (a) Stephen G. Brush (1986) Early history of Selenogony, in: *Origin of the Moon*, W.K Hartmann, R.J. Phillips, and G.J. Taylor, eds., Lunar and Planetary Institute, Houston; (b) David Vaniman, John Dietrich, G.J. Taylor, and G. Heiken (1991) *Exploration, Samples, and Recent Concepts of the Moon*; and (c) *Lunar Sourcebook: A User's Guide to the Moon*, G.H. Heiken, D.T. Vaniman, and B.M. French, eds., Cambridge Press and Lunar and Planetary Institute, Houston.

farther away. In 1873 Edouard Roche, whose formulas still play an important role in understanding planetary ring systems, supported the Laplace theory that the Earth and Moon must have been formed at the same time, in the same neighborhood of the solar system, by "co-accretion". This is the same theory that we use today to explain the formation of binary-star systems. Darwin was forced to admit that the Moon must have been broken up completely as soon as it was separated from the Earth, but nevertheless the fission hypothesis remained popular.

In the early part of the twentieth century, the fission hypothesis was attacked again, this time on mathematical grounds. Forest Ray Moulton argued that according to the mathematics of the stability of rotating fluids, the Earth–Moon system could not have formed by fission. Thomas J.J. See proposed that the Moon was instead a captured satellite. The discovery at that time of retrograde[2] satellites around Saturn and Jupiter suggested that at least some satellites must have been captured. See proposed that the Moon had been formed farther out in the solar system, near the orbit of Neptune. Its orbit had been reduced due to the effects of a "resisting medium" in interplanetary space. However, See's "capture hypothesis" did not gain general acceptance. Although there was nothing illogical about the idea, See was rapidly losing his earlier scientific reputation because of his eccentric behavior. No one seems to have taken his capture hypothesis seriously.

Thus, when manned lunar exploration began in the late 1960s, the three competing theories were unresolved. It was hoped that evidence would be found that would decide the matter conclusively. But, as usual, the scientific data that were returned generated more questions than they answered.

A new hypothesis has been formed, however, which combines the salient points of the above three ideas and brings in a new twist. This is the "planetesimal impact hypothesis", which states that early in the Earth's history, it was struck by another pre-planetary body, roughly the size of Mars (capture hypothesis). The impact expelled large amounts of material (fission hypothesis) that eventually condensed into the Moon (co-accretion hypothesis). This hypothesis reconciles the strong points of the three older hypotheses, while solving the dilemmas in dynamics, chemistry and geophysics that they separately presented. Figure 1.1, from a computer model by Kipp and Melosh,[3] illustrates this new view of the first 12.5 minutes of the impact process.

There are still many open questions regarding the history and formation of the Moon. After the invention of the telescope and before the dawn of the space age, many explanations were proposed, not only for the origin of the Moon itself, but also for the origin of the various features on its surface. The Apollo science investigations put many of these original questions to rest. However, scientists are still debating, because many new questions have arisen even as the old ones were answered.

---

[2] Retrograde: to appear to move backwards from east to west.
[3] M.E. Kipp and H.J. Melosh (1986) Short note: a preliminary numerical study of colliding planets, in: *Origin of the Moon*, W.K Hartmann, R.J. Phillips, and G.J. Taylor, eds., Lunar and Planetary Institute, Houston.

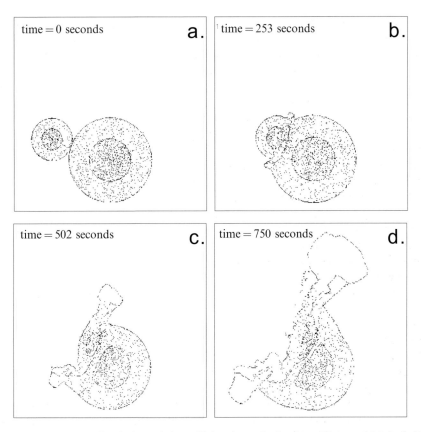

Figure 1.1. Computer simulation of the collision hypothesis, from Kipp and Melosh (1986).

1.2   PHYSICAL FEATURES OF THE MOON

The Moon is almost unique in the solar system in that it is quite large compared with its primary, Earth.[4] The Moon's diameter is 3,476 km, about $\frac{1}{4}$ the diameter of the Earth. By comparison, the satellites of Jupiter and Saturn are about $\frac{1}{100}$ the size of their primary. Table 1.1 shows the overall physical and orbital properties of the Moon. The average density of the Moon is 3.312 (water is 1.0), which is only about $\frac{3}{5}$ that of the Earth. Since the surface rocks are essentially the same, this means that the Moon's interior must somehow be different from Earth's. The Moon has no free water and essentially no atmosphere.

---

[4] The Moon and the Earth are so similar in size that they could easily be classified as a "double planet". Only Charon, Pluto's moon, is of a size-scale so similar to its primary.

Table 1.1 Properties of the Moon, compared with Earth

	Moon	Earth
Density, g/cm^3	3.34	5.515
Gravitational acceleration	1.622 m/sec^2 ($\frac{1}{6}$ Earth normal)	9.8 m/s^2
Length of day	~709 hours (29.5 Earth days)	24 hours
Distance from Earth	~356,000 km–~406,000 km	
Temperature range at equator	−173°C to 127°C (−279°F to 261°F)	0°C to 60°C (32°F to 140°F)
Temperature range at poles	−258°C to −113°C (−433°F to −172°F)	−89.2°C* to −18°C (−128.6°F to 0°F)
Temperature range at latitudes Greater than 80°	−176°C to −73°C (−285°F to −100°F)	−22°C to 6°C (−10°F to 41°F) [Tundra regions, latitude range: 60° to 75°N]
Diameter	3,476 km	12,756 km
Mass	0.7 × 10^{23} kg	59.8 × 10^{23} kg
Escape velocity	2 km/sec	11 km/sec
Atmospheric pressure at surface	1 × 10^{-12} torr	760 torr

* Vostok, Antarctica, coldest temperature ever recorded on Earth

The Moon is tidally locked to the Earth, such that we always see the same side. Thus we refer to the "near side" and the "far side".[5] In order for tidal locking to occur, the Moon must rotate on its axis at the same average rate that it revolves around the Earth.

The Moon has a very low albedo, which means that it reflects very little of the light that falls on it. It appears bright in the sky, but that is only because of the contrast between the lunar surface and the deep blackness of the night sky. In reality, the surface of the Moon is about as bright as coal dust.

Temperatures of the regolith at the Moon's equator range from 127°C (261°F) at midday to −173°C (−279°F) shortly before lunar dawn. The extreme temperatures are due to the fact that there is no blanket of air or water to moderate the temperatures. Surface temperatures at the lunar poles are much colder, ranging from −113°C to −258°C due to the very low incidence angle of sunlight. Thus, although it is much colder at the poles than at the equators, there is also less temperature variation, a fact which may prove useful for the operation of machinery and habitats.

The Moon, like the Earth, has mountain ranges and valleys. However, the origin of these features was very different for the Moon than for the Earth.

1.2.1   Mountain ranges – highlands and basin rings

On Earth, mountain ranges are caused by plate tectonics – two massive chunks of crust pushing against each other. If both of the plates have the same (or very similar) density, everything moves upward and mountain ranges form (such as the Himalayas). On the other hand, if one plate is more dense (sea floor plates are more dense

[5] People sometimes refer to the "dark side" of the Moon. This is a misconception due to a popular song from the late 1960s. There is, in fact, no permanently dark "side" of the Moon, although there are areas in the polar regions that are probably sunless. More on this later.

Figure 1.2. View of the Moon from the Apollo 17 spacecraft, on its return to Earth.

Figure 1.3. Lunar far side.

than continental plates), the oceanic slab will move down and the continental slab will move up, forming ocean trenches with volcanic mountains along the coast.

In contrast with the Earth, the lunar highlands did not result from an active plate tectonic process; instead, they simply represent a lack of destruction by giant meteorite bombardments, such as those that created the lunar basins. Compared with the deepest basins, the highlands regions are as much as 16 km higher in elevation. The lunar highlands are composed mainly of anorthosite, a type of rock also found on Earth, which is made up largely of the mineral feldspar. Samples from some of the Apollo landing sites showed that within the highlands, the surface rocks were often a jumbled mixture of several kinds of rocks – a breccia.[6]

The highlands cover so much of the Moon that it is not easy to separate them into mountain "ranges" as is done on Earth. However, there are other features that are more similar to mountain ranges. These are the circular and semi-circular basin ring structures.

1.2.2   Basins and basin rings

Basins are the largest features of the Moon, and near-side basins can easily be seen with the unaided eye from Earth, appearing as relatively dark circles on the surface. Basins are different from craters mainly with respect to their size. Basins are lunar impact excavations with diameters of at least 300 km, while craters are smaller than 300 km. This is the size at which central crater peaks are observed to change into central rings.

Three important basins on the Moon are the Orientale Basin, the Imbrium Basin, and the South Pole–Aitken Basin. The Orientale Basin was not filled with lava, so its interior structure is visible (Figure 1.4). Four well-defined rings can be seen, the outermost having a diameter of 930 km. Within this ring is another, 620 km in diameter, whose peaks are more jagged and stand higher than those of the outer ring. Within these two rings is a third ring with a diameter of 480 km, and within that, the central ring with a diameter of 320 km.

The Imbrium Basin, which is easily seen on the lunar near side, is filled with lava and thus appears as a dark circle on the face of the Moon (Figure 1.5). Most of the large lunar basins are very old, indicating that in the distant past, impacts of very large objects were much more common than they are today. In fact, the basins of the Moon may come from a time very early in solar system history, shortly after the planets and satellites first formed by the accretion of planetesimals.[7] It is surprising that the planet was not broken completely apart by some of these impacts, considering the size of the basins that were created.

---

[6] Breccias (from the Latin word for "broken") are complex assemblages of rock fragments cemented together by heat, pressure, or both.

[7] Planetesimal: bodies from millimeter size up to hundreds of kilometers in diameter that formed from the ring of dust that encircled the early Sun. Most of these accreted to form the planets, and the rest were eventually ejected from the solar system.

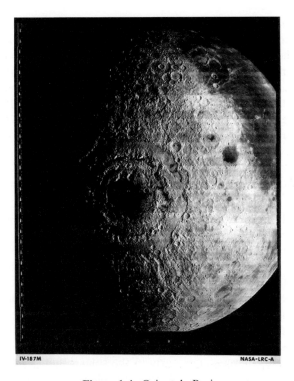

Figure 1.4. Orientale Basin.

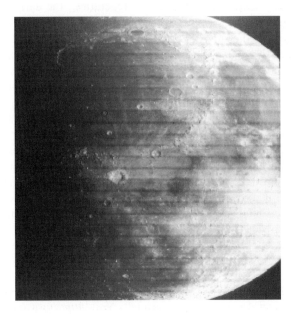

Figure 1.5. Imbrium Basin (upper middle), is easily seen from Earth with the unaided eye.

The South Pole–Aitken Basin is a large impact crater with a diameter of 2,500 kilometers and a maximum depth of 12 to 13 kilometers. It extends from the south polar region to the Aitken Crater on the far side of the Moon, and is the largest known impact crater in the solar system. The extent of the basin was not known until it was photographed by the Lunar Orbiter Satellite. Further imaging and data of the basin were obtained by the Clementine and Galileo missions. The impact event that produced the basin may have exposed the upper mantle of the Moon, but confirmation of that possibility awaits surface analysis.

1.2.3   Craters[8]

The Earth, the Moon, and every other planet in the solar system is continually bombarded by meteorites and comets, which are probably left-over planetesimals from the formation of the solar system. These objects are usually quite small, and on Earth they almost always burn up in the atmosphere before reaching the surface. However, when they do survive the journey, the results are spectacular. Meteor Crater, Arizona, is thought to have been caused by a meteorite 50 meters in diameter (Figure 1.6).

Comets and asteroids strike the Moon with an average speed of 20 km/sec. These impacts produce craters that are ten to twenty times larger than the impacting object. Thousands of millions of years ago, early in the solar system's history, there were probably more meteors and comets striking all the planets, and they were likely much larger than most of the ones observed today. Because the Earth is geologically active, with volcanoes, plate tectonics, erosion and vegetation, the early craters on Earth have long since been erased.

However, much of the surface of the Moon has remained unchanged for 4,000 million years. Thus, the solar system's early bombardment history is still recorded by relatively intact craters of every size, from thousands of kilometers in diameter (basins) down to "zap pits" (micrometer-sized craters in lunar soils and rocks that are observed with scanning electron microscopes).

The internal form of the crater depends on its size. Small craters (less than 15 km in diameter) are simply bowl-shaped depressions. Medium-sized craters, with diameters from 20 to 175 km, have flat floors and a central peak. Craters with diameters greater than 175 km have somewhat complex central structures, and rings begin to replace peaks as the diameter increases. See Figures 1.7, 1.8, 1.9, and 1.10.

When material is ejected from a crater, it follows a hyperbolic path, such that most of the material lands nearby, with decreasing amounts landing farther away. If large boulders are ejected, they may create secondary craters when they land. Secondary craters sometimes occur in straight lines that point to the primary crater.

The material that is not ejected may be dramatically altered as well. The surface

---

[8] Much of the material for this section has been paraphrased from the web page "Exploring the Moon – Impact Crater Geology and Structure". Visit *http://www.1pi.usra.edu/expmoon/ science/craterstructure.html*

Figure 1.6. Meteor Crater, Arizona.

Figure 1.7. Moltke is an example of the smallest class of craters; a simple bowl shape (from the Lunar and Planetary Institute web page, "Impact Crater Geology and Structure", *http:// www.1pi.usra.edu/expmoon/science/craterstructure.html*).

may be broken into rubble, which may then be welded together by the material that has been melted by the energy of the impact. This is one type of breccia. Going deeper below the surface, the rocks are cracked and deformed, but the deformation is less severe at depth.

Figure 1.8. Euler, 29 km in diameter, shows a flattened floor and a central peak (from the Lunar and Planetary Institute web page, "Impact Crater Geology and Structure", *http:// www.1pi.usra.edu/expmoon/science/craterstructure.html*).

Figure 1.9. Schrödinger is 320 km in diameter, really large enough to be called a basin rather than a crater. It has an inner ring that is 150 km in diameter. Schrödinger is one of the youngest impact basins on the Moon (from the Lunar and Planetary Institute web page, "Impact Crater Geology and Structure", *http://www.1pi.usra.edu/expmoon/ science/craterstructure.html*).

1.2.4 Maria

Ancient observers thought that the round, dark, low areas on the face of the Moon were seas, which they called maria, surrounded by land, or terrae, and those terms are still used today. The lunar basins which contain the maria were probably formed by giant impacts very early in the Moon's history. Somewhat later, when volcanic activity began, many of them were filled with lava.

On the near side of the Moon, the visible maria cover 16 percent of the surface. However, there are other areas which have the same type of rock, but their surfaces have been blanketed by the ejecta from later impacts. These are called cryptomare. The total amount of mare (volcanic) material on the lunar near side is about 30 percent of the surface. The rocks that make up the maria most closely resemble terrestrial basalts, a type of dark, fine-grained volcanic rock. The volcanic activity that produced the maria was of a type that is not often seen on Earth flood basalts.[9] They must have come from fissures that were many kilometers in length, and it is thought that the molten material had very low viscosity[10]– similar to that of motor oil.

Associated with many lunar maria are gravity anomalies called mascons (short for mass concentrations). These were discovered by examination of the tracking data of the lunar orbiters and confirmed by the Apollo and Clementine missions. It was found that the spacecraft would accelerate as they neared the maria, then decelerate as they moved beyond them.

The strongest mascon anomaly is associated with Mare Imbrium, although they also occur with Serenitatis, Crisium, Humorum, Nectaris, and Orientale. Clementine found additional evidence for other gravity anomalies associated with ancient, nearly obliterated basins. Although mare basalt has greater density than highland materials, this difference is insufficient to explain the size of the gravity anomaly. It is likely that the lunar crust is thinner below the maria than in other areas, and that dense mantle[11] material is closer to the surface as a result. The combination of the increased density of both the surface basalt and the subsurface mantle material is the most likely cause of these features.

1.2.5 Ridges, lava tubes and rilles

Arcuate ridges, also called wrinkle ridges, are found parallel and interior to the margins of many of the mare basins. They are irregular and extend brokenly for great distances. Their origin is uncertain, but they may represent compression features that result from the differential compaction of layers of rock over subsurface ring systems. They might also represent doming of the crust over shallow intrusions, or

---

[9] Flood basalts are seen in a few locations on Earth, such as the Columbia River and Snake River plateaus of the western United States, and the Deccan basalts of India.

[10] Viscosity: a property of fluids that causes them to resist flowing as a result of internal friction from the fluid's molecules moving against each other.

[11] Mantle: the zone of a planet below the crust and above the core.

Figure 1.10. Lunar Orbiter 3 oblique view of Murchison Crater (foreground), 58 km across and Ukert Crater (back right-center) on the Moon. The southern edge of Mare Vaporum is visible in the background. Note the irregular features and rilles on the floor of Murchison. North is at 10:30 (Lunar Orbiter 3, frame M-85).

accumulation of extrusive lava flows along a linear volcanic fissure system. Ground-penetrating radar data of the southern Mare Serenitatis Basin suggests that differential compaction is the most likely origin for that particular ridge.

Lava tubes are among the most interesting features of the Moon, especially from the standpoint of human habitation. Lava tubes are natural caverns that are the drained conduits of underground lava rivers. Their existence is inferred from the inspection of sinuous rilles that appear to be discontinuous on the surface. Parts of these rilles may be "roofed over" to create a structure that is similar to the lava tubes found on Earth.

The inside dimensions of lava tubes may be tens to hundreds of meters, and their roofs are expected to be greater than 10 meters thick (Hörz, 1985), thus offering an environment that is naturally protected from the hazards of radiation, meteorite impact, and temperature extremes. Figure 1.11 shows a lava tube with segments of collapsed roof. According to Oberbeck *et al.* (1969), basalt "bridges" spanning a few hundred meters are possible on the Moon provided they are at least 40 to 60 meters thick. This is in good agreement with the fact that uncollapsed segments of lava tubes

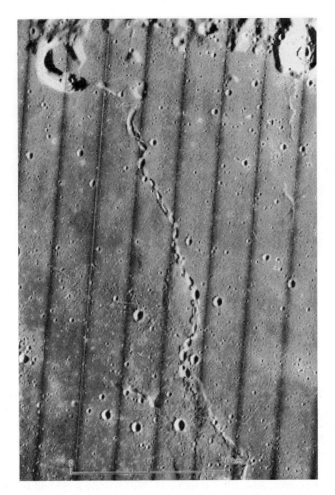

Figure 1.11. A lava tube with segments of collapsed roof (Lunar Orbiter 5, frame 182).

display impact craters a few tens of meters across, occasionally as large as 100 m. These diameters correspond to crater depths of up to 20 meters (Gehrig, 1970). Roof thickness would have to be twice that value (40 meters) in order to withstand the impact (Hörz, 1985).

The upper five to ten meters of any lava roof has probably been eroded by micrometeorite impact into a fine-grained lunar soil. Cracks associated with this regolith are probably a factor of three to five deeper (Pohl *et al.*, 1977). It is difficult to predict the structural integrity and exact thickness of a lava tube roof with great precision (Hörz, 1985). However, an obvious strategy would be to select roofs or roof segments that have only suffered relatively small cratering events. Additionally, future robotic rovers could use ground-penetrating radar to determine the structural integrity of the roof section.

Figure 1.12. View of Hadley Rille from the surface.

Hörz (1985) concluded that natural caverns of suitable size to house an entire lunar base could exist on the Moon. Roof thickness in excess of ten meters would provide safe, long-term shelters as receptacles for modular habitats. It would be very unlikely that the lava tube itself could be modified to create a habitat, however. Site preparation inside the lava tube would consist of leveling the floor with lunar soil.

Hörz (1985) also pointed out that the strongest advantage to using lava tubes is that the sheltered environment allows the use of extremely lightweight construction materials (inflatable habitats, etc.). None of the components would require shielding. Use of thin foil materials would be possible not only for the habitat modules, but also for ducts, storage tanks, and other structures. Not having to shield these mechanical systems from meteorite impact would be a great advantage when it becomes necessary to inspect, repair, or replace them.

Another strong advantage to housing lunar bases inside lava tubes is the relatively constant temperature ($-20°$C), due to the fact that one is underground and some tens of meters removed from the lunar surface (Hörz, 1985). This would be much easier to work with than the constantly fluctuating temperatures ($-180°$ to $+100°$C) on the surface, where complex thermal insulation and control systems are unavoidable. In contrast, the relatively constant, benign temperature within a lava tube, coupled with the freedom from ultraviolet and infrared radiation, would allow the use of common materials such as plastics.

A drawback to using lava tubes is that steep slopes may limit their accessibility. However, relatively shallow rilles do exist. When the Apollo 15 crew visited the edge of Hadley Rille (Figures 1.12 and 1.13), they thought that their Lunar Rover could have negotiated its slopes (Hörz, 1985). Hörz also notes, "Location of a lunar base [in a lava tube may not seem] very economical from an energy point of view, because the mass will have to be lowered and especially raised when needed on the lunar surface and when being readied for export to LEO or GEO.

These energy considerations are, however, a matter of degree, because most large-scale industrial operations rely heavily on gravity for material transport. Some modest elevation difference between the source of raw materials and the processing plant is desirable even for such simple operations as sieving and magnetic separation. For this reason, a lunar base may be more functional if located at the base of some slope such as a sinuous rille/lava tube. Chutes or pipes may be laid out such that they terminate inside the lava tube at exactly that station where the high-graded raw materials are needed.

A great deal of work remains to be done to determine the feasibility of using lava tubes for lunar habitats. Detailed study of data from the Clementine and Lunar Prospector missions is needed to improve our knowledge of the inventory of lunar lava tubes and to determine their spatial distribution. These studies will also give us improved information on roof thickness and dimensions. The *in-situ* examination of lava tubes by tele-operated robots will be among the most interesting missions of the next stages of lunar exploration.

1.3 EXPLORATION OF THE MOON

Since 1959, there have been 52 missions[12] to the lunar surface or to cislunar space, of which 48 were at least a partial success in terms of mission objectives and data return. On 31 January 1966, the Russian Luna 9 spacecraft made the first soft landing of a spacecraft on the Moon. Since that time, substantial *in-situ* analysis of the lunar surface has been made. The bulk of lunar rock samples that were returned to Earth were from the Apollo manned missions. Recent missions have indicated the possible presence of water (hydrogen concentrations) – an unexpected resource that will substantially increase the opportunity for the permanent return of humans to the Moon.

1.3.1 The Apollo experiments

On 11 December 1972, the sixth and final lunar surface mission of the Apollo Program (Apollo 17) landed in a valley near the edge of Mare Serenitatis (Figure 1.13). Astronauts Eugene Cernan and Harrison Schmitt spent 72 hours at the site, named Taurus-Littrow (after the mountains to the north). The site was geologically diverse, with the mountain ring of the Serenitatis Basin nearby, and lava

---

[12] Missions to the lunar surface or cislunar space from 1959 to 2006: 27 Russian spacecraft (24 Luna and 3 Zond), 23 American spacecraft (5 Ranger, 6 Surveyor, 9 Apollo, 1 Pioneer, 1 Galileo, 1 Clementine, and 1 Lunar Prospector), 1 Japanese spacecraft (Muses A), and the European Space Agency's SMART-1 orbiter.

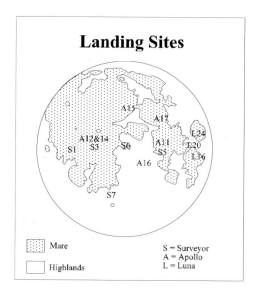

Figure 1.13. Apollo landing sites.

filling in the valley lowlands. The main objective of the mission was to collect samples of these different types of rock.

The crew spent more than 22 hours on the lunar surface, using the rover to traverse the mare plains. The traverses totaled more than 30 km, and nearly 120 kg of rock and soil were collected – the largest total sample mass of any Apollo mission. Many of the rock samples have since been studied in detail.

A mass spectrometer was placed on the lunar surface during the Apollo 17 mission. It provided data on the distribution of many types of rarefied gases, including argon-40 and helium-4, and very small amounts of argon-36, methane, ammonia, and carbon dioxide at sunrise, as well as neon-20, neon-22, and molecular hydrogen. Essentially, all the argon-40 on the Moon comes from the decay of potassium-40 in the lunar interior. Variability in the amount of atmospheric argon (a 6- to 7-month periodicity was observed) suggests the presence of a partially molten core, which is interesting from both scientific and resource utilization standpoints. If deep vents that periodically release gases onto the surface are found, these gases could be trapped and used for manufacturing, as carrier gases in pneumatic conveyor systems, and for other uses.

A similar instrument, the suprathermal ion detector, had, on an earlier mission, documented the presence of a cloud of hot solar-wind electrons near the terminator. This cloud was found to vary with changes in the solar wind and extreme ultraviolet flux. In addition, one other interesting event was found. This event, principally water vapor, must remain suspect because of its proximity to the Apollo 14 mission. However, the large magnitude and the long time duration of the event argue against a mission-related source.

1.3.1.1   *Apollo seismic experiments*

A widely-spaced network of seismometers ("passive seismic experiment") was emplaced during the Apollo missions to determine the characteristics of the lunar interior by monitoring meteorite impacts, impacts of discarded lunar modules, and natural seismic events ("moonquakes"). In the active seismic experiment, geophones were placed on the surface to determine local subsurface structure by recording arriving seismic waves caused by detonating small explosives.

Lunar seismic signals are quite different from terrestrial ones because the Moon's crust is a single plate, whereas the Earth's crust is divided into many plates. The Moon is said to "ring like a bell"; strong signals may last for several hours. Seismic signals on Earth are damped when they reach a plate boundary, but no such damping occurs on the Moon. At the Apollo 17 site, several seismic layers were detected in the subsurface, with the strongest change in seismic signal occurring at a depth of 1,385 meters. This is in agreement with other estimates for the thickness of the volcanic layers at the Apollo 17 site. The material below this seismic boundary is probably chemically similar to the surrounding highlands rocks.

The first few Apollo landings had primarily been engineering experiments. The details of how to land safely on the Moon and survive there for a very short time were of paramount importance. It was only in the last few Apollo missions that scientific questions predominated over engineering questions with respect to choice of landing site, EVA activities, and so on. Many important scientific discoveries resulted from the Apollo missions to the Moon.

1.3.1.2   *Apollo findings*

The top ten discoveries of the Apollo Missions are as follows:[13]

1. The Moon is not a primordial[14] object, but rather an evolved "terrestrial" planet with internal zoning similar to that of Earth. Before Apollo, the state of the Moon was a subject of almost unlimited speculation. We now know that the Moon is made of rocky material that has been variously melted, erupted through volcanoes, and crushed by meteorite impacts. The Moon possesses a thick crust (60 km), a fairly uniform lithosphere[15] (to a depth of approximately 1,000 km), and a partly liquid asthenosphere[16] (1,000–1,740 km). A small iron core at the

---

[13] *http://www.nasm.si.edu/collections/imagery/apollo/apollotop10.html*

[14] Primordial means primitive, fundamental, or original. Some of the asteroids may be primordial objects; appearing now as they were originally formed, and not subject to the processes of change (heating, volcanic activity, and mountain building) that have affected Earth, Mars, and the Moon.

[15] Lithosphere: in plate tectonics, a layer of strength relative to the underlying asthenosphere. It includes the crust and part of the upper mantle. On Earth, the lithosphere is about 100 km thick. On the Moon, the lithosphere may be much thicker. Some research remains to be done to establish this.

[16] Asthenosphere: the layer or shell below the lithosphere, which is relatively weak and in which magmas may be generated and seismic waves are strongly attenuated.

bottom of the asthenosphere is possible but unconfirmed. Some rocks give hints for ancient magnetic fields, although no planetary magnetic field exists today. Compasses won't help you on the Moon.

2. The Moon is ancient and still preserves a record of the first 1,000 million years of its history. That same history must be common to all terrestrial planets. Craters and their relationships to each other, when calibrated using absolute ages of rock samples, provide a key for unraveling time scales for the geologic evolution of Mercury, Venus, and Mars based on their individual crater records. Photo interpretation of other planets is based largely on lessons learned from the Moon. Before Apollo, however, the origin of lunar impact craters was not fully understood and the origin of similar craters on Earth was highly debated.

3. The youngest Moon rocks are about the same age as the oldest Earth rocks. The earliest processes and events that probably affected both planetary bodies can now only be found on the Moon. Moon rock ages range from about 3,200 million years in the maria (dark, low basins) to nearly 4,600 million years in the terrae (light, rugged highlands). On Earth, active geologic forces, including plate tectonics[17] and erosion, continuously repave the oldest surfaces, whereas on the Moon, old surfaces persist with little disturbance.

4. The Moon and Earth are related and formed from different proportions of a common reservoir of materials. The distinctively similar oxygen isotopic[18] compositions of Moon rocks and Earth rocks clearly show common ancestry. Relative to Earth, however, the Moon is highly depleted in iron and in volatile[19] elements that are needed to form atmospheric gases and water.

5. The Moon is lifeless; it contains no living organisms, fossils, or native organic compounds. Extensive testing revealed no evidence for life, past or present, among the lunar samples. Even non-biological organic compounds are amazingly absent; traces can be attributed to contamination by meteorites.

6. All Moon rocks originated through high-temperature processes with little or no involvement with water. They are roughly divisible into three types: basalts, anorthosites, and breccias. Basalts are dark lava rocks that fill the giant basins; they generally resemble, but are much older than, lavas that comprise

---

[17] Plate tectonics: a theory that the Earth's crust is divided into a series of vast, plate-like parts that move or drift as distinct masses. Sometime referred to as the theory of "continental drift".

[18] Many elements have isotopes – individual atoms that have a different number of neutrons in the nucleus. Typically, an oxygen atom has eight protons and eight neutrons, which makes the most common isotope, oxygen-16. Sometimes there are nine neutrons, which makes oxygen-17, while ten neutrons makes oxygen-18. There are "natural abundances" of these isotopes on the Earth, such that an "average mass" can be calculated for the oxygen atom. It was learned from the Viking missions to Mars and from sampling meteorites that there are different oxygen isotope ratios on different planets. In fact, the isotope ratios of many elements vary from one planet to another. The planetary origin of a rock can thus be determined from the isotope ratios of oxygen, which makes up about 40% of all rocks everywhere.

[19] Volatile: elements or compounds that evaporate quickly or are given off as vapor during combustion. These include light elements, as well as the typical gases and liquids such as water, ammonia, carbon dioxide, carbon monoxide, methane, etc.

the oceanic crust of Earth. Anorthosites are light rocks that form the ancient highlands; they generally resemble, but are much older than, the most ancient rocks on Earth. Breccias are composite rocks formed from all other rock types through crushing, mixing, and sintering during meteorite impacts. The Moon has no sandstones, shales, or limestones because water is required for the formation of these rock types.

7. Early in its history, the Moon was melted to great depths to form a "magma ocean". The lunar highlands contain the remnants of early, low-density rocks that floated to the surface of the magma ocean. The lunar highlands were formed about 4,400 million to 4,600 million years ago by flotation of an early, feldspar-rich crust on a magma ocean that covered the Moon to a depth of many tens of kilometers or more. Innumerable meteorite impacts through geologic time reduced much of the ancient crust to a thick layer of boulders and rubble.

8. The lunar magma ocean was followed by a series of huge asteroid impacts that created basins, which were later filled by lava flows. The large, dark basins such as Mare Imbrium are gigantic impact craters, formed early in lunar history, that were later filled by lava flows about 3,200 million to 3,900 million years ago. Lunar volcanism occurred mostly as lava floods that spread horizontally. However, in some cases volcanic fire fountains produced deposits of orange and green glass beads.

9. The Moon is slightly non-symmetrical in shape, possibly as a consequence of the Earth's gravitational influence. The lunar crust is thicker on the far side, while most volcanic basins – and unusual mass concentrations – occur on the near side. Mass is not distributed uniformly inside the Moon. Large mass concentrations ("mascons") lie beneath many large lunar basins which may represent thick accumulations of dense lava. Relative to its geometric center, the Moon's center of mass is displaced toward Earth by several kilometers.

10. The surface of the Moon is covered by a rubble pile of rock fragments and dust, called the lunar regolith, that contains a unique radiation history of the Sun, which is of importance for understanding climate changes on Earth. The regolith was produced by innumerable meteorite impacts through geologic time. Surface rocks and mineral grains are distinctively enriched in chemical elements and isotopes implanted by solar radiation. As such, the Moon has recorded 4,000 million years of the Sun's history to a degree of completeness that we are unlikely to find elsewhere.[20]

The Apollo Program marked the beginning of a global program of human exploration; the comprehensive *in-situ* investigation of the maria, mountains, rilles, lava tubes, and craters awaits the permanent return of humankind to the Moon.

---

[20] Since the Moon has no atmosphere and no magnetic field, solar radiation can reach the surface of the Moon much more easily than it can reach the surface of Earth, Mars, or Venus.

1.3.2   Recent missions to the Moon

The Galileo spacecraft, on its way to Jupiter, captured some new images of the Moon. These images were useful, both to provide more detail about parts of the Moon that had not been mapped before, but also to test the instruments aboard Galileo. In this sense, the Moon is a "reference planet" – an accessible, basic example that can be observed from space and compared with what is already known from studying the lunar samples.

A lot is known about a few places on the Moon, such as the Apollo and Lunokhod landing sites. However, very little is known about the rest of the Moon – for example, the polar regions and the far side. Thus, it is likely that future missions will emphasize global, rather than local, studies, and for this reason, lunar orbiter missions will be important. Galileo is a good example of the usefulness of global imaging missions, as illustrated by Galileo's images of the South Pole–Aitken Basin on the far side. Galileo images suggest that the area has volcanic deposits that have been covered over by later ejecta deposits, creating "cryptomare". This is an important discovery, because it reveals that the Moon was subject to much more widespread volcanic activity than was originally thought. While Galileo's images of the Moon had 5.5 km resolution, future lunar orbiters will have much better resolution for global coverage, and that will allow even more discoveries to be made.

The Clementine mission: In 1994 the U.S. Department of Defense sponsored the Clementine mission to the Moon[21] (Figure 1.14). This was a partnership with NASA in which both groups gained important information. The Department of Defense wanted to know how well their new space-based detection instruments could survive the space environment, particularly the harsh radiation found in the Van Allen belts and beyond. NASA was able to use those instruments to explore the Moon in much greater detail than had been done previously. The Clementine spacecraft was placed in a lunar polar orbit for two months. During that time, it obtained detailed images of almost the entire surface of the Moon, in eleven spectral bands.[22] Scientists have been using these images to learn more about the chemical composition of the entire Moon, and to perform detailed studies of many interesting areas that could not be examined previously, because of a lack of information.

---

[21] Visit *http://www.1pi.usra.edu/pub/expmoon/clementine/clementine.html* for more images and further details on the Clementine mission.

[22] A spectroscope acts like a prism, in the sense that it shows us the different components of the light that would otherwise reach our eyes as a single "color". When light passes through a prism, there are at least seven colors that the human eye can clearly pick out. A spectroscope, however, can see in much more detail, and can pick out millions of tiny variations in color. When rocks and minerals are studied in the laboratory, the spectroscope provides a pattern of intensities of the various "colors" that are unique to each mineral that we observe. The known spectrum of the mineral olivine, for example, is identical to spectra taken of the central peak of the crater Copernicus, so it is known that there is some olivine on the Moon. A great deal of work must still be done to sort out the signatures of various minerals from each other to understand the composition of the lunar surface.

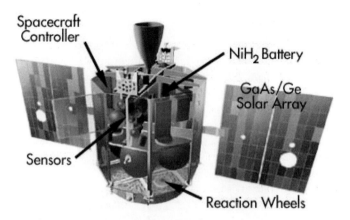

Figure 1.14. The Clementine spacecraft.

By studying various rocks and minerals, it has been learned that there are certain wavelengths (colors) that are characteristic of certain substances. Cameras cannot "un-blend" colors, but they can be made to pick up only the "colors" (spectral bands) that give the most information. Eleven particular spectral bands are known to provide a great deal of information about the composition of rocks and soils, and these color filters were used on the Clementine mission cameras.

The Clementine spacecraft was built and operated by the Naval Research Lab. It was the first of a new class of small, low-cost spacecraft that use very lightweight instruments and hardware. During its two months in lunar orbit, Clementine mapped 38 million square kilometers of the Moon, taking nearly two million images. It also carried a laser ranging instrument to collect accurate data on the topography of the Moon. It was learned that the surface of the Moon is even rougher than had been previously thought.

One of the most newsworthy discoveries from the Clementine mission was evidence for possible water-ice at the lunar poles.[23] If this discovery is verified, it would mean that future lunar manned exploration and eventual colonization will be much easier and sooner than previously estimated.

Lunar Prospector: The "Lunar Prospector" satellite, an inexpensive, long-awaited scientific mission to the Moon, was launched on 6 January 1998, and placed in a 100-kilometer orbit around the Moon (Figure 1.15). It carried five science instruments,[24] including a neutron spectrometer that was designed to look for evidence of water on the surface of the Moon.

Neutron spectrometers measure the energy of neutrons that emanate from the lunar surface as the result of cosmic ray bombardment. Hydrogen in the lunar

---

[23] Visit *http://nssdc.gsfc.nasa.gov/planetary/ice/ice_moon.html* for more information.
[24] The five instruments are a magnetometer, electron reflectometer, gamma-ray spectrometer, neutron spectrometer, and alpha-particle (alpha particles are Helium-4 nuclei) spectrometer.

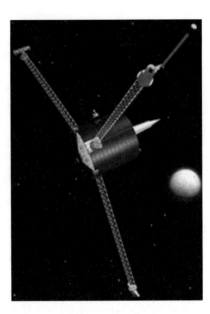

Figure 1.15. Lunar Prospector satellite.

regolith decreases the energy of neutrons on the lunar surface in a characteristic manner. When a concentration of hydrogen is present on the Moon, a spectrometer in lunar orbit will detect that concentration by identifying the characteristic decrease of neutron energy.

The neutron spectrometer on the Lunar Prospector did, in fact, record characteristic dips in neutron energy over the polar regions of the Moon, indicating concentrations of hydrogen in those locations. Because elemental hydrogen is a gas, concentrations of hydrogen on the Moon would have to be present in the form of a stable molecule, and the most likely molecule, based upon physical, chemical, and temperature factors, is water-ice in the permanently shadowed floors of craters.

Initial data from Prospector indicated that from 10 to 300 million tons of water-ice are present in the polar regions of the Moon, with a greater amount of water (hydrogen) in the north polar region than in the south. More recent data now indicate the presence of between 1 and 10 billion tons of water. Because the spectrometer is only able to detect hydrogen to a depth of 0.5 meters, water may be present at depths much greater than those from which the Prospector derived its information. The significance of finding concentrations of hydrogen on the Moon is that hydrogen is a highly valuable element for biological systems, rocket fuel, and industrial purposes. With the hydrogen resources that are indicated by the Lunar Prospector, the time-table for the permanent return of humans to the Moon will be significantly shortened.

In addition to the discovery of hydrogen concentrations in the polar regions, the Lunar Prospector provided data on the lunar surface distribution of other elements such as potassium, rare-earth elements, phosphorus, and iron, and the interaction of

Figure 1.16. European SMART-1 spacecraft.

the Moon with the solar wind. Data is also being used to generate gravity maps of the Moon. Later in its mission, the orbit of the Prospector was lowered to 10 kilometers above the lunar surface so that more detailed information could be obtained.

When the spacecraft had exhausted its fuel, a final experiment was performed: the spacecraft was de-orbited into a permanently-shadowed region of the south pole, to determine if the impact would create water vapor that could be detected by Earth-based instruments or by the Hubble Space Telescope. Spectrometers were used to look for the signature expected from the hydroxyl (OH) molecules that should be a by-product of the impact. Several explanations are possible for the lack of detection of the hydroxyl molecules, so the experiment was not conclusive. Visits to the lunar polar regions by landers, equipped with geologic tools and sensors, will be required to learn if ice (or any other type of hydrogen-containing substance) exists there.

SMART-1: ESA's first spacecraft in its Small Missions for Advanced Research in Technology program (SMART-1) arrived in lunar orbit on 15 November 2004, using solar-electric propulsion and carrying a battery of miniaturized instruments (Figure 1.16). Originally scheduled to end in 2005, the success of the probe later warranted an extension of the mission until 3 September 2006 when the spacecraft impacted the lunar surface in the Lacus Excellentiae region.

The spacecraft's instruments included an ultra-compact electronic camera, AMIE, which surveyed the lunar surface in visible and a near-infrared light, and a near-infrared point-spectrometer (SIR) for mineralogical investigations. It was hoped that the SIR instrument could peer into permanently-shadowed craters using only the light that is reflected from nearby sunlit peaks. However, no definitive results have been reported to date.

1.4 SUMMARY

The Moon is an interesting place; it has a storied past and innumerable places to explore. The first series of data-gathering and human explorations were highly successful. The next step will be to carry out a program of surface explorations and experiments with autonomous and tele-operated robotic devices, in preparation for the return of humans.

2

Science opportunities – engineering challenges

2.1 INTRODUCTION

The transformation of the Moon into an inhabited planet offers many potential benefits to the peoples of the Earth, including access to the vast, low-cost energy and material resources of space. Although these resources hold the promise of substantially increasing the standard of living and quality of life for all of the people of the Earth, a potentially greater benefit of lunar development will be the knowledge that is gained in the basic and applied (engineering) sciences. Among the basic sciences, planetary geoscience and astronomy can expect the greatest advances. Significant growth of knowledge will also result from the engineering efforts that are made to adapt human civilizations to a new planet.

The knowledge of the physical world that has been accumulated over time, and the technologies that are based upon that knowledge, are now sufficient to make the permanent migration of humankind to the Moon possible. The additional knowledge and technological gains that will be realized by going to the Moon will enhance humankind's ability to accomplish ever-more complex objectives, including the human migration to more distant regions of the solar system.

2.2 GEOSCIENCE

There are two important reasons to explore the Moon. The first reason is to know more about the Moon itself; its origin, history, composition, and geologic diversity. The second reason is to apply that knowledge to a useful purpose – the future colonization of the Moon. Exploration will be a significant effort. It will require the services of engineers, scientists and robotic devices working for decades to survey and analyze the major areas of interest on the Moon. The global geologic exploration of the Moon (or of any other planetary body) may be divided into reconnaissance (performed by global remote sensing) and field work, which is the detailed analysis of specific surface and subsurface regions of the Moon. The scientific objective is to understand the processes that produced the observed geologic features.

2.2.1   Geologic reconnaissance missions

Reconnaissance is performed by orbiting satellites that are used to map and characterize global geographic features, provide data on surface minerals and chemistry, and identify areas where resource concentrations may exist. The operations of the Clementine and Lunar Prospector Moon-orbiting satellites are examples of global reconnaissance missions. These missions have done much to categorize lunar surface features from polar orbit. Maps of the regional surface mineralogy of the Moon have been generated, and, among other findings, their data indicate that water may exist in the lunar polar regions.

A unique reconnaissance mission of the Moon was performed during Apollo 17, in which radar data were collected from lunar orbit. The results, although difficult to interpret at the time, nevertheless demonstrated that radar is a very effective method of geophysical investigation on the Moon. Depending upon the wavelength that is used, various depths of the subsurface can be explored with ground-penetrating radar (GPR). The highest frequencies allow detailed exploration of the shallow subsurface, while lower frequencies give less detailed, but deeper information (up to 2 km below the surface). For lunar base construction, exploration of the upper two meters of the regolith will be desirable for delineating areas that are free of boulders.

Areas such as Sulpicius Gallus are thought to contain an abundance of loosely-consolidated glass that is relatively free of boulders. If this can be verified by GPR, then the Sulpicius Gallus formation materials would be ideal for shielding habitats, as well as for extracting useful resources. Moreover, other regions are known (from the Clementine data) to be likely candidates for pyroclastic deposits similar to Sulpicius Gallus. Low-orbiting reconnaissance spacecraft, equipped with radar sounding equipment, could rapidly characterize these regions and select the best candidates for follow-on robotic-rover exploration.

Another important use of GPR (as mentioned in Chapter 1) will be to search for and evaluate lava tubes. An orbital survey of the entire subsurface from low orbital altitude would provide definitive information on the existence and location of these interesting and potentially useful structures. Because space "weathering", including meteorite bombardment, weakens the roof sections of lava tubes, their evaluation by radar will be essential. After a lava tube has been located by low-orbital GPR, rover-mounted, high-resolution surface units can be used to determine the structural integrity of its roof section.

2.2.2   Field work

Field work is the *in-situ* investigation of surface and subsurface geologic structures. It entails imagery, collecting samples, on-site analysis, and documentation. The goal of field work is to gain an understanding of geological processes in detail, which expedites the search for valuable resources such as water-ice, other volatile materials, and areas where other resources are concentrated.

The return of humans to a permanent base on the Moon will inaugurate an extensive program of geologic investigation. Some field work will be done on-site by humans, but the majority of work will be done by geologists using robots that are tele-operated from the Earth or from a lunar base.[1] Surface-roving, tele-operated robots are remarkable tools that will greatly multiply the efficiency and cost-effectiveness of field operations. Technology now makes possible unprecedented realism in the recreation of distant scenes for field geologists through the technique of virtual presence. High-resolution 3-D images can be presented to a pair of eyes (actually, multiple pairs of eyes) at the same time that tactile feedback is imparting touch sensations to the operator's hands. This technology has been spearheaded by medical/surgical procedures and by underwater robotics research. The advantages of robots for work on the Moon include the following:

- Virtually all robots will be sent to the Moon on one-way missions, so there will be no transportation costs for their return to Earth.
- They require no life-support systems.
- They can operate continuously, via direct tele-communication links from Earth.
- They can carry more sensors and larger sample-collection payloads than humans.
- They have greater range and duration than humans.
- They can be made stronger than humans.
- They can work in areas that are too hazardous for humans.
- They are expendable.
- Their parts can be recycled.

The robots that assist with field work will be equipped with cameras, hammers, drills, anthropomorphic arms with tactile feedback, compartments for sample storage, and a means of locomotion (wheeled, walking, or tracked vehicles). They will also carry spectrometers that can perform on-site mineral analyses of samples.

Robots can be designed to make long-distance traverses across the lunar surface. However, these "solo" excursions (such as the Russian Lunokhod rover) are basically one-way missions that end when the power supply or other critical system of the robot fails. More typically, future geologic expeditions will be round-trip missions that investigate a designated region in the vicinity of a lunar base. Each tele-operated robot will travel under its own (battery and solar) power from the lunar base to the area being investigated, conduct its studies over a period of days or weeks, and then return to the base. Excursions employing pairs of robots may be a desirable scenario. Each device can provide images of the other, for augmentation of their control. Like the "buddy system" humans use in hazardous environments, each might be able to

---

[1] Popular illustrations of geologists performing field work on the lunar surface are misleading. The hazard of cosmic radiation will limit human surface activity on the Moon to short-duration, high-priority tasks.

assist the other with various difficult or hazardous chores. The collected samples and data from the mission will be analyzed, and the robots will be serviced and repaired, if necessary, in preparation for the next expedition. In this manner, robots can be reused for multiple missions.

In addition to their use in field work, robots will be used for the transportation and placement of sensors on the lunar surface. For example, geophones can be placed at designated areas to gather seismic information which, when coupled with other geophysical techniques, can provide information about the depth of distinct layers in the lunar subsurface.[2] Robots can also be used to conduct extensive ground-penetrating radar surveys, as discussed previously.

Mass spectrometers can be placed in regions of the Moon where outgassing is suspected to have occurred in the recent past. These areas include the Aristarchus Plateau, the crater Linne, and other sites where so-called "lunar transient events" have been reported (see Appendix M). While most scientific investigators doubt such occurrences, it would nevertheless be useful to document the existence (or lack thereof) of such sporadic events, which, if true, may indicate subsurface deposits of useful volatile materials.

2.3   ASTRONOMY FROM THE MOON

The Moon has many advantages as a location for making astronomical observations and measurements, and it is very likely to become humankind's principal scientific base for astronomy. For this reason alone, the Moon should be given a high priority for human development.

From the time of Galileo until the beginning of the twentieth century, the primary means by which astronomers derived knowledge of the universe was by making telescopic observations of the visible light that was emitted or reflected by objects in space. Significant advances in astronomy were made from these telescopic observations; however, the data that were gathered were incomplete because visible light represents only a small fraction of the electromagnetic spectrum from which the universe may be observed.

In the twentieth century, technical advances have allowed astronomers to expand their investigations to include the entire range of the electromagnetic spectrum, thus greatly expanding our knowledge of the universe. The divisions and characteristics of the electromagnetic spectrum from which astronomical observations may be made are listed in Table 2.1.

---

[2] On Earth, mechanical vibrators are often used instead of explosives because they create a repeatable and well-characterized signal. However, these mechanical vibrators may not be as cost-effective as old-fashioned explosives for the early seismic exploration on the Moon.

Table 2.1: Properties of the electromagnetic spectrum

Photon category	Wavelength (cm)	Photon energy (eV)	Objects studied
Radio	>10	<10^{-5}	Gaseous nebulae; Search for Extraterrestial Intelligence (SETI)
Microwave	10 to 0.01	10^{-5} to 10^{-2}	Pulsars; dusty regions of galaxies
Infrared	0.01 to 7×10^{-5}	0.01 to 2	Cool stars; infrared galaxies; nebulae
Visible light	7×10^{-5} to 4×10^{-5}	2 to 3	Planets; moons; galaxies
Ultraviolet	4×10^{-5} to 10^{-7}	3 to 10^3	Young, hot stars; galaxy evolution
X-ray	10^{-7} to 10^{-9}	10^3 to 10^5	Black holes; hot gases
Gamma ray[3]	<10^{-9}	>10^5	Supernovae; solar flares

2.3.1   Earth-based astronomy

Astronomical observations that are made by Earth-based telescopes are limited by the atmosphere, which distorts images and attenuates portions of the electromagnetic spectrum. The ionosphere reflects long-wavelength emissions in the radio spectrum back into space, thus preventing them from reaching Earth-based detectors. Water in the atmosphere absorbs infrared radiation, clouds block visible light, and the upper atmosphere – particularly the ozone layer – absorbs most of the wavelengths that are shorter (more energetic) than visible light. In addition to the atmosphere, light and radio "pollution" from natural and manmade sources also interfere with observations. Because of these factors, virtually no segment of the electromagnetic spectrum is totally free from distortion or interference, and astronomical observations that are made at the Earth's surface are correspondingly limited.

Some of the undesirable atmospheric effects may be overcome by the use of distortion-correcting optics or by placing telescopes on aircraft or mountain-top locations that are above a significant portion of the Earth's atmosphere. Even with these improvements, however, there is still much to be gained by removing all the effects of Earth's atmosphere and light and radio pollution.

2.3.2   Astronomy from Earth orbit

The most effective method for eliminating the deleterious effects of the atmosphere on astronomical observations is to move telescopes into space. Because technological advances have made access to space possible, telescopes dedicated to the

---

[3] By convention, gamma rays are photons that have energies of greater than 100 thousand electron volts (100 keV). However, by strict definition, gamma rays are photons that originate from the nucleus of an atom, and X-rays are high-energy photons that originate from the interactions of electrons with energy fields or with other matter. The standard convention of separating gamma and X-rays at the 100-keV energy threshold rather than by their origin is observed in the present text.

observation of every major segment of the electromagnetic spectrum have been placed in Earth orbit, creating whole new bodies of knowledge about the universe. For example, orbiting observatories such as the IRAS and Compton Gamma Ray telescopes have made extensive observations of the sky in the infrared and gamma-ray segments of the electromagnetic spectrum, respectively, that would not have been possible by means of Earth-based observatories. The Hubble Space Telescope operates primarily in the visible-light region of the electromagnetic spectrum; it has greater resolving power[4] than, and its images far exceed the quality of, Earth-based telescopes.

However, Earth orbit is again not the ideal location for an observatory. A telescope in Earth orbit is not a completely stable platform for making observations because it moves through, and is displaced by, the Earth–Moon–Sun gravitational fields. Furthermore, it is subject to other disturbing forces such as sunlight and the tenuous atmosphere of the Earth that extends several hundred miles into space. Although the errant motion of orbiting telescopes can be minimized by devices such as gyroscopes or reaction control jets, the resolving power of Earth-orbiting telescopes is compromised by motions induced from destabilizing and corrective forces. Observations of celestial objects are also disrupted whenever the orbit of the telescope causes the Earth to eclipse the object that is being studied.

Telescopes in Earth orbit are difficult to maintain and their useful lifetimes are limited, especially if they use consumable materials such as liquid helium[5] for their operations. They are also subjected to damage from micrometeorites, thermal stress (moving in and out of the Earth's shadow), cosmic rays,[6] solar flares, and high-energy photons (X-rays and gamma rays). Finally, it is expensive to transport telescopes into space, and the number of telescopes that are available for making observations from space is consequently far fewer than desired.

2.3.3 Moon-based astronomy

The Moon is the ideal site in the solar system for the study of astronomy. It has a very low level of seismic activity, and is thus a much more stable platform for astronomy than either Earth-based or Earth-orbiting observatories. Because there is negligible atmosphere on the Moon, the entire electromagnetic spectrum that arrives through

---

[4] The resolving power of a telescope is its ability to distinguish or resolve fine details in the image it produces, such as the ability to distinguish between the separate stars in a binary-star system. The units of resolving power are arc-seconds, which are $\frac{1}{3600}$ of one degree of the angular measurement of a circle. Telescopes such as the 200-inch Palomar Observatory Telescope have a theoretical resolving power of 0.1 arc-second; however, atmosphere distortion and light pollution reduce that value to approximately 1 arc-second. The Hubble Telescope in Earth orbit has a much higher resolving power of approximately 0.02 arc-seconds.

[5] Liquid helium is used to maintain a low-temperature background for the operation of infrared telescopes.

[6] Cosmic rays are charged atomic nuclei that are of galactic and extragalactic origin. They have high energy levels (high velocity) that pose a significant radiation hazard in the space environment.

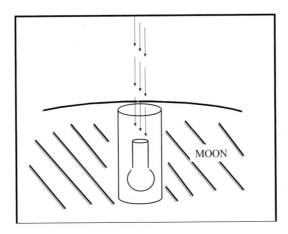

Figure 2.1. Viewing configuration of X-ray and gamma-ray telescopes (Sharpe/Schrunk).

interstellar space is available for observation from the Moon without distortion or attenuation.

The regolith of the Moon can be used to shield telescope instrumentation from temperature extremes, micrometeorites, and ionizing radiation. Shielding can virtually eliminate thermal stresses[7] in telescopes, except for those telescopes that directly view the Sun. X-ray and gamma-ray telescopes, which, by definition, are designed to detect high-energy photons, can be placed several meters below the level of the lunar surface to eliminate the background radiation of space. In this manner, below-ground telescopes will "see" only the primary X-ray and gamma-ray radiation which comes from the small solid angle of the sky that contains the object being observed (Figure 2.1); all other primary and secondary[8] ionizing radiation will be attenuated by the Moon.[9]

A limitation of placing telescopes below the level of the lunar surface is that they will be able to visualize only a small segment of the sky at one time. The rotation of the Moon will thus move the telescope out of viewing range of a given object of interest for an entire lunar month. To overcome this limited observation time, multiple telescopes can be placed at regular intervals around the circumference of the Moon at the equator, at other latitudes, and at the poles, so that any given object in the sky will always be within viewing range of at least one telescope. In this manner, continuous observations of rapidly changing celestial objects will be possible.

---

[7] The temperature at approximately one meter below the lunar surface is a constant minus 20 degrees Celsius.

[8] Secondary radiation is mostly comprised of high-energy photons (X-rays and gamma rays) that are produced by the collision of cosmic rays with the lunar surface. Several meters of lunar regolith are required to attenuate secondary radiation to background levels.

[9] X-ray and gamma-ray telescopes that are positioned below the lunar surface will be subjected to the minimal, natural background radiation of subsurface lunar materials.

One of the biggest advantages of the Moon is that the techniques of interferometry can be fully exploited. Interferometry is the imaging process by which two or more separate telescopes obtain simultaneous[10] images of a celestial target. Optical train elements are purposely introduced to compensate for wave front delay from the different telescopes. The separate images, with optical waves arriving in step, are then combined to form a single image of higher resolution. Optical interferometers now operational in the U.S. include those at the Keck Observatory in Mauna Kea, Hawaii, the Palomar Testbed Interferometer (PTI) at Mt. Palomar, and the Center for High Angular Resolution Astronomy (CHARA) at Mt. Wilson in California. They are capable of milli-arc second resolution, and are routinely able to resolve images of stellar disks.

The advantage of interferometry is that the resolution of the reconstructed image is equivalent to the resolving power of a telescope whose diameter is equal to the distance that separates the telescopes; resolving power of orders of magnitude[11] greater than any single telescope can thus be achieved with interferometry. The seismic stability of the lunar surface and the lack of an atmosphere make the Moon an ideal site for interferometry.

The significantly increased resolving power of telescopes in an interferometry grid on the Moon (Figure 2.2) will permit the observation not only of planets in orbit about nearby stars, but also of large geographic features of those planets, such as mountains and oceans.

The Moon offers astronomers another advantage for making telescopic images. In addition to creating high-resolution images with interferometry techniques, telescopes on the Moon will be able to collect light for longer periods of time, thus creating images of high detail as well as high resolution.[12] A telescopic image is the sum of the photons that are collected from an object of interest; its creation is dependent upon the light-gathering power of the telescope[13] and the exposure time that is taken to make the image.

[10] The technique of interferometry requires that the wavefront of the image that is received by one telescope must arrive within several wavelengths of the same image that arrives at the other telescopes (ideally, the same wavefront is received simultaneously at every site). Because the wavelength of radio waves are longer (and are thus less challenging for making interferometry images) than all other photons, interferometry techniques were first developed for radio astronomy. Images based upon interferometry have now been successfully accomplished with microwave, infrared, and visible-light telescopes.

[11] Resolution on the order of milli-arc-seconds to micro-arc-seconds (10^{-3} to 10^{-6} arc-seconds) may be expected from the first interferometry telescope observatories to be placed on the Moon.

[12] The information content, or detail, of an image is a function of the number of photons that comprise the image. The overall quality of an image depends upon both the resolving power of the telescope *and* the number of photons that have been gathered to create the image.

[13] One measure of a telescope's ability to generate images is its light-gathering power, which is the number of photons that it is able to collect per unit time. Light-gathering power is proportional to the surface area of the primary mirror of the telescope.

Figure 2.2. Interferometry array of telescopes on the Moon (Pat Rawlings, with permission).

Long exposure times (photon integrated periods) are used to collect a large number of photons. Thus, it is possible to create detailed telescopic images of the faintest, most distant objects in the universe by simply gathering light from those objects for a sufficiently long period of time. On Earth, however, atmospheric distortion negates the advantages of long exposure times. For space-based telescopes, detailed images of faint celestial objects can be obtained by using long exposure times, but the small number and operational expense of space telescopes limits the time that is available for creating images of any one object.

The solution to the need for creating images of very faint objects is to place multiple telescopes on the Moon. These telescopes will not need to have greater light-gathering ability than telescopes on the Earth or in Earth orbit. However, each telescope on the Moon can be dedicated to gathering light from a given celestial object of interest for as long as desired – for years or decades – until sufficient photons have been gathered to create an image of suitable detail. In this way, it will be possible to obtain highly detailed images of the most distant objects in the universe. In addition to making long exposures of faint objects, the ability to make continuous observations from the Moon will also be applied to observations of dynamic celestial events, such as supernovas, gamma-ray bursts and the motion of planets around stars.

Permanently shadowed craters in the north and south polar regions, with temperatures as low as 30–40° Kelvin, are ideal for the operation of infrared telescopes. The far side of the Moon is also free from all radio interference from the

Earth, and it is therefore the ideal site in the solar system for the operation of radio telescopes.

The Moon, despite its advantages, may seem to be too far away to become a major site for astronomy. It is a fact that it is more expensive to launch a telescope from the Earth to the Moon than to Earth orbit. In addition to the expense of transportation, there is a problem with power and communications for telescope operation. Solar panels that are attached to a telescope on the Moon cannot provide power to the telescope during the (fourteen Earth-day) lunar night, which is the desired time for making visible-light telescopic observations. Telescopes that operate on the far side of the Moon also have no convenient method for transmitting their data back to Earth.

The solutions to these particular problems are addressed in later chapters of this book. In essence, the cost of transporting telescopes from the Earth to the Moon will become zero, because telescopes will be manufactured on the Moon from lunar materials. The primary support structures of radio telescopes, for example, can be made from lunar iron or aluminum, and mirrors for visible-light telescopes can be made from lunar glass and sand. Surface transportation services, telecommunications, and continuous electric power will be available at any desired site on the Moon through a global utilities infrastructure grid (see Chapter 7), so that telescopes may be assembled, operated, and maintained anywhere on the lunar surface.

The advent of continuous power, communications, transportation, and manufacturing on the Moon, all of which can be made available with existing technology, will enable lunar industries to mass-produce telescopes so that the Moon, in the twenty-first century, can be transformed into a coordinated global astronomical observatory. Thousands of lunar-made telescopes can be placed at both north and south polar regions, at regular intervals along the equator – for example, every 150 kilometers (150 kilometers = five degrees of circumference at the equator), and at 45 degrees north and south latitudes, as depicted in Figure 2.3.

With this arrangement of telescopes, virtually the whole universe will be open to continuous, high-resolution telescopic observation. At every observatory, telescopes that are sensitive to each separate segment of the electromagnetic spectrum can be placed side by side so that they can simultaneously observe any given object of interest in the universe. Because the Moon is a stable platform, each telescope site may also be connected to a local interferometry grid. When a global network of telescopes is in operation on the Moon, the studies described below become possible.

2.3.3.1   The Sun

As the Moon rotates on its axis, the telescopes that are placed at regular intervals at the equator will each make observations of the Sun as it comes into view. The Sun will thus be under continuous observation at every wavelength of the electromagnetic spectrum. All of the features of the Sun (such as sunspots and corona events) will be investigated continuously over each eleven-year solar cycle. Solar flares will also be observed as they occur, thus giving advance warning of impending increases in

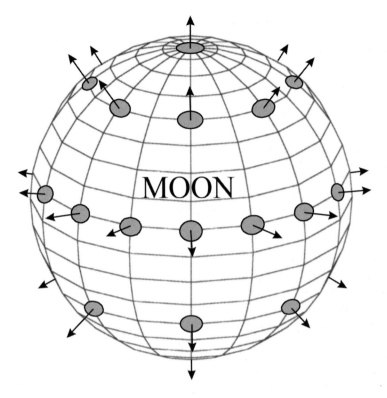

Figure 2.3. Global placement of telescopes on the Moon (Sharpe/Schrunk).

radiation levels to people and sensitive equipment that are working on the Moon, in Earth orbit, and elsewhere in the solar system.

2.3.3.2   *The planets*

Except for those occasions when they are eclipsed by the Sun or the Earth, all of the planets and moons in the solar system will be subject to continuous evaluation by telescopes that are placed at the equator of the Moon. Of particular interest will be synoptic studies of the structure and dynamics of the atmospheres of every celestial body in the solar system that has an atmosphere, including the Earth. Knowledge of atmosphere dynamics on other planets will add to the understanding of Earth's weather patterns, and will aid in planning future manned missions to the planets and moons. For example, avoidance of global dust storms will be a high priority for both manned and robotic missions to Mars. The high resolution of visual images that will be made possible by interferometry will permit detailed examination of the topography of distant objects such as Pluto and its three moons, Charon, Nix, and Hydra.

2.3.3.3   *Asteroids, comets and NEOs*

Telescopes based on the Moon will be able to provide a comprehensive survey of the asteroid belt between Mars and Jupiter and of "near-Earth objects" (NEOs), which are the asteroids and tailless (or "burned-out") comets that approach or cross Earth's orbit. A survey of the asteroids and comets in the Kuiper Belt (which extends from the orbit of Neptune, at 30 AU to 50 AU[14] from the Sun and contains tens of thousands of celestial bodies) and the Oort Cloud (which extends from 50 AU to three light years from the Sun) will also be possible.

The NEOs are of vital interest for two reasons. First, the collision of a one-kilometer diameter or larger NEO (of which there are an estimated two thousand) with the Earth could conceivably destroy all human life here. Therefore, it is impera-tive that the orbits of all the larger NEOs be known with certainty. If it is determined that the Earth is threatened by the pending collision of an asteroid or comet, meas-ures could then be taken to alter the orbit of the threatening object so that a collision can be avoided. The second reason for the interest in NEOs is that they are composed of valuable resources including metals, hydrocarbons, and water. When mining operations of NEOs become possible in the twenty-first century, the Earth and the Moon, for practical purposes, will have access to a virtually inexhaustible supply of raw materials (see Chapter 10).

2.3.3.4   *Stars*

Stars, of which the Sun is a typical example, are large self-luminous bodies of gas that vary widely in size, brightness, and temperature. They have sufficient mass to produce the conditions that sustain continuous nuclear fusion (mostly the fusion of hydrogen into helium) at their center of gravity. The "burning" core of every star produces dynamic surface changes, and some stars, such as Cepheid variables, undergo per-iodic changes in their size and temperature.

Stars undergo several stages of evolution from their "birth" to their "death". Depending upon their size when they are formed, they may have a long, relatively unremarkable existence, or a shorter, more turbulent one which concludes with a supernova explosion that generates (for a brief period) more energy than is produced by an entire galaxy. To provide a record of the birth of stars, a number of telescopes on the Moon can be dedicated to observations of "star factories" such as the Orion (Figure 2.4), Eagle, and Lagoon Nebulae.

Likewise, continuous observations of the dynamics of stars such as red giants and variable stars, and of the death of stars that produce white dwarfs, planetary nebulae, supernovae, neutron stars, and black holes will all be possible. For example, the supernova 1987A in the Large Magellanic cloud is growing brighter by 20 percent per year and the shock wave from the supernova has begun to encounter – and light up – an earlier-expelled outer shell of gas (Figure 2.5). Telescopes on the Moon will be able to provide continuous observations of this unfolding dynamic process over the entire range of the electromagnetic spectrum. When telescopes are available in number on

---

[14] AU = Astronomical Unit equal to the distance from the Sun to the Earth.

Figure 2.4. Star factory in Orion Nebula.

Figure 2.5. Supernova 1987A. Telescopes on the Moon will make continuous observations of Supernova 1988A, and other supernova as they are discovered.

the Moon, other supernovae will also be recorded in real time, adding significantly to the database from which dynamic models[15] of supernovae are constructed.

The observations that have been made of star-forming regions, principally in the millimeter-length and infrared segments of the electromagnetic spectrum, have indi-

---

[15] With the use of multiple telescopes on the Moon, highly detailed models of the location, motion, temperature, size, brightness, dynamic nature, and composition of the observable stars in our galaxy, as well as neaby galaxies.

cated that disks of material – the building blocks of the planets, moons, comets, and asteroids – form around nearly all stars at the time of their birth (Frank, 1996). The implication of these findings is that the majority of stars may have planetary systems analogous to our solar system.

Efforts are underway to discover planets around nearby stars (there is indirect evidence of several hundred planets around nearby stars). The means for detecting planets is by indirect methods such as observing the slight motion, or "wobble", of a star that is caused by the gravitational field of a large orbiting planet. When more sensitive telescopes become available, evidence of planets may be obtained by observing the slight decrease in the brightness of a star that is produced when a planet eclipses the star. Ultimately, when interferometers in the infrared and visible-light spectrum begin operations on the Moon, direct imaging and spectral analysis of planets around nearby stars will become possible.

2.3.3.5   *The search for extraterrestrial life*

The Moon will be a prime location from which to search for extraterrestrial life. If a planet in a nearby solar system were found to have water vapor, oxygen, and small quantities of carbon dioxide in its atmosphere, it would become the object of intense scientific and societal interest because such a planet might harbor life.[16]

Water is necessary for life as we know it, and plants "breathe" atmospheric carbon dioxide. Oxygen is a highly reactive element that readily combines with other elements and molecules to form new molecules; free oxygen is thus quickly removed from a planetary atmosphere. If more than trace amounts of oxygen were to be detected in the atmosphere of a planet in a nearby solar system, it could only mean that new oxygen is constantly being supplied. Because the most likely mechanism for the production of atmospheric oxygen is the biological process of photosynthesis, the discovery of oxygen, water, and carbon dioxide in the atmosphere of another planet would indicate the presence of plant life. Using the Earth as a model, the indications of plant life would also suggest the presence of animal (i.e., potentially intelligent) life as well.

The spectral emission "signatures" of oxygen, water vapor and carbon dioxide are all in the infrared part of the electromagnetic spectrum. The search for evidence of extraterrestrial life will be facilitated by the lunar infrared telescopes placed in the permanently-shadowed regions of the Moon, where continuous cooling of the detectors may be possible without the use of cryogenic liquids.

2.3.3.6   *The search for extraterrestrial intelligence*

If intelligent life forms exist outside our solar system, it is possible that they will use radio waves as a form of communication. "Alien" radio signals might be propagated,

---

[16] As of this writing, carbon and oxygen have been detected in the atmosphere of planet HD 209458 b, which orbits the sun-like star HD 209458 in the constellation of Pegasus, some 150 light-years from our solar system

intentionally or unintentionally, into space where they may then be detected by means of radio astronomy. Radio signals that are used for communication are entirely different from radio waves that are generated by natural means, such as electrical storms. Automated radio reception equipment can sort out any radio waves that are not of natural origin, and thus indicate the presence of extraterrestrial intelligence. Unfortunately for radio astronomers, the Earth is saturated with "intelligent" radio signals from a variety of human-created sources such as radio stations and communications satellites. Even if radio waves from extraterrestrial sources were to reach the solar system, those signals would be overwhelmed by the radio "noise" of Earth sources.

However, the far side of the Moon is a convenient site[17] that is constantly free from radio interference from the Earth. The mass of the Moon attenuates all radio signals from the Earth, and its far side is thus in pristine radio space, where signals from intelligent sources, if they exist, will not be confused with Earth-generated radio noise. The discovery of extraterrestrial intelligence would rival any other significant event in human history.

2.3.3.7   Galaxies

The visible universe is comprised of billions of galaxies[18] that exhibit a wide variety of sizes, structures, and light-emission characteristics. Although millions of galaxies have been imaged and thousands have been studied in some detail, the vast majority of galaxies that populate the universe remain unexamined.

In December 1995, the Hubble Telescope took multiple exposures of a small region of deep space over a ten-day period, creating a single image termed the Hubble "deep field"[19] (Figure 2.6). The composite image contains over 3,000 galaxies, including hundreds of faint galaxies that had never been visualized before. Many of the newly observed galaxies were found to have large red shifts,[20] indicating that they formed within one billion years after the "big bang" that created the visible universe.

---

[17] The only other region of the solar system that is free from radio interference from the Earth is the region of space that is on the opposite side of the Sun from the Earth (180 degrees angular displacement from Earth's orbit).

[18] Galaxies are systems of billions of stars that are bound together by their own gravity.

[19] The Hubble Telescope required 150 orbits over a 10-day period to produce the composite image. Telescopes on the Moon will be able to make images of much higher resolution (with interferometry techniques) for much longer periods of time (months–years).

[20] "Red shift" is the elongation of wavelengths of light (thus shifting visible light towards the red, or longer, segment of the electromagnetic spectrum) caused by the expansion of the universe. The degree of red shift of an object is proportional to the distance of the object from the observer (Hubble's Law). Objects whose emitted light is characterized by the highest red-shift values are therefore at the far limits of the visible universe.

Figure 2.6. Hubble deep field of galaxies.

When large numbers of (lunar-manufactured) telescopes are placed in operation on the Moon, it will be possible for dedicated telescopes to make long-exposure images of these primordial galaxies as they appeared shortly after their formation.[21] Other phenomena that are associated with galaxies and deep space will be studied at high resolution and detail from the Moon, including black holes, which

[21] The continuous observation of hundreds of galaxies that lie at the edge of the visible universe will predictably detect type Ia "standard candle" supernovae in many of these galaxies, thus providing greater accuracy in the estimation of the age and size of the universe.

are thought to lie at the center of most galaxies, quasars,[22] cosmic jets, supernovae, new star formation, gravitational lensing, and gamma-ray bursts.

2.3.3.8   *The structure of the universe*

Telescopes on the Moon will be used to expand the Hubble "deep-field" study into a whole-sky survey of galaxies.[23] By plotting the coordinates, including the red shift (both time and distance separation) of each galaxy in the survey, a comprehensive four-dimensional model of the universe can be created. From this survey of galaxies, the age, origin, evolution, size, structure, and ultimate fate of the universe will be better understood.

2.4   OTHER SCIENCE OPPORTUNITIES

In addition to geoscience and astronomy, the Moon offers unique opportunities to gather knowledge in many other fields, such as cosmic radiation, particle physics, psychology, and sociology.

2.4.1   Cosmic radiation and the solar wind

The lack of an atmosphere on the Moon permits the cosmic radiation and solar wind plasma to reach the lunar surface unimpeded. Thus, analogous to the advantages of lunar-based observations of electromagnetic radiation, detectors can be placed on the lunar surface to study the origins, character, and directions of cosmic radiation. Similarly, measurements of the solar wind can be made directly from the lunar surface.

The interaction of the Earth's magnetic field with the solar wind creates Earth's "magnetosphere", which is a teardrop-shaped cavity in the solar wind that encloses the Earth. The elongated tail of the magnetosphere extends beyond the orbit of the Moon in the anti-sunward direction so that the Moon enters, interacts with, and exits the magnetosphere once every lunar month. Because of the unique properties of the Moon's orbit about the Earth, the structure and dynamics of Earth's magnetosphere can be studied directly from the lunar surface.

---

[22] Quasars, or quasi-stellar objects, are thought to be primordial galaxies that have very large black holes in their center. Quasars emit far more energy than ordinary galaxies, and their emission spectra have large red shifts, indicating that they are among the earliest structures to form after the "big bang", and that they lie at the edge of the visible universe.

[23] The effort to produce a whole-sky survey of galaxies will be immense: the Hubble "deep-field" image comprises a region of the northern sky that would be subtended by a small coin placed at 75 feet (25 meters) from the telescope.

2.4.2   Particle accelerators

The orderly development of mining and manufacturing facilities on the Moon will make possible the construction of large structures such as the lunar railroad system (Chapters 6 and 7). One large-scale scientific project that can be built on the Moon is a super-conducting particle accelerator ("atom smasher") for studying subatomic particles. A 20 to 100-km (or larger) diameter structure can be built just below the lunar surface to take advantage of regolith shielding and constant temperatures. Operation of the collider will be greatly simplified, in comparison with Earth-based colliders, because of the natural vacuum of space that exists on the Moon (see Chapter 8). When it becomes operational, it will be larger, and thus capable of yielding more knowledge on the fundamental structure of matter, than any Earth-based particle accelerator.

2.4.3   Psychology and sociology

By definition, the other-worldly environment of the Moon will present a unique laboratory for the study of individual (psychology) and group (sociology) behavior. The first humans to re-visit the Moon will be subjected to the stresses of confined quarters, isolation, and limited supplies, similar to "wintering-over" scientific expeditions at bases on Antarctica (see Appendix J). The lunar base will thus be a natural scientific laboratory for the investigation of psychological responses to the stressful environments on the Moon. As the size and number of lunar bases increase, the focus of studies will shift to the adaptation of humans to a "new" planet whose habitats have been designed to optimize living and working conditions. The results of those studies will, in turn, be used to create an even more ideal environment for the humans who migrate to the Moon.

Similar scientific studies will be conducted on group behavior – on the interactions of people from diverse cultural, religious, ethnic and other backgrounds, who will work together in the exploration and development of the Moon. The gradual build-up of the lunar population will provide social scientists with ample time to make measurements of group behavior. The results of ongoing studies may be used to modify the pattern of human settlement on the Moon, to take advantage of forces that increase the synergy of lunar development efforts. To that end, one of the conditions for being selected as an emigrant to the Moon may be the willingness to participate in regular surveys that measure the parameters of group behavior. The psychology and sociology of humankind's migration into space are the subject of the book, *Living and Working in Space*, by Dr. Philip Harris, published by Wiley-Praxis, 1996 (2nd edition).

2.4.4   Physiology

Microgravity – the virtual lack of gravity that exists in spacecraft in Earth orbit and free space – produces deleterious physiological effects on humans, including cardiac decompensation, changes in the immune system, and loss of bone mass. The Moon's gravity ($\frac{1}{6}$ that of the Earth) is much more Earth-like than orbiting space stations and is predicted to be less deleterious on humans than microgravity, but the effects of lunar gravity are difficult to estimate. A whole new field of biological investigation related to the effects of reduced gravity on human physiology will begin when humans return to the Moon.

2.5   ENGINEERING CHALLENGES

While the task of basic science is to produce knowledge, the task of the disciplines of engineering is to apply knowledge in the creation and improvement of useful products or tools, such as computers, pharmaceuticals, metal alloys, robots, and life-support systems. When a new product is created, the performance assessment of that product adds to the store of scientific knowledge, which, in turn, broadens the base from which more advanced tools can be made.[24]

The Moon will challenge the engineering disciplines to create new technologies that will operate effectively there, with minimal expense and side-effects. The knowledge that will be gained from creating new technologies will be used for human settlements elsewhere in the solar system, and will be applicable to the solution of problems on the Earth as well. The following is a partial list of the engineering challenges to the human settlement of the Moon.

2.5.1   Robots and tele-operations

The development of remotely-controlled robots on Earth has not been as dramatic as other technologies, such as computers, because they have been mostly limited to dangerous or highly repetitive manufacturing tasks. However, many projects on the lunar surface will be hazardous to humans, dictating that tele-operated robots will be the primary source of labor on the Moon for the foreseeable future.

Tele-operated robots will be used for scientific missions and for mining, manufacturing, construction, and transportation tasks at the first lunar base. To satisfy these roles, substantial advances in robotic engineering will be required. An Earth-based tele-operation service industry that works three eight-hour shifts per day will also be needed to direct the operations of the lunar robots, whose number may rise to several hundred before the first humans return in numbers. In an idealized scenario, human-operated robots will install the utilities infrastructure (communication, power, and transportation) at the first lunar base, mine the regolith, recover oxygen and metals from the regolith, and begin to manufacture solar cells and construction materials. They will install and test scientific instruments, manufacturing equipment, and human-habitable structures as this equipment is delivered from the Earth. When the first humans arrive at the lunar base, tele-operated robots will transport them

---

[24] The combined operations of science and engineering use knowledge and tools to produce more knowledge and tools – an exponential relationship that predicts increasing rates of growth of knowledge and of the number and proficiency of tools.

from the rocket landing site to their fully operational, human-rated underground habitats.

The use of robots during the initial phases of lunar development will gradually be expanded to larger and more complex tasks, such as the construction of the global lunar utilities system (see Chapter 7). Emphasis will also be placed on autonomous operations, where little or no human control of robots is necessary. The lessons that are learned from the lunar experience will be applied to the exploration and development of other areas of the solar system, such as Mercury and Mars.

2.5.2   Chemistry

The desire to "live off the land" will give rise to many innovations in chemistry. Organic (carbon-based) chemical compounds that are abundant on the Earth are not known to exist on the Moon.[25] However, carbon and other light elements such as hydrogen and nitrogen are delivered to the lunar regolith by the solar wind (see Chapter 3). As the regolith is mined and processed for metals and oxygen, carbon and other solar-wind elements will also be harvested and become the feedstock materials for the nascent lunar chemical industries.

Through a series of steps involving carbon (C), oxygen (O), and hydrogen (H), it is possible to make simple organic compounds such as formaldehyde (HCHO) and ethylene (C_2H_4). These two compounds can then be used to create basic food groups (sugars, fatty acids, and amino acids) and plastics. By this chemical engineering means, an organic chemistry/biochemistry/plastics industry can evolve, albeit on a very small scale initially due to the low concentration of carbon, hydrogen, and other light elements on the Moon. However, mining operations of the lunar regolith will continue to add to the store of elements that are needed for the chemical industries. Moreover, almost everything will be recycled, including septic waste, so that as humans begin to occupy the Moon, they become an additional source for organic compounds. When agricultural operations begin on the Moon, crops such as sugar (from sugar beets) and latex (rubber trees), as well as the biomass "waste" of all plants, will provide the feedstock of the organic chemistry industry.

At a later stage of lunar development, mass drivers (and other launch and recovery systems) will provide the Moon with access to the resources of near-Earth objects (see Chapter 10). NEOs have relatively high quantities of hydrocarbons and other light elements, and they will supply the Moon with all of the basic elements and compounds that are needed for the lunar chemical industries.

2.5.3   Health-care challenges

The people who will be living and working on the Moon during the first decades of lunar development will be healthy, young, or middle-aged adults whose medical needs

---

[25] The meteorite remnants of carbon-bearing asteroids (e.g., carbonaceous chondrite asteroids) and comets that are imbedded in the lunar surface may contain useful quantities of carbon compounds.

will be mostly limited to acute care conditions.[26] Until the population reaches several thousand people, it will be impractical to staff a full-service hospital at all times. For this reason, advances in tele-medicine services will be needed, so that small teams of physicians and technologists on the Moon will be able to work in real time with specialists on the Earth via tele-communication links to perform surgery and other complex tasks. Advances in automated health-monitoring systems will also be needed to keep the number of medical personnel at a minimum.

A major effort will be made to prevent communicable diseases from infecting the biosphere(s) of the Moon. To that end, diagnostic technologies that are capable of detecting infectious diseases will be used to screen individuals and cargo prior to their departure from the Earth. To minimize the chance of the spread of disease, human-habitable and agricultural areas will be constructed so that they can be isolated (quarantined) from one another whenever necessary.

When lunar bases are able to support thousands of people, it may become desirable to transport people with physical disabilities to the Moon, where the 1/6 gravity will permit greater mobility than on Earth (see Appendix J). The Moon will thus challenge, and spur advances in, the state of the art of both diagnostic and therapeutic medical procedures.

2.5.4   Life-support systems

Human settlements on the Moon will require advances in control mechanisms and monitors for the long-term control and maintenance of regenerative air, water, agricultural, and waste management systems (controlled ecological life-support systems, or CELSS). Partially regenerative, closed habitat tests with human volunteers have been conducted for as long as 90 days on Earth, but reliable life-support systems that can operate indefinitely in the lunar environment will require substantial engineering. With existing life-support technology, human missions to the Moon of greater than 60 to 90 days will require the periodic delivery of consumable materials from the Earth, analogous to the needs of Earth-orbiting stations.

A significant challenge for the design of CELSS will be the establishment of agricultural facilities on the Moon. The growth of plants from seeds and other agricultural experiments have been conducted in the microgravity environment of space stations, but no food crop cycle (seed-to-seed) has been accomplished in space. An extensive program of lunar agricultural experiments will be necessary to allow a self-sustaining colony to exist there. Multiple variables, such as humidity, temperature, atmospheric composition, air currents, lighting, and nutrients must be tested for growing each of a wide variety of food crops in the lunar soil. The effects of the Moon's gravity ($\frac{1}{6}$ of Earth's gravitational force) on crops must also be determined.

The mining and processing of the lunar regolith during the initial unmanned phase of lunar base development will yield quantities of carbon, hydrogen, oxygen,

---

[26] As the population of people on the Moon increases over time, the scope of medical care will need to be expanded to include the whole gamut of medical conditions, including obstetrics, pediatrics, and the problems of aging.

nitrogen, phosphorus, sulfur, and other elements that are essential for life. They will be used to create the nutrients and the atmospheres for the first agricultural habitats, in conjunction with the development of human habitats. It will then be possible to begin agricultural experiments with food crop cuttings and seeds that are delivered from the Earth to the Moon. Pilot development trials of combinations of plants and animals living continuously in controlled ecosystems will become an ongoing investigation. Which species belong together, which are the optional choices, and which must be kept apart? Which sets of lighting, temperature, humidity, atmospheric composition, and air flow produce the best results? Answers to these questions will form the basis for biological aspects of habitat architectures for all subsequent destinations in space.

The successful adaptation of food crops to the lunar environment, and the development and integration of technologies for the control of air, water and waste-recycling systems, represent significant engineering challenges. However, no absolute barriers are foreseen. When reliable life-support systems become operational on the Moon, the full-scale migration of people from the Earth to the Moon, and to other sites in the solar system, will be possible. The recycling technologies may also be applicable for air, water, and waste management systems on Earth.

2.5.5   Mining and manufacturing operations in the lunar environment

Initially, the mining and manufacturing equipment that is needed on the Moon will be imported from Earth. However, the lack of an atmosphere means that there will be no protection from hazardous radiation and meteorites, that waste heat cannot be dissipated by atmospheric convection, and that lubricants such as oil and grease will evaporate in the lunar vacuum. The radical (in Earth terms) temperature extremes on the lunar surface will also subject equipment to significant thermal stresses. These harsh conditions will require substantial design changes in virtually all equipment if it is to operate on the lunar surface.

Another problem for lunar operations is the abundance of dust. Lunar dust is an electrostatically charged abrasive powder that clings to spacesuits and almost all other surfaces, including equipment and the seals around air locks. Dust contamination is a threat to machine operations because it can obstruct and abrade moving parts, and interfere with visual tele-operations. During the Apollo missions, the seals on spacesuits were compromised (in some cases becoming unsafe after only two uses). Dust coated the instruments and machinery and was very difficult to remove. The astronauts tried to brush the dust off each other before re-entering their surface module, but the brushing only embedded the particles more deeply into their suits. Moreover, one of the astronauts suffered hayfever-like symptoms during the return journey; likely caused by the dust that was brought into the cabin on suits and boots. Explosives that are used in mining operations and the exhaust from rockets that land and take off from the Moon will carry the dust at high speeds for long distances.[27]

---

[27] A significant advantage of mass drivers (see Appendix K) as compared with rockets is that their operations will not disturb the lunar dust.

Lunar dust will be a constant companion of machines and humans on the lunar surface, and its negative features must be understood and overcome.

2.6 SUMMARY

The extension of human presence to the Moon will lead to significant advances in human knowledge. The Moon will, in effect, be used as a laboratory from which a host of scientific investigations can be conducted. The process of building that laboratory will be accompanied by increases in engineering knowledge – knowledge that will be made available for the solution of problems on Earth and for the further extension of human presence to more distant reaches of the solar system.

3

Lunar resources

3.1 INTRODUCTION

The Moon has more than sufficient resources for all phases of lunar development. It also has access to the vacuum of space from the lunar surface, low gravity (as compared with the Earth), and other unique features that will enhance exploration and development projects on and from the Moon. The present chapter reviews the resources that the Moon offers for future exploration and settlement.

3.2 ELEMENTS

The cost of transporting payloads from the Earth to the Moon is greater than U.S. $20,000 per kilogram. Lunar activities will be greatly facilitated if the materials that are needed for exploration and settlement are obtained directly from the Moon. The Moon possesses all of the elements that are found on Earth. From these elements a global infrastructure that supports all human activities on the Moon can be constructed.

The elemental composition of the Moon is usually described in two categories – major elements and trace elements. In general, a major element is one that constitutes more than 1 percent of the total. Table 3.1 lists the major elements found on the Moon, based on analysis of lunar samples.

Trace amounts of other important elements that are present in the regolith are listed in Table 3.2. Many of the lighter elements in Table 3-2 are delivered to the lunar regolith by the solar wind.

As previously mentioned, iron, aluminum, and titanium will be used for construction materials, while silicon is used in the production of computer chips, photovoltaic (solar) cells, fiber-optic cables, mirrors, and lenses.

Table 3.1: Major elements in the lunar regolith (from Turkevich, 1973)

Element	Percent of atoms		
	Mare	Highland	Average surface
Oxygen (O)	60.3 ± 0.4	61.1 ± 0.9	60.9
Silicon (Si)	16.9 ± 1.0	16.3 ± 1.0	16.4
Aluminum (Al)	6.5 ± 0.6	10.1 ± 0.9	9.4
Calcium (Ca)	4.7 ± 0.4	6.1 ± 0.6	5.8
Magnesium (Mg)	5.1 ± 1.1	4.0 ± 0.8	4.2
Iron (Fe)	4.4 ± 0.7	1.8 ± 0.3	2.3
Sodium (Na)	0.4 ± 0.1	0.4 ± 0.1	0.4
Titanium (Ti)	1.1 ± 0.6	0.15 ± 0.08	0.3

Table 3.2: Trace elements of the lunar regolith (derived from Haskin and Warren, *The Second Conference on Lunar Bases and Space Activities of the 21st Century*, W.W. Mendell, ed., 1991)

Element	Grams per cubic meter of lunar regolith (approx.)
Sulfur (S)	1,800
Phosphorus (P)	1,000
Carbon (C)	200
Hydrogen (H)	100
Nitrogen (N)	100
Helium (He)	20
Neon (Ne)	20
Argon (Ar)	1
Krypton (Kr)	1
Xenon (Xe)	1

The trace elements will be used to produce fatty acids, amino acids, vitamins, and sugars that are needed for life-support systems,[1] as well as plastics. The inert elements such as helium and neon will be used for compressed gas and other applications. The lighter isotope of helium (He-3) is present in greater concentrations on the Moon than on the Earth, and it has significant economic potential as a fuel for future atomic fusion reactors (see Appendix H).

---

[1] Each cubic meter of typical lunar regolith contains, " ... the chemical equivalent of lunch for two large cheese sandwiches, two 12-oz sodas (sweetened with sugar), and two plums, with substantial N and C left over" (quoted from Haskin, p. 393, 1988; N = nitrogen and C = carbon).

3.3 THE LUNAR REGOLITH

The lunar regolith is a resource in its own right because it can be used as a building material. Once a roof has been placed over a habitation module or a robotic manufacturing facility, the lunar regolith can be placed on top of the roof to form a protective shield against ionizing radiation, temperature extremes, and meteorites.

The regolith can also be piled into a berm that acts as a support for solar arrays and other structures, and as a shade that minimizes thermal expansion and stress in metal rails for the lunar railroad (see discussion, Chapter 6). Through a sintering (heating and compressing) process, regolith can be made into bricks for construction projects. Bulk regolith is a resource that will be used early in lunar development because it does not require a substantial infrastructure of technology, factories, or processing plants. Through the adaptation of mature terrestrial rock processing technologies, Moon rock may be quarried, hewn and machine tooled to produce a variety of valuable building materials. See Appendix U on Lunar Rock Structures.

3.4 WATER

As discussed in Chapter 1, the Clementine and Lunar Prospector missions both found evidence for hydrogen in the polar regions of the Moon. Although water can be created by combining the hydrogen and oxygen that are presumed to be present at virtually all sites in the lunar regolith, it will be much simpler to recover water in the form of ice or hydrated minerals if they are present in the north and south polar regions as suspected. Water will not be needed for initial unmanned base operations because the first scientific instruments, tele-operated robots and other equipment will require neither water, nor the oxygen and hydrogen that are water's elemental constituents. As mining and manufacturing at the unmanned base become more sophisticated, water will be needed as a solvent, and both hydrogen and oxygen will be extracted and stored for human use.[2]

Water is essential for manned missions. Controlled ecological life-support systems (CELSS) and experimental agricultural programs will require substantial quantities of water. Hydrogen and oxygen will also be used as the fuel for rockets that return humans to the Earth, and for the operation of fuel cells. If water ice is readily available at the lunar poles, the establishment of permanent human settlements on the Moon will be possible sooner than would otherwise be the case.

---

[2] Manned operations on the Moon are not expected to commence until several years after the first unmanned base has been established. A significant proportion of operations at the unmanned base will be dedicated to the creation of human-habitable facilities, a lunar utilities infrastructure, and the recovery and storage of water and oxygen in anticipation of piloted (human-tended) missions.

3.5   SUNLIGHT

Sunlight that reaches the lunar surface is constant, intense, and virtually inexhaustible. It delivers $1.365 \, kW/m^2$ to the lunar surface when the Sun is directly overhead (orthogonal to the lunar surface). Since photovoltaic (solar) cells that have an energy conversion factor of 20 percent or more can be made on the Moon from lunar regolith materials, all of the electrical power needs of lunar exploration and development projects can be satisfied with solar electric power. A lunar-based power system that is composed of lunar-made solar cells could potentially supply all of the energy that is needed for lunar development and for all of the future energy needs of the Earth (see discussion of a lunar-based solar electric power system in Chapter 10).

Sunlight can be used for agricultural applications on the Moon and it can be concentrated with mirrors to create the high temperatures that are needed for raw material processing and manufacturing. The only problem with sunlight is that it is not available for power generation during the lunar night. However, that problem can be overcome by constructing a solar-powered electric grid around the circumference of the Moon. A circumferentially-placed solar power grid will always be energized, thus providing abundant, constant[3] electric energy for all lunar projects (see discussion in Chapter 7).

3.5.1   Sun-synchronous operations and "Magellan routes" of discovery

The relatively slow rate of rotation of the Moon about its axis offers the opportunity for Sun-synchronous travel on the surface of the Moon (9.6 miles/hr = 15.4 km/hr at the equator and proportionately less at higher latitudes). For example, a solar-powered wheeled vehicle could move west continuously at 1.3 km/hr at 86° south latitude, and maintain the Sun at the same angle in the sky. This route of travel would enable the vehicle to receive sunlight for solar power continuously during its journey around the circumference of the Moon. Such pathways are referred to as "Magellan routes" after the explorer Ferdinand Magellan who headed the first circumnavigation expedition of the Earth.[4]

Similarly, Sun-synchronous railroad operations would allow solar-powered mobile factories to be transported on a lunar railroad at a speed that matches the rotation of the Moon, thus maintaining the Sun in the same position for constant solar power and the operation of solar ovens. The relatively slow rate of rotation of the Moon would allow the train to stop temporarily to load raw materials and to off-load finished products.

A negative feature of sunlight is that it contains ultraviolet radiation, which can

---

[3] Energy storage devices will be needed when the Earth eclipses the Sun.
[4] Magellan routes of discovery can also be exploited on the Planet Mercury. Sun-synchronous routes would allow robotic devices to match the surface speed of rotation of Mercury and thus maintain a constant, optimum Sun angle for solar power and for photographic perspectives, with shadows cast forward along the circumferential route of exploration.

degrade some materials. However, repair and replacement of components must be planned for in any case, and this issue would be handled in the same way that industries on Earth address the problem. Sunlight is also accompanied by the solar wind, which, particularly during solar flares, contains ionizing radiation that is harmful to instrumentation[5] and to biological organisms.

3.6   VACUUM

The vacuum of space extends to the surface of the Moon, and is a valuable asset that has multiple applications, including:

- As explained in Chapter 1, telescopes that are placed on the lunar surface are "space" telescopes that not only have the advantage of the lack of an attenuating atmosphere, but also have a stable platform (the Moon) that makes them easier to operate and maintain.
- Certain manufacturing processes benefit from the presence of a vacuum, such as the separation of elements by ion sputtering (see Appendix E). Manufacturing processes such as the production of microelectronics, micro-machines, carbon nano-tubes, and thin film solar cells also require a vacuum.
- The production of highly-refined metals, glasses, and other materials will be facilitated on the Moon because the lunar vacuum will allow trapped gases to escape from molten materials far more readily than woud occur on Earth, and there will be no atmospheric gases to contaminate or combine with them during the refining processes.
- Rusting of (external) iron and other metal structures will not occur.
- Buildings and towers will not be subjected to aerodynamic forces, and will thus require far less structural strength than their Earth counterparts.
- Rockets and mass drivers will be able to deliver payloads between the lunar surface and lunar orbit without concern for weather delays or aerodynamic loads on launch vehicles (see Chapter 10).
- Particle accelerators ("atom smashers"), require a vacuum for their operations. The Moon will thus be an ideal place to build[6] them.
- Laser communication and power-beaming systems[7] will be able to operate effectively at all times on the Moon because of the lack of atmospheric distortion and attenuation.

[5] To the greatest extent possible, all man-made implements and structures will be placed underground for protection from radiation, meteorites, and temperature extremes.
[6] When a permanent lunar utilities infrastructure and mature manufacturing facilities are in place on the Moon, it will be possible to build and operate a large nuclear collider whose diameter may be as large as the diameter of the Moon.
[7] Optical communication systems are desired for wireless communications because they have high bandwidth capability and produce no radio interference (radio transmissions are prohibited on the far side of the Moon to prevent interference with radio telescope operations).

3.7   TEMPERATURE PROFILE

The temperature of the lunar surface at the equator varies between $+127°C$ during the lunar day and $-173°C$[8] during the lunar night and is colder with increasing latitude. This large temperature difference makes it possible to generate power by the operation of heat engines, such as Stirling engines, that convert temperature differences into mechanical energy. In the polar regions, towers whose upper structures are always in sunlight and whose bases are in permanent shadow could thus be used for continuous heat engine operation. On a global scale, pipelines that transport gases around the circumference of the Moon could be used for thermal management (heating or cooling) at desired locations.[9]

A negative feature of temperature extremes is that, depending upon their composition, structures such as exposed solar power arrays could undergo considerable thermal stress during the transition between lunar night and day. However, at a depth of approximately one meter and below, the temperature is a fairly constant $-20°C$ at the equator. This is advantageous for minimizing thermal stresses. To the greatest possible extent, lunar habitats, structures, equipment, storage facilities, and instruments will be placed below the lunar surface. A simplified representation of the temperature variation on the Moon at the equator is depicted in Figure 3.1.

3.8   PHYSICAL MASS OF THE MOON

Lunar gravity allows materials and fluids to be separated by gravity-gradient processes (a distinct advantage over zero-gravity operations on orbiting space stations). Although only $\frac{1}{6}$ that of the Earth, lunar gravity will also allow humans to move with the same bipedal posture and locomotion that they are familiar with on Earth. The physical mass of the Moon can be used to protect human habitats from cosmic and solar radiation, as well as meteorites. The Moon will provide more protection, and will be more Earth-like, than an orbiting space station.[10] Finally, the absence of subsurface water (hydrostatic pressure) combined with the inherently-low seismic activity of the Moon means that engineering requirements for structural foundations, footings, and tunneling are less stringent than on Earth.

[8] Temperatures may be as low as $-230°C$ in permanently-shadowed craters in the north and south polar regions of the Moon.

[9] Thermodynamic processes require a means to reject or dissipate heat. On the Moon, radiation of heat into space will be the primary method of heat rejection for the first stages of lunar development. However, radiation (as compared to conduction and convection) is an inefficient means of heat rejection, and the challenge will be to design thermodynamic processes that are effective on the Moon.

[10] When human populations are able to live permanently on the "space station" Moon, the human species will be protected against extinction by a catastrophic event on the Earth, such as a nuclear or biological war or the collision of a large asteroid with the Earth

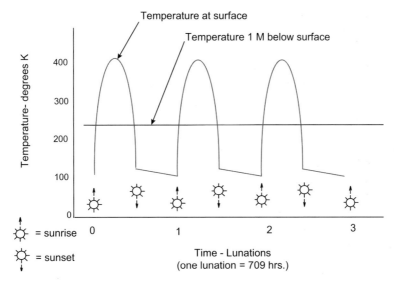

Figure 3.1. Typical diurnal temperature profile at the Moon's equator (Sharpe).

3.9   TOPOGRAPHY

The varied topography of the Moon offers opportunities for diverse human activities. In the polar regions, high elevations are well suited to the establishment of long-distance point-to-point communications that are desired for the first lunar base, as well as facilities for generating electricity from solar power. The mountains, plains, lava tubes, and craters of the Moon will be of considerable interest to geologists and tourists,[11] and lava tubes may be the first sites for human habitats. The relatively flat mare plains will facilitate surface and surface-to-orbit transportation.

3.10   STERILE ENVIRONMENT

Because the Moon is devoid of life (including bacteria and viruses), spacecraft that are intended for exploration of other regions of the solar system can be constructed on the Moon with less concern for forward contamination.[12] Likewise, specimens

---

[11] Observation towers at the north and south poles of the Moon will provide tourists with an unobstructed view of the Earth, Sun, planets, and the northern and southern galactic hemispheres, respectively.

[12] There are two types of contamination that are of concern to scientists: forward contamination occurs when terrestrial bacteria travel aboard our spacecraft to other planets, after which it may be difficult to determine whether microbial life that is found elsewhere is truly indigenous to that other location. In backwards contamination, toxic life forms from other planets might be accidentally brough to Earth.

from other regions of the solar system such as Mars or Europa can be brought back to laboratories on the Moon where they can be extensively examined without the possibility of contamination of the Earth's biosphere. Special laboratories that are isolated from other regions on the Moon can be used to conduct biologically-hazardous experiments without endangering life forms elsewhere, and they can be used to prototype various controlled ecosystem configurations in anticipation of later human occupation.

The isolation of the Moon from the terrestrial biosphere also means that "ideal" lunar habitats for human and agricultural uses can be constructed without the threat of pests and pathogens that exist on the Earth. For example, habitats on the Moon can be free from tsetse flies, mosquitoes, malaria parasites, plague, tuberculosis, poisonous snakes, cockroaches, and scorpions.

3.11   LOW GRAVITY

In addition to the relaxed requirements for structural strength on lunar buildings, the low gravity of the Moon has other advantages. The energy that is required for the launch of spacecraft from the surface of the Moon to escape velocity from the Earth–Moon system will be far less[13] than from the Earth. The use of the Moon as the launch site for spacecraft is the subject of Chapter 10.

The ease of physical movement for humans on the Moon as compared with the Earth may be expected to have recreational applications, such as human-powered flight and the development of games that are unique to the lunar environment. The Moon will also become a center for the study of the physiological effects of low gravity, and it may become a rehabilitation center for individuals with physical limitations.

3.12   ORBITAL MECHANICS

The Moon always presents the same "near-side" face to Earth (see Chapter 1), it is close to Earth, and it has an orbit with relatively low eccentricity. This combination is advantageous for the construction of permanent Earth-facing radio and optical-transmission communication stations, as well as Earth-observation telescopes. Tidal locking also produces a radio interference-free zone on the far side of the Moon that is ideal for the operation of radio telescopes (discussed in Chapter 2). The orbit of the Moon about the Earth adds to the velocity of satellites that are launched from the Moon, and permits spacecraft trajectories to be optimized for launch to specific areas of the solar system. The Earth-Moon distance is small enough to permit Earth-based direct command and control of devices located there (a "two-way light time" delay of

---

[13] It requires 22 times less energy to propel a payload to escape velocity (from the Earth–Moon gravitational system) from the Moon than from the Earth. The Moon will become the principal site for the manufacture and launch of spacecraft (see Chapter 9).

less than three seconds), an advantage that will greatly simplify the operation of lunar-based systems.

3.13 LAGRANGE POINTS

Lagrange points (named after the mathematician, Joseph Louis Lagrange) are points in space where gravitational forces between celestial bodies are more-or-less balanced. As a result, it requires less fuel to maintain a spacecraft at those locations over time. The Lagrange points of the Earth–Moon system are illustrated in Figure 3.2. In the figure Lagrange points L-4 and L-5 are ahead of and behind the Moon in its orbit around the Earth, and their positions form the points of equilateral triangles with the Earth and the Moon. Both L-4 and L-5 are more stable than the other Lagrange points; objects located at L-4 and L-5 will remain in place more easily than objects paced at other points. In addition, there are more stable "Lissajous"[14] orbits around L-4 and L-5.

Lagrange points are significant for lunar development in that raw materials, such as near-Earth objects (NEOs), can be safely stored at L-4 and L-5 prior to mining and processing. Eventually, large spacecraft may be assembled at these points prior to departure to other points in space (as suggested by Gerard O'Neill).

L-1 is strategically positioned between the Earth and the Moon and is a likely candidate for the storage of fuel and other materials for transit vehicles. L-1 may also be the site for a hotel/transfer station, and for a space elevator that extends to the

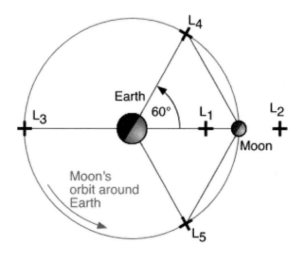

Figure 3.2. Earth–Moon Lagrange points.

[14] Lissajous orbits are periodic orbits in which there is a combination of planar and vertical movements. They require less station-keeping effort than halo orbits, in which the satellite follows a simple circular or elliptical path around the Lagrange point.

surface of the Moon. L-2 and L-3 may be used for observation and communication stations for the Earth–Moon system.

3.14 SUMMARY

For the next stage of human advancement on the space frontier, the Moon is an ideal objective. It has an abundance of resources and unique features that favor human exploration, scientific studies, and settlement.

4

Lunar robotic and communication systems

4.1 INTRODUCTION

It seems inevitable that humankind will someday settle the solar system. The process will begin with small steps and at first we will stay close to home – the Moon. Not only will the techniques we develop for fashioning the Moon to our purposes apply to other places – asteroids, other planets, and their moons – but the Earth itself will benefit. On the Moon and elsewhere we will, over time, find more effective ways to harness the renewable energy of the Sun to provide power, develop improved methods for recycling, and convert the local raw materials into tools and supplies. Many of the resulting processes and techniques will find their way back to Earth. The technology for establishing the first unmanned lunar outpost is available now. This chapter discusses the systems and devices that can be placed on the Moon to establish these first lunar scientific and industrial capabilities.

4.2 OBJECTIVES

4.2.1 Continued science investigations

Over the last half-century, manned and robotic spacecraft have been sent to the Moon – some flew by, some orbited, and some landed. The scientific objectives associated with these missions included terrain photography and mapping, soil and rock composition determinations, particles and fields studies, lunar geophysical investigations, and many others. However, there are still unanswered questions that need additional data. Camera-equipped roving vehicles that can investigate features not observable from orbit can obtain some of this information. These rovers will search for lunar resources; find geologically interesting materials; and explore the far side of the Moon, which until now has only been viewed from lunar orbit. Beyond these objectives there exist many others that fit into the categories of applied science and engineering.

4.2.2   Continuous operations

For long-term, continuous development of the Moon, equipment must operate reliably and continuously. While it may not be possible to develop equipment that operates forever, systems that are placed on the Moon must be built with the goal of minimizing maintenance. Moreover, systems planning should include the goal of being able to recycle equipment or material that has been placed on the Moon and has outlived its usefulness. Eventually, we may see machines that can fabricate replicas of their own components, and robotic devices assembling and disassembling those machines as well as each other.

4.2.3   Mining and manufacturing

To reduce the cost of developing the Moon, locally-produced rocket fuel for transportation and electrical power for lunar surface systems are the highest priority. These capabilities will allow an order-of-magnitude reduction in the cost of transportation to and from the Moon. When rocket fuel for the return trip does not have to be carried up on the outbound vehicle, additional payload can be sent. When lunar factories can run on local power, nuclear reactors will not be required and the size of batteries will be reduced to what is needed for keep-alive power during the lunar night. Many concepts for regolith mining and manufacturing processes that turn raw materials into finished products have been proposed. Eventually, it will be possible to produce a majority of the items needed on the Moon from the material that is already there. These items will be made of glass, metals, ceramics and other materials, and formed into solar cells, building materials, beams, tubes, wires, mirrors, and lenses. Appendix D gives examples of how the resources of the Moon can be extracted and processed into finished products such as tools and construction materials.

With power and communications resources available on the lunar surface, vehicles that travel to the Moon will not have to bring all of their own power; nor will they need their own complete communications systems – they can connect to the power and communications "grid" that is already in place.

4.2.4   Preparing for the return of humans

Eventually, humans will return to the Moon. Based on our experience with initial robotic operations, we will be able to develop a better understanding of how much advance preparation can be completed by using robots and lunar materials. It will be possible, for instance, for robots to prepare areas for habitation by moving regolith into berms. Robotic operations will also help to find ways to mitigate the problems associated with lunar dust, vacuum, ionizing radiation, and triboelectric[1] charging.

4.3   COST AND SCHEDULE ISSUES

The final design of the robotic equipment to be placed on the Moon will depend on when the process begins, as well as how much funding is available (and from whom). Historically, space systems and spaceflight operations have been so expensive that only governments of world powers could afford them. The $30 billion cost (1960s' dollars) of Apollo paid for only six Lunar Modules, which visited the lunar surface for a total of 300 hours. This works out to $100 million per hour. If we only consider the 160 man-hours of time spent suited up and working outside on the lunar surface, then the cost was about $180 million per man-hour.

Prior to the "space race", space flight technology was approximately equivalent to that of the aviation industry 100 years ago, which consisted of some hot air balloons, science fiction, and a couple of talented bicycle mechanics[2] who proved that heavier-than-air flight by humans was possible. In the 1960s, space technology became more equivalent to the aviation industry of World War I – airplanes were used almost exclusively by governments. Commercial use of aircraft for cargo and passenger transport did not become widespread until after World War II.

National space agency plans for lunar development focus on sustainability and affordability, which in turn require that private industry and international partners become involved to the greatest extent possible. It makes financial sense in the long run, therefore, to take the following factors into account for the design of lunar systems, in order to minimize and offset expense:

1.   Utilize unmanned (tele-operated and autonomous) robotic devices in lieu of humans wherever possible. Machines do not require food, air, water, or a ride home when they are done; and capabilities for transparent tele-presence[3] have increased tremendously.

[1] Triboelectric effect: the generation of static electricity by friction between different materials. The airless lunar environment presents unusual challenges in this area.
[2] Orville and Wilbur Wright: Americans generally credited with making the first controlled, powered, heavier-than-air human flight on 17 December 1903. In the two years afterward, they developed their flying machine into the world's first practical fixed-wing aircraft, along with many other aviation milestones. Previous to their aviation success, they operated a bicycle repair and sales shop.
[3] Tele-presence is a human/machine system in which the human uses head-mounted displays and body-operated remote actuators and sensors to control distant machinery. It provides a virtual environment for humans to control devices (e.g., robots) in a hostile or remote environment. Transparent tele-presence is the experience of being fully present at a live, real location that is remote from one's own physical location. Someone experiencing transparent tele-presence would therefore be able to behave, and receive stimuli, as though at the remote site.

2. Use robotic devices to prepare habitations and resources for humans, out of locally-available materials, rather than shipping them from Earth.
3. Wherever possible, make use of commercial off-the-shelf (COTS) hardware and software, rather than designing from scratch. (However, adapting existing systems for use in the lunar environment may not be cost-effective in many cases.)
4. Develop a robust permanent infrastructure for basic utilities such as power and communications, so that these basic needs will not have to be replicated and tested every time a new spacecraft is landed on the Moon.
5. Use incremental design and development techniques (build–test–build) wherever feasible. In addition to lowering the risk that a major development might not work right, this method allows time for commercialization (and lower cost) of new technologies.
6. Offset expenses by incorporating activities on the Moon that offer paybacks on investment, building toward eventual economic self-sufficiency. One example would be to include small items (such as pictures) that would be purchased by private individuals.[4] Alternatively, if a data-collection system were funded by a private enterprise, there would also be a potential for selling the information to other users, similar to the way in which commercial remote-sensing satellites are operated on Earth. Intellectual property rights for new processes which are developed by private industry would be another source of income.
7. Examine the possibilities for long-range financing by investment consortia,[5] which are formed by contracts between governments, industry, non-profit organizations, or any combination of these. This option offers so many possibilities that a separate section (following) has been written to describe them further.

4.3.1 Investment consortia: interdisciplinary and inter-institutional cooperation

Consortia are common in the non-profit sector. For example, Five Colleges Inc. is one of the oldest and most successful consortia in the United States. An example of a for-profit consortium is Airbus Industrie ("Airbus"). Formed in 1970, Airbus is one of the world's premier manufacturers of civilian airliners. Another successful consortium was Six Companies Inc., formed by six smaller general contractors in order to submit a bid for the Hoover Dam contract (which they won).

Project Apollo was a politically-rooted program with an exceptionally skillful management approach to coordinating the required engineering resources, then appending some significant science. But it was not a self-sustaining venture, any more than were the pyramids. No single business, government, or educational interest

---

[4] A recent example of this is the partial funding of the Genesis I inflatable Earth-orbiting spacecraft, conceived and implemented by Bigelow Aerospace.
[5] A consortium is an association of two or more individuals, companies, organizations, or governments (or any combination of these entities) with the objective of participating in a common activity or pooling their resources for achieving a common goal.

– science, tourism, education, or entertainment – presently possesses the wherewithal to develop the Moon by itself. In the absence of an overriding political rationale, an endeavor of such magnitude must be a combined effort.

The consortium approach presents the opportunity for a "virtual" community of participants from the worlds of business, education, government, and technology to combine their resources, wisdom, and experience. Planning activities could make use of Internet-based tools for conducting meetings, video conferencing, and for collaborative systems engineering. A way can thus be found to start from a small basis and expand the effort in many directions. The consortium (or consortia) could find "bootstrap" opportunities[6] to incrementally fund additional lunar development.

4.4 BOOTSTRAPPING LUNAR DEVELOPMENT

Humankind presently has the means to develop lunar commerce with minimal modification or growth of existing technologies.[7] Within a few years we could place solar panels, robots, scientific instruments and other enabling components at our preferred location on the Moon. *In-situ* experiments can begin that will enable mining, processing, and manufacturing.

Robotic mobility systems are currently available that are capable of supporting the exploration of relatively flat areas of the Moon. It may not be necessary for rovers and instruments to be capable of working throughout the complete day/night cycle – the rover could land at lunar dawn, shut down for a few tens of hours during the lunar midday for cooling, work through the lunar evening, then shut down again during the lunar night. One great benefit of a new rover mission would be that the mission could be viewed by millions of people in real time via the Internet. An example of such a system is the Mars Exploration Program. Two rovers, *Opportunity* and *Spirit*, have explored the Martian surface for two years (Figure 4.1). They each weigh about 175 kg and can travel at speeds up to 30 meters per hour. They are solar-powered, with batteries for keep-alive thermal power during the night. Their solar arrays provide about 150 W of electrical power, and the mobility system uses about 100 W. The body section provides room for scientific and engineering equipment.

The mast provides a location for two sets of stereo cameras, and is roughly the same height above the ground as a human's eyes. The single manipulator arm includes a microscopic camera. The rovers have autonomous capabilities to detect and avoid driving hazards.

---

[6] Bootstraps are companies that are started by their founders with their own internal funding, without any external financing. While it involves a significant risk for the founders, the absence of any other stakeholder gives them complete freedom and helps them realize their vision without any external interference.

[7] Since humanity has already visited the Moon with robots and humans, we clearly have the technology to get there. The current technological challenge is not so much getting there as it is learning how to remain there (addressing dust mitigation, resource extraction and utilization, ionizing radiation protection, and so on).

Figure 4.1. Artist's rendering of Mars Rover at work (courtesy NASA-Jet Propulsion Laboratory).

In comparison with lunar mission operations, Mars operations are more complicated, require more timing precision, and have higher risk in many areas. If these rover designs were modified for use on the Moon, they could be in communication with Earth 24 hours per day, and because the Moon is very close to Earth in comparison with Mars, they could be operated more easily. On the Moon, the three-second communication delay would permit near-real-time driving, which is not possible with the Martian scenario, where radio signals require 7 to 44 minutes to make the round trip between the two planets (Section 4.6.1 discusses this latency issue in more detail).

The Mars Exploration rovers are a good example of the state of technology that exists for planetary surface exploration today. Some other applicable technologies that are evolving rapidly are those of wireless digital communications, which holds promise for interconnecting devices around a lunar base without cables; and of rapid prototyping, wherein machine parts can be formed from layers of various powder materials using devices employing computer numeric control and lasers. The technologies of rovers and robotics that have potential near-term payoffs are discussed in the next section.

4.5 ROBOTICS TECHNOLOGY FOR LUNAR BASE DEVELOPMENT

Robotics technology is evolving as an increasingly important component in the exploration of space. The next stages of exploration of the Moon will see robots used independently in the first stages of the development of a lunar base and, with the return of humans to the Moon, in synergistic roles to augment human capabilities. Here we examine the role of robots in the next phases of exploration and development of the Moon.

4.5.1   Introduction

People sometimes debate the relative merits of human versus robotic exploration. Proponents of robotic exploration say that it is far less expensive to send robots into space than it is to send humans, and that robots, as extensions of our senses, are our "virtual presence" in the solar system. They say that robots can do anything that humans can do, and they can do it cheaper. On the other hand, people who have worked in human spaceflight argue that humans have unique capabilities for on-the-spot thinking, observing, and reasoning. They list examples of how humans can fix things that break, improvise when plans change, and take advantage of unexpected opportunities. Historically, this discussion has focused on the "either/or" line of reasoning. How do we reconcile these two viewpoints, which both hold value?

4.5.2   Apollo experience with robots for exploration

Tele-robotic devices were successfully operated on the lunar surface in the 1960s and 1970s during the Luna, Surveyor, and Apollo programs. Spacecraft instruments were commanded mainly in real time by operators on Earth, and information was received about the results via the tools of the day – strip charts, printouts, and images.

Controlling lunar surface devices was relatively cheap and simple, compared with other space operations. The Moon is a cooperative target for Earth-based antennas in terms of its position in the sky and near-unchanging distance; and pointing antennas on the Moon's surface back toward Earth is also relatively simple. The physical environment is predictable, and in itself contains nothing requiring fast or complicated reactions. There is no atmosphere or rain to cause rust or corrosion, the main design considerations being the long-term effects of radiation, temperature extremes, and dust. A new generation of devices could be made to last for a much longer period than did the original instruments, if we focus on maintainability and reliability.

4.5.3   International Space Station experience: robots for science, assembly, and maintenance

The complexity of the International Space Station (ISS) has resulted in a large number of hours spent on assembly and maintenance. NASA and its international partners are developing maintenance robots that are either operated by the crew onboard the station or by ground control personnel, to reduce the need for human spacewalks (extra-vehicular activities, or EVAs).

Space walks performed by humans take much longer, and require more resources, than would spacewalks performed by robots. A human who prepares for a spacewalk might spend up to six hours in the airlock, as the atmosphere slowly

Figure 4.2. Special-Purpose Dexterous Manipulator (SPDM/Dextre), designed for replacing components on the exterior of the International Space Station (courtesy of Canadian Space Agency).

bleeds away, to avoid decompression sickness (the "bends"). Also, because there are many potential hazards related to working in the space environment (a slip of a tool could cause a cut that opens and depressurizes the spacesuit), humans always work in teams. They must use the "buddy system", in which two individuals go out to the worksite together. A third crew member is then required to monitor the activities from inside, providing extra camera views, communication with the ground, and (in many cases) operation of the remote manipulator system. The Space Station Remote Manipulator System (SSRMS) is a large "crane" that is used to position hardware at the work site, move large assemblies,[8] and provide a stable position for EVA crew via foot restraints. It is also designed for use with the Special-Purpose Dexterous Manipulator (SPDM/Dextre) a two-arm dexterous system for performing parts replacement tasks on the exterior of the Station (Figure 4.2).

There will be other robots in use on the International Space Station, such as

---

[8] Remember that even though objects have negligible weight in microgravity, they still have mass, and, consequently, inertia.

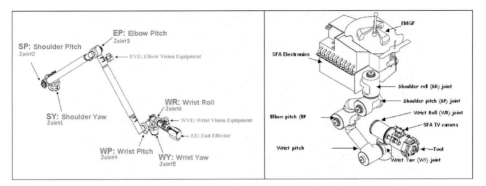

Figure 4.3. Robotic systems for the Japanese Experiment Module on the International Space Station.

the Japanese Remote Manipulator System and the Japanese Small Fine Arm (Figure 4.3). These devices will be used to move experiments to and from an external deck. During the assembly of the Japanese Module, the SSRMS will hand items over to these robots so that they can be attached to their designated positions.

Robotic systems on the ISS are used to assemble large modules and structures, and to offload work from the on-orbit crew. In addition to being hazardous, human EVAs are also costly. Using robots for tedious, strenuous, or dangerous tasks allows more crew time for conducting science experiments and other high-value activities.

4.5.4   Deep-ocean experience – production robots

In the last 20 years, industry has learned that the optimum method of working in the sub-sea environment involves both humans and robots. Robots have a much greater depth range than do human divers, working in depths of the ocean that would otherwise be inaccessible. However, the robots are not sent out alone. Humans on the ship or platform are available to pilot the robots, repair things that the robots cannot, react to surprises, and keep a close watch on what is happening.

In deep-sea drilling operations, there is frequently a need to repair equipment that is on or near the sea floor. Divers who make these repairs are often accompanied by remotely-operated vehicles, whose principal job is to fetch and carry tools, and to provide a view of the operations for the people at the surface. The petroleum industry must answer to its shareholders and show profitability, thus they aggressively look for the most efficient and economical way to do things. The fact that they use robots is strong evidence for the value of this technology.

4.6   TECHNOLOGICAL CHALLENGES

Robotic technology has made tremendous advances in the last 30 years; and many universities now have robotics research departments. However, there are still technological challenges to be addressed if we want to use robots on the Moon. These challenges include latency, bandwidth limitations, and the need for an infrastructure in which they can operate.

4.6.1   Latency

Radio signals take about three seconds to make a round trip from the Earth to the Moon and back. This amount of time lag, called latency, is how long it takes an Earth-bound operator to experience the results of an action (e.g., throw a light switch, and three seconds later see the light come on). In highly dynamic situations such as driving a vehicle at high speed, latency may be a problem. Operators could quickly find themselves in a disorienting situation, potentially damaging the equipment. There are two ways of handling the latency problem: first, proceed slowly; and second, use automated subroutines.

4.6.1.1   Proceed slowly

Going slowly requires us to design operations so that the situation does not outpace the operator's ability to control it. On the Moon there is nothing in the natural environment that will require fast response to an unanticipated event. Quick reactions would only be required in contingency situations (if something unexpected happened), and keeping a slow pace will reduce the chances of such occurrences. The martian rovers (*Sojourner*, *Spirit*, and *Opportunity*) have been successfully operated using a combination of moving slowly and predictive modeling, despite the much longer latency of signals between the Earth and Mars.

Going slowly has other benefits. In the case of mobility devices, less dust (a nemesis of thermal control, bearings, gears, and mechanisms in general) will be disturbed and re-deposited where it is not wanted. Going slowly requires less bandwidth for control and (in some cases) uses less energy for mobility. It also reduces the work required to dissipate the heat that is caused by expending energy. (In the Moon's vacuum, the only practical way to get rid of excess heat is by radiation.) However, a slow pace of operations does not always imply low power or slow movements. For example, when climbing a steep slope, high energy (torque) is required, and if the robot is at risk of tipping over, rapid action would be required. Moreover, a stalling motor may overheat faster than a motor running at a high speed.

The only disadvantage to working slowly may be a human one: impatience with a scheme of things that does not match the way we would do things on Earth. Slow-paced lunar operations might be boring to watch, and there could be a tendency to think of them as inefficient – not getting things done "fast enough". But the Moon has been waiting for us for four billion years. We can proceed slowly and methodically, keeping our tools and equipment in working order for a longer period of time. Less maintenance reduces the overall operational cost, which is an important consideration for commercial enterprises.

4.6.1.2   *Automated subroutines*

It will be useful to include processors and software in the lunar equipment that can control precise or repetitive operations, so that the role of the Earth operator would be more supervisory. This method would be similar to the way many factory robot devices work on Earth.

Examples of tasks that could be automated to some degree include excavating, traversing, grading a pre-defined area (perhaps using boundary beacons), or making bricks. Process automation would also include collecting samples based on human identification and direction. Robots can be "taught" how to perform a task before the actual work begins. Thus, a general-purpose robot becomes capable of specialized tasks.

4.6.2   Bandwidth requirements

Another aspect of tele-operation involves controlling devices that replicate the hand-eye coordination of a skilled technician. This is the kind of situation presently associated with surgical telemedicine. Two manipulators for hands, and two cameras for stereo vision, along with some upper-body mobility, would allow a doctor or nurse to control the robot in a manner analogous to a human caregiver. The same device, under the control of a specialist in another profession (e.g., a master craftsman of some kind), would instantly have a new job description.

Sensory immersion in operator control stations requires high-bandwidth, high-resolution systems with imaging and haptic[9] control.[10] Higher bandwidth does not eliminate the latency problem, but would provide more information that could be used for making decisions. Device control as well as video and data could be stored as multi-channel records, which could be used later for analysis, training, and general distribution.

With today's available and emerging technologies, Earth-bound operators and audiences can have a far greater sense of participation. They could "work on the Moon" every day, and be home for dinner. This sense of being on the Moon could eventually be distributed throughout the world with a virtual viewing-room environment.

4.6.3   Infrastructure requirements

In the early phases of lunar development, none of the robotic devices will be entirely autonomous. Rather, they will be under the direct or indirect control of humans

---

[9] Haptic: of or relating to or proceeding from the sense of touch.
[10] Eventually, when pressurized environments exist on the Moon, stereo audio will also be useful.

on Earth. Tele-operated devices require continuous direction, but more autonomous devices have some capacity to function on their own, with only general supervision and guidance. They will be acting as surrogates for the humans on Earth, who will be specialists sharing the job of controlling the progress of the lunar operations.

A facility to control lunar devices might initially resemble other mission control centers. There would be command and display equipment for operators, as well as a network of computers, data lines, and transmitter–receiver stations at various places on Earth. The communications equipment would be used for both robotic control as well as control and monitoring of lunar satellites.[11] Work stations for operator control of devices (both "mobility" and "skilled technician" types) would be present, providing a virtual lunar environment.

Safeguards for both physical security and prevention of operator errors before they are made (via simulation and command checking) would be included. "Pilots" in the ground control facility will benefit from immersive visualizations, as well as the ability to model and practice operations virtually, before they are performed on the Moon. In this way, actions that might damage the lunar hardware would be simulated before they were performed, and revised if needed.[12]

It will also be useful to distribute the tele-operation capability to many user locations: international partner space agencies, corporations, universities, and individuals. Groups and individuals could then be assigned time slots to be "on the Moon" in a virtual sense, conducting activities of their own design, for their own purposes.

Internet technology will allow people in all parts of the world to share in these tasks as operators, analysts, and spectators. Video from ongoing operations can be streamed real-time to the World Wide Web. Control of the devices can be handed over from specialist to specialist across continents, so that more people will participate and all of them can work within the hours that are natural for their time zone. Multiple devices can be operated together in synergistic tasks.

The three-second delay between the time when an action is commanded and the time that results are seen is addressed by operational and technical methods, described in Section 4.6.1. Of all the natural objects in space, only the Moon allows this near-real-time tele-presence by controllers on the ground.

Given a lunar rover with stereo vision, manipulators, and a set of field tools, Earth-bound geologists could conduct field trips to various sites via tele-presence. Mapping, sample collection, and some *in-situ* analyses could be conducted. The "reality" of the representation of the site to the humans on Earth is essentially unlimited by technological constraints; and the data could be distributed worldwide by television and Internet. This is similar to the method by which the ongoing Mars Rover programs have been conducted, as discussed in Section 4.6.1.1.

---

[11] The relay satellites would allow communication with the lunar far side.
[12] Examples include commands that would cause a camera's image sensor to be exposed to direct sunlight, or cause a mobile robot to drive off a cliff.

4.6.4 Lunar surface infrastructure requirements

Development of tele-operated devices on the lunar surface requires line-of-sight communications with Earth. Continuous power for solar cells will also help by alleviating the need for batteries while taking advantage of the abundant solar power.[13] In Chapter 5, we explain why the summit of Mons Malapert, in the south polar region of the Moon, offers the optimum location for direct and continuous Earth–Moon communications, long periods of continuous sunlight, and proximity to regions of scientific interest. For the first robotic lander missions to the Moon, an area near Mons Malapert should be given consideration as the location of the first sunlight-dependent robotic base.

Eventual human exploration and development of the Moon will require robotic precursor missions to the base site, to set up the habitation modules, get the oxygen plant running, get the power plant running, and so on. Then, these robotic systems must keep everything running (for perhaps several years) until the humans arrive.

4.7 THE ROBOT ASSISTANT

On the surface of the Moon, humans face challenges that are distinctly different from spacewalking on orbit. First, there is the partial-Earth-gravity environment. Now, instead of just adding inertia, the spacesuit adds weight, and this affects how one moves and performs tasks. It also affects the length of time which can be spent inside a spacesuit, because carrying the weight of the suit and additional life-support equipment on one's back causes fatigue to set in sooner than it otherwise would. Second (unlike orbital operations), there is a sense of up and down. One cannot change one's orientation randomly to pick something up. If there is an interesting rock on the ground, one must bend down to get it, and then worry about maintaining one's balance (we learned from Apollo that this can be very challenging in a 200-pound – 90-kg – spacesuit). Third, because the gravity on the Moon is less than the gravity on Earth, other problems occur that would not be experienced in a terrestrial setting. An example is the hazard caused by cables that were used to deploy the Apollo surface experiments: there was insufficient gravity to keep them lying flat, and this resulted in a tripping hazard.

There are useful things that can be done with robots that will offset the disadvantages that humans face when trying to work in spacesuits. These "assistance" tasks are fairly well understood because of our experience with robots in industry.

---

[13] Due to the lack of an appreciable atmosphere, approximately eight times as much solar radiation reaches the surface of the Moon, compared with the amount that reaches the surface of the Earth.

4.7.1   Evaluating robotic technology for the Moon

There are a number of potential tasks that may lend themselves to robotic automation or tele-operation. To find the easiest and most useful tasks, we must compare the
difficulty of roboticizing each task with its potential benefit.

We identified a number of tasks and subtasks that would be associated with
surface excursions, because traverses by humans will be a primary science activity.
A complete listing of the activities is given in Appendix A. Our list was evaluated for
aspects of difficulty as well as benefit. The difficulty scores for the task were added
together, as were the benefit scores. Things that are easy for robots to do will have low
numbers on the difficulty scale. Things that are most beneficial will have high
numbers on the benefit scale. The scoring factors are described in the following
sections.

4.7.2   Difficulty aspects

4.7.2.1   Technical heritage

How much new design and development work will be required? For some tasks, such
as inspection of a vehicle exterior, systems exist that are already space-proven. The
heritage risk is thus low. For other systems, such as emplacing a lunar surface habitat,
the hardware is still in the early development phase, and the technical risk is comparatively high.

4.7.2.2   Complexity of the task

Some tasks are simple – they only require one or two steps; there may be only one or
two objects to be manipulated, and the robot would not have to be overly dexterous
to do it. Simple tasks are easier to roboticize than complex ones.

4.7.2.3   Robotic compatibility

This is the general question of how well a robot could perform a task. For example,
inspecting debris shields for micrometeoroid damage is very compatible with robotics,
whereas repairing those shields is not. On a planetary surface mission, observing a
sample with a microscope is a robot-compatible task, but deciding which sample
should be placed under the microscope requires human judgement.

It is difficult to put a metric on the robotic compatibility factor. Instead, we rely
on the judgment of robotics experts to provide their opinions on which tasks are, or
are not, robot-compatible. It is noteworthy that robot compatibility can be built into
a design, to a certain extent. For many of the planetary surface tasks, we determined
that the task could be robot-compatible, but only if designers intentionally make it so.

4.7.2.4   Task criticality

What are the consequences of a failure of the robot to accomplish the task? If it is
only an inconvenience, less redundancy is required than would be the case if the
failure could lead to a catastrophe.

4.7.3   Benefit aspects

4.7.3.1   Task duration

How much spacewalking or surface excursion time might be saved if a robot could do the task? Tasks that require more than six hours to accomplish are prime candidates for robotic development, because spacesuits are limited by the amount of breathing air and battery power that can be conveniently carried along. Shorter tasks that occur frequently are also prime candidates.

4.7.3.2   Task frequency

How often would that block of time be saved? If the task requires five hours, but only happens once during the entire mission, then that task is not the most beneficial one to roboticize. It would be more useful to focus on tasks that require an hour every day.

4.7.3.3   Task pervasiveness

If a task is performed at many different places, it is probably more beneficial to roboticize it than it would be to roboticize a task that is only performed at a single location. For example, breaking a rock in half, in order to examine the pristine interior, or to acquire a sample and leave half behind as a controlled specimen, will occur in many places throughout a geological surveying sortie. On the other hand, loading samples into a spacecraft for return to Earth would only occur at one (or a few) locations.

4.7.3.4   Human factors

Something that the crew would consider as challenging, interesting, and fun would not be as beneficial to roboticize as something that was tedious, boring, strenuous, or dangerous. For example, collecting lunar regolith and putting it in bags, then transporting the bags to the site where an above-ground habitat is being built would be a good robotic task from the perspective of human factors. Exploring the crash site of a Ranger[14] or Surveyor[15] spacecraft would be something that humans would like to do,[16] and there would be minimal benefit from roboticizing it.

[14] From 1964 to 1965, NASA sent four successful Ranger missions to the Moon, which were designed to relay pictures and other data as they approached the Moon and finally crash-land into its surface. One less-than-successful mission, Ranger 4, disabled, crashed just out of sight on the far side of the Moon. The Ranger 8 crash site was located near to the Apollo 11 landing site. Ranger 6 impacted on the eastern edge of Mare Tranquillitatis, and Ranger 7 crashed south of the crater Copernicus. Ranger 9 impacted in the crater Alphonsus.
[15] Surveyor 2 was the only vehicle of the Surveyor series to crash into the lunar surface. Surveyor 3 was visited by the Apollo 12 astronauts.
[16] There are a total of 66 man-made objects on the Moon. The locations of 16 of them are unknown.

4.7.3.5   *Distance from safety*

Finally, how far away from the airlock must a human travel if they perform the task? If the task is very close to the airlock, then the crew spends less time getting there, and it is quicker to get back to the airlock in an emergency. All other things being equal, it would be more beneficial to roboticize tasks that take place farther from the airlock.

4.8   PERVASIVE SUBTASKS AND CAPABILITIES

4.8.1   Acquiring imagery

When a repair is completed, it is useful to take pictures of the finished work. These are typically referred to as "close-out" images. They allow the crew to evaluate their work after they are back inside. If they determine that additional work needs to be done based on this analysis, plans can be made for an additional EVA or resupply of parts as appropriate.

Imagery is also used to improve situational awareness. Working in a spacesuit not only limits peripheral vision, but also forward vision and range of motion. It is more likely for a space-suited crew member to bump into something, or get snagged on something that was not noticed, than it would be for a person working in a shirt-sleeve environment to have a similar accident. NASA requires that two people go out on each spacewalk – one has a buddy in case of trouble. Having a robot that can function as a "buddy", watching out for problems (perhaps being controlled by a human who remains inside) can be useful.

The ability to see the work site makes a critical difference in the difficulty of doing the task (try using a screwdriver in total darkness!). Robots need visual information to an even greater extent than do humans, because they generally lack the tactile feedback that a human would have.

When astronauts are performing maintenance during a spacewalk, having a camera on the end of a manipulator would allow them to peek around corners or under things. A "flying eyeball" (Figure 4.4) for inspection of the exterior surface of the spacecraft or habitat helps identify problems, assess work sites, find leaks, and in general allows better planning before the extra-vehicular activity (EVA) begins. On a planetary surface, having a camera on the end of a manipulator might allow the crew to choose samples for inspection prior to bending over (or deploying a specialized extention tool) to pick them up. This would eliminate the amount of repetitive activity and decrease fatigue (many of the rocks that seem interesting enough for a closer look often turn out to be similar to a sample that has already been collected, in which case the new sample is discarded).

This imagery capability is also central to landing site survey and qualification. We must see what the ground looks like so that we can plan ahead for challenges, or even decide to pick another location for the initial base camp. Site surveys require mobility and the ability to capture images. Since they do not involve manipulating objects, they are among the easiest things to do with robots. Appendix A lists additional pervasive capabilities that should be developed early.

Figure 4.4. AERcam-Sprint. The ball in the left panel is a robotic camera designed to float about a space shuttle and the International Space Station and take pictures. The right panel shows a model of AERcam-Sprint in use in the Shuttle payload bay. AERcam-Sprint is NASA's first Autonomous Extra-vehicular activity Robotic Camera (AERCam).

4.8.1.1 *Manipulation and mobility*

Some tasks involve manipulation in addition to mobility. Among these, the highest-payback tasks involve maintenance. When we begin designing the systems that will allow humans to live on the Moon, we should remember to add robotic interfaces to things that could break, and to include robotic compatibility in the design. Things that are robotically compatible should be human-compatible as well, adding redundancy to the system.

4.8.2 Lessons learned from shuttle, space station, industry and medicine

Planning ahead will allow us to use robots as efficiently as possible. Lessons from the International Space Station and from terrestrial robotics applications, about what works and what doesn't work so well, are all applicable to lunar development:

- *Successful work systems rarely double as technology demonstrations.* Usually, the prototype is a way to learn how to do things better, and we do not expect the prototype to stay on the job for years. We need to understand each robotic system and have some operational experience with it, then develop more advanced versions based on our experience.
- *Design tasks for single-arm operations.* In space, a crew member must hold on to whatever they are working on. On a low-gravity planet, one may still need to hold onto something to maintain one's balance. Although a robot may have more than two arms, the task should be operable by either human or robot; thus, single-arm operations are still important.
- *Minimize fasteners and actuators.* Find a way to use only two bolts instead of twenty, only one switch instead of ten, and so on.

- *Provide clear visual and physical access and visual cues,* and design those visual cues with the imaging system in mind. It is not very useful to make an orange-and-white target for a monochrome (black-and-white) vision system. Make sure that alignment guides are located where the vision system can see them.
- *Minimize motion requirements.* To turn a valve, build a robotic tool with a motor in it, instead of only having a steering wheel or lever on the equipment (see Figure 4.5 for an example).
- *Allow for proper tolerances between parts.* There is a balance between too tight a tolerance and too loose a tolerance. Either extreme can cause problems – especially on the Moon, where abrasive micrometer-size dust particles are abundant.
- *Provide status indicators for robot-activated functions.* Robots usually have no tactile feedback system, so they need some other way of knowing when they have made a connection, tightened a bolt sufficiently, and so on.
- *Provide clear identification of objects and directions.* The perceptions of "up" and "down" are workable for humans, but they do not apply to a robot, which has no neuro-vestibular system.
- *Provide operational margins.* Design for the mid-range of capabilities. One may expect a mechanism to work continuously for fifteen years, but it should be easy to replace nonetheless.
- *Limit the amount of force required to operate mechanisms.* This makes the task more human-compatible, and more robot-compatible at the same time.

4.9 RECOMMENDATIONS

4.9.1 RoboTractor

The development of a robotic excavation/soil-moving capability is a critical component of an overall robotic exploration system. Whether habitats are built above ground or in lava tubes, significant (and tedious) effort will be required to construct a livable space. Although lava tubes are shielded from some of the ionizing radiation that would cause cell damage (and eventual cancer) in humans, habitats in lava tubes would still need protection from meteor strikes, which, if they occur near or directly above the habitat, could cause a cave-in. Protection from meteor strikes might include concrete- or composite-reinforced walls. Above-ground structures will require that bags of regolith be placed over them to protect the occupants (human or electronic) from radiation damage.[17] Telescopes will require a flat, level, and solid substrate upon which they can be mounted, and if we wish to land a vehicle in a highlands area, landing site preparation must be performed. Even when operating in

---

[17] The amount of regolith shielding that will be needed is still being debated, but will probably be greater than early estimates to account for the secondary radiation that is produced by cosmic rays.

a smooth mare area, berms may be needed to shield other outpost components from the blast of dirt and dust that will accompany each rocket lift-off.

Almost any facility that is expected to remain stable for a long duration on the Moon will require tedious and strenuous effort to create. Such efforts are far more suitable for robotic tractors, trenchers, and scrapers than they are for humans with picks and shovels. Appendix A includes a case study of how such a vehicle might be developed.

4.9.2   Engineering standards

Lunar systems designed to operate indefinitely should be based on standards, conventions, and engineering codes which assure a high level of compatibility and interoperability for future systems. Commonality standards are needed to add flexibility and robustness, so that every piece of equipment sent to the Moon now and in the future can be interfaced (added to, cannibalized, or reconfigured) with other equipment.

Adherence to such standards will increase efficiency and safety in the decades to come. One such standard is the metric system; another is the adoption of tolerance and dimensioning standards that account for the dust and temperature extremes of the Moon.[18]

4.9.3   Incorporate a tele-robotic control mode into autonomous systems

A fully autonomous rover might be developed to explore areas of the Moon that are too dangerous or too far away from the habitat, with instructions to return pictures and information to an orbiting lunar satellite that relays the information to the human outpost. However, such a robot might still get into trouble, and would need assistance from a human who can assess the situation and give commands that result in the robot being able to be "driven out" of a situation in which it otherwise might have been intractably jammed.

4.9.4   Design early for robots

Even if robots are not developed during the early planning stages, provisions should be made to include robot capabilities as part of the overall planetary work system. Incorporating robot-specific interfaces allows tasks to migrate to robotics as the capabilities are fielded. Figure 4.5 shows a microconical tool. It is used on the International Space Station as an attachment device, and was designed to be operated either by a human or a robot – even though there were no robots on the ISS that could use it at the time of its development. When a human uses it, the "steering-

---

[18] Currently the U.S. is the only spacefaring nation that continues to use the English measurement system. This is not difficult to correct from a technology standpoint, because metal-working tools worldwide are programmed and designed to use either type of measurement. The only obstacle is habit.

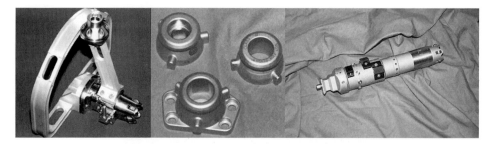

Figure 4.5. The microconical fitting and tools, developed by Oceaneering Space Systems in support of the International Space Station (ISS). The "steering-wheel" tool is used by an EVA astronaut. The cylindrical tool is used by a robot.

wheel" device on the left is connected to the microconical (center), and the human can easily turn it. When a robot is working, the robot uses the rotating tool shown on the right.

4.9.5 Plan ahead

Identify overhead tasks associated with the primary job so that timelines accurately reflect the required work. When a more accurate timeline is established, the need for robot capabilities and interfaces becomes apparent. When the need is defined, a more efficient system (combining human and robotic tasks) can be developed within a context that best supports the overall mission. An example of the top-level planning that is required is given in Appendix A.

A lack of understanding of the kind of work robots can actually perform in space may lead to a reluctance to place them in the critical path of assembly or maintenance activities. Consequently, robotic accommodations might not be widely incorporated into the architecture due to cost, weight, volume, and so on. However, as development continues and new tasks are discovered, it becomes difficult and even more expensive to retrofit equipment for robotic operations. A similar phenomenon was seen in the U.S. automotive industry, which lost ground to Japan and other countries that had begun to make extensive use of robotics and automation in their assembly lines as early as the 1970s.

4.9.6 Build a little, test a little

Testing and analysis of each successive prototype should be used to define the specifications of the next prototype. This allows partial functionality to be fielded earlier, producing results that encourage additional development. Robot prototype demonstrations early in the planning stages will instill confidence and showcase capabilities to planners and managers. Robot accommodations and tasking will occur when the understanding exists about what robots can do and what is required (accommodations) for them to perform the tasks.

Figure 4.6. Robonaut is a humanoid robot designed by the NASA Johnson Space Center in a collaborative effort with DARPA. The Robonaut can function as an astronaut surrogate. Robonaut is designed to be used for "EVA" tasks – that is, those which were not specifically designed for robots (text and image courtesy of NASA Johnson Space Center's Robonaut website).

4.9.7   Develop robot surrogate capabilities

There may be situations at a lunar outpost (such as a depressurization event) in which there is an urgent and immediate need to "go outside" – which humans cannot do, even in a spacesuit, without a lengthy pre-breathe (refer to Section 4.5.3, ISS experience). A robot surrogate provides a unique capability to respond to contingency and emergency situations when suited human crew members cannot. This is the most complex and sophisticated class of robots, exemplified by "Robonaut" (Figure 4.6). Robonaut is an anthropomorphic device which can completely replace a human in some situations. It is operated via virtual-reality devices and controls. Work at JSC has focused on the mechanical aspect of Robonaut and enhanced tele-operation capabilities. DARPA partnered with JSC to develop an autonomous operation capability for the system. Robonaut eliminates the robotic "scars" (special robotic grapples and targets) and specialized robotic tools of traditional robots. However, it still keeps the human operator in the control loop through its tele-presence control system.

Tracked vehicles, such as "Matilda" are already in use in dangerous areas on Earth. (Image courtesy of Mesa Robotics, Inc.)

Archetypal robotic rover: Lunokhod 1 was the first roving remote-controlled robot to land on another world.

Walking Mobility. An eight-legged Scorpion robot prototype is being evaluated at NASA Ames Research Center, where scientists are analyzing how similar robots someday may explore planets. (Image and text courtesy of NASA)

Snakebots can weave through narrow passageways, inspect hard-to-reach areas, coil around pipes, and climb from one structure to another. In rugged terrains where wheels would be impractical, snakebots would not tip over or be easily stuck. (Image courtesy of Hirose Robotics Lab, Japan; text courtesy of NASA).

Figure 4.7. Examples of mobility methods in robots.

4.9.8 Mobility is key

Many of the tasks we examined for robots include a requirement for mobility. Technologies that implement or facilitate robot mobility are critical to future planetary exploration. On the Moon, we may envision free-flying robots (such as AERcam-Sprint, Figure 4.4), tracked vehicles, wheeled vehicles, ambulatory (walking) vehicles, and many other methods of locomotion (Figure 4.7).

4.10 LAUNCH VEHICLE CAPABILITY

While no system is currently capable of placing humans on the Moon and returning them safely to the Earth, there are some existing launch vehicles that are capable of placing payloads on the Moon, with masses ranging from tens to thousands of kilograms. They give us the capability for placing many of the components of a robotic base on the lunar surface. During the time that it takes to establish a robotic outpost on the Moon, heavy-lift and human-rated rocket systems will be developed that permit permanent human habitation.

4.11   CONCLUSION

Robotic technologies are key to future lunar development, and should be incrementally developed by starting with those tasks that are the most beneficial and the least difficult to roboticize. When these technologies are proven, they will signal the beginning of the true "space age" – large-scale space operations comprised of a mix of human, tele-operated, and autonomous robotic participants.

The following chapters discuss four phases of lunar development, from the first robotic landers to the fully inhabited "Planet Moon".

5

The first lunar base

5.1 INTRODUCTION

The transformation of the Moon into an inhabited sister planet of the Earth (the "Planet Moon Project") is divided into four phases. Phase 1 concerns robotic activities on the Moon prior to the return of humans. Phase 2 (discussed in Chapter 6) describes the return of humans. Phase 3 (Chapter 7) explains how power, transportation, pipeline, and communications networks around the circumference of the Moon in the polar regions provide the basis for autonomy of lunar industrial and economic operations. Phase 4 of lunar development (Chapter 8) concerns the global human exploration, development, and settlement of the Moon.

As shown in the previous chapters, the Moon has the energy and material resources that are needed to sustain permanent human colonies, and the technology exists to undertake large-scale lunar development projects. The question of whether it is possible to colonize the Moon has been answered – in the affirmative. The first step on the road to a globally-inhabited Moon will be to establish an unmanned robotic outpost. From these, extensive analysis and research (including mining and manufacturing experiments) will be performed and the first elements of a global utilities infrastructure will become operational. These activities will prepare the way for colonization to follow.

The polar regions of the Moon are probably the best locations for initial lunar base activities. Evidence of hydrogen (implying water-ice deposits) was discovered by the Lunar Prospector mission. Due to favorable topography, Earth–Moon–Sun geometries, and access to resources, Mons Malapert in the south polar region is an excellent location for the first lunar base.

5.2   LUNAR BASE SITE SELECTION CRITERIA

There are several desirable criteria for the successful installation and sustenance of a permanent settlement on the moon. Among them are the following:

1.   Availability of abundant, safe, economical, and uninterrupted power.
2.   A reliable, dedicated, and uninterrupted Earth–Moon communications link. The mascons on the Moon create a non-uniform gravitational field that causes variability in satellite orbits and results in a need for continuous reboost and orbital adjustment. A communications link that does not rely exclusively on a satellite is preferred.
3.   A site that is in close proximity to areas of scientific interest.
4.   A site that provides some degree of natural protection from the harsh environment. By locating the base in a crater, rille or lava tube, it may be possible to maintain constant temperature with the use of regolith shielding.
5.   Transportation and logistics to and from the base should be relatively simple. This requirement could involve the use of guidance-related equipment on the lunar surface or in lunar orbit, unambiguous terrain cues for use by crews or automated equipment, or direct guidance control from Earth.
6.   A location from which a crew evacuation operation may be conducted without difficulty.

5.3   MONS MALAPERT

Mons Malapert[1] in the south polar region of the Moon, may be the best location for the first lunar base. This feature is located at longitude 2° east, latitude 85.75° south (see Figures 5.1 and 5.2). The vertical distance from terrain near the base of the mountain to the summit may be as much as 8,000 meters. However, the summit of the mountain is calculated to project approximately 4,700 meters above the 1,738-km lunar reference radius of the Moon. The difference in these elevations is explained by the fact that the base of the mountain lies approximately 3,000 meters below the lunar reference radius of the Moon. Moreover, an unnamed crater immediately south of the peak is a possible "cold trap" that may contain water-ice (Margot *et al.*, 1999).

Mons Malapert the most favorable conditions for Earth visibility for a direct communication link, and long periods of sunlight (predictions of 87% full sunlight and 4% partial sunlight throughout the lunar year) for solar-electric power genera-

[1] This feature has not been assigned a formal name by the International Astronomical Union. We refer to it as Mons Malapert because it is (a) adjacent to Malapert Crater, (b) lunar mountains are traditionally referred to as "Mons" or "Montes" (plural).

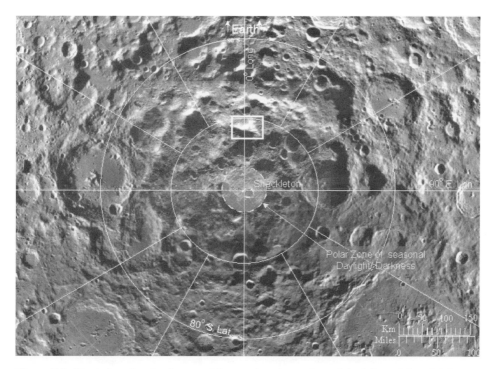

Figure 5.1. Earth-based radar image of the south polar region of the Moon. Mons Malapert, located at longitude 2°E and latitude 85.75°S, is highlighted (image courtesy of Margot *et al.*, 1999).

Figure 5.2. Clementine 750-nm image of Mons Malapert (detail from Figure 5.1). Note the small crater at the left (west) side of the summit.

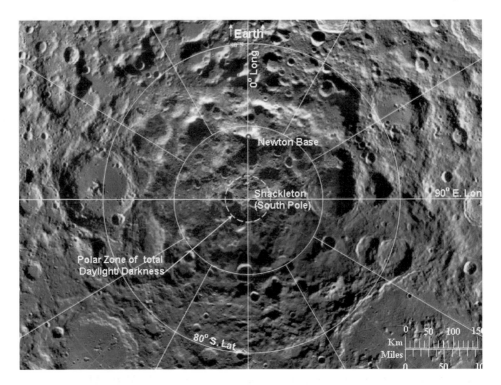

Figure 5.3. Newton Base at the summit of Mons Malapert (Margo *et al.*, 1999; Sharpe).

tion. The summit is also the "high ground" that dominates the south polar region of the Moon.[2]

Preliminary evidence also suggests that the rim of Shackleton Crater (at the geographic south pole of the Moon) may be visible from the peak of Mons Malapert, a little over 100 km away. If confirmed, this would provide a line-of-sight link for power and communications. The higher relief may also enable local line-of-sight communications across the entire region, without the need to erect towers and support structures. Using optical communications, bandwidth will be available for a variety of uses, both locally and globally. The first lunar base site of interest (Site "1", or "Newton Base") is at the summit of Mons Malapert (see Figure 5.3).[3]

---

[2] Although a "Peak of Eternal Light" (at which sunlight is always available) has been identified at the lunar north pole by the SMART-1 orbiter, we still prefer the south pole because of the surrounding topography. The larger craters in the south pole area also provide more possibilities for larger deposits of water ice.

[3] Considerable uncertainty – plus or minus a kilometer or so – still exists as to large-scale absolute elevations of features in the lunar polar regions, and smaller but still significant uncertainties exist with respect to relative elevations. The most conclusive way to determine these relationships will be to go there and take sightings.

5.3.1 Site characteristics and mission opportunities

There are interesting scientific and logistical aspects to sites in the south latitudes from 83° to 86° degrees. These latitudes correspond to locations that are between 125 and 220 km (75 to 135 miles) from the poles. First, these are essentially polar regions, and the surface and subsurface soil temperatures will be only slightly higher than those at the geographic pole. The physical and chemical characteristics of the lunar regolith at these latitudes will probably be similar to regolith characteristics at the poles.

Second, 83° is the latitude above which the Earth will set once a month on the near side of the Moon due to the slight tilt (1.6°) of the Moon's axis of rotation with respect to its orbit around the Sun. Thus, sorties into areas closer to the poles must either be timed to occur while the Earth is visible or they must use satellite relay communication. Simultaneously, sunlight must be available for use as a power source for continuous operations.[4] The periods of sunlight and of direct view of Earth are asynchronous. With careful scheduling, it may be possible to design a robotic-rover traverse profile that will include a dash to the south pole, reconnaissance, and return to the base site – all in sunlight, and all with Earth visible. To do this, the rover would need an average speed of 2–3 km/hr, and surface features on the horizon must not obscure either the Earth or the Sun, both of which will be very low in the sky.

5.4 INITIAL ROBOTIC OPERATIONS AT THE LUNAR BASE

Once a location for a permanent base has been selected, one of the first activities of a robotic lander would be the installation of a continuous, wide-band Earth–Moon telecommunications link. The Earth-synchronous rotation of the Moon (that makes the same side of the Moon always face the Earth) allows for continuous operations anywhere on the Earth-facing side of the Moon below 83° north and south latitude. This system will be the channel through which much of the robotic activity will occur during the initial phases of lunar base development.

5.5 THE LUNAR "SEED"

Before there can be a lunar global civilization, there must be self-sufficient lunar cities. Before there can be self-sufficient lunar cities, there must be bases; before the bases, outposts; before the outposts, expeditions. Before these, there must be smaller endeavors. This is analogous to a "seed" from which sprouts the entire process. Each

---

[4] If sunlight is not available, robotic equipment will probably have a "sleep mode" for use while waiting for sunlight to return to the area. Although it is possible that batteries could sustain operations through the lunar night, the mass of such batteries would increase the launch mass (and consequently the cost) of this option.

stage of growth should be self-sustaining and fiscally viable (fundable on a pay-as-you-go basis).

The expansion of the physical infrastructure and amenities required to change a campground into an outpost, an outpost into a base, and so on, must rely on the use of local materials as the main ingredient of construction elements, rather than depending on supplies from Earth to fulfill all possible needs. However, for all but the simplest uses, the right subsets of raw materials must be found, collected, and processed. The products must then be assembled and placed at the site of their intended use, tested, and commissioned into use. The entire process must be failure-tolerant and amenable to updating, while at each step it must produce either goods or information that justify the overhead costs of sustaining and "growing" the project. The minimum amount of mining, processing, and manufacturing equipment that can accomplish the above tasks is the lunar "seed" that will initiate mankind's permanent return to the Moon (see Appendix I).

5.6   A CANDIDATE LUNAR SEED: THE SELENO-LAB

One payload that would satisfy these objectives would be an unmanned engineering laboratory, containing durable, precision equipment and a stock of laboratory materials.

5.6.1   Mission concept

A concept for the lunar seed mission is to emplace a robotic payload consisting of at least one rover and a fixed-base central station/utility facility at a lunar base site. The two portions would be able to operate independently, as well as cooperatively, and both could be controlled by operators on Earth. The rover would be sent as a separate payload in advance of the lab so that it could set up the lab when it arrives. The rover would derive most of its energy from solar power, and spend the 350 hours of lunar daylight exploring, surveying, and doing field geology tasks. The controls would provide a sense of "being there" for the operators. The system could provide tactile feedback to the operators, who might themselves be future crew members.

Some space hardware systems can be scaled down to micro-size and work just as well as larger ones; others cannot. Computers, for instance, can be made very small, whereas antenna sizes are dictated by the wavelengths of the signals they are transmitting, and rover wheel sizes must be based on terrain roughness and distance-dependant camera visibility requirements. Thus, the rovers will need to be as large or larger than the Apollo Lunar Roving Vehicles and Lunokhods.

5.6.2   Design requirements

The Seleno-Lab's main objective would be to demonstrate and refine many presently-existing candidate techniques for identifying, collecting, and processing lunar materials. The laboratory would be landed near known or suspected resources. It would remain in constant two-way communication with one of several Earth-based transmitting and receiving stations throughout its lifetime. Its design would include features to protect against lunar night and day temperature extremes, thermal shock conditions caused by terminator and eclipse events, and gradual dust accumulation. The rover discussed above would explore the region, collect resources, and return them to the lab for analysis or processing. The lab would be small enough to be flown as the payload of a single launch vehicle, yet be large enough to include all of its necessary consumable materials and operating hardware, including a means of surviving within the power and thermal constraints of lunar night.

5.6.3   Fixed-base facility

The fixed-base facility could be contained in, or be part of, the lander mechanism. It, too, would be mainly solar-powered. Like the rover, it would have manipulation and visualization capabilities also tied to Earth operators, but only limited mobility. During lunar night, the rover could return to the fixed base, where it would be placed in a powered-down, thermal conservation configuration. Interactive maintenance and repair tasks between the roving and fixed-base systems could be performed. With a sufficient budget, additional rovers and fixed-base facilities would be desirable. This way, more possibilities for synergistic interaction between them would be available.

5.6.4   Design process

The design of the laboratory equipment, like that of any other complicated, unique project, will be the product of numerous option decisions, tradeoffs, and compromises. What may distinguish this particular design process from previous ones is a new capability, recently spearheaded by the aerospace industry, of collaborative 3-D software modeling of each physical component. Using these tools, all involved parties can instantly share details of configuration design, detect problems, and propose and implement solutions and improvements. The same tools used to design, document, and test the laboratory can also be used in its operation.

5.6.5   Investigation objectives

The laboratory will produce new knowledge relating to materials and processes. Which materials and processes can be used to replace the heavily water-dependent processing methods that evolved on Earth? How can the vacuum and low-gravity environment of the Moon be used to advantage? Can devices employing principles of superconductivity, for instance, be used during the lunar night, to utilize the scarce electrical power more effectively? Depending upon the size of the lab, multiple operations might occur simultaneously, in different areas of inquiry.

5.6.6   Operations

The lab itself might be the property of a government or governments, industry, universities, or a consortium. Actual operations could be conducted by these entities, or by separately-contracted specialty organizations. Data dissemination and planning coordination, as well as actual command and control activities, would benefit by using the capabilities of the internet.

5.6.7   Production

After a sufficient number of extraction and manufacturing processes have been proven, additional equipment, robotic devices, and supplies will be delivered to the Moon to begin the lunar base operation. The next iteration of the lab will begin turning out small-scale pilot production runs of materials. As the lunar base develops, equipment can be landed that will produce tubing and containers for routing and storing fluids and gases, as well as the fluids and gases themselves. Structural materials will also be produced, including aluminum or iron raw stock and the products of basalt melts (glasses and ceramics). Cables and wires for conducting electricity will be fashioned. Problems will be identified and solved in areas of computer chip manufacturing, electronic components, and antenna production. Other research will be focused on manufacturing the fundamental constituents of a global lunar infrastructure: the solar cells[5] and solar mirrors that will power the new "planet". Finally, the tasks will turn toward building machine tools which, when directed from Earth, can replicate the suite of hardware originally landed to leverage production capacity. The tele-operation of robotic devices on the Moon from the Earth is depicted in Figure 5.4.

5.6.8   Growth of the lunar base

Over a period of several years, re-supply missions will deliver additional consumables, replace or upgrade old equipment, and add new capabilities, until full-scale industrial operations are established. Ideally, all new equipment would be designed with interchangeable parts so that they can eventually be cannibalized for use in other applications. Getting to the Moon costs a lot, but operating things that are already there should cost far less. Once things are there, they do not go away, so recycling and refurbishing the old tools and materials will be an important element in the lunar machine life cycle. The increasing capability and sophistication of industrial activities will enable operators on Earth to direct the construction of lunar pathways and shelters in preparation for the return of humans to the Moon.

---

[5] Lunar-made solar cells will be added to the initial electrical power supply at the lunar base, and the output of electrical power will increase over time.

Figure 5.4. Tele-operation of robotic devices on the Moon (Thangavelu/DiMare).

5.7   DASH TO THE POLE

Simultaneously with the build-up of lunar base operations, wheeled robotic vehicles will embark on exploratory missions in the south polar region. With increasing capability and experience, excursions to the south pole from Newton Base at Mala-pert will be possible. The Moon, like the Earth, experiences annual seasons, but the Moon's seasons are far less pronounced than Earth's mid-latitude summers and winters. Seasons result when a planet's axis of rotation is not vertical with respect to the plane of its revolution around the Sun. For the Earth, this angle is 23.5° from vertical; for the Moon it is only 1.6 degrees. (Another way of stating this is that on the Earth, the Arctic and Antarctic Circles, or latitudes above which constant daylight or night can occur, are 23.5 degrees away from their respective poles. The equivalent lunar Arctic and Antarctic circles are only 1.6 degrees, or roughly 50 km (30 miles), away from each pole. So, once a year on the Moon at each pole, a circle 100 km (60 miles) across experiences total darkness; and roughly 6 months later that same area experiences total daylight.)

The "dash to the pole" must occur during the southern hemisphere's midsum-mer, when the Sun will have its highest elevation (1.6 degrees above the horizon), and during the approximately two-week period that the Earth appears in the sky, rising up seven degrees then setting back down again. Assuming a trip of 250 km each way (an allowance is added for driving around obstructions), the rover would have to travel at a constant 1.5 km/hr without stopping. Alternatively, the rover might be parked for two weeks at the pole in the sunlight until the Earth returns in the sky, before making the return trip. If the rover has left visible tracks (which it almost certainly would), these will expedite the return.

It is interesting to observe that just a few years ago, accurate determination of these traverse opportunities could only be made using ephemeris programs running on large institutional computers. Now, anyone with a personal computer and some inexpensive, commercially-available software can routinely compute and compare these and many other such cases. This decrease in cost of goods, and concurrent increase in availability of information, puts the Moon within the reach for both traditional and non-traditional space operations.

5.8   CIRCUMPOLAR "MAGELLAN ROUTES" EXCURSIONS

Once the south pole has been reached, the horizon surveyed, and confidence in the performance of roving vehicles has been achieved, missions can begin to branch out and explore more of the polar regions. The asynchronous appearances of the Sun and Earth at various points within the Moon's polar regions permit some interesting excursions, using the same basic methodologies as described for the polar mission.

Over a period of two or three years, it would be possible to execute a series of traverses wherein most of the topographically-accessible terrain above 84° latitude could be visited. This latitude represents the approximate limit of constant visibility from the Earth. It also provides an interesting mnemonic: the circumference of the lunar 84th latitude measures 709 miles and the duration of the lunar day–night cycle is 709 hours. This means that if one were standing at 84° south latitude, facing due west, and began traveling forward at exactly one mile per hour (1.6 km/hr), the Sun would always maintain its position in the sky relative to the traveler.

If a solar-powered rover were to embark upon this westward journey at one mile per hour, it would be able to accomplish a lunar circumnavigation until, 709 hours after it began its trip, it appeared in the east, returning toward its point of departure and completing a "Magellan route of discovery" at 84° south latitude. The rover would spend at least half of its journey out of direct contact with the Earth and would need to store data for later transmission. If it continued its westward journey it would be able to make multiple circumnavigations, and by making minor alterations to its route it would identify optimum pathways for the placement of circumferential utility networks (discussed in Chapter 7).

5.9   INTERNATIONAL COOPERATION AND
COMMERCIAL PARTICIPATION

As of this writing, the exploration of the Moon has been the exclusive province of individual nations and space agencies. That paradigm will change, however, with the

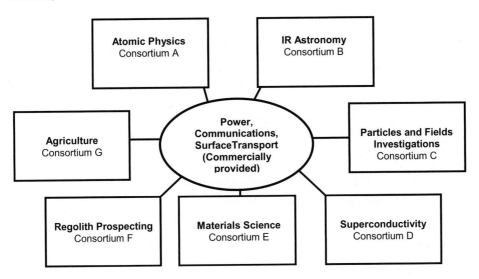

Figure 5.5. A model of the nascent lunar economy.

next stage of lunar development. The creation of the first lunar base and the natural progression of the transformation of the Moon into an inhabited sister "planet" of the Earth will provide opportunities for exploration, research, commerce, education, travel, and recreation by a wide range of interested parties. The most efficient path for each step of lunar development will use the "division of labor" that is possible when involved parties engage in cooperative efforts to provide the segments needed to achieve a common goal.

The national and international space agencies, for example, have traditionally focused on areas of scientific interest such as astronomy and geology, which are financed through (limited) taxpayer funds. To increase the funds available for science projects, investor-supported commercial entities will construct and operate infra-structure projects such as electric power grids and communication and transportation networks (see Appendix L). By this strategy, a viable lunar economy will come into existence where commercial enterprises supply goods and services (at a profit) to customers such as consortia of national space agencies that conduct scientific research projects (see Figure 5.5).

5.10   THE EMBRYONIC CIRCUMFERENTIAL INFRASTRUCTURE

Over a period of several years, terrain surveys of the south polar region, including the coordinated exploration by "swarms" of robots (see Figure 5.6) will have produced accurate maps of the south polar region and a list of promising sites for commun-ications relay stations and power generation towers on high ground. Manufacturing efforts will have been focused on the mass-production of solar cells, wires, insulators,

Figure 5.6. A swarm of robots explores the south polar region (Thangavelu/DiMare).

and other elements of the lunar electric power system. It will thus become possible to emplace a small network of simple signal repeater devices and solar-power generation stations at these high locations around the south polar region. If half of the relay points are in the sunlight, energy can be beamed to those that are not.

With the gradual build-up of power and communications systems in the vicinity of the first lunar base, the entire south polar region will gain access to abundant electrical power and continuous high-bandwidth communications. This network will provide navigational information to rovers, command and telemetry relay functions, and continuous power in the lunar electric grid, and will constitute the first elements of the global lunar infrastructure. In addition, autonomous and tele-operated mining and manufacturing projects will construct shelters, such as arch-supported buildings placed at the bottoms of lunar rilles and covered with regolith to provide radiation shielding and thermal protection, in preparation for human occupancy.

5.11 SUMMARY

The establishment of the first lunar base in the south polar region will provide the knowledge and experience that are needed to begin large-scale projects based upon *in-*

situ resource utilization. The south polar region of the Moon will be explored in depth and the incremental build-up of industrial and scientific capabilities at the first base will prepare the way for humans. International cooperation and participation by commercial enterprises will establish the beginnings of the lunar economy, creating a sustainable and affordable path to the goal of the permanent human settlement of the Moon.

6

Return of humans to the Moon

6.1 INTRODUCTION

When a round-trip Earth–Moon transportation capability becomes operational in the second decade of the twenty-first century, humans will return to the Moon, this time to stay. The arrival of humans will initiate the second phase of lunar exploration and development, and the entire pace of lunar activities will accelerate.

6.2 PRECURSOR MISSIONS

During the first stage of lunar development, discussed in Chapter 5, robotic devices will have conducted extensive surveys of the Moon (e.g., at Mons Malapert and environs). Engineering studies of mining, processing, and manufacturing will have verified a number of lunar-based technologies that are needed for lunar development. These technologies include the extraction of oxygen, silicon, iron, and aluminum from the lunar regolith and the production of solar cells, glass, bricks, wires, and structural beams (Appendix D).

Autonomous and tele-operated robots will have prepared the way for the return of humans by installing navigation/landing beacons, constructing the beginnings of communications and solar-powered electric networks, and by creating one or more underground shelters. Robotic geoscience expeditions in the south polar region will be routine and several large telescopes will be either operational or in the advanced planning stages, including infrared telescopes in permanently-shadowed craters.

While lunar robotic exploration and experimentation are underway, development of spacecraft that are capable of delivering a crew of humans to the Moon will be complete. The Orion Crew Exploration Vehicle (CEV) spacecraft that is being designed by Lockheed-Martin, for example, will be able to deliver a crew of four people to the Moon for mission durations of several weeks, and return the crew to

Figure 6.1. Launch of the Lockheed Martin Orion Crew Exploration Vehicle (CEV).

Earth (see Figure 6.1). Other governments such as China, India, Japan, and Russia may also develop spacecraft capable of delivering humans to the Moon.

6.3   RETURN OF THE HUMANS

The return of humans to the Moon, with multiple crews and eventual permanent habitation, will increase the cost and complexity of lunar operations. An unmanned lunar base can get by with little more than a communications link and a regular supply of electric power, but a sheltered base that is suitable for human habitation will require facilities that can support the physiological, psychological, and other needs of humans. There are many reasons for returning humans to the Moon, not the least of which is that they are indispensable for solving complex problems, conducting science and engineering experiments, and supervising construction projects.

Several challenges must be addressed and overcome if long-term human settlements on the Moon are to occur. The surface of the Moon is hostile to life. Its environment is hazardous and it lacks the basic requirements of life such as food, water, and an oxygen-bearing atmosphere. If humans are to exist for extended periods on the Moon, then shelters with reliable life-support systems must be in place to provide for their safety, as well as for their physiological and psychological needs.

6.4 HUMAN FACTORS

Prior to the arrival of the first human settlers, the tools, equipment, and consumables needed to support them can be delivered to the lunar base. Habitats will be assembled robotically – at the Earth-facing slope of Mons Malapert near the summit, for example – that provide protection from cosmic radiation, micrometeorites, and temperature extremes. One scenario for habitation is to deliver inflatable habitats (see Figure 6.2) to the Moon and install them, using tele-operated robots, in the bottoms of trenches or natural channels, then cover them with loose regolith. Two or three meters of regolith surrounding a structure will decrease the thermal extremes and also protect against ionizing radiation.

Figure 6.2. Depiction of the Bigelow Aerospace Inflatable Habitat in low Earth orbit. These inflatable structures could be placed in sheltered underground locations on the Moon and provide a long-term habitat for the next generation of lunar explorers (Image courtesy of Bigelow Aerospace).

Figure 6.3. The MALEO office on the Moon (Thangavelu/DiMare).

6.4.1   The MALEO site office

An alternative form of housing for the initial crews can be achieved by deploying a MALEO[1] spacecraft base (see Appendix R). The rationale for MALEO stems from the fact that putting a lunar base together on the lunar surface entails much astronaut-supervised activity in a high-risk environment. Also, without any support infrastructure during the initial stages of evolution, lunar dust and gravity can hamper construction activities in many ways.

In the MALEO strategy, the modular components are separately flown into Earth orbit where they are assembled. The components that make up the MALEO base are comparable with a space station, including habitat modules, laboratory equipment, power generators, attitude control and communications systems. Assembling the components would entail operations similar to the assembly of the International Space Station. A superstructure is added to take the thrusting loads during transportation from Earth orbit to the Moon. Once all the systems are in place and checked out, the entire assembly is flown to the Moon and landed as a complete, functioning lunar base that can be commissioned shortly after touchdown (Figure 6.3).

When humans arrive on the Moon, they will disembark from their spacecraft and

---

[1] MALEO: Modular Assembly in Low Earth Orbit. MALEO is proposed by one of the authors (Thangavelu) as a strategy for assembling a lunar base in low Earth orbit and then transferring it to the Moon (see Appendix R).

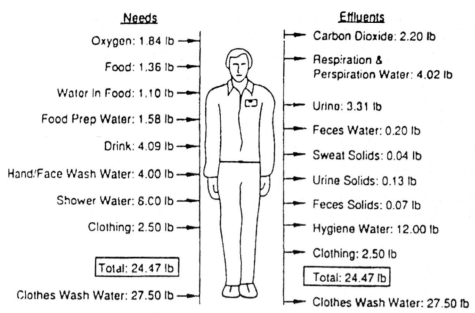

Needs

Oxygen: 1.84 lb

Food: 1.36 lb

Water In Food: 1.10 lb

Food Prep Water: 1.58 lb

Drink: 4.09 lb

Hand/Face Wash Water: 4.00 lb

Shower Water: 6.00 lb

Clothing: 2.50 lb

Total: 24.47 lb

Clothes Wash Water: 27.50 lb

Effluents

Carbon Dioxide: 2.20 lb

Respiration &
Perspiration Water: 4.02 lb

Urine: 3.31 lb

Feces Water: 0.20 lb

Sweat Solids: 0.04 lb

Urine Solids: 0.13 lb

Feces Solids: 0.07 lb

Hygiene Water: 12.00 lb

Clothing: 2.50 lb

Total: 24.47 lb

Clothes Wash Water: 27.50 lb

Figure 6.4. Human system input/output (Hypes and Hall).

proceed to the habitats at Mons Malapert. From their secure and well-provisioned base, they will be able to explore and conduct a wide range of experiments and construction projects.

6.4.2 Environmental control and life-support system (ECLSS)

Reliable life-support systems must be developed for permanent human habitation. They need air, food, and water to survive, as well as regenerative waste management systems (see Figure 6.4). Manned space missions such as the International Space Station are partially-closed life-support systems, in which water and oxygen are recycled, but food is periodically supplied from Earth and solid wastes are stored for later removal. However, it will be easier to create self-supporting human habitations on the Moon than in Earth orbit! The reason is that the Moon has local material resources that can be used to support human missions, whereas Earth-orbiting stations are dependent upon re-supply missions from the Earth.[2] As noted earlier, the lunar base will also be safer than Earth-orbiting stations because the habitats on the Moon can provide higher levels of protection for the crews from the hazards of space.

As the term implies, ECLSS provides the crew with the basic conditions for life functions and comfort – parameters that directly impact crew survival and produc-

---

[2] Eventually, the Moon will be the major source of food, water, oxygen, and other resources for Earth-orbiting stations.

tivity. Therefore, a safe and reliable environmental control and life-support system is essential to the successful operation of a crewed lunar base. Two approaches exist in the design of ECLSS to date. They are physio-chemical systems and biological–biospheric systems.

6.4.3 Physico-chemical (non-biological) systems

Expendable or regenerative chemicals can be used to recycle air and water. By circulating the effluents through a variety of chemical filters and processes, it is possible to maintain the atmosphere and the water in the proper constitution for life. The operational characteristics of these systems are well understood and routinely employed in spacecraft today. They are dependable, show highly predictable behavior, and are rather easy to operate. However, for permanently-manned lunar bases, the material resources needed to operate such systems must be replenished at considerable expense. While small physico-chemical systems are adequate for short-term missions and activities, more capable regenerative systems will be needed to sustain the operations of a lunar base that supports tens to hundreds of people over long periods of time.

6.4.4 Biological–biospheric systems

Biospherics is a newly-evolving discipline that studies and emulates the environmental, ecological, and life-support system of planet Earth. Advances in biospherics show much promise in the development of CLSS that can be applied to long-duration habitats on the Moon, and eventually Mars.

By introducing biological systems such as plants and animals into a cycle that resembles an ecological system on Earth, a symbiotic, balanced, and efficient ECLSS can be evolved. Such systems can be designed to regenerate food supplies (something that physico-chemical systems do not do) through complex feedback loops that are being developed and tested now with encouraging results. The result is a system that can imitate the functions of the Earth by regenerating the air, water, and food within the enclosure with only minimal additional input. Such a system is referred to as a sustainable ecological life-support system (SELSS).

Current experimental life-support systems show promise but are quite complex in their layout and function, and their performance has not been consistent. However, as experience with these systems increases, so will reliability and ease of operations. The long-term goal of the ECLSS community is to produce a fully-closed system that can regenerate all of the water, air, and food without adding anything to the system after startup. This ideal[3] system is often referred to as the closed ecological life-support system or CELSS.

---

[3] The "ideal" of a lunar (or orbiting) base with completely-closed regenerative life-support systems is not possible because such a system will always have losses (with the ingress and egress of equipment, people, etc.). The design goal, therefore, is a stable controlled system that approximates the "ideal" closed system.

6.4.5   Medical care

Humans will need medical care on the Moon, and the first crews to return to the Moon will include physicians. The crews that are selected for trips to the Moon will have no pre-existing medical conditions and will be in excellent physical condition. Their health status will be monitored regularly, as in human spaceflight today, by Earth-based medical specialists on a minimum interference basis. However, there is an inevitable element of risk of injuries and other acute medical problems on lunar missions, and plans must be made for emergency medical treatment.

The Moon is only three days' travel time from the Earth and a crew member who develops appendicitis or other serious medical conditions can be flown back to Earth for treatment. For less serious conditions (e.g., infections and minor injuries), medical problems will be handled on the Moon by the crew, using limited medical facilities and supplies. As more tools and equipment are delivered to the Moon with each crew, however, an increasingly-sophisticated on-site medical diagnosis and treatment capability will evolve, and it will become possible to treat more serious problems such as burns and broken bones on the Moon without curtailing lunar missions. Telemedicine techniques will also allow Earth-based physicians to assist in medical diagnostic and treatment procedures on the Moon (Figure 6.5).

6.5   LEA (LUNAR EXCURSION ACTIVITIES)

Moving about on the lunar surface in a spacesuit will be necessary for tasks that are difficult or impossible to accomplish with autonomous and tele-operated robots alone. It is a hazardous procedure that must be minimized because of cosmic radiation and micrometeorites. LEA will consist of a group of crew members in spacesuits (portable life-support systems). Spacesuit designs now provide reasonable ease of donning and doffing, dexterity, and comfort during long tasks. Special measures are needed to keep lunar dust from contaminating the "in-suit" astronaut environment and to prevent back-tracking of material into the habitats.

A great deal of effort is ongoing to build the next generation of spacesuits. Depending on the task to be performed, ranging from emergency operations to supervision missions, they will have varying designs and performance characteristics. Safety and comfort are foremost in the minds of spacesuit designers as LEA missions will last several hours. Spacesuit concepts and engineering models being developed and tested range from the soft, multilayer kevlar spacesuit similar to the Apollo era and the shuttle EVA suit, through the semi-partial pressure hard suit, to the full hard suit. The hard suit option allows the crew to quickly initiate a surface EVA, as it operates at the same pressure as the lunar base interior. This eliminates the need for the lengthy pre-breathing procedure that is used to prevent the "bends" caused by gaseous nitrogen bubble buildup in the blood stream. A recent concept is the mechanical counter-pressure suit, the skin of which adheres firmly to the body, providing the required pressure. In the partial pressure suit, the torso is rigid and the pressure is

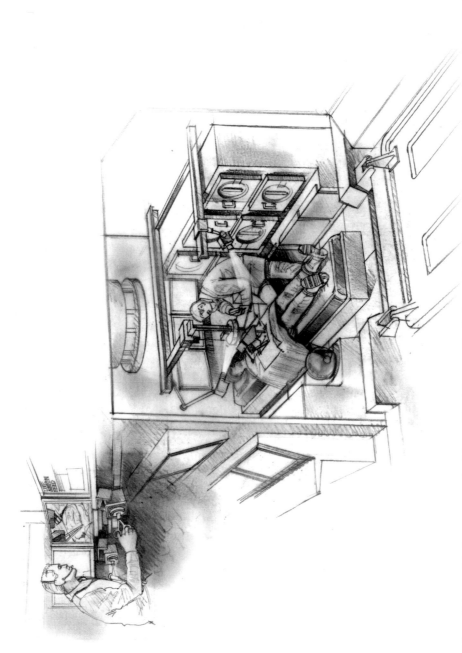

Figure 6.5. Telemedicine assisted surgery on the Moon. Physicians on Earth (upper left) will use telemedicine techniques to assist medical personnel on the Moon (center right) (Thangavelu/DiMare)

slowly reduced as the crew arrives at the location of the mission, thus decreasing the pre-breathing procedure period. Examples of spacesuits are shown in Figure 6.6.

For each LEA, the crew members will don their suits, exit the base through airlocks, and be transported to the work site via rovers. The roof of such vehicles could be made from solar panels so that the rover receives a charge "on the run", and the crew would have some protection from the Sun and micrometeorites. Although the range of battery or fuel cell-powered rovers is limited, the radius of operation of rovers can be extended by placing solar-powered battery-charging stations at out-lying locations[4] (see Figure 6.7). When the crew arrives at a scientific or construction site, they perform work in conjunction with autonomous or tele-operated robots that are controlled from the lunar base or from the Earth.

6.5.1   Nomad Explorer

A large robotic/crewed rover, termed the "Nomad Explorer", with a range of several thousand kilometers may also be landed during the early stages of development (Appendix S). Such a vehicle can assist in large-scale construction activities that are foreseen throughout the early development of the Moon. By adding several systems that can assist in base assembly activity, such a rover might be able to carry out development work in remote regions that are the ideal locations for observatories and other scientific facilities (see Appendix S).

One of the myths about space activities is that there are two clearly separate philosophies in the design and architecture of lunar base building systems and their operations: purely robotic and purely crew-operated activities. Nothing could be further from the truth. The most effective methods of exploration and construction operations are designed around a synergistic man–machine architecture. Robots work best under highly-predictable conditions, while humans are able to adapt to an unpredictable or constantly-changing environment. Using a human–robot buddy system, it will be possible to exploit the best attributes of both robots and humans. While it is possible to maximize robotic activity during the initial stages of lunar base and utility infrastructure construction, human crew intervention will likely be required for optimum operations (Figure 6.8).

6.6   GROWTH OF THE FIRST LUNAR BASE

Using *in-situ* resources, human crews at the base (alongside robotic devices) will install additional habitable facilities and agricultural enclosures, and both pressurized and unpressurized structures for manufacturing, storage and maintenance operations. The continuing expansion of habitable structures will permit ever-increasing numbers of people to live and work at the lunar base. The lunar base will thus become

---

[4] For safety purposes, vehicles will initially be confined to a radius of operations equal to the walking distance from the vehicle to the lunar base.

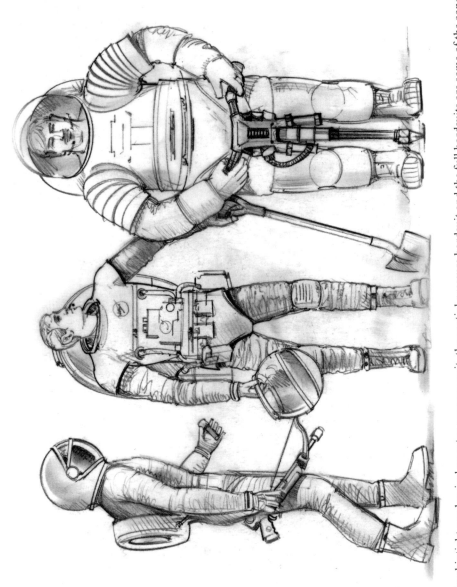

Figure 6.6. The skintight mechanical counter-pressure suit, the partial pressure hardsuit and the full hardsuits are some of the concepts being explored for a range of lunar excursion activities (Thangavelu/DiMare).

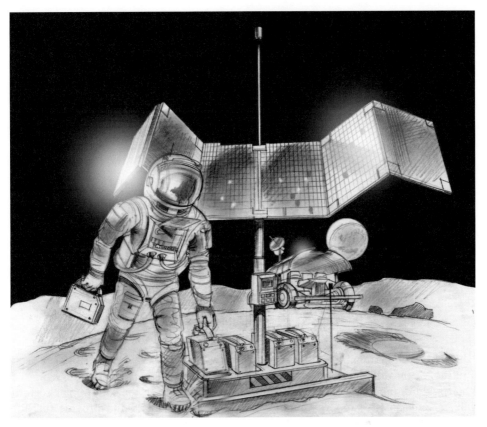

Figure 6.7. Battery-recharging station for Lunar Excursion Vehicle (in background) (Thangavelu/DiMare).

an ongoing, adapting, and incrementally-growing facility whose end-point is global human settlement (see Appendix T).

Cargo missions by space agencies and commercial enterprises from Earth to the Moon will supply the first lunar base with consumables, mining equipment, tools, scientific instruments, and the first experimental agricultural modules. The underground complex at Mons Malapert will be expanded and additional inflatable habitats will be delivered and commissioned. Crews from multiple nations will travel to the Moon for increasing durations and, eventually, there will be a continuous human presence on the Moon.

6.7 EARTH–MOON TRANSPORTATION SYSTEMS

During the first few decades of development, significant advances will take place with *in-situ* resource utilization for the support of habitats and the construction of utility

Figure 6.8. Human/robot synergy on the Moon (Thangavelu/DiMare).

networks. At the same time, transportation systems that operate between Earth orbit and the Moon will be developed. The production of hydrogen and oxygen from the lunar regolith (and from the mining of putative water-ice in the south polar region) will be sufficient to support human crews at the lunar bases and provide fuel for spacecraft that return to Earth orbit.

For example, after a spacecraft delivers its cargo to the lunar surface, it will be refueled and fitted with an aero-braking heat shield that has been manufactured on the Moon (from lunar titanium dioxide, for example). With a cargo (such as water), it will then be launched from the Moon on a return trajectory to the Earth, use its aeroshell to brake into Earth orbit, rendezvous with an orbiting station to transfer cargos, and the cycle will then be repeated. In this manner, a stream of cargoes, including human crews, will be delivered to and from the Moon, assuring the continued step-wise growth of lunar exploration and development activities (Figure 6.9).

6.8   THE LUNAR RAILROAD

Small, unpressurized excursion vehicles will serve the needs of local transportation for crew and cargo (Figure 6.7). The Lunar Excursion Modules (LEMs) used in the Apollo program served this purpose very well. However, these vehicles had limitations in range, safety, and cargo capacity. They also disturb and accumulate the abrasive lunar dust, which is a hazard for both machines and humans (see

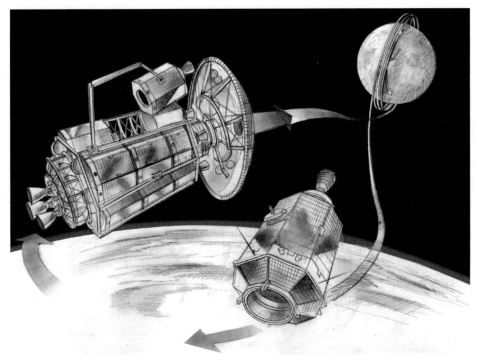

Figure 6.9. Earth–Moon transit vehicles. Lunar-made heat shields are attached to the return stage (the spacecraft on the left) for aero-braking into Earth orbit. (Thangavelu/DiMare).

Appendix J). One solution to the need for effective, safe, and high-speed long-distance transportation on the Moon is a railroad system. The "lunar railroad" will greatly facilitate the exploration, development, and human settlement of the Moon.

During the early phases of lunar development, a pathway for the rail system will be surveyed and, after the right of way has been approved by the lunar government (see Chapter 9), the rail bed will be prepared. The lunar regolith that is removed in preparation of the rail bed can be amassed into a berm[5] at the side of the railroad to serve as a support for the solar panels and as a sunshield for the rails. The lunar dust that is removed from the rail bed is also a source of solar wind gases and other elements, and it can be processed *in-situ* by a mobile factory robot.[6]

After the rail bed has been prepared, lunar-manufactured rails and cross ties will

---

[5] Before the regolith is moved into a berm, high-priority elements such as hydrogen and helium-3 will be recovered.

[6] The regolith contains hydrogen, carbon, nitrogen, and other elements that have been delivered to the Moon by the solar wind, as explained in Chapter 2. The lighter elements will have high value on the Moon because they can be used for life-support systems and for chemical processes such as the production of plastics. The Helium-3 that is recovered will also have a potentially high value as a fuel for future nuclear fusion reactors – both for rocket propulsion and for surface-based power plants.

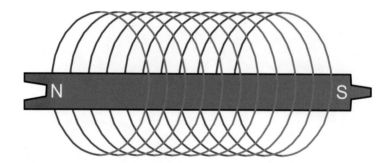

Each segment is made by melting regolith inside an electric field. Iron particles align themselves in the electric field. When the segment cools, it is a permanent magnet. Each segment will then interlock naturally with the segments in front and behind. Bolts and drilled bolt holes would not be needed.

Figure 6.10. Smart railroad track segments (David McKay, pers. commun., 1994, and *Space Resource News*, Vol. 4 No. 2, 1995).

be emplaced. A three-car system, consisting of a power cart, a hauling cart, and an erector cart, can be placed on the tracks. Assisted by multipurpose rovers in the beginning, the hauling cart will be loaded with lunar-manufactured track and delivered to the site of construction of the rail line.

Alternatively, the tracks could be made out of sintered basalt instead of metal. A factory-bot could then move along the intended track pathway and build the track segments more-or-less in place (if sufficient basaltic material is in the vicinity). By using sintered basalt, "smart" tracks could be made that lock themselves in place without bolts. This would by done by sintering the regolith in a reducing environment inside a magnetic coil. The reducing environment is needed to convert FeO to Fe (see Figure 6.10).

The tele-operated[7] erector cart would then align each rail section and drop it into place, where it naturally aligns itself in relation to the adjacent section. The system will then move forward along the new rail line to lay more track (Figure 6.11).

One of the challenges associated with the lunar environment is thermal expansion and contraction during the lunar day/night cycle. To minimize these thermal stresses, the solar arrays can be placed at the sunward side of the rail bed and thus keep the rails in perpetual shadow.[8]

At first, the railroad will be a simple two-track rail system, and the trains will be composed of automated and tele-operated cars that have been delivered to the Moon from the Earth. As lunar manufacturing becomes more sophisticated and proficient,

---

[7] The operation of trains will be mostly automated, with little or no need for control by Earth-based tele-operators. For the completion of more complex tasks, a team of tele-operated devices and on-site construction workers may be necessary.
[8] To minimize thermal shock on solar panels, resistive heating elements can be added to the electric grid to supply heat to the panels in advance of the lunar sunrise.

Figure 6.11. The first lunar railroad (Thangavelu/Schrunk/DiMare).

railroad cars and electric motors will be made on the Moon from lunar materials, and when practicable the railroad will be electrified, deriving its power from the electric grid.

6.9   POWER TOWERS

Power for the lunar base will be principally obtained from solar cells, analogous to the power supply for the International Space Station. At first all of the power requirements needed for lunar activity will be derived from photovoltaic panels and solar dynamic generators that are delivered to the Moon from Earth. (Solar dynamic generators use solar concentrators to drive a high-efficiency mechanical turbine system, generating kilowatts of power. Their thermodynamics employ Stirling and Brayton cycles and their power-to-mass ratios are typically much higher than solar photovoltaic arrays. Recent advances in system design, serviceability and materials offer great promise of this technology for long-duration operations.) Eventually, indigenous materials will be used as feedstock for the creation of silicon solar-cell arrays. The manufacture of solar cells from lunar materials will be an important application of *in-situ* resource utilization.

As power supplies increase and experience is gained with the building of the lunar railroad, it will become possible to build power towers that support large arrays of solar cells. Towers that are placed at high elevations in the south polar region will be able to supply the electric grid with continuous megawatt levels of power (Figure 6.12). Multi-purpose tower structures for use in the polar regions will be

Figure 6.12. Multipurpose power towers (Thangavelu/DiMare).

easier to build on the Moon than on the Earth since there are no wind forces or eroding ground-water conditions on the Moon. However, procedures will be needed to combat the temperature differential that will occur on such a structure when exposed to sunlight on one end and the cold of deep space on the other.

One method would be to hang curtains of solar photovoltaic panels in such a way that the core structure is always maintained in the shadow of the panels. Each tower could serve as an antenna mast, a solar power station, a camera-mounted observation post and, eventually, a support for cable car systems. Large lights and sunlight diffuser "sails" hooked to these tall structures could then be used to create ambient lighting conditions at the base of the towers via mirrors and windows. The first tower, if built at the summit of Mons Malapert, for example, will be tall enough to receive sunlight continuously – and therefore provide a continuous supply of electric power to the lunar base. The large temperature differential between sunlit and shadowed areas, sometimes more than 300 degrees Celsius, will allow the operation of thermo-couple devices for power generation.[9] Initially, while continuous power[10] is still unavailable, power storage may be accomplished through fuel cell technology.[11] At later dates, energy may be stored in superconductor rings, or as kinetic energy in a cluster of flywheels in a "flywheel farm" (discussed in Chapter 7).

6.10   RAIL LINK TO THE SOUTH POLE

Assuming a construction system that can produce and lay as much as a kilometer of track (i.e., 10 pairs of 10-m tracks) during one Earth day, a single rail line from Newton base at Mons Malapert to Shackleton Crater could be built within a year (see Figure 6.13). The railway system will haul tools, structural material and consumables to the geographic south pole at Shackleton Crater in preparation for the construction of the second human-habitable base and supporting infrastructure elements.

6.11   THE SECOND LUNAR BASE

The rail link to the south pole will provide the principal base at Mons Malapert with access to the floor of permanently-shadowed craters that have increased concentra-tions of hydrogen (most likely water-ice). Water in the form of ice would be a highly-valuable resource for sustainable human habitations, and will be of considerable interest to geologists because its analysis could yield clues to the early history of the solar system. Once the railway track from Newton Base to Shackleton is in place, robotic devices that were used to explore Mons Malapert and environs would be delivered to points along the rail line, and the south polar region would then be explored in depth. Crews from Newton Base would take the train to Shackleton Crater to oversee the construction of the second base on the Moon (Figure 6.14).

---

[9] Problems associated with maintaining temperature differentials and with heat dissipation in the lunar environment will need to be solved for thermoelectric power generation.

[10] The Newton base site may, by virtue of its topography, receive as much as 340 days of sunlight per year for power generation.

[11] The putative water-ice in the polar region would provide all the fuel and oxidizer necessary for this storage system.

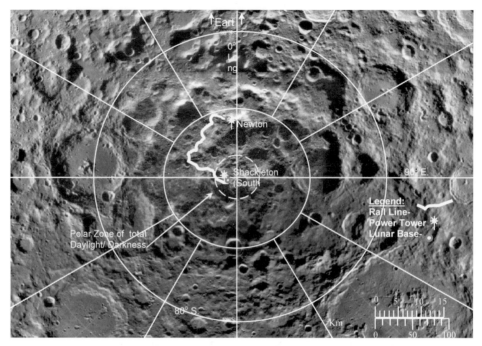

Figure 6.13. Depiction of the initial rail line, bases, and power towers in the south polar region (Margot, *et al.*, 1999; Sharpe/Schrunk/Thangavelu).

The facilities at Shackleton Crater will include the power tower, a heliostat that will monitor solar activity around the clock, a series of science and astronomy stations, and resource processing, manufacturing, and construction facilities. With these efforts and the expanded radius of operations, combined with continued importation of manufacturing equipment and other tools from the Earth, the industrial bases on the Moon will be capable of producing a wide variety of increasingly-sophisticated products. Eventually, these might include electric motors, construction hardware, computers, and telecommunications equipment.

6.12   SUMMARY

The return of humans to the Moon will lead to the establishment of inhabited bases in the south pole region, thus marking a significant milestone in lunar development. The first elements of a permanent lunar utilities infrastructure will provide electric power, transportation, and telecommunications services to government and commercial users. A lunar economy will come into being, based initially upon mining, manufacturing, and the production of basic goods and services. From that beginning, permanent human residence on the Moon will be possible. The stage will be set for the development of the entire south polar region of the Moon.

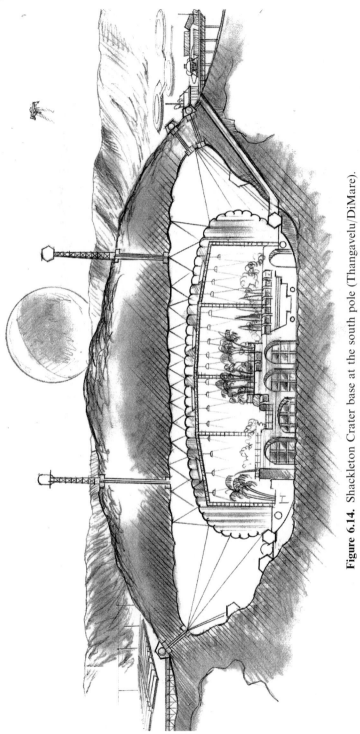

Figure 6.14. Shackleton Crater base at the south pole (Thangavelu/DiMare).

7

Circumferential lunar utilities

7.1 INTRODUCTION

The global settlement of the Moon will require electric power and communications as well as transportation networks (the lunar utility infrastructure), on and below the lunar surface. Without these elements in place, the exploration of the Moon and other large-scale tasks would be difficult. When *in-situ* resource utilization capabilities begin producing infrastructure components, such as solar cells, bricks, metal structures, and electric cable, the placement of a permanent global utilities infrastructure on the Moon will commence. Although technological advances and innovation are expected as a by-product of lunar development, virtually all of the infrastructure needs of the Moon can be satisfied by simply adapting existing Earth-based technologies to the lunar environment. There is no need for technological breakthroughs.

7.2 LUNAR ELECTRIC POWER

Electric power will be as important for the development of the Moon as food and shelter were for the spread of civilization across the Earth. Since there is no electric power system on the Moon right now, every early lunar mission must carry its own power supply, and when that power supply becomes exhausted, the mission must end. The dependency on self-power results in high transportation costs and limits the scope and duration of every mission.

When continuous and abundant supplies of electric power become available, the full scientific and material benefits of the Moon will be realized. Lunar base development and lunar exploration will proceed apace with little or no requirement for material support from the Earth. Missions to the Moon will no longer require spacecraft to carry their own power supplies with them – they will simply "plug in" to the existing lunar electric grid when they arrive on the lunar surface.

The highest priority of the Planet Moon Project will be given to creating the means to support exploration and development projects. During the first phases, greater emphasis will be placed on constructing the lunar electric power and transportation systems, rather than conducting scientific experiments. The rationale for this approach is simple: first build the laboratory, then conduct the experiments. Once the "laboratory" has been built, virtually unlimited energy and material resources will become available, and global scientific and other human activities on the Moon will be possible on a large scale. There are two principal options for providing electric power on the Moon: nuclear reactors and photovoltaic (solar) cells. Both of these sources of power have been deemed to have limitations that will delay large-scale lunar missions well into the future. However, solar cells will be the likely source of electric power for both near- and long-term lunar exploration and development projects.

7.2.1 Nuclear power

Nuclear reactors are often mentioned as a likely source of electrical power for future lunar bases, for several reasons. First, a significant body of knowledge has been accumulated on the design and safe operation of nuclear reactors. That body of knowledge can be applied directly to the design of a nuclear reactor that can operate safely and effectively on the Moon. Second, nuclear reactors can provide continuous electrical power at levels (100 kW–1 MW range) that will be needed for initial lunar base requirements. Third, all of the components of a nuclear reactor can be built and tested on the Earth, and transported to the Moon with existing rocket technology.

Unfortunately, the advantages of nuclear reactors are offset by significant negative factors. The design, testing, and construction of a flight-worthy nuclear reactor would be expensive and time-consuming. The delivery costs would also be high. But the biggest problem for nuclear reactors is the potential for political reaction, particularly in the United States, against the launch of nuclear reactors from the Earth and their use in space.

The launch and operation of nuclear reactors present legitimate concerns for public safety and for the environments of the Earth and the Moon. There are definable risks of the failure of a launch of a rocket containing radioactive elements, and of the contamination of the lunar environment resulting from the operation of a reactor on the Moon.

While these risks may be reasonable and acceptable, and appropriate means for protecting the Earth and Moon environments can be developed, the level of risk *perceived* by the public may be sufficiently high that the nuclear option will not be acceptable. Political forces have been successful in blocking or curtailing the construction in nuclear reactors in many countries, and for attempting to prevent the launch of space probes that contain small amounts of plutonium, such as the Galileo mission to Jupiter and the Cassini mission to Saturn. It is probable that those same forces would also attempt to impede the development and deployment of a nuclear power source for the Moon. Although nuclear power has several positive features for lunar applications, its costs, long development time, and potential for political road-blocks limit its promise as the principal source of power on the Moon.

7.2.2 Solar electric power

The second option for the production of electric power on the Moon is photovoltaic (solar) cells. The Moon has an abundant supply of energy in the form of sunlight; it is constant,[1] unobstructed, and virtually inexhaustible, and it can be converted into electric power by solar cells. Solar cells have been a source of electric power for long-duration spacecraft for several decades, and they have been used successfully on the Moon. The first stages of a functioning lunar electric power supply will therefore be established by simply transporting "off-the-shelf" solar arrays from the Earth to the base camp on the Moon.

One potential drawback with sunlight as a source of energy is that it is not available during the lunar night.[2] Consequently, some form of energy storage device,[3] such as batteries or flywheels, will be needed to provide power for nighttime lunar operations. The incorporation of energy storage devices substantially increases the mass, cost, and complexity of solar power systems, and for this reason solar cells have been regarded as less attractive as a primary power source for lunar operations than nuclear reactors.

Nevertheless, as noted in Chapter 3, the lunar regolith is an abundant supply of materials that can be used as feedstock for the Moon-based manufacture of solar cells. If a number of Earth-made solar arrays were placed on the lunar surface and connected into an electric power grid, the grid would supply the power required for solar cell fabrication equipment to make solar cells from the lunar regolith. Another possibility is a solar-powered mobile robotic factory that "paves" the lunar surface with additional solar cells (see Appendix A). The lunar-made solar cells would be added to the electric grid, and steadily increasing power levels on the Moon would thus be realized. The combination of inexhaustible sunlight that is unobstructed due to the lack of an atmosphere (and only rarely obstructed by the Earth during an eclipse of the Sun) and the local availability of lunar regolith feedstock for the

---

[1] Sunlight on the Moon is only interrupted during lunar eclipses, when the Earth is interposed between the Moon and Sun. Also, because there is no atmosphere to attenuate it, sunlight that arrives at the lunar surface is about eight times as intense as the sunlight that reaches the surface of the Earth.

[2] The Moon rotates on its axis once every revolution around the Earth, and half of the Moon is always in daylight and half is in darkness. Images from the SMART-1 spacecraft indicated that a crater rim at the north pole of the Moon may receive sunlight continuously; all other areas of the Moon experience periods of darkness.

[3] Unmanned (robotic) lunar bases may not need continuous power. They can "power down" to a limited extent during the lunar night and resume operations during the lunar day, thus obviating the need for large power storage devices. However, they typically require "keep alive" power to prevent damage to sensitive electronics caused by the extreme cold of the lunar night.

construction of solar arrays has given rise to a substantial body of literature on the subject of generating electrical power on the Moon.

7.3   THE LUNAR POWER SYSTEM (LPS)

The most ambitious plan for generating electrical power on the Moon with solar cells that are made from lunar regolith materials has been put forward by Criswell and Waldron. They have patented and published articles on a plan for supplying electric power to the Earth at low cost from a Lunar-based Solar Power system (LPS). They envision the construction of large solar collecting bases on opposing (east and west) limbs of the Moon. The reason for the placement of bases on opposite sides of the Moon is to assure that one or the other of the two bases is always in sunlight, thus providing power to the Earth continuously.

The power that would be generated at each of the two lunar base sites of the LPS would be converted to microwave energy, beamed to receiving stations on Earth, reconverted into electricity, and then fed into electric power grids on the Earth (Figure 7.1). The estimated power that could be generated on the Moon by the

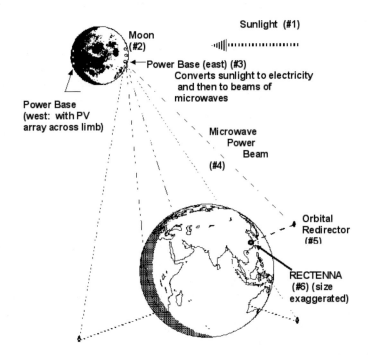

Figure 7.1. The Lunar-based Solar Power system (LPS) (courtesy of Dr. David R. Criswell): 1, sunlight; 2, Moon; 3, sunlight is converted to electric power by solar arrays on the Moon; 4, electric power is converted to microwave energy on the Moon and beamed to the Earth; 5, relay stations in Earth orbit send microwave energy to the Earth; 6, rectennae convert microwave energy to electric energy and deliver it to power grids on the Earth.

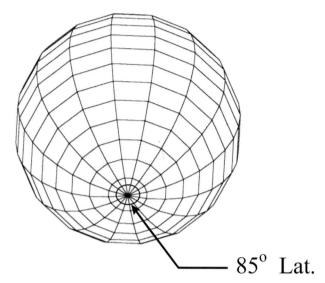

Figure 7.2. Circumferential electric grid at 85° south latitude. Solar power stations that are positioned at regular intervals along an electric grid at 85° south latitude provide continuous electric power to the grid (Sharpe).

LPS is in the multi-terawatt range (one terawatt equals one trillion watts, or one million megawatts). The projected power levels from the LPS will thus exceed the present combined output of all electric-generating power plants on Earth (Criswell, 1994), and will be able to supply Earth with all of its future energy needs. The lunar power system has the potential to be highly profitable (see Appendix L). If that economic potential is realized the Earth will be the recipient of an abundance of clean, low-cost electric power.

7.4   CIRCUMFERENTIAL ELECTRIC GRID

The LPS is intended to satisfy the future energy needs of the consumers of electric power on Earth, but it would also, be the source of electric power for all future lunar missions. For the first permanent lunar base, the LPS concept could be modified by creating a solar-powered electric grid that courses around the circumference of the Moon. If solar panels were placed at regular intervals along a given circumference of the Moon, then 50 percent of the panels would be powered at any one time, and continuous[4] electric power would be available in the grid for all lunar operations (Figure 7.2).

The circumferential LPS can be created by delivering the manufacturing equipment for the production of solar cells and other electric power components from the

---

[4] Some form of power storage will still be needed to provide power during eclipses of the Sun by the Earth.

Earth to the first lunar base. Solar cells can be fabricated from lunar regolith feed-stock (using autonomous and tele-operated robots) and then connected to the lunar electric grid. As more solar cells are added, the grid would grow in east and west directions from the first lunar base until a solar-powered electric grid is constructed around the circumference of the Moon. When the solar-powered circumferential electric grid is in place, continuous electric power in the multi-megawatt range[5] will be available for all lunar exploration and development needs. Further growth of the electric power grid will produce more power than is needed on the Moon and the excess can then be exported, for a profit, to Earth or other sites in the solar system, as envisioned by Criswell.

The solar-powered circumferential electric grid concept has the following advantages:

- It can be created in a number of years from lunar regolith materials.
- Most of its components can be manufactured and assembled robotically.
- It can grow incrementally to any desired size.
- Its construction is based upon the adaptation of existing technologies to the lunar environment; no "breakthrough" technological developments will be required.
- It will produce electric power during every stage of its construction.
- It will eliminate the need for nuclear reactors.
- It will produce abundant and continuous electric power.

For these reasons, we advocate a solar-powered circumferential electric grid as the source of electric power on the Moon.

7.5   CONSTRUCTION OF THE UTILITIES INFRASTRUCTURE

During the first two phases of lunar development, a continuing stream of tools and equipment will be delivered to the Moon. The number of people living on the Moon will gradually increase in step with the construction of habitats and life-support systems. This increasing capability to do useful work will enable commercial entities and the international lunar community to undertake the construction of the circum-ferential utility networks. As mentioned in Chapter 5, the production of solar cells from lunar regolith materials can proceed on a nearly continuous basis at Mons Malapert because the summit area receives sunlight for as much as 90 percent of the lunar year.[6]

---

[5] The overhead Sun delivers slightly more than 1.3 kilowatts of energy per square meter to the surface of the Moon. Assuming a solar cell efficiency of 20%, the output of a solar array that is 4 meters wide would produce approximately one megawatt of electrical power per kilometer of length when the sunlight is orthogonal to the array. A 950-km circumferential grid at 85° south latitude would thus generate an estimated 300 megawatts of electric power continuously.

[6] Subsequent analyses may well favor the north polar region as the site of the first lunar base. The above discussion is applicable to either the north or the south polar region.

The challenge of building the circumferential electric grid will be similar to that of building the lunar railroad (Chapter 6), and both construction projects can be undertaken simultaneously. For example, the mining and manufacturing processes that are used to produce iron and aluminum support structures for the railroad will also be used for the construction of the electric grid.

The construction of the electric grid can begin by adding lunar-made solar cell components along illuminated segments of the right of way of the lunar railroad that extends from Mons Malapert to the south pole. A system of governance of the Moon (see Chapter 9) will be required to sanction the construction of the utilities system. The government will also need to establish and guarantee property rights to construction companies and other institutions that utilize regolith materials and build the infrastructure elements along the right of way. Sites of scientific, esthetic, or other interest will be avoided so that they are not disturbed by the construction project.

As the grid grows lengthwise around the circumference of the Moon, the number of days of continuous power will increase until the grid extends more than halfway around the Moon. At that point, some portion of the electric grid will always be in sunlight, and the goal of continuous power on the Moon will be realized.

A problem with communications will arise as the construction of the rail line and electric grid proceeds around the circumference to the limbs and far side of the Moon. Line-of-sight communications for the direct control of robotic devices from the Earth will not then be possible and a system of relays for communications will be necessary. A high-bandwidth telecommunication system for the transmission of voice and data will be added in parallel to the rail line and electric grid.

7.5.1   Lunar telecommunications network

Telecommunication equipment such as fiber-optic cable can be produced at the same industrial park from the same lunar regolith materials that are used to make solar cells. Thus, as solar cells and railroad tracks are added to the growing ends of the lunar railroad and electric grid, a telecommunications network that parallels them will also be installed. This will enable voice and data communications links to be made from the Earth and from lunar bases to any location on the utility network.

7.5.2   Lunar pipeline system

The fourth element of the circumferential utilities system is a pipeline network that will be used for the global transportation and storage of liquids and gases. The mining and processing of lunar regolith materials for the production of silicon, aluminum, iron, and other elements will also yield oxygen and solar wind gases such as hydrogen, nitrogen, and helium, including potentially valuable Helium-3 (see Chapter 3 and Appendix H). All of these gases must have a means of storage and transportation.

For storage, oxygen produced from the regolith (see Appendix E) will be combined with solar-wind hydrogen (also present in the regolith) to form water, and the remaining oxygen and other gases be stored in liquid or gaseous form.[7] For transportation, a global pipeline network that parallels the other elements of the circumferential utilities grid could be installed. The pipelines will be constructed from regolith materials, such as iron, glass, or cast basalt, at the industrial base. In addition to the transportation of liquids and gases, the pipeline system may also be used for active thermal management purposes. For example, gases and liquids in the pipeline can be used to remove waste heat (e.g., by means of phase changes or closed-cycle convection) from industrial operations, and to deliver heat to and from agricultural and human habitat areas. The heating and cooling cycles of a global pipeline/thermal management system may also be used to drive heat engines (e.g., Stirling engines) that derive energy from the temperature difference between sunlit and dark areas of the Moon.

7.5.3 Flywheel farm

Flywheels manufactured on the Moon could provide energy during periods when solar energy is at a minimum. Because there is no atmospheric drag, storing energy in a flywheel would be more efficient than would doing so on Earth. The flywheel farm will be useful during the early stages of construction of the circumferential grid, before continuous power is available (see Figure 7.3).

7.5.4 The Newton–Shackleton cable car system

Another element of the utility infrastructure is a cable car system that stretches between the bases at Newton and Shackleton. The cable cars will carry both crew and payload, thus providing an alternate transportation means between the two base locations. Towers can be built along suitable peaks between the two bases to support these cars over long, uninterrupted stretches of lunar-made cable (Figure 7.4).

Loaded with cargo and personnel, a logistics system attached to the cable will provide services to the region by dropping off and picking up crew and equipment all along the cable path using tethered elevators and pressurized gondolas. The cable car would service points of activity, such as infrared telescope operations on the lunar surface, and it eventually would be a tourist attraction as well.

7.6 DEVELOPMENT OF THE SOUTH POLAR REGION

After the first circumferential railroad, electric grid, and associated utilities are completed,[8] interconnecting infrastructure elements will be added and the entire

---

[7] Cannibalized fuel tanks from rockets that have landed on the Moon can serve as cryogenic storage tanks.

[8] A milestone of the construction project will be realized when the two ends of the first circumferential rail line are connected on the far side of the Moon, analogous to joining the ends of the first transcontinental railroad project of the United States in the 19th century.

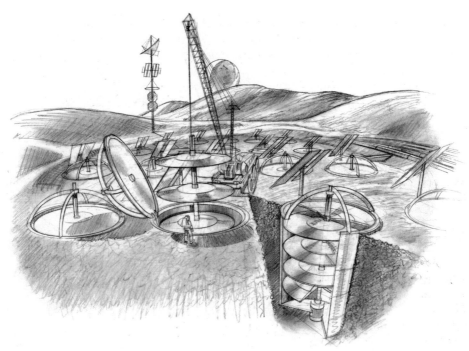

Figure 7.3. A flywheel farm used to store energy (Thangavelu/DiMare).

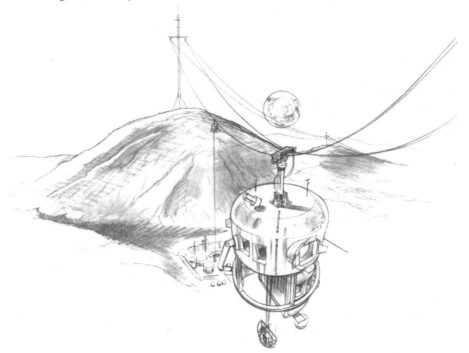

Figure 7.4. The Newton–Shackleton cable car system (Thangavelu/DiMare).

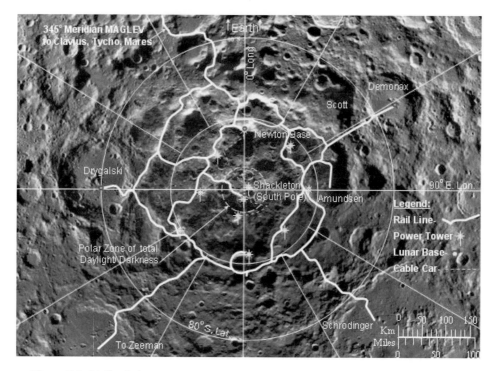

Figure 7.5. Utility infrastructure grid at the south pole (Sharpe/Schrunk/Thangavelu).

south polar region will become open to exploration, development, and human settlement (Figure 7.5). The continuously energized electric power grid will make power available to any location in the south polar region, and the power supply will continue to grow as new solar cells are added to the grid.

The manufacturing capability of the industrial base of the Moon will evolve to a level of sophistication equal to that of Earth-based industries, and plans will be made to extend the utilities infrastructure network to all other areas of interest on the Moon. Ready access to all areas by the lunar railroad (see Figure 7.6) will enable the scope of human activities to increase, including education, tourism, arts, and sporting events.

7.6.1   Lunar agriculture

The experience that has been gained from early agricultural experiments will be used to increase the acreage and efficiency of food crop production. The lighter elements that are harvested from the regolith[9] will be added to the lunar soil to support the

[9] The regolith, as noted in Chapter 3, has small quantities of elements that are essential for the growth and maintenance of life systems such as carbon, hydrogen, nitrogen and sulfur. These nutrients will be available for agricultural projects when horticultural operations begin.

Figure 7.6. Railroad operations in the south polar region (Thangavelu/DiMare).

growth of a variety of plants for consumption. The availability of continuous and abundant electric power will permit the operation of greenhouses whose lighting, humidity, temperature, and other operating parameters are optimally tuned to the production of specific agricultural products.

At first, all foodstuffs that are cultured on the Moon will be plant life.[10] Some plant life, such as fruit trees, will be woven into habitat spaces, and the lunar dwellers will thus live within luxuriant vegetation that serves both esthetic and useful purposes. As experience is gained and available power levels continue to rise, food production will be expanded to include fish and livestock.

7.6.2 Emergence of lunar cities

The continued growth of infrastructure and human-habitable facilities on the Moon will lead to the development of multi-level villages that can accommodate 100 to 1,000 people in comfort and safety (Figure 7.7). These early lunar towns will be the prototype for large-scale human settlements on the Moon that have fully regenerative

[10] Honeybees will be imported from the Earth for use with crops that require pollination.

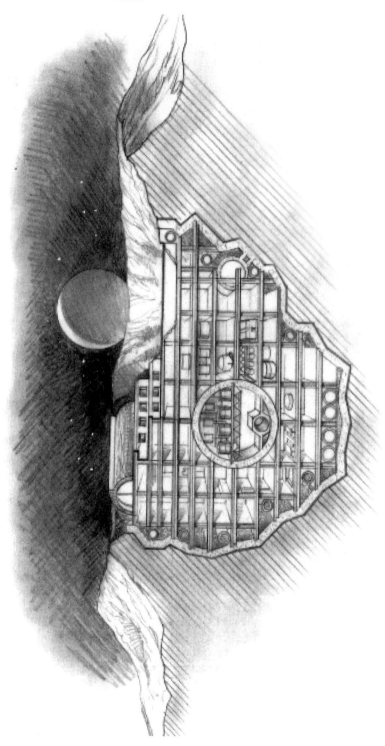

Figure 7.7. Lunar base (Thangavelu/DiMare).

life-support systems, a continuous supply of electric power and communications, and access to other areas of the Moon by the lunar railroad.

One important consideration for early human habitation is lava tubes, which may be as much as 30 meters below the lunar surface and tens of kilometers in length. Lava tubes have inherent advantages for human habitation in that they afford protection from ionizing radiation, provide a more benign thermal environment, and are oriented horizontally. After a lava tube has been structurally reinforced, it can be used for early human habitats and can grow, with the creation of connecting tunnels, to accommodate much larger populations in the future.

7.7 SUMMARY

Virtually all of the infrastructure needs of lunar development can be satisfied with the adaptation and nominal growth of existing technologies, combined with *in-situ* lunar resource utilization. The creation of a circumferential utility network in the south polar region of the Moon will provide communications and transportation services, and continuous and abundant electric power for a wide range of human activities on the Moon. The stage will then be set for global human habitation of the Moon – for the transformation of the Moon into an inhabited "sister planet" of the Earth.

8

The Planet Moon

8.1 INTRODUCTION

The history of human exploration forecasts that, following the commissioning of the first lunar "outpost", the entire globe of the Moon will eventually be explored, and multiple permanent human settlements will be established. The future Moon will be as comfortable and secure a place for humans to live, work, study, and play as anywhere on Earth. Its technological capability will equal that of the Earth, and our Moon will have access to the virtually unlimited energy and material resources of space.

8.2 GLOBAL DEVELOPMENT

After a circumferential infrastructure becomes operational in the south polar region, it can be expanded to other regions, and global human settlement will be possible. Self-sufficient lunar industries will be able to produce virtually all of the goods and services needed for development of the lunar frontier and the only cargoes that will need to be imported from the Earth will be the humans who wish to live, work, study, and tour the Moon.

8.3 CISLUNAR TRANSPORT AND LOGISTICS SYSTEMS

A variety of human-rated launch vehicles, orbital depots, and lunar-landing facilities will be developed for the movement of people between Earth and Moon. The economics of cislunar[1] spaceflight suggest that four separate nodes or transit points

[1] The term "cislunar" refers to the region of space encompassed by the orbit of the Moon around Earth.

Figure 8.1. Earth spaceport (Thangavelu/DiMare).

will be used. They are the Earth Space Port, the Earth Orbital Station, the Lunar Orbital Station and the Lunar Base Space Port. There are three distinct vehicle types that will ply between these transit points: the Earth-to-orbit shuttle, the cislunar orbital transfer vehicle and the lunar lander shuttlecraft. In each of the above categories both robotic and crewed vehicles are feasible. Cargo carriers that are robotic[2] and their systems will not need to be safety rated for passenger transport. For this reason, they will be far more economical to build and maintain.

8.3.1   Earth spaceports

Many places on Earth will host spaceports, first at locations close to existing launch facilities like Kennedy Space Center in Florida, Baikonur Cosmodrome in Kazakhstan, Tanagashima Island in Japan, Kourou in French Guiana, and Sriharikota Space Center in India. Eventually, spaceport terminals will appear as extensions of airports all over the world, with vehicles, passengers and baggage being serviced at those locations (see Figure 8.1). A wide range of vehicles will be used for transportation between the Earth and the Moon, including expendable vehicles and two-stage fully reusable systems (Figure 8.2).

[2] The technologies that are now used for guidance and control of robotic aircraft will be adapted to cargo spacecraft. Automated (robotic) berthing of spacecraft has already been demonstrated with the (U.S) Orbital Express system

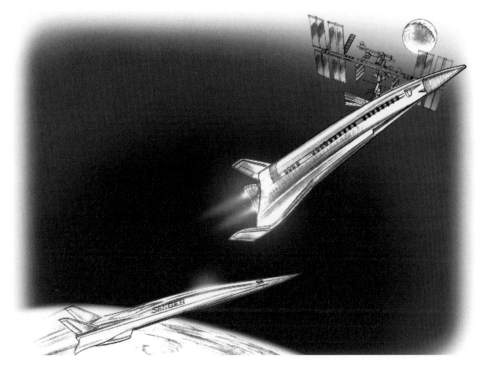

Figure 8.2. Two-stage Earth to LEO transport (Thangavelu/DiMare)

8.3.2 The Earth orbital station

Skylab was America's first space station, in operation from 14 May 1973 until 8 February 1974. The Russian Mir space station (derived from their Salyut series of stations) was in operation for fifteen years, from 1986 until 2001. The current space station, which has been occupied since 2000, is the International Space Station (Figure 8.3). All of these vehicles were designed for scientific research activities in microgravity. The experience and data gained from operating these stations will be used to build larger stations for commercial operations. Several stations could be placed in various orbital inclinations to optimize orbital plane alignments between the Earth and the Moon, providing multiple launch windows between the two in a given period.

8.3.3 Cislunar vehicles

A new generation of vehicles will evolve for hauling crew and cargo between the Earth-orbit and lunar-orbit stations (the two spacecraft at the middle and left side of Figure 8.4). Since they are designed purely to function in space and operate under the thrusting loads of small engines, they will not need to have the sturdy skeleton of an atmospheric reentry vehicle or a lunar lander. As their flight times are limited to a

Figure 8.3. International Space Station after its assembly is completed (∼2011) (image courtesy of NASA).

Figure 8.4. Lunar orbital and landing craft (Thangavelu/DiMare).

maximum of three Earth days to and from Earth or lunar orbit, the environmental control and life-support systems for crew are sized accordingly, with replenishments available first only at the Earth station but eventually at both lunar and Earth destinations. (See Section 8.8 for a discussion of a design of a large cislunar cruiser that perpetually cycles between the Earth and Moon.)

8.3.4   Lunar orbital station

Both an equatorial and a polar orbiting station will be useful for lunar development; they will be used as fuel and cargo depots and as transfer stations for passengers and crews. After insertion into lunar orbit, arriving crew, passengers, and cargo would transfer from the cislunar vehicle to one of the orbiting stations. Subsequently, they would enter a lunar lander shuttlecraft whose destination will be a lunar port city. The polar-orbiting station will serve the entire globe, aligning itself with different locations around the lunar globe once every fortnight. It could also serve as an orbiting hotel for tourists.

8.3.5   Lunar lander craft

In-situ resource utilization (ISRU) processes will be used to extract oxygen (see Appendix E) and hydrogen on a high-priority basis from the lunar regolith, and possibly also water ice from the polar regions.[3] The harvested oxygen and hydrogen will be valuable for life-support and industrial processes, but perhaps one of the most important uses will be as fuel for rockets that transfer people and equipment to and from the Moon. This reduces the need for Moon-bound vehicles to carry fuel for the return trip. The mass and volume of the fuel that would have been used for the return trip can now be used to carry cargo from the Earth to the Moon.

After the passengers and cargo are transferred to the lunar lander shuttlecraft (the vehicle on the right in Figure 8.4), the lander would undock from the station and descend to the Moon. The lander would be refueled at the lunar base with lunar-derived liquid hydrogen and oxygen, and then deliver Earth-returning passengers and cargo back to lunar orbit for rendezvous with the lunar orbital station. The cycle of transportation between the lunar surface and lunar orbit would then be repeated at whatever frequency is economical at the time. When mass drivers become operational (Appendix K), water ice if it exists will be launched from the south polar region to the lunar orbital station, where it will be electrolyzed into its hydrogen and oxygen components and used to refuel the lunar lander.

[3] The Lunar Prospector satellite discovered increased concentrations of hydrogen at the lunar poles, and radar imaging by the Clementine satellite and Earth-based facilities suggest that this hydrogen may be in the form of water ice. These findings will be verified through sample analyses of the regolith in the polar regions.

8.4 LUNAR BASE PORT

The lunar base port will be the gateway to Planet Moon, serving functions analogous to any international air or seaport on Earth. Initially, a single base port will be the hub of lunar development projects. Eventually, several ports will be built at many locations on the Moon. The port may need a lunar quarantine facility to assure that Earth–Moon mutual contamination is kept to a minimum. Crews and passengers will be "acclimatized" to $\frac{1}{6}$ gravity in special training quarters and oriented to the lunar environment and activities.

Newton Base at Mons Malapert is used as an example of a lunar port city. It will be the center of lunar construction activities and will have access to communications, rail transportation, and abundant electric power from the circumferential utility network. Tunneling and excavation efforts at Mons Malapert will expand the size of Newton Base by creating large interconnecting underground spaces. This underground complex will be sealed (e.g., with a glass or metal lining) to contain an atmosphere, and developed into a comfortable and safe, self-sufficient lunar city[4] (Figure 8.5).

The city will not only be the center for scientific, manufacturing, and construction activities, it will also be a major tourist destination. The interior will be designed to be esthetically pleasing, with large open spaces and abundant vegetation. The diversions that will be possible at Newton Base include sporting events (in $\frac{1}{6}$ G) and human-powered flight (see Figure 8.6).

8.4.1 Retirement and rehabilitation

Advances in nutrition and medical care have led to greater longevity of the population of the Earth, and people now enjoy longer productive and professional lives than was possible in the past. Worldwide living standards have also been steadily increasing in many parts of the world. The aging population has more wealth than ever before in history, and a significant fraction of that wealth is directed to tourism and the treatment of the physical infirmities of aging. In view of these facts, the "Planet Moon Project" will be a magnet for the aging population for several reasons:

1. Participation in lunar development projects will be an exciting task for people of all ages.
2. Lunar development projects will require expertise from a wide range of professions and management that retirees can provide.
3. The Moon offers the ultimate tourism experience.
4. And the lower gravity ($\frac{1}{6}$ G) will allow people with physical limitations to enjoy greater levels of mobility than are possible on Earth.

People who have retired from employment on the Earth can extend their productive lives by providing their services on the Moon. They would also be able to enjoy the

---

[4] The previously used inflatable habitats may be moved to new areas under development or they may be used for other purposes.

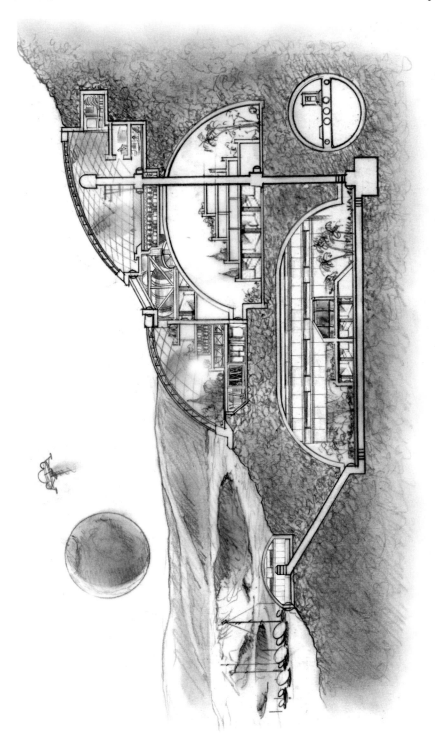

Figure 8.5. Newton Base at Mons Malapert (Thangavelu/Pimenta/DiMare).

Figure 8.6. A mall at Newton Base. The interior will be esthetically pleasing, with vegetation, waterfalls, and open spaces (Thangavelu/DiMare).

Figure 8.7. High-speed maglev transport on the Moon (Thangavelu/DiMare).

adventure of living and working on a new world and they could take advantage of the reduced gravity environment that permits greater ease of mobility than is possible on Earth. The common elements among these ideas suggest that it may be worthwhile to create lunar retirement and rehabilitation facilities on the Moon. These facilities promise to be viable economic projects that meet the needs of the older population who will potentially play a vital role in development projects on the space frontier.

8.5 THE MAGLEV RAIL SYSTEM

The utility network of the south polar region will likely be extended over the entire Moon as the lunar civilization and culture develop. The simple two-track rail system can then be augmented by the development of a magnetic levitation ("maglev") rail system, which will provide high-speed transport of humans and cargo on the Moon (Figure 8.7). It will be highly efficient because there is no air resistance on the Moon and it may be able to travel at a speed of one kilometer per second.

The growth of the railroad toward the equatorial regions will lead to the creation of new cities, scientific laboratories, astronomical observatories, and tourist destinations. At the equator, the track will provide access to the landing site of the first spacecraft ever to land on the Moon, as well as the Apollo landers where humans first explored the Moon. The available electric power in the lunar power system will also grow substantially, and there will be an abundance of power available for all scientific and commercial projects. Excess energy will be beamed to energy markets on Earth and other destinations in space as described in Chapter 7, and will be a substantial component of the lunar economy.

8.5.1   The Sun-synchronous railroad

In addition to its ability to carry cargoes, the circumferential rail line will offer the opportunity to experiment with "Sun-synchronous trains." Sun-synchronous trains move west continuously on tracks around the latitudinal circumference of the Moon at the same speed as the rate of rotation of the Moon, and thus remain stationary with respect to the Sun. If a train were to travel west at 85° south latitude, at the rate of 1.4 kilometers per hour, it would maintain the Sun at a constant angle overhead. At the equator of the Moon, the rate of westward movement required for Sun-synchronous operation would be 16 kilometers per hour.

One application of a Sun-synchronous train could be experiments with greenhouse cars that grow food supplies in continuous sunlight.[5] If such experiments were successful, an entire train of greenhouse cars, a "farm on wheels", could be operated to take advantage of biological processes that thrive on continuous sunlight. Another application could be a "factory on wheels" train. Mirrors on the train could focus sunlight and thus create the constant high temperatures that are needed for raw material processing and manufacturing operations (Figure 8.8).[6]

8.5.2   The 345th meridian magnetic levitation high-speed transport train

The Maglev rail line from the south pole to the north pole would follow a path that causes the least disruption to the lunar environment. It would course across the smooth mares for most of the distance between the poles and would end in loops around the 85°N and 85°S parallels. The proposed route of the rail line will extend from the south polar region to the north pole along the 345° meridian longitude (the "345 Meridian Route") and would thus intercept many of the landmarks on the Moon that are clearly visible from the Earth, making it a particularly delightful experience for lunar tourists (Figure 8.9).

A pole-to-pole connection across the Earth-facing side of the Moon will open up the Moon to global exploration and development. It may provide access to water resources in both the south and north polar regions and give rise to the expansion of utility networks around the circumference of the Moon at the equator and other

---

[5] Because the plants would require adequate protection from cosmic and solar radiation, the sunlight might need to be delivered to the plants by a system of mirrors, windows and UV filters.

[6] The average speed of the train would be Sun-synchronous; however, the speed would be varied to accommodate the loading and unloading of raw materials and finished products.

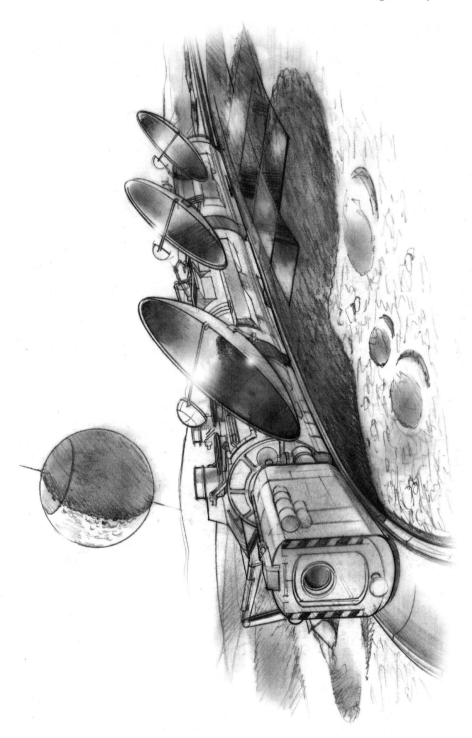

Figure 8.8. The Sun-synchronous railroad (Thangavelu/DiMare).

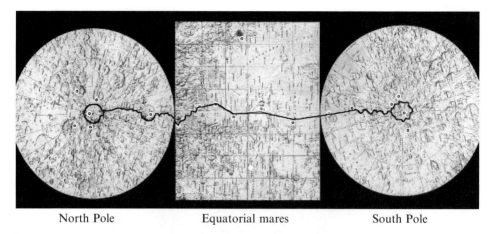

| North Pole | Equatorial mares | South Pole |

Figure 8.9. The 345th meridian rail network (Thangavelu).

latitudes. Terawatt levels of electric power will then be generated by the lunar power system, global geoscience expeditions will be possible, and the first of several thousand lunar-made telescopes will be placed in a global astronomical observatory network. Cities will be constructed at multiple sites along the route, including the seat of the lunar government,[7] people will be able to migrate to and settle on the Moon, and the new, self-sustaining *Planet Moon* will become a reality.

8.5.3   The ballistic cargo delivery system

Because there is no atmosphere on the Moon, the trajectories of payloads that are launched by mass drivers (Appendix K) can be calculated with precision. As a result, the accelerating and reverse-mode braking components of mass drivers may be built at fixed locations on the Moon for sub-orbital flights. Using the power of the electric grid that is continuously available, these mass drivers will be able to launch cargo-filled containers on parabolic trajectories to various locations, where they would be captured by receiving mass drivers and either brought to a halt or accelerated on a trajectory to the next receiving station. By mechanically adjusting the azimuth and elevation angles of these launchers, it will be possible to fly mail and other cargo to every location on the Moon that has a launch–capture mass driver system (Figure 8.10).

8.6   SCIENCE PROJECTS

The extension of utility networks to all areas of interest on the Moon will enable scientists to construct unique scientific facilities, including a global, coordinated

---

[7] It would be appropriate, perhaps, for the seat of the lunar government to be located at the Crater "Plato", near the 345th meridian in the northern hemisphere of the Moon.

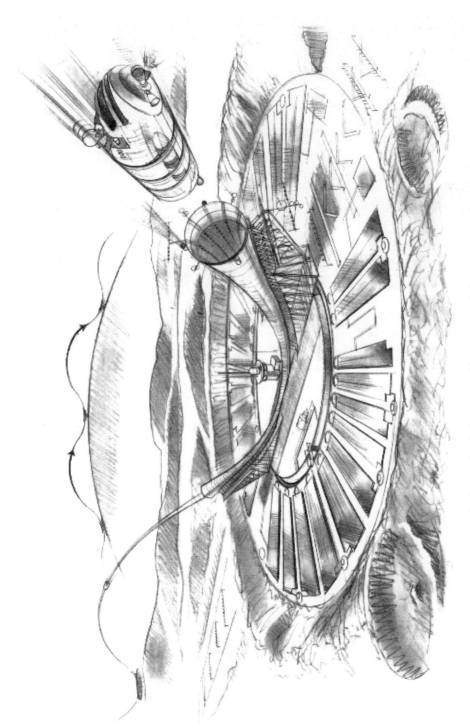

Figure 8.10. The ballistic cargo delivery system (Thangavelu/DiMare).

Figure 8.11. Optical interferometry at Copernicus Crater (Thangavelu/DiMare).

astronomical observation capability, a biohazard experimentation and containment laboratory, and a superconducting "atom smasher".

8.6.1   Astronomy

The Moon is an ideal platform for astronomical observations. As soon as practicable, the lunar industrial base will build telescopes and deploy them at strategic positions around the Moon as discussed in Chapter 2. Separate telescopes that are sensitive to a specific electromagnetic range (i.e., range of wavelengths) will be arranged so that the entire electromagnetic spectrum from any given astronomical object of interest can be studied continuously (unless briefly eclipsed by the Sun, Earth, or other planets).

In addition to the ability to collect photons over long periods of time, many telescopes will be connected together in an interferometry grid, thus providing high-resolution images. The high rims of craters such as Copernicus would allow very long rim-to-rim baselines, and the crater peak would be used for the beam integration facility (see Figure 8.11).

This system configuration of telescopes holds the promise of resolving details of planetary bodies around nearby stars and of studying their spectra to determine if the conditions of life exist, thus providing important data for planning future interstellar missions.

The large number of observatories thus established would eventually allow every astronomical institution, university, school, or interested individual to have continuous observation capability through the internet. Such remotely operated observatory models already exist on Earth, but lunar-based facilities promise to be the premier location in space for astronomical observations.

8.6.2   The quarantine and biohazard experimentation facilities

Lunar development will inevitably lead to an extensive program of solar system exploration (see Chapter 10). Consequently, spacecraft that are launched from the Moon will gather and return samples of asteroids, comets, planets, and moons from every region of the solar system. Protecting the Earth's biosphere from contamination by these samples – and protecting the samples from Earth contamination – will thus be an important issue.

As the lunar manufacturing base matures, it will become possible to build sophisticated analytic biohazard laboratories on the Moon (Figure 8.12). Since the Moon has no atmosphere, it will be an ideal place to operate scientific facilities for the investigation of materials from other regions of the solar system. Sample return vehicles would land at these isolated, specially designed facilities, where the samples would be analyzed tele-robotically by scientists on the Earth and the Moon. Once the samples are determined to be safe, they could be transported to other laboratories on the Moon, or flown to the Earth for further investigation.[8]

Such facilities will also be ideal for conducting experiments that are considered to be too risky for Earth-based operations. Experiments involving recombinant DNA technologies or biohazardous materials, such as bacteria, viruses, and toxins, could be conducted safely at the facility without endangering environments or life forms at other locations. In the event of contamination or other adverse event, several termination procedures would be possible. For example, venting the laboratory to the lunar vacuum could be used to destroy bacteria or viruses; alternatively, as a "last resort," the entire facility could be simply annihilated ("melted down") and rebuilt.

8.6.3   Hadron collider

By colliding atomic particles that have been accelerated to relativistic speeds, scientists are able to gain knowledge of the fundamental properties of matter. However, this form of study on Earth has its limitations, including the high-energy requirements, the need to conduct experiments in a vacuum, and the need for long acceleration pathways. To overcome these limitations, scientists will have the opportunity to construct a large particle accelerator ("hadron collider") on the Moon. An acceleration track

---

[8] As awareness grows of the abundance of organic material in meteorites that survive the fall to Earth, it has become increasingly likely that we have already been contaminated by every potential life form in the Solar System. Concerns for backward contamination have decreased somewhat, along with the knowledge that the vacuum of space may not be sufficient to destroy life forms.

Figure 8.12. Biohazard laboratory (Thangavelu/DiMare).

could be constructed around the circumference of the Moon and, using superconduct-ing magnets and energy from the lunar power system, experiments with the interaction and annihilation of highly energetic atomic particles could be conducted. Being lunar-based will allow scientists to conduct these experiments more safely and at higher energy levels than would be possible on Earth.

8.7   TOURISM ON THE LUNAR CONTINENT

Tourism is one of the biggest industries on Earth. When lunar cities become self-sufficient and reliable Earth–Moon transportation systems come into being, a robust tourism industry of the Moon will evolve.

The Moon is an interesting destination for tourists from the Earth. It is an entirely "new" world whose surface area is as large as the continent of Africa, and there is much to explore. Moving about in the $\frac{1}{6}$ gravity will be a unique experience, and it will be possible to participate in human-powered flight in large specially-designed air excursion spaces.[9] The perspective of space (e.g., of the Earth) from the lunar surface will be spectacular, particularly when there is a special event such as an eclipse of the Sun by the Earth.

For a typical tour of the Moon, tourists will arrive at Newton Base via tethers, a

---

[9] Routine advances in flight controls will prevent novice pilots from flying into obstacles and will assure a safe landing on every occasion.

space elevator from the L-1 Lagrange point (tethers and space elevators are discussed in the following section), mass drivers (Appendix K), or chemical rockets. If they opt to spend time at the L-1 point, they will have the opportunity to visit the L-1 Point International Museum of Space Artifacts, where space stations, great space observatories of yore, and other artifacts are cocooned after they have been decommissioned (Appendix Q).

Upon arrival at the lunar surface they will be able to enjoy one or more sessions of human-powered flight, observe or partake in $\frac{1}{6}$-G sports, and take a sightseeing expedition by rail to the south pole and the far side of the Moon. They can journey north to the equatorial mares regions to visit one or more of the original Apollo landing sites. From there they proceed to the north pole and stay in a hotel that provides a continuous "upright" view of the Earth or one in which the Earth is never visible at all. Figures 8.13, 8.14, and 8.15 depict the maglev sightseeing train on a tour of the Moon.

8.8   NEXT-GENERATION TRANSPORTATION SYSTEMS

Continued growth of the size and sophistication of the lunar industrial base, combined with access to abundant supplies of materials and energy, will lead to the development of new transportation systems. Among the promising technologies that may find application for the transportation of cargos are Earth–Moon "Cyclers", tethers, space elevators, and mass drivers. The sooner such technologies are developed, the less pollution of the lunar atmosphere will be caused by chemical rockets.[10]

8.8.1   Earth–Moon Cycler

The gravitational forces and dynamics of the Earth–Moon system allow spacecraft to follow a figure-eight "cycler" path around the Earth and Moon. With proper maneuvering, the spacecraft can be made to swing around the Moon and the Earth continuously in a figure-eight orbital path. The advantage of this orbital configuration is that the spacecraft requires relatively little energy to remain in orbit, and it can be used to ferry humans and cargoes between the Earth and the Moon. The design of "cycler" spacecraft would be based upon space shuttle-derived technologies (see color plate section). When the cycler approaches the Earth, a spacecraft in low Earth orbit could be accelerated (e.g., by chemical rockets or space tether) to match the velocity of the cycler. The crew would then transfer to the cycler for the trip to the Moon. Upon reaching lunar orbital space, the crew would transfer to a lunar lander vehicle that would then take them to the lunar surface.

---

[10] Apollo experiments have shown that exhaust contamination from a single vehicle is only detectable for about four hours after landing; however, the accumulated rocket exhaust from regular traffic to and from the lunar surface would interfere, to some degree, with telescopic operations on the Moon.

Figure 8.13. The maglev sightseeing train (Thangavelu/DiMare).

Figure 8.14. View of the lunar landscape from the club car (Thangavelu/DiMare).

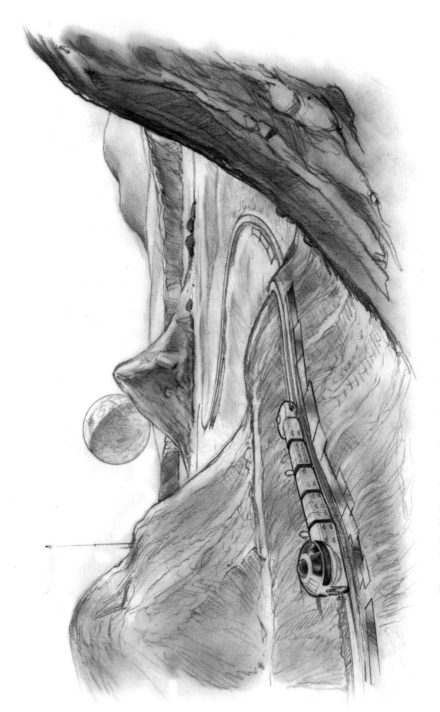

Figure 8.15. Sightseeing maglev train at the south pole (Schrunk/Thangavelu/DiMare).

When an industrial base becomes functional on the Moon, the components of a large (e.g., one or more kilometers in diameter) Earth–Moon Cycler can be fabricated on the Moon from lunar materials and launched into lunar orbit (or a Lagrange point such as L-1 or L-5) for final assembly. The Cycler can then be placed in a high-inclination orbit that loops around both the Earth and the Moon. The Cycler will support a population of hundreds of people and will need sufficient shielding to protect the crew and passengers from the ionizing radiation of space. After being maneuvered into orbit, it will require very little energy to remain in perpetual orbit. The advantage of the Cycler is that it will serve as a safe, comfortable, and efficient crew and cargo transfer vehicle between Earth and Moon orbits – and it would be a very popular tourist destination. If proven successful, the Earth–Moon Cycler would be a model for Earth–Mars Cyclers.

8.8.1.1   Rapid prototyping and manufacturing

As lunar missions grow longer and more complex, a variety of failure modes of systems and hardware caused by anomalies, accidents, abnormal stresses, and even nominal wear and tear can be expected. Since it is not possible to carry spares for every system, and there are limits to redundancy, what can be done to repair or replace non-functioning tools and equipment?

The answer is rapid prototyping and manufacture (RPM). RPM is evolving technology that will have a significant impact on repair and re-commissioning of broken hardware (see discussion in Section 4.4 and Appendix V). It involves the creation of a solid computer model of the failed part and its duplication with high fidelity using raw materials or reprocessed materials from the failed part. A failed component such as a valve, bearing, impeller, or thruster nozzle may thus be re-fabricated at a lunar base or onboard a spacecraft such as the Earth–Moon Cycler, as depicted in Figure 8.16.

8.8.2   Tethers

A space tether is a long cable or wire, typically deployed from a spacecraft. NASA has conducted four successful space tether experiments, and two that were not successful. Unfortunately, the two unsuccessful experiments received much publicity whereas the successes did not; thus, people tend to discredit them. However, space-based tether systems promise to be very useful. Multi-kilometer-long rotating orbital tethers, made of high-tensile-strength material such as carbon nanotubes (currently in development stages), with guidance, navigation, and control attachments, will be able to move payloads from one orbit to another orbit (of higher or lower energy) or from lunar orbit to the surface of the Moon without the need for propellants.

Tethers operate on the principle of conservation of energy. When a payload is attached and then released by the tether, the energy gain (or loss) of the payload is matched by an equal and opposite energy change of the tether. With payload systems attached at either end, in controlled, very low lunar orbit, a tether could

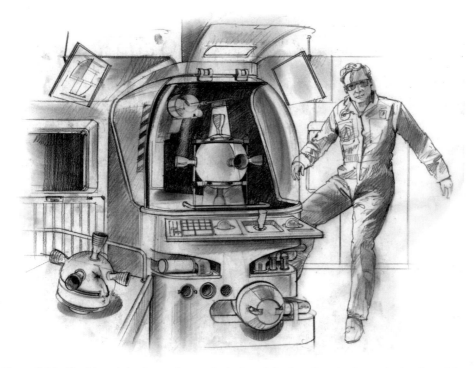

Figure 8.16. Rapid prototyping and manufacturing. A broken thruster is repaired onboard the Earth–Moon Cycler (Thangavelu/DiMare).

skim the lunar surface at precise elevated locations and pick up and drop off cargo (Figure 8.17).

Large tethers in Earth orbit and in lunar orbit could be used in conjunction with the Earth–Moon Cycler to deliver passengers between the Earth and the Moon. For example, a rotating tether operating in Earth orbit could capture a passenger craft in low Earth orbit and release it (after 180° of rotation) at a higher velocity that matches the velocity of the approaching Cycler, and the passenger craft would then rendezvous with the Cycler.

When the Cycler reaches the Moon, a second tether system in lunar orbit would grapple the passenger craft and release it at a lower velocity so that the passenger craft would enter lunar orbit and then descend to the lunar surface. Alternatively, a combination of tether systems could take the passenger craft from the Cycler and deposit it directly on the lunar surface. The combination of tether systems and Cycler would be able to move passengers and cargos between the Moon and Earth orbit safely and comfortably without a net expenditure of energy.

8.8.3 Space elevators

A space elevator is, in effect, a long cable whose orbital period around a planetary body is equal to the rate of rotation of that body. The orbital mechanics of this kind

Figure 8.17. Operation of an orbital tether cargo system (Thangavelu/DiMare).

Figure 8.18. Terminus of the L1/lunar space elevator at the surface of the Moon (Thangavelu/DiMare).

of system enables one end of the cable to remain stationary with respect to a point on the surface of the planetary body at its equator. It is then possible to take an "elevator" along the cable path to move from the surface of the planetary body to a point in space – at a significant saving in energy costs. Space elevators have been proposed for Earth applications; for the Moon, the center of mass of a space elevator could be placed at the Earth–Moon Lagrange point (L-1). The terminus of the cable would be at the equator of the Moon, and would enable a low-energy means of transportation between the Moon and cislunar space (see Figure 8.18).

8.8.4   Mass drivers

A lunar spacecraft-manufacturing industry will be established to manufacture both manned and unmanned spacecraft for launch on exploratory missions throughout the solar system (see Chapter 10). To launch spacecraft from the Moon on trajectories to destinations in space, a series of large electromagnetic mass drivers (which

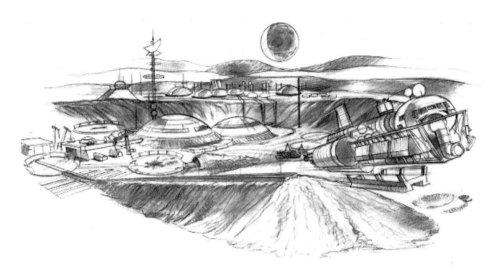

Figure 8.19. Launch of manned spacecraft from the Moon (Thangavelu/DiMare).

use the same magnetic propulsion system as the maglev railroad) can be constructed (see Appendix K).

Manned missions (e.g., to the Earth or Mars) could then be launched (and recovered) by mass drivers on the Moon. The launch system for human operations will be designed so that accelleration ("G") forces of the mass driver are within human tolerable limits. A bonus of the operation of mass drivers is that they can be designed to be "energy-neutral" (similar to tethers). The electrical energy used to launch a payload into space is recovered when the mass driver operates in reverse to capture payloads returning from cislunar space to the lunar surface. Mass drivers will also decrease the need for rockets – which will eliminate, in turn, the consumption of fuel and (equally important) will reduce the contamination of the lunar environment by rocket exhaust. Figure 8.19 depicts the launch of a manned spacecraft from the lunar surface on to another destination in space.

8.9 SUMMARY

After the first robotic missions demonstrate that lunar resources are readily accessible, can be applied to useful purposes, the permanent human settlement of the Moon will begin. Astronaut crews will then return to the Moon, at increasingly frequent intervals, to create self-reliant human colonies. From the first base, an extensive program of exploration will commence, additional colonies will be established, and a utility infrastructure network will be extended over the global structure of the Moon. With tourism and the continued migration of people from the Earth, the Moon will become a part of humanity's economic sphere of influence and the transformation of the Moon into a vibrant and productive inhabited sister "planet" of the Earth will then become a reality.

9

Governance of the Moon

9.1 INTRODUCTION

A system of governance that provides a rule of law is essential for the stability of every social organization. Without a rule of law that stipulates and enforces the rights and obligations of individuals and institutions, a modern complex society could not possibly exist: there would be no guarantee of individual rights and no security of person, property, or contracts. A system of governance must be created for the peoples and institutions that are now planning to explore and develop the Moon. The structure of the lunar government and the timing of its creation are of critical importance, because the success of the *Planet Moon Project* will depend directly upon the success of the government.

9.2 LUNAR GOVERNMENT ORGANIZATION

As presently envisioned, the social structure of the Moon will begin as a series of separate lunar bases, and will eventually evolve into a global network of interconnected municipalities. The simplest governmental organization that would meet the needs of this postulated lunar society would be two-tiered, with a single central government and a second level of multiple local governments, as depicted in Figure 9.1. The role of the central government would be to solve societal problems and provide administration for global matters of the Moon, such as the assignment of mining rights and the utilization of lunar orbits. The role of local governments would be to solve societal problems at the municipal level. The central lunar government would be the sovereign government of the Moon, with a higher level of authority on the Moon than any other government, including those based on Earth.

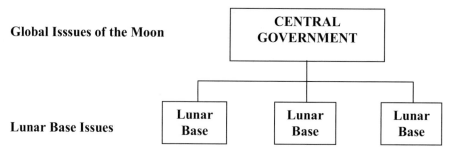

Figure 9.1. Government organization of the Moon.

9.3   LUNAR GOVERNANCE AND THE OUTER SPACE TREATIES

The exploration and use of outer space by governments and other institutions has been conducted under the direction and principles of the outer space treaties of the United Nations. On this basis, the outer space treaties will provide guidelines for the foundation of the new government of the Moon. The United Nations Committee on the Peaceful Uses of Outer Space (UNCOPUOS) has produced five major treaties:

- Treaty on Principles Governing Activities of States in the Exploration and Use of Outer Space, including the Moon and other Celestial Bodies, 1967 ("The Outer Space Treaty").
- Agreement on the Rescue of Astronauts, the Return of Astronauts, and the Return of Objects Launched into Outer Space.
- Convention on International Liability for Damage Caused by Space Objects, 1973.
- Convention on Registration of Objects Launched into Outer Space, 1976.
- Treaty Governing the Activities of States on the Moon and Other Celestial Bodies, 1979 (The "Moon Treaty").

The first four UNCOPUOS treaties have been ratified by a majority of nations, including the major spacefaring nations, but the fifth treaty, the "Moon Treaty", has been ratified by only eight nations, none of which is a spacefaring nation. The outer space treaties are declarations of broad principles regarding the conduct of nations and other institutions in the space environment. They emphasize international cooperation and have established the following principles for the exploration and use of outer space:

- Space is the province of all mankind.
- Space is the common heritage of all mankind.
- Space is to be used for the benefit of all mankind.
- Governments, other institutions, and individuals have the right of innocent passage through space, and the right to explore and use space for peaceful purposes.

- Military forces and weapons of mass destruction in outer space are prohibited.
- Earth-based governments are prohibited from appropriating territory in outer space under a claim of national sovereignty.
- Governments are liable for damage that they cause to others in the conduct of space activities.
- Protection of the environment of space and celestial bodies is to be observed by all parties.

The ratified outer space treaties have been observed by all spacefaring nations to date. Hopefully, the trends of international cooperation and peaceful pursuits in the exploration and use of outer space have been permanently established. However, the treaties have several inherent weaknesses that limit their direct applicability to a new lunar government.

The first problem with treaties is that their provisions are purposely broad so that they will be acceptable for ratification by the greatest number of nations. As a result, they are vulnerable to disparate or even contradictory interpretations, which degrade their effectiveness. Second, a treaty only applies to those nations that have ratified it; the people of those nations that have not ratified the treaty may thus be excluded from the benefits of the treaty. (To satisfy the spirit of the outer space treaties, the benefits of lunar development should accrue to all humankind, not just to those people whose governments have ratified the treaties.) The third problem is that treaties are inflexible documents that often become outdated, especially in the environment of rapid technological change that may be expected in the development of the Moon.

Finally, treaties are agreements, not laws, and they are thus unable to provide enforceable remedies for those nations or other institutions that violate treaty provisions. Since agreements cannot be effectively enforced, international treaties can only be used as *guidelines* for the governance of the Moon, not its *laws*. A global rule of law of the Moon can only be realized, therefore, when a sovereign lunar government creates an enforceable body of laws for the people and institutions of the Moon.

One significant deficiency of treaties involves the utilization of space resources, such as the lunar regolith, for commercial purposes: Treaties have not provided a legal definition of how governments or other institutions or individuals can "use" resources for their separate purposes. For example, suppose that a valuable resource such as a lake of frozen water is discovered in a crater in the south polar region of the Moon. If a space agency such as ESA, NASA, or JAXA were to establish a base at the south pole and gain access to that resource, would that space agency be allowed to use all of the water for its own scientific and industrial purposes, or should some or all of that resource be preserved for others to use? What if three or four different space agencies were to establish their separate bases at the crater and stake overlapping claims for the use of that resource? What if one or more of the lunar bases caused pollution of the resource? In whose court and under whose system of governance would these types of disputes be adjudicated?

These questions – and hundreds more – regarding property rights, mining, processing, transportation, and sale of resources are far beyond the scope of treaties, and can only be answered by a lunar government that has the authority to provide a rule of law *for* the Moon. In recognition of the limitations of treaties, the Moon Treaty of 1979, Article XI, Item 5, specifically provided for the development of a lunar government, an "international regime", to deal with the exploitation of lunar resources when such exploitation becomes feasible:

> *Article XI, 5   Parties to this Agreement hereby undertake to establish an international regime, including appropriate procedures, to govern the exploitation of the natural resources of the Moon as such exploitation is about to become feasible.*

9.4   SPACE LAW

Since 1979 governments have extended the rule of law into space through the creation of "space laws"[1] for the governments and commercial institutions that conduct business in space. Space law is created by national governments under the aegis of international treaties. It applies, for example, to the liability issues that are raised when a satellite decays from orbit and crashes back to Earth. If a national space agency were to establish a base on the Moon, the body of laws of that base would be provided by that space agency's government. It might seem logical that the central government of the Moon would also be an extension of one of the existing Earth-based governments. However, the existing outer space treaty agreements prohibit a claim of sovereignty over bodies in space, including the Moon, by existing Earth governments, as expressly stated in Article II of the "Outer Space Treaty of 1967":

> *Article II   Outer space, including the Moon and other celestial bodies, is not subject to national appropriation by claim of sovereignty, means of use or occupation, or by any other means.*

Thus, the Outer Space Treaty of 1967 establishes a limit to the level of government that can be transferred from the Earth to the Moon. Earth-based governments are permitted to establish municipal, or local, governments for their individual lunar bases but they are not permitted to become the sovereign, or central, government of the Moon (see Figure 9.1). The United States, for example, can provide the governance for a lunar base that it has constructed on the Moon, but it would not be allowed to convert that base and adjacent lunar territory into the fifty-first state of the United States.

The prohibition against establishing sovereignty over any territories of the Moon by Earth governments creates a legal void at the top of the system of governance of

---

[1] Space laws apply to the interactions of people who conduct business in space, but who reside on the Earth. They are thus distinguished from astro laws, which apply to the interactions of humans who reside in an extra-terrestrial setting.

the Moon. Without a lunar government that has the authority to address and solve societal problems at the global level of governance, there can be no legal certainty of the rights and obligations of any institution that aspires to explore and develop the Moon. The question thus arises: If a central lunar government does not now exist, and if existing governments do not have sovereignty over a territory of the Moon (other than a lunar base), how can a rule of law that addresses the global issues of lunar development be established?

9.5   CREATION OF A NEW LUNAR GOVERNMENT

There are two options that may be taken to satisfy the need for a lunar government. The first option is the *a-posteriori* approach, which is to do nothing now and simply wait for a global lunar government to evolve as a result of the interactions and conflicts between the bases that are placed on the Moon by the several national space agencies and by commercial businesses. The second option is to create a system of lunar governance now, in advance of the next phase of lunar exploration, the *a-priori* approach.

9.5.1   The *a-posteriori* approach to lunar governance

The current practice for creating space law is *a-posteriori*, or "after the fact". That is, space laws are created *after* technological advances have extended humankind's reach into a new territory of space. This is the way that space laws were written for events in Earth orbit. Similarly, a new lunar government can also be created, *a-posteriori*, in response to the need for a rule of law after permanent human settlements have been established on the Moon.

Under the existing treaty agreements, each spacefaring nation can continue its individual exploration efforts of the Moon. A logical result of this nation-based pattern of lunar exploration would be the establishment of multiple national bases at separate geographic locations on the Moon, analogous to the system of national scientific bases that is now employed to explore the continent of Antarctica (the "Antarctica model").

If the Antarctica model were to be adopted for lunar exploration and development, bases could be established by different national governments or consortia on the Moon. Each base would have its own municipal system of governance (and customs, standards, measuring system, language, etc.) that is provided by the "mother country" that built the base. Such a system of governance would not violate the treaties that prohibit a claim of sovereignty over territories on the Moon by Earth governments, because treaties do not apply to events that occur within a space vehicle (and, by extension, a lunar base). Thus, the people and institutions at a lunar base could operate under the set of laws of their national government, just as the laws that are in place at scientific bases in Antarctica are the laws of the sponsoring country. In this manner, the rule of law at a Japanese lunar base would be Japanese – an extension of the government of Japan; the rule of law at a Russian base would be Russian, and so on.[2]

Now suppose that the Europeans, the Chinese, the Americans, the Japanese, and the Russians all establish separate lunar bases, and experience (inevitably) conflicts over the appropriation and use of lunar regolith resources and other related matters, such as environmental pollution. To resolve their conflicts, they decide to create a higher level of governance – a lunar government – which mutually satisfies the needs of the five governments. In this way, a sovereign, central government of the Moon will eventually evolve from the Antarctica model, as anticipated by the Moon Treaty. However, this *a-posteriori* approach to the creation of a new lunar government has several negative aspects:

- It is a prescription for national competition, balkanization, and confrontation.
- The new government would be created after lunar bases have been constructed, when the need for conflict resolution becomes imperative, and when compromise is much more difficult, as opposed to advanced planning and cooperation.
- The uncertainty that is caused by the initial lack of a uniform code of law will prevent commercial enterprises from investing in lunar development projects.
- Nation-based exploration programs would tend to be competitive instead of cooperative, which would result in the unnecessary duplication and expense of lunar development efforts.
- The new government would be created by the principal spacefaring nations, and would tend to exclude the non-spacefaring nations, thus breaching the spirit of outer space treaties, which seek to include all of humankind in outer space affairs

Although the creation of a lunar government after lunar bases have been established on the Moon is feasible, *and is the option that is now being followed*, such an approach is full of limitations and uncertainties. It is also possible that the resulting lunar government would be unsatisfactory to most Earth governments, particularly those governments that are excluded from participating in the design and operation of the new government.

9.5.2   The *a-priori* approach to a new government

The alternative to creating a government after lunar bases have been established is to establish a limited system of lunar governance now, in advance of the next significant stage of lunar exploration and development. The creation of a new government at this time would greatly facilitate the cooperative international development of the Moon, and it would have none of the disadvantages of the *a-posteriori* approach to lunar governance.

[2] In the worst case scenario, national spheres of influence could lead to conflicts, and one nation could establish dominance over the other bases of the Moon in defiance of international treaties. Unfortunately, the latter is the historic norm for governments of the Earth.

One of the benefits of the *a-priori* creation of a lunar government would be the inclusion of non-spacefaring nations in the strategic planning of lunar missions and projects. The lunar missions that are currently planned only involve those nations that have the technological, financial, and managerial means for exploring the Moon. However, many of the tasks of lunar exploration do not require even a partial spacefaring capability. For example, the tele-operation of robots, the creation of software, and the design of lunar infrastructure components – all of which will be needed – can be accomplished by people who reside in non-spacefaring countries. A strategic plan for missions to the Moon can thus be organized so that all peoples, not just the people in the spacefaring nations, will have the opportunity to share in the planning and operation of lunar projects.

The ability of humankind to explore, develop, and colonize the Moon is directly related to the tools (e.g., rockets, robots, financial institutions, life-support systems, computers, etc.) that are now at our disposal. By analogy, the new lunar government, will be another tool that is created to facilitate humankind's migration into space. The new lunar government will be the offspring of the outer space treaties; how well it functions will be directly related to the quality of its design.

The goal of the lunar government will be to facilitate the exploration and development of the Moon under the guidelines of the outer space treaties of the United Nations. The new central government could begin as a simple port authority and then grow with each phase of lunar development. The ultimate stage of maturation of the lunar government would be a separate, sovereign government that would be dedicated to the best interests of the peoples and institutions of the Moon.

9.6 PORT AUTHORITIES

The first stage of lunar governance could begin now with a simple port authority for the Moon. Port authorities are quasi-public entities whose limited governmental powers include the authority to issue interest-bearing bonds, to police the area under its jurisdiction, to award contracts for services, and to lease property. They derive their powers from a host government, such as an existing national government or an international consortium of governments.

The advantage of port authorities is that they are relatively simple governments that have a limited scope of action within a defined territorial jurisdiction. Examples of existing authorities that serve as models for the port authority of the Moon include, in the United States, the New York Port Authority, the Denver International Airport Authority, and the Tennessee Valley Authority.

9.6.1 The LEDA model of lunar governance

A prototype port authority for the Moon has been developed by the United Societies in Space (USIS) and its affiliated World-Space Bar Association (WSBA). It is termed the Lunar Economic Development Authority (LEDA) and, if adopted, would act as an internationally sanctioned port authority for the Moon. The LEDA form of governance is discussed in detail in Appendix F. An existing national government or a consortium of governments could serve as the host nation for LEDA; alternatively, an entirely new space government called "Metanation" has also been proposed as the host government for LEDA, also discussed in Appendix F.

As it is now envisioned, LEDA would be an interim governmental unit for the early development stages of lunar development. It would function under the guidelines of the outer space treaties and its purpose would be to provide a rule of law for the responsible, peaceful, and effective exploration and development of the Moon. Later, when the Moon has a sizable population, LEDA would sponsor a referendum to establish a permanent sovereign government for the people of the Moon. The management and advisory structure of LEDA would include representatives of participating governments (potentially all national governments) and public and private space agency and interest groups. As proposed, LEDA would have the following powers and responsibilities:

A. *Security and administration.* LEDA would create rules, perform inspections, set standards and safety rules, resolve disputes, and sanction law-breakers in a court system. To assure that the rules serve the best interests of all involved parties, quality standards for the design and follow-up evaluation of rules would be observed (see Appendix G).
B. *Coordinate competing interests.* LEDA would facilitate or mediate competing interests of international projects on or near the Moon by national space agencies, scientific organizations, universities, and private corporations or consortia.
C. *Economic assistance.* LEDA would control an authority bank that sponsors a fiscal and monetary system on the Moon. It would arrange relevant financing out of bond revenues for developers who care to participate in construction of these facilities and utilize the Authority's bank.
D. *Property management.* LEDA would maintain a property-leasing system and site permitting activity as municipalities are constitutionally expected to do. This authority would serve to facilitate those who want to build habitats, mines, ports, factories, large ships, catapults, or company towns.
E. *Environment.* LEDA would oversee the protection of the lunar environment. Venue-wide rules should be tendered before the settlers, developers, tourists, and workers arrive on the Moon.

With these powers and mandates, LEDA would oversee the long-term planning and coordination of lunar activities; the management of lunar resources; the creation of standards for lunar development; and financing for exploration and development projects.

9.7   LONG-TERM PLANNING AND COORDINATION OF LUNAR SCIENCE AND DEVELOPMENTAL PROJECTS

A lunar government such as LEDA would provide the forum for the creation and execution of a long-term strategic plan of lunar development. There is, as yet, no agreed-upon master plan for the exploration and development of the Moon. What exists, instead, are exploration programs that are planned by the individual national space agencies without reference to an international cooperative strategy for lunar development. This piecemeal and fragmented approach by individual space agencies, based upon national plans and priorities, is not in accord with the outer space treaties' vision of international cooperation. Furthermore, it limits everyone's ability to view their efforts as an integral segment of a larger international plan of global lunar exploration.

The need for a master plan for lunar development was recognized by representatives of national space agencies and other scientific institutions who convened the first International Lunar Workshop (ILW-1) in Beatenberg, Switzerland, in June 1994. The workshop discussed plans for the strategic phased development of the Moon based upon international coordination and cooperation; the first workshop concluded with the "Beatenberg Declaration" which stated that the time had arrived to:

- Begin the first phase of the lunar (exploration and development) program.
- Prepare for future decisions on later phases.
- Implement international coordination and cooperation.
- And establish, at a working level, a mechanism for regular coordination of activities

Based upon the Beatenberg Declaration, the International Lunar Exploration Working Group (ILEWG), comprised of existing and emerging national space agencies and other lunar study groups, was formed. ILEWG recognizes the need to develop an international strategy for the exploration of the Moon, to establish a forum and mechanisms for the communication and coordination of activities, and to implement international coordination and cooperation.

Since the first meeting, lunar workshops have been conducted in multiple countries on an annual basis. The list of nation-participants is now greater than thirty, and includes all of the major spacefaring nations. LEDA would be the natural home base for ILEWG and other scientific groups (such as the Lunar Base Subcommittee of the International Academy of Astronautics) which are dedicated to lunar exploration. It would provide a single forum for the discussion and planning of lunar missions. In this manner, the expertise, resources, and mission plans of all interested parties could be coordinated for the common goal of lunar exploration and development.

9.7.1   Lunar resource management

If a governmental authority such as LEDA becomes operational, in coordination with an international exploration group such as ILEWG, it will have jurisdiction over the entire territory (and orbits) of the Moon. It will have the power to assign property rights and grant exclusive, limited-term leases of lunar territory and orbits to governments and other institutions for such purposes as mining and manufacturing operations. By awarding and upholding contracts, LEDA will create legal certainty for the use of lunar resources.

Because the outer space treaties mandate that the exploration and use of space is the common heritage and province of all mankind, the resource management role of LEDA would include the authority to protect the lunar environment in conjunction with development projects.

9.7.2   Standards for lunar development

Lunar development will be impeded if each nation that explores the Moon has its own unique set of tools, hardware, communications systems, and other facilities that do not interface with the tools and systems of other nations. The lunar government would facilitate lunar activities, therefore, by establishing and enforcing hardware, software and operational standards for scientific exploration and infrastructure development. It would adopt the metric system of measurement, for example, and require that all new hardware that is brought to the Moon must interface smoothly with the infrastructure that is already in place there.

The creation of a new government for the Moon will also offer the opportunity for experimentation with quality standards for the design, evaluation, and disposition of laws. Quality standards for laws hold the promise of causing the lunar government to be self-correcting in the direction of optimum service to the people as a whole (see Appendix G).

9.7.3   Fundraising for the lunar development

The lunar government would derive revenue (and thus be self-supporting) by charging fees to governments and other institutions for its role in the management of lunar resources. It would also, as a port authority, have the power to issue bonds[3] for public sale. The monies raised by bond issues could then be used for development projects, such as a lunar industrial park, and to award contracts to corporations for the construction of a lunar utilities infrastructure, including electric power, communications, and transportation services. The government would repay the bonds and interest from the user fees and tax revenues that are generated from the use of these services.

While it is possible that lunar exploration and development could be undertaken by the efforts of national governments alone (as was done with the United States' Apollo missions to the Moon), the optimum use of human and financial resources would involve both governments and corporations. This would allow the national space agencies and other government institutions to concentrate on exploration and

---

[3] The bonds could be double or triple tax-exempt (i.e., exempt from taxation by two or three levels of government) with the approval of sponsoring national governments.

other scientific investigations, while the corporations could focus on providing the goods and services that support those investigations. Because governments have limited resources, and because it is advantageous for governments to have investors (rather than taxpayers) assume the financial risks, it may evolve that corporations will contribute more funds for lunar development than do governments[4] (see Figure 5.5).

However, a "catch 22" condition is currently in effect for the Moon. Governments now limit their financial and manpower commitments to scientific missions that must have their own self-contained power and communications equipment, which are highly expensive to transport from the Earth to the Moon. Likewise, corporations are reluctant to place electric power generation stations or communication systems on the Moon because there is currently no market for those services (in addition to their concern of the lack of a rule of law that defines property rights on the Moon).

The lunar government could overcome this problem by using its bond money and other assets to contract for infrastructure projects, and, in partnership with space-faring nations, guarantee a market, at a reasonable rate of return on investment, for the commercial providers of services. For example, the new lunar government and the national space agencies could agree to purchase communication services from the lunar telephone company, transportation services from the lunar railroad company, oxygen from the lunar oxygen company, and electric power from the lunar electric corporation. If such guarantees were to be made available to commercial vendors, they would then be able to raise funds from investors for lunar development projects. Corporations could then establish profitable commercial operations on the Moon and the spacefaring governments would have access to goods and services at considerable savings for their individual scientific missions (see discussion of economics, Appendix L).

9.8   GROWTH OF THE LUNAR GOVERNMENT

A port authority of the Moon based upon the LEDA model can be designed so that it will be able to grow in step with the governance needs of the lunar society. The first stage of lunar development will be robotic missions that are directed from Earth, and the governance of the Moon will likewise be directed from the Earth. Because succeeding stages of development of the Moon (see Chapters 6–8) will involve an increasing number of permanent human residents, the port authority will evolve from a limited Earth-based government to a fully sovereign Moon-based government.

[4] The principal reason that the Moon Treaty has not been ratified by the major spacefaring nations is the concern that the treaty will limit or prevent for-profit institutions from undertaking lunar development projects. The significant benefits of the Planet Moon Project for all people (see Chapter 8) will be lost if for-profit institutions are not allowed to contribute to lunar development. It will therefore be incumbent upon the lunar government to use all of the tools (including for-profit corporations) that are available for the efficient, responsible, and peaceful development of the Moon.

9.9   SUMMARY

The technological capability for establishing a permanent human presence on the Moon is now available, but the system of governance that is needed to facilitate that presence does not yet exist, and needs to be created. One option for creating a lunar government is to do nothing (the current *a-posteriori* approach), and wait for the need of a new government to present itself. The *a-posteriori* approach to the creation of a new lunar government is a recipe for confusion, unproductive competition, and delays in lunar development, and is at odds with the spirit of the outer space treaties that seek international cooperation in the exploration and use of outer space.

The second option is to create an interim system of lunar governance, such as a lunar port authority, in advance of the next significant phase of lunar exploration and development. The Lunar Economic Development Authority (LEDA) has the potential to facilitate the orderly and effective transformation of the Moon into a functioning sister "planet" of the Earth, in harmony with the scope and intent of the outer space treaties.

10

Endless frontiers

10.1 INTRODUCTION

The interactions of investigative science (science) and creative science (engineering) are synergistic, and their outputs – scientific knowledge and technology, respectively – are growing exponentially. When the first permanent base is established on the Moon, the ever-growing technological prowess of humankind will become linked to the unlimited energy and material resources of space. The Spacefaring Age will then come to fruition: large-scale exploration and development activities in space (e.g., planetary engineering) will be possible; an abundance of energy and material wealth will be delivered to the Earth; the solar system will be explored in depth; and missions to the stars will be initiated.

10.2 "SPACEPORT" MOON

The transformation of the Moon into a permanently-inhabited planet will lead to a fundamental change in our perspective of the roles of the Earth and the Moon in the exploration and utilization of space. It is natural for present-day Earth-bound peoples to regard space activities only in terms of Earth-based programs. But, as humans establish a permanent presence on the Moon, Earth-centered thinking will give way to the reality that the Moon has become humankind's principal base for the exploration and utilization of space.

The Moon will be the hub of space activities for several reasons. First, the industrial base of the Moon will evolve to the point that it will be able to produce all of the hardware that is needed for the physical exploration of space.[1] Mass-production techniques and micro-/nano-technologies will be used to create all of the components of manned and unmanned spacecraft, such as computers, cameras, sensors, rockets, aerobraking heat shields, and solar panels, from the lunar regolith. Except for occasional special circumstances, the space exploration and development industries will be shifted from the Earth to the Moon.

Second, the Moon is an ideal site from which to launch and recover spacecraft. Exploratory missions to any site in the solar system can be launched from the Moon with much less energy requirements than from the surface of the Earth, and there is no atmosphere to block or delay the launches. Thousands of very small, single function probes can be launched from the Moon at high accelerations (100 to 1,000-G force, or higher at velocities 10 to 100 times greater than is now possible with rocket propulsion). These probes could be sent on high-velocity fly-by missions to any area of interest in the solar system, such as Pluto or comets at the aphelion[2] of their orbits. Associated with the fly-by missions, larger spacecraft can be launched from the Moon to a given object of interest, such as Mercury or a moon of Jupiter, where they will be maneuvered to desired orbits or surface locations. For larger spacecraft, the acceleration force of the launch would be tailored to the mission and size of the spacecraft. The ability of mass drivers to operate "in reverse" will also enable both manned and unmanned spacecraft to de-orbit and land on the Moon without the use of chemical rockets (see Appendix K).

The reason for using the Moon for the production and launch of spacecraft is dictated by economics: it will cost far less to place a satellite into Earth orbit, for example, by means of mass drivers on the Moon than by a rocket that is launched from the Earth. Also, rockets that are launched from the Earth present environmental and safety risks that will be avoided by Moon launches. At some point in the future, the only cargo that will be lifted from the surface of the Earth into space will be humans; virtually all other space cargo will be produced at off-Earth sites.

The third reason is economics. The cost of launching a space probe from the Earth to other areas in the solar system is now (at the beginning of the twenty-first century) on the order of U.S.$10,000.00 per pound. A major reason for this high cost is that over 90 percent of the launch vehicle is comprised of chemical fuel and rocket components that are expended in launching the payload. In contrast, the launch of the same payload from the Moon by mass driver may be no greater than U.S.$0.01 to $0.10 per pound because the "propellant" of mass drivers is the electrical energy in the lunar electric grid. No rocket components will be consumed or expended in the launch process because the "rocket" is the mass driver that is fixed to the lunar

---

[1] The sophistication of the manufacturing base of the Moon will approach that of the Earth when continuous electric power levels in the electric grid exceed 1,000 megawatts (i.e., shortly after the first functioning circumferential grid is in place in the south polar region). At that point, all of the resources of the Moon will be available to the lunar industrial base and it will be possible to manufacture virtually any tool on the Moon that can be made on the Earth.
[2] Aphelion: the point in the orbit of a planet or comet that is at the greatest distance from the Sun.

surface; the only moving part is the payload itself. Thus, as compared with rocket launches from the Earth, the use of mass drivers to launch – and recover – payloads at the lunar surface (see Appendix K) may result in a reduction of launch costs by five to six orders of magnitude.

The cost of space probes that are produced on Earth is also high because spacecraft are custom-made for their missions and require years of design and testing prior to launch. Spacecraft that are manufactured on the Moon, on the other hand, will be less expensive because they can be mass-produced. Testing spacecraft on the Moon will be simplified because the testing can be performed on the lunar surface, where the vacuum of space begins. Space missions that were launched from the Earth to Jupiter and Saturn (the Galileo and Cassini missions), for example, cost well over U.S.$1 billion. For that same cost, thousands or tens of thousands of spacecraft can be manufactured and separately launched from the Moon with a far greater volume of data return and less risk of mission failure. For these reasons, the Moon will become the principal base for the planning, design, manufacture, launch, and operation of scientific missions to every point of interest in the solar system.

10.3   THREE EMERGING TECHNOLOGIES

In addition to the lunar utilities infrastructure, mass drivers, and industrial base that produces spacecraft, several technologies will be further developed so that the advantages of the Moon for the exploration of the solar system and beyond can be fully exploited. Three of these technologies merit further discussion: solar sails, energy transmission, and solar-power satellites.

10.3.1   Solar sails

Solar sails may become the predominant form of interplanetary transportation in space in the twenty-first century (Figure 10.1). Solar sails are highly efficient because the source of their energy is sunlight.[3] The sails only need to balance the forces of gravity and sunlight (using control surfaces to attain the proper alignment with the Sun) to produce the thrust vector that propels them from one part of the solar system to another. The pressure of sunlight on a sail decreases in proportion to the square of the distance of the sail from the Sun and for this reason solar sails have much greater performance in the inner solar system (Mercury–Venus–Earth–Mars) than in the outer solar system. Although solar sails will be able to move large payloads over

---

[3] The propelling force on a solar sail is the sum of the non-ionizing photons (sunlight) that impact and are reflected by the sail. The photons of sunlight are different from the solar wind, which is comprised of particles (mostly electrons and protons).

Figure 10.1. Solar sail (Jerome Wright, courtesy of Jet Propulsion Laboratory).

interplanetary distances at high speeds,[4] they have large surface areas[5] and cannot withstand aerodynamic forces. For this reason, they must transfer their cargoes to short-range transportation systems when their final destination is a planet or moon that has an atmosphere.

One advantage of solar sails is that they have the ability to "hover" over the Sun or a planet or a moon by using the sunlight pressure on their sails to counteract gravitational forces. In this manner, a data-relay solar sail can be made to hover over the north or south pole of a planet or moon. Another advantage of solar sails is that laser beams can be used to augment their propulsion (Figure 10.2).

A laser located on the Moon could be used to add propulsive forces to a solar sail and thus decrease the transit time for high-priority missions, such as the transportation of astronauts from the Earth–Moon system to Mars. Of particular interest is a "ping pong" arrangement, where one laser is located at the launch site to augment

---

[4] "A solar sail of 700 meters could transfer 5 tons from the Earth/Moon orbit to Mercury orbit in approximately 15 months" (from Wright, J., *Space Sailing*, Gordon & Breach Science Publishers, 1992, p. 215.)

[5] The sail on a typical interplanetary sailing ship will be on the order of one square kilometer or larger in size.

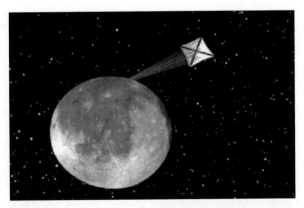

Figure 10.2. Laser beam propulsion of a solar sail from the Moon (compiled from NASA images by Oceaneering Space Systems).

Figure 10.3. Laser propulsion of solar sail for Moon–Mars transit missions. A laser on the Moon drives the solar sail to Mars for the first half of the journey and a laser on Mars decelerates the sail as it approaches Mars. This process is reversed for the return voyage from Mars to the Moon (compiled from NASA images by Oceaneering Space Systems).

propulsion forces and the second laser is located at the final destination (e.g., Mars) for braking, as depicted in Figure 10.3.

Solar sails can be made from very thin (two to four microns) sheets of aluminum. Because aluminum and other construction materials are present on the Moon, all of the components of solar sailing cargo ships can be manufactured on the Moon and launched by mass driver into space for final assembly.[6] A fleet of solar sailing ships

---

[6] The L-3 or L-5 Lagrange points of the Earth–Moon system are balanced-gravity sites for the assembly of large space structures.

will provide the Earth–Moon planetary system with a highly efficient and capable interplanetary transportation system.

10.3.2   Energy transmission from the Moon

It is possible to generate terawatts of electric power on the Moon from solar cells that are made from lunar-derived materials. The power levels will meet or exceed all of the needs of the lunar civilization, and excess energy will be available for export to other areas in the solar system, including the Earth. The method for transferring energy from one site to another is to convert the electric energy in the lunar power grid into microwave or laser beams, which are both forms of electromagnetic radiation.[7] The beamed energy is then reconverted into electric energy at the receiving site. As explained in Chapter 7, nominal advances in beaming technologies are expected to allow terawatts of microwave energy to be delivered from the Moon to the Earth to supply a major proportion of the future energy needs of the Earth.

Lasers convert electric energy into coherent beams of light. The laser beam is able to transmit energy over interplanetary distances at high efficiencies.[8] The Moon will be an excellent site for the operation of lasers that transmit power to other sites in the solar system. This will make it possible to supply the electric power that is needed for exploration and development projects, such as the exploration of the moons of Jupiter and mining operations on near-Earth objects (NEOs).

In addition to the transmission of electric energy, the laser beam can be used as the energy source for solar/laser thermal rockets[9] and for augmenting sunlight propulsion for solar sails on interplanetary missions. Laser-augmented solar sails will be the high-performance interplanetary transports of the twenty-first century; they will be able to carry payloads from the Earth–Moon system to Mars, for example, in less than a month.[10] When the lunar industrial base is able to construct lasers for power transmission, the high levels of electric power in the lunar electric

---

[7] The wavelength of the light of microwave radiation is tens of millimeters whereas the wavelength of visible-light laser radiation is in the micrometer range.

[8] The advantage of laser beams is that they maintain their beam size over very long distances and thus can be used to transmit energy throughout the solar system. Potentially, they can be used to transmit power to satellites on interstellar missions. Laser beams are subject to attenuation and dispersion by the atmosphere of a planet, whereas microwave beams can deliver energy through a planet's atmosphere with little interference. As a generalization, lasers will thus be used for point-to-point transmission of power in space, and microwave energy will be used to supply power from space to the surface of planets such as Earth and Mars.

[9] Solar thermal rockets use the Sun to heat propellant such as water or liquid hydrogen into gas. The gas is then exhausted through a rocket nozzle to provide thrust. Solar thermal rockets can also use lunar-based laser beams as their power source to provide much higher levels of performance than can be provided by sunlight alone.

[10] A 46-billion (46 gigawatt) watt laser could (theoretically) drive a 50-meter solar sail with a 10-kg payload to Mars in 10 days (report in *Science Magazine*, Vol. 281, 7 August 1998, p. 765, from *Workshop on Robotic Interstellar Exploration in the Next Century*, sponsored by NASA's Jet Propulsion Laboratory).

grid will be made available for extensive exploration programs throughout the solar system.

10.3.3   Solar power satellites

Solar power satellites are space-based power supply systems. They are composed of arrays of solar panels that convert sunlight (or photons from other sources such as lasers) into power and then beam that power by microwave (or laser beam) to a distant site, such as a receiving station on Earth or Mars. Considerable literature has been written about solar power satellites; they promise to deliver large amounts of electric power to distant sites at low cost.

The lunar industrial base will be used to build the solar panels and other structural components of solar power satellites. The separate components of the solar power satellites will then be launched by mass driver from the Moon to their destination in space. They will be maneuvered into their final orbit (by tethers, rockets, etc.) and assembled by means of autonomous or tele-operated robotic devices. Large orbiting solar power satellites can then provide continuous power in the 100–200 (or more) megawatt range[11] – for example, for the global exploration and development of planets such as Mars.

10.4   EXPLORATION AND DEVELOPMENT OF THE SOLAR SYSTEM

Based upon an evolved, mature industrial base on the Moon and nominal scientific advances, the following program of exploration and development of the solar system *from the Moon* will become possible.

10.4.1   Mercury

Thousands of small satellites can be constructed on the Moon and launched by mass drivers on high-velocity, data-gathering fly-by missions to Mercury. Particular attention will be given to the polar regions of Mercury, where Earth-based radar imaging has indicated that water- or sulfur-ice may be present (Figure 10.4).

Following the global survey, larger lunar-fabricated spacecraft will be launched from the Moon on a trajectory to Mercury. Upon reaching Mercury, the spacecraft will enter orbit around the planet, and surface-roving robots will be soft-landed on the surface for *in-situ* analysis and global exploration.[12] Mercury revolves about its axis of rotation once every 88 days. This relatively-slow rate of rotation will allow a fleet of autonomous wheeled robots to explore the entire surface area of Mercury by

[11] Assuming a solar cell efficiency of 20%, a solar power satellite that has an area of one square kilometer would produce as much as 260 megawatts in Earth/Moon orbit.
[12] Orbital-maneuvering and landing rockets could use a combination of liquid oxygen and powdered aluminum or liquid hydrogen, or solid propellants – all of which will use lunar-derived resources.

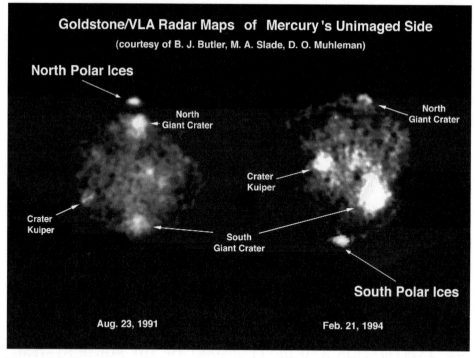

Figure 10.4. Radar maps of Mercury. Radar reflections indicate the presence of ice in the polar regions of Mercury.

moving across the terrain on "Magellan routes" of discovery[13] in synchrony with the day–night terminator.

Solar sails will have greater capability when they are closer to the Sun, and at Mercury they will receive approximately seven times more energy per unit surface area than at the Earth–Moon system. They will have sufficient (sunlight) thrust to enter into orbit around Mercury without the need for chemical rockets. Orbiting survey instruments can thus rely completely on solar power. If, as theorized, the north (and south) polar regions of Mercury are found to have water- or sulfur-ice and other valuable resources, an autonomous industrial base and circumferential utilities will be established in the polar region, analogous to the program of lunar development. Lunar-made mining and manufacturing equipment will be launched by mass drivers from the Moon to Mercury. Solar sails, with cargoes that include lunar-made solar-power arrays, will also be launched and maneuvered into a hover position above the north pole of Mercury, as depicted in Figure 10.5.

---

[13] Globe-circling expeditions are termed "Magellanic routes". They allow robotic devices to match the surface speed of rotation of planetoids and thus maintain a constant, optimum Sun angle for solar power and photographic perspectives along a circumferential route of exploration.

Figure 10.5. A lunar-made solar sail satellite maneuvers to a hover position above the north pole of Mercury (compiled from NASA images by Oceaneering Space Systems).

The mining and manufacturing equipment will be landed on the surface of Mercury in the polar region, and accompanying construction robots (e.g., automatons and tele-operated devices) will assemble the first elements of the industrial base and infrastructure utilities grid of Mercury. Continuous electric power from the orbiting solar arrays will be transmitted via microwave or laser beam to the electric grid of the industrial base. The solar sails that hover above the polar regions will also be relay stations for the Moon- or Earth-based remote tele-operation of robots at the industrial base. After an industrial base has been established at either pole of Mercury, work will begin on the robotic construction of a circumferential utilities infrastructure,[14] as well as mass drivers and power-transmission lasers.[15] The global exploration and development of Mercury will thus parallel the lunar experience, conceivably including a permanent, self-governing human population. Mercury could therefore become an important source of resources (e.g., heavy metals and silicates) and laser power for use in other areas of the solar system.

10.4.2 Venus

Spacecraft that are manufactured on the Moon, including solar sails, solar power satellites, balloons, and various landing craft, can be launched by mass driver to Venus. Satellites will be inserted into orbit, balloons placed in the atmosphere, and autonomous robots will be delivered to the surface of Venus by a combination of

---

[14] The problems of thermal stress and heat rejection will be much more severe on Mercury than on the Moon. Active thermal control systems, such as a circumferential pipeline system, will be critical to the operation of equipment on the sunward side of Mercury, and it may be necessary to move all equipment to underground locations at "high noon".

[15] The proximity of Mercury to the Sun makes it an ideal site for the operation of multiple lasers that beam power to other locations in the solar system.

Figure 10.6. Solar sail shade for cooling Venus (Sharpe/Schrunk).

aerobraking techniques, gliders, parachutes, and lunar-made chemical rockets. The hostile environment of Venus will limit initial scientific missions to orbital radar imaging, atmospheric probes, and short-duration robotic surface excursions.

The atmosphere of Venus is mostly carbon dioxide (CO_2) and large balloons will be inserted into the upper atmosphere (where temperatures are lower) for processing CO_2 and other components of the atmosphere. Energy for aircraft propulsion in the Venusian atmosphere, and for the operation of chemical processing plants, will be supplied by orbiting solar power satellites. The harvested carbon and oxygen, sulfur (from sulfuric acid), nitrogen, and other elements in the Venusian atmosphere would then be delivered to the Moon and other sites in the solar system.

The evolution of solar sail technologies will enable very large solar sails to be constructed on the Moon (and Mercury), and placed in Venus' orbit and at the Venus–Sun L-1 Lagrange point. These solar sails could be positioned so that a significant fraction of the incident sunlight energy on Venus would be blocked (Figure 10.6). As a result, the temperature of the atmosphere would gradually decrease to the point that liquid water could exist on the surface. It would then be possible to begin a vigorous program of converting Venus into an inhabitable planet, including the delivery of water to Venus from donor sites in the solar system, such as comets and the moons of Jupiter. Such a program would probably require centuries or millennia to complete, but the ever-improving ability to conduct effective and responsible planetary engineering projects would eventually lead to the transition of Venus into a habitable planet.

10.4.3 Near-Earth objects (NEOs)

Sharing the region of Earth's orbit around the Sun are billions of planetary bodies called near-Earth objects, or NEOs. The NEOs are a mix of asteroids and "burned-out" comets that have migrated to the inner solar system from the asteroid belt (between Mars and Jupiter), and from the Kuiper Belt and Oort Cloud in the outer reaches of the solar system. It has been estimated that 100 million Earth-crossing NEOs are greater than 10 meters in diameter, and an estimated 2,000 of them are larger than one kilometer in diameter (Lewis, 1996). The NEOs have been classified according to their orbital path:

- *Amor*: "Earth-grazing" objects whose closest approach to the Sun (perihelion) is between the orbits of Earth and Mars.
- *Apollo*: Objects whose orbits cross Earth's orbit, but whose orbital period around the Sun is greater than one year.
- And *Aten*: Objects whose orbits may cross Earth's orbit, but whose orbital period is less than one year.

NEOs are the principal source of meteorites that strike the surface of the Earth. The larger NEOs are a threat to the Earth–Moon planetary system because a collision of one of them with the Earth would be catastrophic.[16] A number of the largest NEOs *will* collide with the Earth[17] within the next several million years, with the potential of destroying a significant fraction of human life on Earth. However, the NEOs also contain a wealth of raw materials. The *in-situ* analysis of meteorites, and the astronomical observations and spectroscopic studies of orbiting NEOs, have disclosed that they contain resources such as water,[18] hydrocarbons, nitrogen, sulfur, phosphorus, and strategic and noble metals that can be used for all phases of solar system development projects.[19]

Virtually any of the NEOs can be reached by spacecraft that are launched from the Moon by mass driver and maneuvered to a rendezvous using solar sails, tethers, or chemical rockets. A high priority for the first lunar telescopes will be to conduct a survey of the size, number, spectroscopic characteristics, and trajectories of NEOs. Simultaneously, scientific probes will be constructed on the Moon and launched on trajectories that intercept and land on the more accessible NEOs; solar-sail and associated lander/retrieval vehicles will also study NEOs on fly-by missions and will return samples and data. As the NEOs are explored and analyzed, a program for the utilization of their resources can be initiated.

---

[16] The date of collision of the next large NEO with the Earth is unknown because a complete survey of the NEOs and their orbits has not yet been made.

[17] An argument can be made that humans should establish a permanent presence on the Moon so that human life will be preserved even if a large asteroid collides with and destroys life on Earth.

[18] The "burned-out" comets that comprise 50% of the NEOs no longer have tails because their surface volatile materials (mostly water) have evaporated. Nevertheless, their remaining, intact mass is estimated to be approximately 50% water, which may be the most valuable resource in space.

[19] The Amor asteroid, Eros, for example, contains billions of tons of carbon, hydrogen, nitrogen, and metals.

Figure 10.7. Photograph of the main-belt asteroid Gaspra (NASA).

Figure 10.7 is a photograph of the large (19 km long, 12 km diameter) S (stony) asteroid, Gaspra, which resides in the asteroid belt between Mars and Jupiter. Most NEOs are smaller than Gaspra, but the majority of them are thought to be similar in composition and appearance. If trajectory analysis indicates that a large asteroid will collide with the Earth or the Moon, the threat could be eliminated by altering the orbit of the asteroid. Methods for altering the orbit of an NEO include the following:

- Deflection by high-energy laser beams based on the Moon and/or Mercury.
- Explosives that are planted on their surface.
- Space "tugs" such as solar sails or chemical rockets.
- Mass drivers (that impart small changes in velocity to the NEO in one direction by accelerating small fragments of the NEO in the opposite direction).

Alternatively, robotic mining equipment could be delivered from the Moon to the NEO and its mass removed and transported, block by block, to the Moon (or to

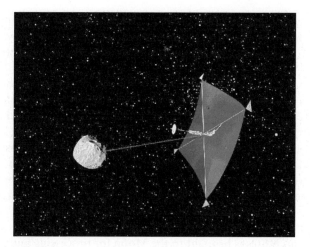

Figure 10.8. A solar sail transports asteroid material to cis-lunar space for processing (compiled from NASA images by Oceaneering Space Systems).

an advantageous location in space such as the Earth–Moon Lagrange point, L-5) for resource utilization and construction projects. The mining process will consume the NEO, thus physically removing it as a threat to the Earth–Moon system. A scenario for a future program of lunar-based mining of NEOs is offered.

10.4.3.1 Mining of NEOs

Robotic-mining and bulk-processing equipment, communications equipment, and solar cell arrays[20] are manufactured on the Moon and launched by mass driver on a trajectory that intercepts the designated NEO. The mining, communication, and power supply equipment are landed and set up on the asteroid, and mining operations of the NEO begin.[21] A pallet of raw material from the asteroid, such as a mixture of ore and water-ice, is prepared for transport to the Moon. Utilizing solar sails, lunar-made chemical rockets, solar thermal rockets, or mass drivers that are placed on the asteroid,[22] the pallet of raw material is delivered from the asteroid to lunar orbit (see Figure 10.8).

Upon entering cislunar space, a solar sail, chemical rocket, or other propulsion device intercepts the pallet and adjusts its trajectory so that it can be captured by a mass driver on the lunar surface (Appendix K). The deceleration and capture of the

[20] The power that is needed to operate mining and other equipment on the asteroid may be supplied by power transmission lasers based on the Moon.
[21] The human tele-operators of the mining equipment will be located on the Earth or the Moon.
[22] A mass driver that is operated from the surface of the asteroid will alter the orbit of the asteroid each time that it launches a cargo pallet into space. This means that the asteroid can be maneuvered into a more desirable orbit (i.e., closer to or farther away from the Earth–Moon system).

pallet by the mass driver converts the orbital kinetic energy of the payload into electric energy, which is added to the lunar electric grid. Alternatively, a tether cargo system in lunar orbit could deliver the pallet of raw material directly to the lunar surface.

After it has been delivered to the lunar surface, the pallet from the NEO is transferred to a processing area where the raw materials are refined. Water and the lighter elements, such as carbon, nitrogen, and phosphorus that have high value on the Moon are retained for lunar applications. Materials that have higher value on the Earth (or other areas in the solar system) than the Moon, possibly including platinum, nickel, gold, cobalt, and palladium, are delivered from the Moon to those other locations (the metals could be shaped into a lifting body, covered with ablative materials, and maneuvered through Earth's atmosphere to a designated landing field).

The above strategy is repeated for hundreds of NEOs so that a steady stream of raw materials from NEOs is brought to the Moon for processing and refining. In this manner, the threat of collision of a large asteroid with the Earth/Moon planetary system is reduced, and the Moon, the Earth, and other sites in the solar system are supplied with an abundance of valuable[23] material resources.

10.4.4 Earth

The Industrial Revolution occurred as the result of the link-up between the human technological expertise of the eighteenth century and the resources (e.g., coal and iron ore deposits) of the Earth. The (ongoing) Industrial Revolution has generated substantial increases in living standards and quality of life for a large segment of Earth's population, but these desirable conditions have led to the consumption of large amounts of energy, as shown in Figure 10.9.

This rate of growth of energy use, and concomitant growth of living standards and quality of life, cannot be sustained from the consumption of declining and increasingly costly Earth resources alone. Furthermore, there is concern about the deleterious environmental effects of the continued use of Earth's energy and material resources, and the resulting measures to alleviate environmental degradation may curtail or even cause a regression of economic activity and living standards. In other words, the prospects for the "closed-Earth" paradigm of the future of humankind, based on dependence upon Earth's resources, are daunting: increasing populations, increasing pollution, and declining resources will lead to lower living standards and quality of life for the people of the Earth on a global scale.

The establishment of permanent human settlements on the Moon will thus have a profound beneficial effect for the people of the Earth. The industrial capability of the Planet Moon will tap into the resources of space and supply the Earth with an abundance of materials and clean energy. The "closed-Earth" paradigm of declining resources will be replaced by the "open space" paradigm of unlimited resources, and

---

[23] The estimated Earth-market value of the (iron, nickel, and platinum-group) metals of the smallest known metallic NEO, 3554 Amun, is U.S.$20 trillion.

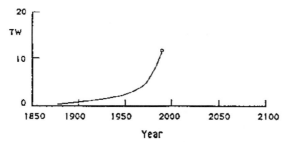

Figure 10.9. World energy consumption in terawatts over time (R. Bryan Erb).

the people of the Earth will be able to enjoy substantial improvements in living standards and quality of life – the "greening of the Earth".

10.4.4.1   Electric power for the Earth

As noted in Chapter 6, the circumferential solar-energized lunar power system will supply the Moon with all of its energy needs and, based on nominal advances in beaming technologies, excess power will become available for transmission to Earth and other locations in space. The growth of the lunar power grid in support of expanding permanent settlements, combined with the incentive of large markets for clean energy on Earth, should lead to the export of terawatts of energy to the Earth by the middle of the twenty-first century.

The delivery of an abundance of low-cost, clean solar electrical power to Earth locations will lead to significant advances in living standards on a global scale and, by eliminating the need to consume fossil and fission fuels, will also reduce pollution. The mining, transportation, and consumption of fossil and fission fuels cause the release of greenhouse gases (e.g., carbon dioxide) and heat into the biosphere.[24,25] The use of the lunar power system (LPS) to replace conventional sources of electric power will therefore be doubly beneficial, because the LPS releases no greenhouse gases into Earth's biosphere and generates no waste heat since the "heat engine" of the LPS is the Sun. In fact, excess energy that is not needed for immediate consumer needs can be used to clean up toxic waste sites.

10.4.4.2   Hydrogen, oxygen, and potable water

The lack of freshwater is a significant problem for the quality of life of a large segment of the Earth's population.[26] When excess energy from the lunar power system

---

[24] The operation of nuclear (fission) power plants releases heat into the biosphere but does not produce greenhouse gases. A significant problem with nuclear-fission sources of power is that fissionable materials can be converted into nuclear weapons – another reason to develop space solar power as the prime source of energy for the Earth.

[25] If a coal-fired electric generating plant is 30% efficient, then 70% of its energy is released into the biosphere in the form of heat.

[26] *Science Magazine*, 25 August 2006, pp. 1067–1090.

becomes available, it will be used to desalinate ocean water and pump that water to arid regions. Excess energy will also be used for the electrolysis of seawater (the breakdown of water into its constituent elements: $H_2O \Rightarrow 2H + O$) to make hydrogen available as a replacement for hydrocarbon fuels.

An advantage of hydrogen as a fuel (e.g., in fuel cells) is that the "waste product" of its use is potable water. One scenario for the use of hydrogen is to cool it to its liquid form (20 Kelvin or below) for use as the coolant in superconductor electrical transmission networks. The transmission network would thus act as a conduit for the delivery of both electrical power and hydrogen.

The oxygen that is obtained from the electrolysis of seawater would be used for multiple industrial purposes, and excess oxygen could be released in river deltas. Many river deltas (e.g., the Mississippi river) have "dead zones" as the result of oxygen depletion. By delivering oxygen to these areas, dead zones would be eliminated and marine life could be re-established. Another benefit of the desalination and electrolysis processes is that valuable minerals and elements (e.g., thorium, uranium, gold, and mercury) would be recovered from seawater.

10.4.4.3 Delivery of space resources to Earth

As discussed with the mining of NEOs, an abundance of space resources (e.g., metals) will be delivered to Earth (and the Moon). The people of the Earth will benefit from the increased availability (and hence lower costs) of materials, and from an improvement in quality of life as mining operations and their associated environmental degradation are curtailed.

10.4.5 Mars

The exploration of Mars, including manned missions, will be a primary goal of space exploration in the twenty-first century. There are compelling reasons for exploring Mars. Of all of the planets, Mars is the most Earth-like (Figure 10.10). It has an atmosphere, and its geology indicates that it once had lakes and rivers of liquid water, which are conditions that are conducive to the evolution of life. If the compelling (but circumstantial) evidence for ancient bacterial life on Mars is confirmed it would imply that the evolution of life is possible for every star that has Earth-like planets. In other words, if life ever existed on Mars, there is a strong possibility that life is ubiquitous throughout the universe.

In addition to the evidence of life on Mars in the distant past, there is also a possibility that microbial life still exists there. If these life forms are found in protected underground areas of Mars, astrobiology will develop into an even more important and well-known branch of science. The science of extraterrestrial life will require an extensive program of exploration of Mars; these investigations will be not only fascinating, but of benefit to biological research on Earth.

Another objective for going to Mars is that it may potentially be transformed into an inhabited planet with large human populations. Exploration of Mars is now underway – satellites and surface rovers are currently investigating Mars, and addi-

Figure 10.10. Mars.

tional missions are planned for the next several years. At some point in the mid-twenty-first century, manned missions to Mars are expected to commence. Sending humans to Mars may not be as cost-effective as robotic missions, but given human-kind's inherent curiosity and historical desire to explore new lands, manned expeditions to Mars may be regarded as inevitable. What program of exploration, then, should be adopted for sending people to Mars?

Many strategies for the human exploration of Mars are now being discussed. A typical plan[27] would be for a crew of four to six people to be launched from the Earth to Mars on a 2–3-yr round-trip mission, which will include a stay of several months on the Martian surface. In advance of the launch of the manned spacecraft, a nuclear reactor would be sent to the surface of Mars to make chemical fuel from the Martian atmosphere for the return trip, and to provide power to the crew when it arrives at Mars on a second spacecraft. Such missions will permit the astronauts to make detailed investigations at each landing site area and to make excursions over a wide territory with rover vehicles. The current Mars plans are an attractive strategy for the initial human exploration of Mars. However, the long-term missions are hazardous for several reasons, including exposure to space radiation, and they involve the development and use of new, large booster rockets and space nuclear reactors that will be expensive.

[27] See *"The Case For Mars"* by Robert Zubrin, The Free Press, New York, 1996.

An alternative approach to near-term human missions to Mars is to give priority to the exploration and development of the Moon before human expeditions to Mars are attempted. The developed Moon would then be used as the principal base (the "stepping stone") from which to explore Mars. While this strategy emphasizes the Moon as the next major space objective, the long-term objectives of current Mars exploration strategies may actually be achieved sooner and at less expense and risk by using this approach.[28] There are several reasons for giving priority to the near-term exploration and development of the Moon:

1. The future industrial base of the Moon will be able to build and test the Mars exploration vehicles and infrastructure on the Moon and launch them to Mars at less cost than equivalent Earth-based missions
2. The life-support and other technologies that will be required for manned Mars missions will be fully developed and commissioned during the course of lunar development.
3. Expeditions to Mars that are initiated from the Moon will not require the (politically tenuous) use of nuclear reactors.
4. The robotic, ISRU, and construction techniques that are developed for the Moon can be exploited to create a permanent Mars utilities infrastructure. When that infrastructure is in place and functioning on Mars, human exploration and development of Mars will become easier, safer, and less expensive.

The following discussion presents a scenario for the exploration and development of Mars from the Moon.

10.4.5.1   *A strategy for the exploration and settlement of Mars*

During the early stages of lunar development, an ongoing program of unmanned exploratory missions are launched from the Earth to Mars. After an industrial base and permanent utilities infrastructure are in place on the Moon, unmanned spacecraft are constructed and tested on the Moon and launched by mass driver on a trajectory to Mars. Imaging and communications satellites are placed in Mars orbit, and tele-operated and autonomous surface-roving robots are placed on the Martian surface by means of aerobraking techniques and lunar-made chemical rockets. Detailed surveys and analyses of Mars are performed.

Based upon the results of the program of robotic exploration, a site for an industrial base on Mars[29] is selected. Components for several solar power satellites are built on the Moon, launched to Mars, and assembled in Martian orbit. Power-

---

[28] The question regarding space priorities is not whether we should explore and develop the Moon ahead of Mars, or *vice-versa*. It is, instead, what are the most expedient means by which the exploration and human settlement of the solar system can be achieved?

[29] The north pole region is suggested as the site of an industrial base because of its proximity to water-ice deposits there. The analyses that are provided by surface-roving robots may subsequently indicate that some other location, such as the equator, is more favorable for the first base.

Figure 10.11. Solar power satellites that have been manufactured on the Moon will supply electric power for exploration and development activities on Mars and other sites in the solar system (NASA).

receiving antennas and robotic mining and manufacturing equipment are manufactured on the Moon, launched by mass driver, and delivered to the industrial site on Mars. Lunar-made solar sails are placed in a hover position above the polar regions of Mars for the command and control of robotic devices on the Martian surface. The solar power satellites deliver hundreds of megawatts of electric power via microwave to receiving antennas that feed into the Martian electric grid at the industrial site (Figures 10.11 and 10.12).

Automated and tele-operated assembly processes, controlled from the Moon and the Earth, would enable the construction of a Martian industrial and scientific base as well as utilities infrastructure, from local resources. An electric grid, which is supplied with power from orbiting solar power satellites[30], a telecommunications and railroad network, and a pipeline system course around and through the north pole region, analogous to the earlier development of the lunar utilities infrastructure at the south pole of the Moon (Figure 10.13).

Drawing upon the Moon experience, human-habitable structures at the industrial/scientific Martian base are constructed robotically from local resources. Greenhouses are erected, and seeds from food crops on the Moon are delivered from the Moon to Mars. A replica of the controlled, ecological life-support system that is used on the Moon is constructed robotically on Mars.

---

[30] Solar arrays could be manufactured at the Martian industrial base and connected into the Martian electric grid. However, atmospheric attenuation of sunlight and accumulations of dust on the solar arrays will limit their effectiveness, and the preferred source of power for the electric grid is a system of solar power satellites that beam (microwave) power to receiving antennas on the Martian surface.

Figure 10.12. Depiction of lunar-made solar power and communication satellites at Mars. The solar power satellite is in orbit around Mars and the solar-sail communication satellite is in a hover position above the north pole (compiled from NASA images by Oceaneering Space Systems).

The components of a human-habitable spaceship-cruiser that is capable of safely housing 50 to 100 people in space is manufactured on the Moon and launched into space for assembly. The Earth–Mars "cycler" will provide the life-support systems, recreation, and radiation protection that are needed for round-trip journeys between the Earth–Moon system and Mars. The (as yet unmanned) cycler is then boosted, with chemical rockets, on a trajectory that "swings by" Venus [31] and the Earth (for gravity assist) on its journey to Mars (Figure 10.14).

Spacecraft containing human crews are then launched by mass driver from the Moon to rendezvous with the Earth–Mars cycler when it passes the Earth–Moon system.[32] After launch, the manned spacecraft are linked to the in-transit cycler that, then completes the journey to Mars.[33]

---

[31] The orbit around the Sun is used to gain speed for the trip from the Earth–Moon system to Mars. The spacecraft swings by the Earth–Moon system to pick up the human crew and to use the gravity of both Venus and the Earth for additional acceleration (gravity assist) on the trajectory to Mars.

[32] The minimum acceleration required to achieve lunar orbit is approximately 1.6 G. Based on test pilot experience, this acceleration would not be harmful to most people. (The maximum launch load experienced by the shuttle is ~4 G.)

[33] An alternative strategy would be to launch a capsule with a small crew from the Moon on a direct trajectory to Mars. Lunar-made chemical or nuclear propulsion systems (which will be "less troublesome politically" because they will not operate within Earth's or Mars' environment) would further boost the capsule as it leaves the Earth–Moon system, and would brake the capsule on approach to Mars. This approach would entail a transit time of several months for chemical propulsion and one to two months for nuclear propulsion systems.

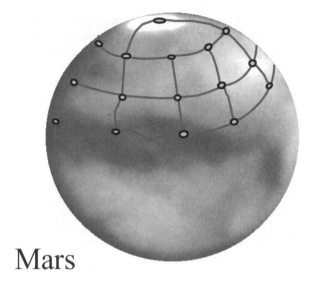

Mars

Figure 10.13. Depiction of the utilities infrastructure grid on Mars (Sharpe).

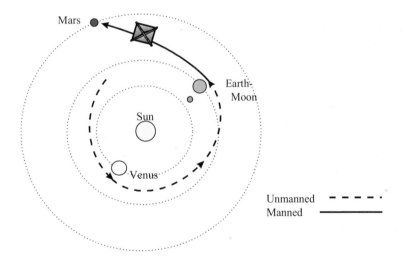

Figure 10.14. Initial trajectory of Earth–Moon–Mars cycler (Sharpe).

Upon reaching Mars space, the crew transfers to the lunar-made descent rockets, which detach from the cycler and deliver the crew to the Martian surface base that has been prepared for them. The cycler sails past Mars and returns to the Earth–Moon system.

The humans who arrive at Mars step out of their spacecraft and proceed to their permanent housing area via the previously-constructed Martian railroad. They will have ready access to completely-autonomous mining and manufacturing facilities, life-support systems, scientific laboratories, and an evolving global rail, communications, and power network. The human exploration and development of Mars then commences on a global scale.

A series of cyclers make regular round-trip journeys between the Earth–Moon system and Mars so that the human population on Mars grows over time. For individuals who wish to return to the Earth–Moon system, chemical rockets or mass drivers that are made at industrial sites on Mars launch them from the surface of Mars[34] to rendezvous with the next cycler that enters Mars orbital space on its regular journey between the Earth–Moon system and Mars. A self-sufficient, inhabited Mars thus becomes a reality.

There are many possible variations of the above "Moon first" scenario for the exploration of Mars, such as the utilization of resources from Mars' moons (Deimos and Phobos) and their conversion into human-habitable orbiting platforms. Mining and processing operations on the Moon will recover fissionable elements (uranium, thorium) that can power nuclear rockets, which would be capable of shorter flight times to Mars than chemical rockets. The important point is that the human development of the Moon can potentially supply the materials, power, and technological advancements that are needed for the global exploration and development of Mars.

10.4.6   The asteroid belt

A program of exploration and utilization of the asteroid belt will be an extension of the plan to explore and utilize near-Earth objects. The resources of the asteroid belt will supply the peoples who live in the inner solar system (on Earth, Moon, Mercury, Venus, and Mars) with all of their material resources for the foreseeable future (see photograph of Gaspra, an asteroid lying in the inner edge of the asteroid belt, Figure 10.7).

10.4.7   Comets

The lunar industry that produces thousands of satellites, atmospheric probes, and robotic surface rovers will have a stockpile of space probes and solar sails that can be made available on short notice for special missions. When a new interloping comet from the outer reaches of the solar system is detected, spacecraft can be taken from existing inventories to launch from the Moon and intercept the comet. Fly-by satellites will make global surveys and images of the comet, probes will analyze its tail, and robotic rovers will land on its surface. If the comet is on a collision course with the Earth or the Moon, its orbit will be altered by the same means that are used to alter the orbital path of near-Earth objects.

---

[34] Mass drivers could be placed at the east and west slopes of Olympus Mons (the highest mountain on Mars) for the launch and recovery of spacecraft, thus obviating the need for chemical rockets. Tethers may also be able to pick up and deliver payloads from the summit of Olympus Mons.

10.4.8   Jupiter

An extensive program of exploration of Jupiter and its moons will be initiated from the Moon. Small fly-by satellites, larger orbiting satellites, atmospheric probes, and surface rovers will be constructed on the Moon and launched by mass drivers to Jupiter and its moons. Images and other investigations of Jupiter's moons, including radar imaging of Europa, Ganymede, and Callisto (to search for liquid-water oceans beneath their ice-covered surfaces) will be made. Multiple robots will be placed on the surface of each of the moons for *in-situ* analyses. Solar power satellites (SPSs) and tethers (for maneuvering satellites in orbit around Jupiter) that have been constructed on the Moon will be launched to Jupiter orbital space. The SPSs will be augmented by laser transmission of power from the Moon and Mercury, and will transmit power to robots in Jupiter space and on the surface of the Jovian moons.

10.4.8.1   Moons of Jupiter

Europa (Figure 10.15) is an example of an intriguing Jovian moon for investigation. Topographical images from the Galileo satellite strongly suggest that Europa has a

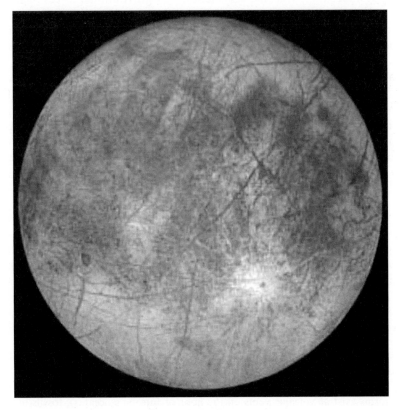

Figure 10.15. Europa.

Figure 10.16. Submarine robotic devices explore volcanic vents and other areas of interest in the oceans of Europa (image from NASA/JPL website).

liquid-water ocean below its icy surface. If Europa has an ocean of liquid water and if it has hot underwater volcanic vents (caused by gravitational forces related to the orbit of Europa around Jupiter), then it is possible that life exists on Europa, just as life forms exist at hot water vents on the floor of Earth's oceans. The value of finding and studying life forms on another world would be incalculable to the biological sciences – and to the whole philosophical subject of life in the universe.

If liquid water is found, a fleet of spacecraft will be launched from the Moon to Jupiter where they will be maneuvered into orbit and onto the surface of Europa. Samples of the surface ice will be analyzed for evidence of life.[35] A submarine probe will then explore Europa's oceans by first melting a path through the surface ice. Once the submarine has reached the ocean, it will begin a search for underwater volcanic vents similar to those on Earth (Figure 10.16). Based upon the experience with the Europa mission, similar missions may be undertaken on Ganymede and Callisto.

10.4.9   Saturn, Uranus, and Neptune

Analogous to the scientific missions to Jupiter, lunar-constructed and launched fly-by, satellite, probe and (where possible) atmospheric, surface and subsurface expeditions will be made to Saturn, Uranus, and Neptune, and their moons.

10.4.10   Pluto, Eris and Sedna

Satellites will be launched from the Moon on fly-by survey missions of Pluto, Eris, and Sedna. Larger satellites also launched from the Moon can be placed in orbit

---

[35] If life forms exist at volcanic vents on the floor of Europa's ocean(s), fossils of those life forms should be present in the ice, including the surface ice, of those oceans.

around these mysterious bodies, carrying surface rovers that perform *in-situ* examinations.

10.4.11   The Kuiper Belt and Oort Cloud

Just beyond the orbit of Pluto is the Kuiper Belt of asteroid-like bodies (planetoids). More than 40 planetoids have been discovered in the Kuiper Belt (Jewitt *et al.*, 1996), and there may be as many as 100 million planetoids that are larger than 10 km in diameter. As the resolution of astronomical observations improves, the number and size of objects in the Kuiper Belt will be defined more accurately. Subsequently, small fly-by satellites will be launched from the Moon to objects in the Kuiper Belt, followed by larger-scale scientific missions.

The Oort Cloud is a suspected collection of planetoids that are thought to be one source of the comets that enter the inner solar system. The Oort Cloud extends two to three light years from the Earth, and it may contain trillions of planetoids. Solar sails that are augmented by lasers from the Moon and Mercury will be sent on missions to explore the Oort Cloud region.

10.5   ROBOTIC MISSIONS TO THE STARS

The experience and technological gains that will have enabled the extensive investigation of the solar system will eventually lead to a program of robotic exploration of nearby stars.

10.5.1   The cosmic "seed"

Soon after the first permanently inhabited base has been established on the Moon, future lunar development will proceed without material support from the Earth. A self-sufficient and self-replicating industrial base will grow indefinitely after an essential core or "critical mass" of robotic mining and manufacturing hardware and technology has been delivered to the Moon from the Earth. The abundant supplies of energy from the Sun, and of material resources from the Moon and NEOs, will provide the needed raw materials, and the only required link to the Earth will be for the transfer of human passengers.

The field of robotics promises to make ever-greater contributions to space activities. Machine intelligence, as measured by instructions per second, is expected to match, and then exceed, that of the human brain by the middle of the twenty-first century.[36] That increasing capability, combined with advances in sensor and manipulation technologies, will lead to the development of autonomous robotic devices ("automatons") that will use space resources to undertake large-scale and hazardous space exploration and construction projects, and to replicate themselves.

---

[36] *Scientific American*, p. 37, January 2005.

The learning curve of space development will advance as each new area of the solar system is explored. The minimum sum of critical tools and knowledge that is needed for the development of an autonomous and self-sufficient base in space, using solar energy and local resources, is designated a "cosmic seed". A study by NASA in 1990 estimated that the "seed" that would be required for the development of an autonomous lunar base would have a mass of approximately 100 tons (see Appendix I). With advances in miniaturization (i.e., nano-) technology, this mass might someday decrease, to the point where the "seed" could be placed within a single spacecraft.

The goal of developing cosmic seeds will be to launch them to the nearest stars, where they will use sunlight energy and local material resources to explore and analyze each star's solar system, including any biological species. The imaging and other data gathered will be transmitted back to the Earth–Moon system. The seed would have the ability to replicate itself and send a duplicate seed on to another star system, and the process would then be repeated.

One problem that must be overcome is the interaction of spacecraft with the interstellar medium at relativistic speeds. The interstellar medium (the "vacuum" of space) contains particles of gas, dust, and cosmic rays at a concentration of approximately one particle per milliliter of space. To reach the nearest stars in less than 100 years, spacecraft must be able to travel at 5 percent of the speed of light or higher ($\geq 0.05c$). At such speeds, the interaction of the spacecraft with the gas and particles of the interstellar medium will produce aerodynamic drag and secondary radiation from collisions with cosmic rays. A means of protecting spacecraft from these hazards must be devised.

10.5.2   Robotic mission to the Proxima Centauri system

If technological advances will enable interstellar travel at relativistic speeds, a program of robotic exploration of the stars from the Moon is offered:

- A "cosmic seed" and solar sail are manufactured on the Moon and launched by mass driver on a trajectory to the Proxima Centauri star system, a group of three stars that are located at a distance of 4.4 light years from the Sun.[37] This process is repeated for multiple solar sails and cosmic seeds that are launched at regular intervals (e.g., monthly) on the same trajectory to receive, amplify, and transmit messages, and thereby maintain a continuous communication link between Earth and the foremost robotic emissary.
- The solar sails are propelled by sunlight, as well as by giga/terawatt lasers located on the Moon, Mercury, and other sites in the solar system. The sails are accelerated to a speed of 0.10 times the speed of light ($0.1c$)

[37] Proxima Centauri is chosen in this example because it is the nearest star system. If telescopic observations from the Moon detect planets capable of supporting life on other nearby star systems, those stars might have a higher priority for exploration than Proxima Centauri.

over a period of weeks to months, which carries them to the nearby star system in fewer than 50 years.[38] Upon reaching relativistic speeds, the sails will be furled, or oriented so that they parallel the direction of flight, thus minimizing encounters with particles in space. During flight, technological advances that can be used by the cosmic seeds are forwarded to the spacecraft.

- As the first probe nears the Proxima Centauri star system, the solar sail uses the sunlight pressure and gravitational and magnetic fields of the stars/planets to decelerate and maneuver, and they then go into orbit around the star Alpha Centauri (Rigel Kent). The light from the star provides the energy for the solar sails to proceed on a program of exploration of the star system. A survey of the stars and their major planetary bodies is performed;[39] photographs and other data of the survey are transmitted back to Earth.

- A candidate planetoid of the appropriate size and location for the establishment of a scientific base is identified, and a cosmic seed is placed on the planetoid by the solar sail (Figure 10.17). The cosmic seed utilizes the locally-available materials and local sunlight to create an industrial base, which in turn produces scientific probes, solar sails, mass drivers, chemical rockets, robots and so on, as well as replicas of itself.

- The knowledge base of the cosmic seed then directs an extensive program of scientific investigation of the planets, moons, and major asteroids of the Proxima Centauri star system. Particular attention is given to any planet or moon that has the potential for life. If, for example, an Earth-like planet with oceans of water and an atmosphere containing oxygen and carbon dioxide (an indication of plant life photosynthesis) is discovered, a search for life (and characterization of that life form, if found) on the planet will be undertaken.[40] The data return from scientific investigations is periodically transmitted to the Earth.

- More cosmic seeds and solar sails are constructed from the industrial base(s) of Proxima Centauri and sent on exploratory missions to more distant star systems. In a similar fashion, cosmic seeds are sent from the Moon to all of the other stars in the local star neighborhood of the Sun (e.g., Barnard's Star, Sirius, 61 Cygni A and B, etc.). All of those seeds establish industrial–scientific bases in their respective star systems, and report their findings to the people of the Earth. Each cosmic seed then replicates itself and sends multiple cosmic seeds to more distant

[38] Controlled nuclear fusion technology may be sufficiently advanced by the mid-twenty-first century that it will be an alternative means of propulsion for interstellar travel. Nuclear fusion rockets would use deuterium (heavy hydrogen) and helium-3 fuels that have been obtained from the Moon. Rocket propulsion would be obtained by modifying the nuclear fusion combustion chamber so that the high temperature products of fusion are allowed to escape through a magnetic "nozzle" at high velocity, thus producing thrust.

[39] Using our solar system and recent discoveries of planets orbiting other stars as a model, there should be some planetary systems in our galaxy where plant life – or even animal life has evolved – but that is just speculation. One of the objectives of the exploration of stars will be to discover if other stars have solar systems similar to ours.

[40] If intelligent life beings exist on a planet in the star system Proxima Centauri, they would have to contend with the "UFO" from Earth that has invaded their planet!

Figure 10.17. Interstellar sail cruiser approaching an asteroid in the Proxima Centuri "solar system" (modified from a NASA image by Adolf Schaller).

stars. Within 100 years, extensive knowledge of the 10 to 20 stars nearest to the Sun will be returned to the Earth, and within 1,000 years, thousands of star systems will have been explored.

- The cosmic seeds will replicate and explore every star system they encounter in the Milky Way galaxy. An ever-widening network of knowledge gathering and reporting in the galaxy will be established (which will escalate the problem of how to handle the large volume of data that will be returned). White dwarf stars, neutron stars, black holes, nebulae, and star-forming regions can all be visited, greatly increasing our understanding of the universe. The *in-situ* exploration of thousands, then millions, of star systems will eventually answer the question of the possible existence of intelligent life elsewhere in the Milky Way galaxy.

This scenario for the exploration of the stars is based upon the development of technologies that permit robotic emissaries of the human race to travel between stars at relativistic speeds. Those technologies are expected to become available from the nominal growth of existing knowledge bases, and when they do, the peoples of the solar system may begin receiving detailed images of planets and moons from neighboring stars, possibly as soon as the end of the twenty-first century.

10.6 HUMAN MISSIONS TO THE STARS

Continuing advances in all sciences related to space exploration and settlement can be expected to lead to the initiation of human missions to the stars. The "pathfinder" missions of the cosmic seeds will identify planetary systems of nearby stars that, even if they do not harbor life, are amenable to the establishment of ecological systems that can sustain human life. Given the size of the universe, the odds of the uniqueness of life on the planet Earth are exceedingly small; however, until life is discovered on other worlds, the only safe assumption is that life is unique to the Earth. On the further assumption that humankind constitutes a net positive force in the universe, it is our responsibility not only to ensure the survival of humankind but to extend that life throughout the universe. When habitable planets of stars are identified, the natural curiosity of humankind, combined with the desire to settle new lands, will lead to the migration of the human species to other habitable regions of the galaxy.

10.6.1   The great diaspora

The planets, moons, comets, and asteroids of the solar system will provide a virtually-endless supply of raw materials from which to undertake not only planetary engineering projects but also the construction of large habitable colonies in space (e.g., at L-5, *à la* O'Neill) capable of sustaining millions of people indefinitely.[41] Advances in physics and space sciences (e.g., matter/anti-matter propulsion technologies) should make it possible to launch self-sustaining human colonies on missions to nearby star systems by the end of the twenty-first century. After a journey of hundreds of years – involving multiple generations of the on-board explorers – the first humans will then set foot on the new host planet and establish a second inhabited solar system in the galaxy. If it is assumed that the journey of a self-contained colony from the Earth to the next star system requires 500 years and that the inhabitants of the new planet require another 500 years to build and launch two new colonies on voyages to neighboring star systems, then the number of inhabited stars systems will double every thousand years. By this doubling strategy, humankind will be able to colonize every inhabitable region of the Milky Way galaxy within a few million years – a "blink of the eye" in geologic time. Thus, assuming that no life of greater intelligence than our own exists in the Milky Way galaxy, it will be possible for humankind to become the dominant species of our galaxy and reach the status of a Kardashev Type III civilization.[42]

10.7   SUMMARY

The transformation of the Moon into an inhabited sister "planet" of the Earth will be a significant step in history – humankind will become a multi-world species. However, the significance of the Planet Moon will not just be that we have created a new world for human habitation, but that we have also become a true spacefaring civilization, and our advancement on the endless space frontier will be assured.

[41] Spacecraft that are capable of making the journey to nearby stars will be the natural successors to the "cycler" spacecraft that were developed for the Earth–Moon system and for the round-trip journeys between cislunar space and Mars.

[42] The Russian astronomer Kardashev denoted three types of civilizations: Type I, II, and III. A Type I civilization controls the energy of its own planet; a Type II controls the energy of a star (e.g., the Sun); and a Type III controls the energy of its galaxy.

11

Conclusion

The Moon is now within our reach, and is the logical next site for human exploration and settlement. The alternative to the migration of peoples into space is to remain forever on our home planet – which is to say that we have explored and learned enough, that we have satisfied our need for advancement, and that we wish to accept our present existence as good enough. This closed-Earth model of the future offers declining resources to serve growing populations – a frightening scenario in which war could become the major option for survival. The migration to and settlement of the Moon breaks the closed-Earth paradigm – it offers unlimited opportunities for the growth of humankind based upon the virtually-limitless resources of space.

The human settlement of the Moon will result in increased knowledge of the basic sciences. Perhaps more importantly, the challenges and problems that must be overcome to make settlement possible will produce advances in virtually every field of human endeavor, to the potential benefit of all people. The exploration and development of the Moon offers the following:

- A whole new world for cultural invigoration and growth.
- Advances in science, engineering, government, and law.
- A peaceful outlet for national competitive energies.
- Room for population growth.
- Expansion of business opportunities.
- Virtually-unlimited material and energy resources.
- The opening of endless frontiers.

The Moon's existence may be regarded as an incredible offer, a gift. It is truly a "stepping stone" on which we will learn to live and work in space, and from which we will explore space. Soon after the first lunar base becomes operational, it will become obvious that endless supplies of solar electric energy are available on the Moon. Based upon that realization, the global transformation of the Moon into an inhabited "sister planet" of the Earth will be inevitable. The benefits of the Spacefaring Age to human existence will thus be realized: the people of the Earth will receive an abundance of energy and material resources from space; the entire solar system will be open to exploration and settlement; and we will begin organizing the first trips to the stars.

Appendix A

Robots on Planet Moon

Bonnie Cooper, Ph.D.

A.1 ROBOTICS TECHNOLOGY FOR THE FIRST AND FOLLOW-UP LANDER MISSIONS TO THE SOUTH POLE

Because the orientation of the lunar spin vector is fixed with respect to the ecliptic, the limb of the Sun rises a maximum of 1.6° above the horizon in the lunar polar regions. Thus, it has long been speculated that there may be craters in the polar regions which, never being touched by the Sun, are extremely cold (~40 K or −233°C) and could trap ice from comet impacts on the Moon. Data from the Lunar Prospector, which orbited the Moon in 1998, indicate that there is a small but statistically significant increase in the hydrogen signature at the lunar poles compared with other areas. Although preliminary interpretations suggested that the north lunar pole had a greater signature than did the south pole, later refinements to the analysis showed that there was slightly more hydrogen in the south polar region (Feldman *et al.*, 1999).

A.1.1 Benefits of sending robotic lander missions to the lunar polar regions

The Lunar Reconnaissance Orbiter (LRO), to be launched in 2008, will provide much additional information about the lunar polar regions. One instrument, CRaTER, will characterize the global lunar radiation environment. DIVINER will measure lunar surface temperatures at scales that provide essential information for future surface operations and exploration. LAMP is an instrument that images in the ultraviolet region of the spectrum, and should be able to "see" in the shadowed regions. The LEND instrument will provide refined information about hydrogen abundance over the entire lunar surface, as a follow-on to the results obtained by Lunar Prospector in 1999. The Lunar Orbiter Laser Altimeter (LOLA) will give us a precise global lunar topographic model and geodetic grid, which will allow construction of a digital elevation model of the polar regions. Finally, the Lunar Reconnaissance Orbiter

Camera (LROC) will address two fundamental requirements: landing site certification; and polar illumination. It will acquire images to assess meter- and smaller-scale features for a hazard analysis of potential landing sites.

However, no matter how promising the LRO results may seem, we will not have direct proof of the existence of ice until a robotic lander obtains physical samples. If water-ice is confirmed to exist there, it is potentially the most readily-available source for consumables. Lunar Prospector data suggest that the ice accounts for 1.5 ± 0.8 percent by weight of the material in the shadowed craters (Margot *et al.*, 1999). By comparison, the hydrogen reduction process can extract up to 5.4 wt% oxygen from lunar pyroclastic orange glass (Allen *et al.*, 1994). We calculate that it would require 5,497 kW-hours to extract a tonne[1] of ice from 67 tonnes of regolith. This is slightly less than the estimated energy required for hydrogen reduction of ilmenite, but more than the estimated energy for magma electrolysis (e.g., Hepp *et al.*, 1994). However, the near-constant sunlight at Mons Malapert would reduce the mass of the equipment that would be needed to produce thermal energy. This fact, combined with the lower technical risk of obtaining water by melting it, would be two good reasons for locating a lunar resource extraction facility in the south polar region.

Estimates of the total extent of shadowed areas poleward of $87.5°$ latitude are 1,030 and 2,550 square kilometers for the north and south poles, respectively. Within the 2,550 square kilometers of permanently-shadowed crater floors of the south polar region, and assuming that water-ice totals 200×10^6 metric tons, there should be approximately 78 kg/m^2 of water-ice. However, the real distribution of any ice in the polar regions is not known. Mobile robots are needed to conduct a broad survey of the area south of Mons Malapert.

A.1.1.1   *Exploration for water-ice*

Because the ice (if it exists) may be widely distributed in the polar regions, more than one robot lander is needed for the survey described above. Multiple robots could survey a much broader area and provide data on the variation in ice content, both laterally and at depth. Calculations suggest that the ice may be buried at a depth from 5 to 50 cm below the surface. If so, the dry regolith must be removed (or penetrated) to access the layer that contains the ice. Robots that can acquire samples at depth will determine the accessibility of this resource.

An end-to-end method for finding, measuring, and storing water can be developed for robotic explorers. The robot (or many copies of one robot design) would explore permanently-shadowed crater floors, and find promising locations for sampling both the surface and intervals into the subsurface. It will extract samples and analyze their chemical content using mass spectroscopy, coupled with complementary techniques to resolve ambiguities. If water-ice is found to be among the constituents of the sample, the robotic system will extract the water by heating, and store the water in a pressure vessel. NASA's RESOLVE project is an example of a

---

[1] Tonne: metric ton, 1,000 kilograms.

rover-mounted payload that can go through this complete cycle from drilling to water storage.

These first robots could be based on existing commercial robotic systems that are used in arid, dusty regions and which are intended for tele-robotic operation. The base system might be a tracked vehicle, which allows the robot to climb steep, uneven grades. Alternatively, a robotic base system such as ATHLETE (described below) might be used.

Extremely robust mobility will not be required at first. The lander must have a smooth surface to land on, and the smoothest surfaces are often found in the floors of large craters. The robot would then deploy from the lander and explore within the confines of that single crater. Because continuous solar power will not be available when the first robots arrive (the infrastructure needed will not yet be in place), they will use batteries or fuel cells. The instruments and tools will be based on flight-proven or commercial systems. A six-degree-of-freedom manipulator arm with stereo vision, onboard lighting, and an end effector will allow multiple views at varying distances. Additional instrumentation might include color cameras with filter wheels to provide spectral information for geologic interpretation. The first robotic explorer might have a capability of handling 10 kg of regolith over a one-month time period, representing up to 100 individual samples.

Permanently-shadowed lunar crater floors are generally not within line of sight of the Earth, thus information cannot be relayed directly. Instead, the robot and lander combination will require an orbital vehicle, or a relay station at a higher topographic point, for communication with Earth. Because the robot is controlled via radio frequency, the lander that carries it to the Moon can function as a local transponder, such that the robot carries a lightweight RF system for short-distance relay to the lander. The lander then communicates with the orbiting spacecraft, which transfers information from the lander to Earth as a part of its periodic data transmittal.

A.1.1.2   Verifying and validating the resource potential

Reliable calculations on the amount of ice that may be available in the polar regions, and the ease (or difficulty) of extracting it are not possible because of several unknowns: (1) the porosity of the regolith with depth (although this could be as much as 40 percent based on Apollo core tube samples); (2) the actual existence of water-ice – in any quantity – in the lunar polar regions; (3) the degree of uniformity of ice distribution (if it exists). Moreover, the market for this consumable is currently non-existent. Consequently, this project would not be a likely candidate for a commercial venture, even if the mineral rights question were solved. However, should the project succeed in finding quantities of water-ice as described here, there would likely be an increased interest on the part of commercial entities to become involved.

Sampling will provide direct information about the matrix material, the volume percentage of ice, and the variation of ice content with depth at many discrete locations. This information allows us to estimate the total reservoir potential of an area, which in turn enables us to place a potential monetary value on the resource. Investment decisions are based on this kind of concrete information. For example, consider a 20-km crater near the lunar south pole, such as Shackleton. A "reservoir" might consist of a layer of icy regolith 1 meter deep and 50 cm below the surface, which covers the entire crater floor. This can be modeled by a disk with radius 10 km and thickness of 1 meter, and would translate to a total reservoir volume of approximately 0.314 cubic kilometers. Assuming an average ice content of 0.25 percent by volume (which is only 25 percent of the current estimate), we would have access to 785,000,000 kg of water ice. Further assuming that the lunar market value of the water is merely 0.1 percent of the launch cost for an equivalent amount of water, it would have a value of about $10 per kilogram. The hypothetical reservoir that we have described thus has the potential to provide an investor with a gross revenue of $7,800,000,000.[2]

A.1.1.3   Technological challenges to be overcome in using lunar resources

To establish the usefulness of water ice (if found), it will be necessary to show that it can be accumulated and stored for long periods of time, awaiting the arrival of humans and the need for consumables.

Batch versus continuous processing
Near-term efforts to utilize lunar resources will rely on batch-processing technology. Small amounts of feedstock will be taken through the processing steps one at a time. Batch processing is usually simpler, but continuous processing is desirable when it becomes technically and economically feasible. With continuous processing, new feedstock is entering the system constantly, and product is constantly coming out, thus larger amounts of product can be manufactured.

Seals and pressure retention
Reliable methods are needed for sealing and re-sealing pressure vessels that are handling regolith, in spite of dust which may collect on seal surfaces. Moreover, each opening and closing of the pressure vessel represents a loss of the volatiles contained within it, which is undesirable. Demonstrating the ability to move regolith into and out of pressure vessels will be useful everywhere on the Moon. A system that is operable in the lunar polar regions can be used in the equatorial regions as well, and also on Mars or other planetary bodies.

Facing the cold
The lunar polar regions are colder on average than areas nearer the equator, and in the permanently-shadowed areas temperatures may reach 40 K ($-233.15°C$). The temperature a few centimeters below the surface is probably more moderate than that at the equatorial regions, and does not experience as much of a temperature change over the course of the lunar day–night cycle. Unfortunately, equipment immediately above the surface will still be subject to the stress that is caused by thermal cycling

---

[2] Assuming, of course, that the resources can legally be extracted and sold for profit, which is not established.

over a range of several hundred degrees. In the same way that temperatures at the International Space Station vary by as much as 500°C as the station moves into the sunlight and back into shadow, temperatures anywhere on the lunar surface will heat up when they are in direct sunlight, and cool off when they are in shadow. The temperature of the lunar subsurface has little impact on the temperature of equipment that rests upon it.

However, we have some experience dealing with very cold temperatures. In fact, cryogenic systems that contain liquid hydrogen are designed to operate at 20 K (−253.15°C), which is 20 degrees lower than the expected 40-K temperatures in the permanently-shadowed regions of the Moon. The challenge will be to expand our capabilities with cryogenics to include systems with more moving parts.

A.1.2   Other ISRU experiments

While the lunar south pole region may be optimum for extracting water from ice, many other items are needed to develop a sustainable and affordable lunar outpost. Currently, we do not know very much about the geology of the polar regions. Remote-sensing instruments that operate in the visible or infrared regions of the spectrum cannot function in a region where no light enters their detectors. Our knowledge at this time is limited to what can be gleaned by radar, laser altimetry, and other "active" remote-sensing methods; to data provided by instruments that detect particles instead of rays, such as the alpha spectrometer, electron reflectometer, and neutron spectrometer aboard the Lunar Prospector; and to methods which measure field strengths, such as Prospector's magnetometer and Doppler gravity experiments.

It has been speculated that the abundance of volatiles (such as carbon, oxygen, helium, and nitrogen) is likely to be highest inside permanently-shadowed regions, because the low temperatures there favor binding of these gases to the lunar regolith. So, while it is reasonable to expect that there may be sufficient quantities of volatiles near Mons Malapert, the abundance of iron, aluminum, silicon, and titanium is not known.

The Lunar Reconnaissance Orbiter (LRO) will provide much-needed additional information, but, ultimately, the most accurate and precise information will come from instruments placed directly on the surface. These experiments will benefit by having mobility, and several robotic development projects are underway to provide robots for exploring the lunar polar regions.

ATHLETE is a test bed lunar vehicle having six legs that terminate with wheels. The wheels are smaller than would normally be found on a robot that is designed to cross rough terrain. This is possible because the "legs" of the robot have six-degrees of freedom (6-DOF), so that if the rover gets stuck, it can "walk" itself out rather than driving out. The legs can also be used as "arms" when needed. For example, two wheels can be used to grasp a bucket excavator. Each leg also has a socket drive in its axle hub to operate rotary tools (such as drills). A habitat module can be built on top of one of these rovers, and transported to its final location at some distance from the

Figure A.1. ATHLETE robotic rover, under development by NASA.

landing area. Three doors at right angles would allow the habitation modules to dock with each other to begin the formation of an outpost.

The Construction Resource Utilization eXplorer (CRUX, Figure A.2) is an integrated suite of instruments and related software designed for resource exploration. It consists of an instrumented drill (Prospector) and surface geophysical and optical mobile sensors (Surveyor), linked with a mapping and decision-support system (CRUX Mapper/DSS). The Surveyor's geophysical instruments will map shallow subsurface regions to help locate optimal drilling sites. The Prospector's drill will carry instruments down-hole to measure site-specific regolith geotechnical properties and detect water ice.

A.1.2.1 Geologic investigations

Proponents of a return to the equatorial region of the Moon have a rationale for their preference: the equatorial regions of the Moon have been available for study for centuries, and much more is known about them than is known about the polar regions. For example, Coombs and others (1987) have identified lunar rilles that appear to terminate abruptly, which suggests that the lava which flowed within them may have solidified at the top, after which the lava flowed on, leaving behind a "lava tube". Lava tubes could be used for habitats, or for emergency shelters from solar magnetic storms (Hörz, 1985).

In addition to the possibility of finding lava tubes, some of the edges of mare areas have deposits of pyroclastic "fire fountain" glass beads. They are rich in titanium and iron silicate and have the highest measured concentration of other volatiles adsorbed to their surfaces (McKay et al., 1974). The technology for extracting all these materials is well understood, and has been demonstrated in the laboratory (Allen et al., 1994). Lunar pyroclastic deposits would provide a resource for the oxygen and other materials that are needed for a lunar outpost. Moreover,

Figure A.2. Concept of operations for the Construction Resource Utilization eXplorer (CRUX).

because the pyroclastic beads are well-rounded and poorly consolidated, they have excellent engineering properties – they can be excavated much more easily than any other type of lunar regolith material (Cooper, 1994).

It would be fortunate if evidence of volcanic activity (especially pyroclastic glass deposits) were found in the lunar polar region, because having iron- and titanium-rich materials nearby would add to the potential usefulness of Mons Malapert as a lunar outpost location. Currently, there is no evidence for lunar pyroclastic deposits anywhere in the south polar regions.[3] However, many unknowns remain about the geology of this area, and it is possible that some volcanic materials could be found. Lunar Orbiter data from the region north of Mons Malapert shows craters with flat floors and other areas that appear to be fairly level. This is suggestive of the presence of basalts – volcanic rocks that may have flooded the low-lying areas billions of years ago.

Whether the flat-floored crater north of Malapert is caused by volcanism or is simply a highland plains region, it offers a favorable location for a landing site. Moreover, it is of scientific interest because its genesis is not known.

---

[3] If mare or volcanic-related materials exist anywhere in the south polar region, they are more likely to occur in conjunction with the South Pole–Aitken Basin, on the far side. However, this basin has very little mare fill in comparison with other basins on the Moon such as Serenitatis or Tranquillitatis.

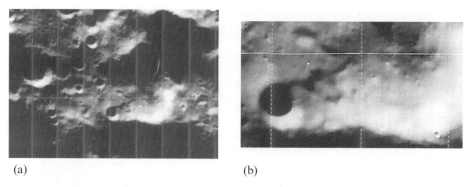

(a)   (b)

Figure A.3. Channel and crater near the peak of Mons Malapert. Detail from Lunar Orbiter IV (1967/05/19), showing Mons Malapert. (a) To the left of the peak is a crater and a channel which appears to be associated with it. The feature, approximately 7 km in diameter, is shown in an enlarged view in (b).

There are other features of scientific interest in this area that could be explored with robotic rovers. A recent analysis of Lunar Orbiter data (Cooper, 2006) shows evidence of a channel near the peak of Malapert Mountain (Figure A.3). Continuing studies are planned using multiple data sets (Clementine, radar data, and additional study of Lunar Orbiter frames) to understand the nature and origin of this feature. LRO data will also be useful. However, the "ground truth" that is available by sample analysis is always preferred, thus a robotic rover would again be invaluable.

Each new mission has generated new questions, and it is likely that LRO will find features of interest that must be explored further with robots. For example, if evidence of lava tubes is found in the polar region, robots will be needed to explore the area to determine their extent, as well as the thickness of the "roof" layer. With this information we will be able to make plans on how to use the lava tubes for radiation shelters or thermal control.

The exploration capabilities developed by these robotic precursor missions will remain applicable throughout the duration of planetary surface exploration. At the beginning, there will be a significant role for humans in the loop: geologic assessment of terrains and surface traverse planning, assessment of outcrops and surface features, and sample acquisition planning; real-time decision-making; troubleshooting, prioritizing tasks, reviewing information, and commanding actions. As the technology develops, these functions will gradually be transferred to the robotic systems.

A.1.3   Robotic development of the south pole infrastructure

The more that can be done by robotic precursor missions, the safer it will be for humans to follow. Risk is mitigated by creating the infrastructure and verifying that it is working properly. The humans who follow will know that there are full tanks of propellant, and a habitat that is protected from ionizing radiation, micrometeorite, and thermal extremes, and which has an Earth-like, breathable atmosphere.

A.1.3.1 Automated solar cell production

Ignatiev (pers. commun., 2006) has shown that lunar resources can be used to fabricate solar cells which will provide energy for the lunar outpost. Because the Moon's surface is already an ultra-high vacuum, the vacuum epitaxy deposition process can be used to make thin-film solar cells directly on the surface of the Moon. All of the elements needed are present – silicon, iron, titanium oxide, calcium, and aluminum.

The components of these solar cells include a substrate, bottom electrode, silicon p–n junction, top electrode, anti-reflection coating, and cell interconnects. The bottom and top electrodes can be made from aluminum or iron silicide, and the substrate can be formed from melted regolith glass. The anti-reflection coating could be created from TiO_2, SiO_2, or evaporated regolith. Thin-film metals (iron, aluminum, or calcium) would serve for cell interconnects.

A mobile robotic system is envisioned (Figure A.4) that would have a launch mass of between 150 and 200 kg. Solar thermal collectors will provide energy of evaporation, and photovoltaic panels will provide power for mobility and control. As the robot moves along, it creates a continuous layout of solar cells that lie directly on the lunar surface.

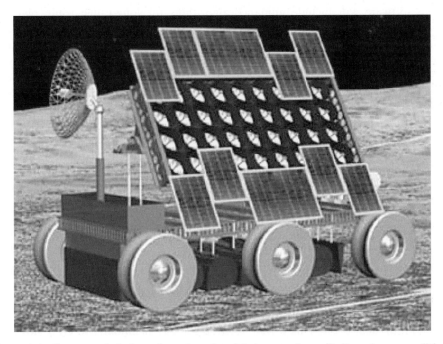

Figure A.4. Conceptual design of a robot that fabricates solar cells from lunar regolith as it moves along in a Sun-synchronous orbit (image courtesy of Alex Ignatiev, University of Houston).

At a velocity of one meter per hour, the robot would travel 672 meters in a lunar "day" (29 Earth days). A robot solar-cell-paving machine that was landed near Malapert Mountain would travel up the side of the mountain to a point about 15 km from the landing site, then begin traversing around the sunlit portion at a constant elevation. The robot could be remotely controlled to ensure that it remains in the sunlight for as much of the time as possible. Converting one square meter per hour, and assuming that the solar cells are 5 percent efficient, a power capacity of 200 kW would exist by the end of the first year.

Adding more solar-cell-production robots would increase the available energy to industrial scales within a short time. For example, 10 rovers as described above could provide from 2 to 4 megawatts per year; 100 rovers could provide up to 40 megawatts per year. Larger-scale or more numerous robots could produce up to 2 gigawatts per year. Abundant energy is needed for a viable lunar economy.

A.1.4   Robotic explorations of the lower latitudes

Robotic rovers that operate with a combination of supervisory and autonomous control will begin to explore the lower latitudes (closer to the equator) on "Magellan routes" – paths that remain in sunlight at all times – so that solar power can be used as the principal power source. They would require more autonomy than past rovers, due to the 14-day traverse of the lunar far side (Teti et al., 2005). Their autonomous system software will include guidance algorithms for long-range route planning (based on LRO data) as well as short-range planning to avoid obstacles detected by sensors. They will be able to recognize features to determine their current position, in a method that may be similar to the way that Clementine's StarTracker cameras provided navigation guidance by recognizing constellations. The software must also manage strategies for re-planning routes when necessary, and for continuously pointing its solar panels towards the Sun.

The robots' vision system will include obstacle detection, and a laser scanning system might be used for redundancy and elimination of false readings. Other components that will be needed include telescoping masts for communication with Earth.

An additional challenge for early robotic rovers will be position determination without GPS. One possible solution would be to remain in continuous contact with the lander. That might seem to limit the ability of the rover to travel anywhere that was not within the line of sight of the lander; however, lunar material is fairly transparent to radar, and radio signals should be able to penetrate to depths of about 10 m (Heiken et al., 1991). This concept would need to be fully tested to determine its limitations, and alternative methods would need to be developed for areas where a larger amount of material (a mountain, for example) stands between rover and lander.

Another method for position determination would be to have a transceiver spacecraft in lunar orbit. This would function somewhat similarly to GPS; however, with only one spacecraft more time would pass between opportunities to access the signal. This is less of a problem on the Moon than it would be for humans on Earth: the robot can continue its tasks while waiting for the next opportunity.

A.2   ROBOTIC TASKS AND SUBTASKS

A.2.1   Robotic tasks for planetary surface missions

As explained in Chapter 4, there are many tasks that could be performed by robots on the lunar surface. Prioritizing those tasks allows development to proceed in a logical and efficient way. Table A.1 lists the tasks that could potentially be performed by robots. The remainder of this section explains how components of the technology can be developed in a logical sequence.

In Tables A.2, A.3, and A.4, the tasks listed in Table A.1 are sorted into categories and ranked according to payback (highest payback first, calculated as "benefit/difficulty"). Depending upon the interests of the project, various subsets of capabilities can be chosen for development. The first subset, "enabling technologies", would be of interest for long-term planning of a habitable lunar outpost. However, if one only wants to plan for a scientific sortie, then the tasks in Table A.3 would be of primary interest. Table A.4 details the capabilities that are not required for the first lunar outpost, but would be useful for expansion of lunar activities.

Enabling technology for planetary surface mission phases includes some daunting tasks. There may be no humans present, and robots must pave the way. Scoring these items involves asking which tasks should be roboticized, which tasks should be developed for hard automation, and which components should be postponed until other technologies are in place.

Because these are enabling technologies for future human missions to the Moon and Mars, they all must eventually be done; but it is nevertheless useful to know that trenching and excavating are higher-payoff tasks than landing site surveys and certification. If you have limited dollars to invest and you want to make progress, then the high-payback tasks should be addressed first. Technologies developed and lessons learned can then be applied to the more difficult, or possibly less beneficial, tasks. Figure A.5 shows the rankings for these tasks.

If we assume that only limited robotic development precedes the next human mission to the Moon, then we would logically also assume that many of the tasks related to planetary surface exploration are likely to be performed by humans, with assistance from a general-purpose robotic assistant. Table A.3 describes these tasks, and their rankings are depicted graphically in Figure A.6. Note that in the alternative scenario we propose, in which many robots are deployed prior to human missions (Chapter 4), some of the assumptions made here would need to be modified.

The highest-payback tasks for the "robot assistant" include such items as EVA monitoring, watchman rounds, EVA preparation, scooping, grasping, and raking. Breaking rocks, observing them in the field with a microscope, and autonomous surface traverses also got high marks. By comparison, loading samples into a return spacecraft is not so useful.

The third and final subset includes tasks that are not critical for a first lunar outpost, but would be part of an ongoing effort to explore the Moon and develop a colony. These are ranked and described below, and their relative scores are illustrated in Table A.4.

Table A.1. Planetary surface mission tasks

Task #	Description
1	Emplace the infrastructure for human exploration: power source, habitat, communication system; check out to ensure that everything is working before the humans arrive
	The details of this task are dependent upon specific mission architectures; thus, it must be broken down into subtasks if it is to be analyzed. Tasks 1b, 3, 37, 38, 41, and 50 are key components of infrastructure emplacement. This task is too broad to be evaluated as an independent item, and thus is not specifically enumerated elsewhere in this appendix
1b	Make plant equipment operational
2	Surface traverses (for science sorties or for delivering materials)
3	Implement a global-positioning network, either satellite-based or ground-based or some combination of both
4	Selecting samples: observing, judging, deciding (knowledge- and experience-based activity)
5	Coring and drilling
6	Scooping, grasping, raking
7	Breaking rocks
8	EVA preparation, worksite preparation
9	EVA worksite teardown and cleanup
10	Splitting and archiving of half of each sample in a facility that is secure from contamination
11	Bagging and labeling of samples
12	Packing samples for transport
13	Loading samples into return spacecraft
14	Sample characterization – evolved gas analysis (derived from heating the sample and measuring off-gassed products)
15	Sample characterization – microscopic observation
16	Sample characterization – wet chemistry analysis
17a	Geophysical exploration – emplacement of geophones
17b	Operate telescopes
17c	Repair telescopes
18	Geophone placement
19	Geophysical exploration – seismic event generation (explosives, vibro-seis, crash landings of spacecraft)
20	Geophysical exploration – ground-penetrating radar exploration along traverses
21	Geophysical exploration – electrical properties measurements along traverses
22	Emplace surface science experiments
23	Surface physics – radiation level analyses at multiple locations and various times
24	Surface physics – measurements of electrostatic charge differential between sunlit and shadowed regions
25	Surface physics – heat flow measurements at multiple locations and over a several month time span
26	ISRU experiments – extraction of water vapor from lunar regolith in the permanently shadowed regions

Task #	Description
27	ISRU experiments – hydrogen reduction of glasses and ilmenite
28	ISRU experiments – sintering/melting of regolith for magma electrolysis and casting bricks and construction materials
29	ISRU experiments – chlorine plasma reduction of basaltic regolith
30	ISRU experiments – vapor-phase reduction of regolith
31	ISRU experiments – ion (plasma) separation of regolith components; ion sputtering
32	ISRU experiments – slurry/solution processes for oxygen recovery, such as HCl dissolution and electrolysis, H_2SO_4 dissolution and electrolysis, HF dissolution and electrolysis, lithium, aluminum or sodium reduction
33	ISRU experiments – making solar cells from silicate materials, especially ilmenite
34	ISRU experiments – extraction of other gases by heating of regolith
35	ISRU experiments – collection and containment of produced gases
36	ISRU experiments – collection and bagging of raw or processed regolith for radiation shielding over habitats
37	Landing site survey and certification for optimum placement of surface modules
38	Outpost maintenance during human absence
39	Trash management
40	Mating/docking of surface modules to each other
41	Storage of cryogenic liquids on the lunar surface – conduct tests of systems to control cryogen boil-off by the use of thermodynamic vents, vapor-cooled shields, low-conductivity supports, and refrigeration and re-liquefaction equipment
42	Watchman rounds (e.g., to look for micrometeorite damage to habitats or equipment)
43	Deploying solar-flare-detection telescopes
44	Operation and maintenance of solar-flare-detection telescopes
45	Situational awareness
46	Worksite assessment
47	EVA monitoring
48	EVA tool retrieval
49a	Brush operation (brush off rocks before beginning spectral analysis, doing "delicate" excavations)
49b	Cleaning dust off equipment (other than with brushes)
49c	Suit cleaning
50	Trenching/excavating to bury cables at the base camp, examine geological sections, dig radiation storm shelters
51	Sawing core samples in half
52	Set up oxygen extraction plant

Note: The numbers in the left column are tags that were given when the task was first enumerated, and have no bearing on the payback score that was eventually assigned. They are retained as key fields for cross-reference to other tables in this appendix.

Table A.2. "Enabling technology": tasks that must be accomplished robotically as precursors for future human missions

Task #	Description	Pervasiveness
50	"Dirt work": trenching/excavating *(used to bury cables at the base camp, examine geological sections, dig radiation storm shelters. – includes, but is not limited to, 29 subtasks in Table A.6)*	Moon and Mars
38	Outpost maintenance during human absence *(includes, but is not limited to, 31 subtasks in Table A.6)*	Moon and Mars
03	Emplace transponder or repeater beacons *(implementation of a global-positioning network, either satellite-based or ground-based or some combination of both – includes, but is not limited to, 24 subtasks in Table A.6)*	Moon, Mars asteroids, Europa
35	Collection and containment of produced gases (ISRU experiments) *(similar to task #P-41, but does not include cryogenics handling)*	Moon and Mars
24	Measurements of electrostatic charge differential between sunlit and shadowed regions (surface physics) *(emplacement of the instrument is the robotic part of the task, and is represented by task #P-22 – the operation of the instrument might be more appropriately classified as automation)*	Moon
37	Landing site survey and certification for optimum placement of surface modules *(includes, but is not limited to, 23 subtasks in Table A.6)*	Moon and Mars
40	Mating/Docking of surface modules to each other *(involves multiple steps and extensive technology development – although automated docking is performed on-orbit at the International Space Station, the conversion of this capability to a planetary surface is not straightforward)*	Moon and Mars
34	Extraction of other gases by heating of regolith (ISRU experiments) *(similar to task #P-14)*	Moon and Mars
41	Setup of cryogenic storage facilities on the lunar/Martian/ planetary surface *(storage of cryogenic liquids on the lunar surface – conduct tests of systems to control cryogen boil-off by the use of thermodynamic vents, vapor-cooled shields, low-conductivity supports, and refrigeration and re-liquefaction equipment – includes, but is not limited to, 15 subtasks in Table A.6)*	Moon and Mars
26	Extraction of water vapor from lunar regolith in the permanently-shadowed regions (ISRU experiments) *(this task involves multiple steps – if extensive robotic precursor missions are not conducted, then this task is more likely to be performed initially within a pressurized habitat, along with other experiments on resource extraction)*	Moon
33	Making solar cells from silicate materials, especially ilmenite (ISRU experiments) *(same notes as P-26)*	Moon and Mars
28	Sintering/melting of regolith for magma electrolysis and casting bricks and construction materials (ISRU experiments) *(same notes as P-26)*	Moon and Mars
1B	Make plant equipment operational. *(includes, but is not limited to, 23 subtasks in Table A.6)*	Moon and Mars

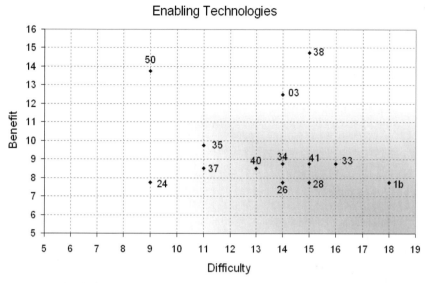

Figure A.5. Payback for long-term, enabling technology tasks. Robot development should focus on the least difficult and highest benefit tasks. The slope of the line from the origin to each point is used to determine relative payback for each task. Refer to Table A.2 for the legend and explanation of each item.

A.2.2 Pervasive subtasks, activities, and capabilities

In fact, some tasks are clearly at a more fundamental level than others. Some are so fundamental that they are more properly classified as "pervasive activities and capabilities". These pervasive subtasks, activities, and capabilities are important for almost all areas of robotic development.

1.0 *Situational awareness*: [photographing samples *in-situ*; compare worksite with pre-existing maps and other data; "verbalize" activities and sample description; photograph areas of interest; select areas of interest; select samples for further evaluation]. Much of this could be accomplished by including a visual inspection system on the end of a manipulator arm, with data fed to a crew person's helmet display, a ground base, or a work station.

2.0 *Mate/Undo connectors*: This is an integral part of 28 subtasks in Table A.6. The ease or difficulty of doing this would depend on the connector. If the connector is robotically compatible (e.g., a bayonet or blind-mate connector) then the task would be straightforward, having been designed for robotics in the first place. In a more complicated scenario, such as a contingency task, a robot could be sent out if the automated method failed. However, it would be relatively complex to make a repair robot-friendly if the system was not originally designed as such. It could require dexterous manipulation, tactile feedback, multiple "hands" for manipulation, etc. This level of technological development is in its infancy (see Robonaut discussion in Chapter 4).

Table A.3. EVA assistance/EVA minimization tasks

Task #	Description	Pervasiveness
47	EVA monitoring (EVA situational awareness) *(very similar to task #P-45)*	Moon and Mars
42	Watchman rounds	Moon and Mars
08	EVA worksite preparation *(included as a subtask of 29 other tasks)*	Moon and Mars
09	EVA worksite teardown and cleanup *(included as a subtask of 29 other tasks)*	Moon and Mars
06	Scooping, grasping, and raking *(includes, but is not limited to, 28 subtasks in Table A.6)*	Moon and Mars
48	EVA tool retrieval *(included as a subtask of 44 other tasks)*	Moon and Mars
22	Emplacement of surface science experiment packages. *(uncludes, but is not limited to, 20 subtasks in Table A.6)*	Moon, Mars asteroids, Europa
49c	Suit cleaning *(includes, but is not limited to, 24 subtasks in Table A.6)*	Moon and Mars
7	Breaking rocks *(includes, but is not limited to, 24 subtasks in Table A.6)*	Moon and Mars
15	Microscopic observation (sample characterization). *(includes, but is not limited to, 7 subtasks in Table A.6)*	Moon and Mars
14	Evolved gas analysis (sample characterization). *(includes, but is not limited to, 7 subtasks in Table A.6)*	Moon and Mars
19	Geophysical exploration – seismic event generation (explosives, vibro-seis, etc.). *(includes, but is not limited to, 29 subtasks in Table A.6)*	Moon and Mars
16	Sample characterization – wet chemistry analysis. *(includes, but is not limited to, 7 subtasks in Table A.6)*	Moon and Mars
10	Splitting and archiving of half of each sample in a facility that is secure from contamination *(not represented by any of the subtasks in Table A.6, because it does not occur during the "generic" EVA sortie, but might occur as a separate, unique EVA or as "cleanup" work at the end of a sortie)*	Moon and Mars
11	Bagging and labeling of samples *(includes, but is not limited to, 6 subtasks in Table A.6)*	Moon, Mars asteroids, Europa
18	Emplacement of geophones (geophysical exploration). *(includes, but is not limited to, 41 subtasks in Table A.6)*	Moon and Mars
51	Sawing core samples in half *(samples might be brought back to the lab for this operation; it may occur in an external lab glovebox, but would typically not be part of an EVA per se)*	Moon and Mars
49b	Cleaning dust off equipment (other than with brushes) *(includes, but is not limited to, 24 subtasks in Table A.6)*	Moon and Mars
12	Packing samples for transport *(includes, but is not limited to, 5 subtasks in Table A.6)*	Moon and Mars
2	Exploratory surface traverses *(includes, but is not limited to, 52 subtasks in Table A.6)*	Moon and Mars

Task #	Description	Pervasiveness
05	Coring, drilling, and trenching *(includes, but is not limited to, 24 subtasks in Table A.6)*	Moon and Mars
46	Worksite assessment *(includes, but is not limited to, 9 subtasks in Table A.6)*	Moon and Mars
20	Geophysical exploration – ground-penetrating radar exploration along traverses *(includes, but is not limited to, 41 subtasks in Table A.6)*	Moon and Mars
21	Geophysical exploration – electrical properties measurements along traverses *(includes, but is not limited to, 21 subtasks in Table A.6)*	Moon and Mars
49a	Brush operation (brush off rocks before beginning spectral analysis, doing "delicate" excavations) *(includes, but is not limited to, 24 subtasks in Table A.6)*	Moon, Mars, asteroids
36	Collection and bagging of raw or processed regolith (ISRU experiments) *(includes, but is not limited to, 27 subtasks in Table A.6)*	Moon and Mars
04	Selecting samples – observing, judging, deciding (knowledge- and experience-based activity) *(includes, but is not limited to, 18 subtasks in Table A.6)*	Moon and Mars
13	Loading samples into return spacecraft *(includes, but is not limited to, 3 subtasks in Table A.6)*	Moon and Mars

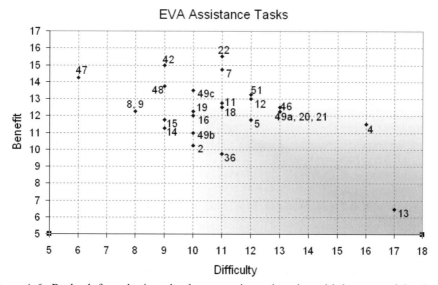

Figure A.6. Payback for robotic tasks that occur in conjunction with human activity. Lower difficulty numbers and higher benefit numbers represent the most useful robotic tasks for early development, and are located in the upper left quadrant. Refer to Table A.3 for the legend and explanation of each item.

Table A.4. Future development tasks – not required for first lunar outpost, but useful for expansion of lunar activities

Task #	Description	Pervasiveness
18b	Geophysical exploration – monitoring of geophones	Moon and Mars
25	Heat flow measurements at multiple locations and over a several month time span (surface physics) *(emplacement of the instrument is the robotic part of the task, and is represented by task #P-22 – the operation of the instrument might be more appropriately classified as automation)*	Moon and Mars
23	Radiation level analyses at multiple locations and various times (surface physics) *(emplacement of the instrument is the robotic part of the task, and is represented by task #P-22 – the operation of the instrument might be more appropriately classified as automation)*	Moon and Mars
17b	Telescope operation	Moon
39	Trash management *(included as a subtask in tasks #P-1a, P-3, P-9, P-17, P-19, P-26 through P-34, P-37, P-38, and P-50)*	Moon and Mars
17c	Telescope maintenance and repair	Moon
43	Deploying solar-flare-detection devices *(similar to task #P-42)*	Moon and Mars
44	Operation and maintenance of solar-flare-detection telescopes *(similar to task #P-17)*	Moon and Mars
29	Chlorine plasma reduction of basaltic regolith (ISRU experiments) *(this task involves multiple steps – if extensive robotic precursor missions are not conducted, then this task is more likely to be performed initially within a pressurized laboratory, along with other experiments on resource extraction)*	Moon
30	Vapor-phase reduction of regolith (ISRU experiments) *(same note as P-29)*	Moon
31	Ion (plasma) separation of regolith components; ion sputtering (ISRU experiments) *(same note as P-29)*	Moon
17a	Deploy telescopes *(includes, but is not limited to, 40 subtasks in Table A.6)*	Moon
27	Hydrogen reduction of glasses and ilmenite *(same note as P-29)*	Moon
32	Slurry/solution processes for oxygen recovery – HCl dissolution and electrolysis; H_2SO_4 dissolution and electrolysis; HF dissolution and electrolysis; lithium, aluminum or sodium reduction (ISRU experiments) *(same note as P-29)*	Moon

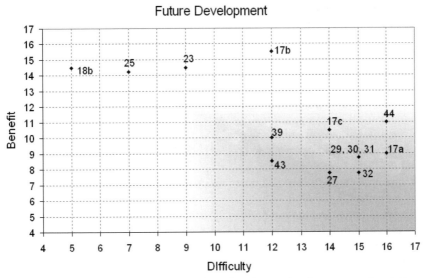

Figure A.7. Graphical display of the rankings for future development tasks that are not performed immediately, but which must be planned for future colonization of the Moon. Highest payback tasks are in the upper left quadrant of the graph. Refer to Table A.4 for the legend and explanation of each item.

3.0   *Remove/replace bolts*: The same issues occur for this task as for mating and undoing connectors. This could be an integral component of the same 28 subtasks mentioned above.

4.0   *Third arm for EVA crew*: This would also be an integral component of the 28 subtasks mentioned above. It might be used while replacing ORUs; deploying antennae/solar arrays; inspection in hard-to-reach areas; making and verification of utility and structural connections; cleaning exterior surfaces; clearing fluid lines; cutting and repairing multi-layer insulation; moving debris shields; coring, drilling, and trenching; scooping, grasping and raking; breaking rocks; splitting and archiving samples; bagging and labeling samples; packing samples for transport; deploying or repairing telescopes; and cleaning dust off equipment.

5.0   *EVA caddy/EVA buddy*: For transferring equipment and tools from an intermediate staging platform (pressurized rover or lunar truck) to the work area and back. This is an integral part of 31 subtasks in Table A.6. The robot should include a power outlet for EVA extension cords, and self-docking batteries. Then, if an EVA crew person needs a power tool that does not have batteries (or whose batteries are depleted), they can plug the tool into the robot. Moreover, the EVA caddy would be an important component for portable life-support system (PLSS) recharging, should the need arise. Studies are currently underway to determine the optimum system designs for planetary surface EVAs, given the fact that spacesuits and life-support packs are both cumbersome and massive, and add to the mental and physical fatigue experienced during long EVAs. One idea being considered is to design the suits to carry a four-hour supply of life-support consumables (making them lighter), and have a robotic assistant along to carry additional consumables for EVAs of longer duration. In addition to the items discussed in 4.0 above, the "EVA buddy" might also be used for many other tasks:

- exploratory traverses;
- microscope operations;
- chemical analyses;
- emplacement of geophones;
- emplacement of surface science experiment packages (similar to the Apollo ALSEP;
- ground-penetration radar or electrical properties measurements along traverses.

6.0   *Verbalize activities and description of area of interest*: This is an activity that occurs continuously in conjunction with an EVA – the crew are talking to each other and to the ground, discussing the situation and exchanging information. If a robot were acting autonomously on a planetary surface, some type of feedback about its activities would be useful; however, the amount of feedback and its quality would depend on the design of the robot. An autonomous, intelligent robot might be making decisions based on the information that it gathers, and modifying its own behavior based on those decisions. In the situation of a robotic precursor mission to the lunar far side, it would not be possible for the robot to interact with those who are monitoring its activities in the same way that a line-of-sight communication would occur. Nevertheless, it would be useful to have feedback from the robot that explains what inputs were used to make decisions and what decisions were made. This would verify the functionality of the autonomous robot and provide additional information about what to expect during the human-piloted phase of the mission.

7.0   *Verify functionality of equipment installation (monitor or test)*: This is a general inspection task, associated with all of the tasks in Table A.2.

A.2.3   Specific high-value robotic technologies for lunar development

A.2.3.1   Telescoping mast

On the Moon, communications must rely on line-of-sight contacts between stations.[4] Because the curvature of the Moon is greater than the curvature of the Earth, the horizon is closer. Table A.5 shows the distance to the horizon on each body, depending upon the height of the observer.

It will be useful to deploy masts with transceivers to facilitate communications

---

[4] Depending on the frequency used, there is a possibility that radio signals can travel straight through the dry lunar terrain, at least for a limited distance. A frequency of 10 MHz would be the optimum choice for subhorizon radio communication through soils (Olhoeft, 1984).

Table A.5. Distance to the horizon on Earth and Moon

Height above surface:	2 meters	10 meters	20 meters	50 meters
Earth, $R = 6{,}371$ km	5.0 km	11.3 km	16.0 km	25.2 km
Moon, $R_{\mathbb{C}} = 1{,}737$ km	2.6 km	5.9 km	8.3 km	13.2 km

between ground stations. A transceiver atop a 50-meter mast would be accessible at distances of 13 km away on the Moon.

Besides transceiver beacons, other instruments could be located atop (or along the length of) masts. These might include reconnaissance cameras, flood lights, and solar power collectors for the lunar polar regions (Figure A.8).

Telescoping masts are routinely used on Earth by radio and television news vehicles. The basis of this technology is mature; however, the telescoping masts are usually deployed with human assistance and they are carried on relatively-large vehicles (vans or trucks) which travel under human guidance, usually on paved roads. Free-standing standard models up to 84 feet (26 m) can be installed on a variety of vehicles or shelters. Guyed models ranging in height from 20 feet (6 m) to 134 feet (40 m) are available for vehicle, trailer, or field installations. Work is needed to

Figure A.8. Example of a telescoping mast used for a communications transceiver (image courtesy of Wil-Burt Company, Orrville, OH).

determine if guy wires would be of sufficient value to warrant the additional complexity involved in deploying them. Although there is no hazard from high winds on the Moon, there may be center-of-gravity issues associated with working on slopes that would require the use of guy wires to stabilize the mast.

Power for telescoping masts on the Moon must also be considered. On Earth they are typically operated via hydraulics (as is the case for several other "heavy" equipment items discussed in this section). Additional study should be undertaken to determine if an alternate power system would be more appropriate for the Moon and Mars.

A.2.3.2   *Trenching tool*

Trenching is an activity that will be required repeatedly during planetary surface missions. Site preparation will require trenching to bury power and communication lines (e.g., Sullivan, 1994). Trenching is also used in geological investigations of near-surface features (e.g., Spudis and Taylor, 1990). Finally, trenching or digging equipment may be required on a pressurized rover, so that "field camps" can be established that offer protection from radiation storms, in the form of dugouts. Trenching is a good candidate for robotic activities because it is pervasive, time-consuming, dangerous, and labor-intensive. In fact, trenching is recognized by OSHA as one of the most hazardous construction operations. Commercial-off-the-shelf (COTS) trenching machines include smaller models that can be powered by a tractor the size of a large riding lawnmower. These machines also offer the capability of changing out tools. For example, several of the COTS models have interchangeable attachments so that the machine can be operated as a trencher, a backhoe, a snow plow, and so on. This technology should be considered as a starting point for robotic capability development.

A.2.3.3   *Picking up rocks*

When a terrestrial geologist is doing field work, a frequent activity is picking up rocks to examine them closely. This activity helps the geologist to decide which samples warrant further study, and over half of the rocks which are picked up are promptly discarded. Picking up rocks while wearing an extra-vehicular mobility unit (EMU) and gloves, however, is difficult. This activity requires that the crew person bend over or kneel down, both of which can present problems (e.g., Eppler, 1999), especially if repeated many times during the course of the sortie. Picking up rocks corresponds to subtask #5F3, "Operate sampling tool (acquire sample)" from Table A.6, although in many cases the sampling tool might be a gloved hand (or a robotic grasping tool).

Two approaches to minimizing this task may be envisioned. In one approach, a robotic system allows close examination of the sample without picking it up. This may be done with a portable microscope which uses a CCD camera and a video link to a monitor. The monitor may be carried on the rover at the approximate height of the crew member's eye, or fed to a monitor in the suit helmet, so that initial examinations of many rock samples can be done without the need to bend down and pick up the rock.

Figure A.9. "EZ-Reacher" pick-up tool (image courtesy of Arcoa Industries, Inc.).

However, a second approach is also needed, because samples must eventually be collected anyway – the above system does not eliminate the need for a grasping arm. Moreover, it is likely that the ability to pick up rocks will be useful during the precursor missions, for base site preparation. Cobble-size rocks may interfere with the emplacement of habitat and physical plant modules, in which case it would be necessary to remove them. Clearly, picking up rocks is a pervasive task which would be a good candidate for robotics. Pick-up tools that are used on Earth are available which could be modified for EVA use. For example, one COTS pick-up tool (Figure A.9) is said to be capable of picking up a dime or a five-pound brick, and features a locking handle. It is available in a variety of sizes from 20 inches to 35 feet.

A.2.3.4   Robotic brush

Although it is not useful for cleaning spacesuits, brushing (perhaps using an electro-static or magnetic brush) may be effective for cleaning other items, such as tools and equipment (e.g., solar power cells or other hard, smooth surfaces). Dust was a serious problem during the Apollo missions (Sullivan, 1994). Dust got into areas where the release bolts were located and made it difficult to deploy some experiments. It is likely that dust caused some of the equipment to fail early, and it is known that the accumulated dust on surfaces caused a gradual increase in the operating temperatures of the instruments (Bates *et al.*, 1979). A brush end effector mounted on a mobile robot would be able to clean equipment on the lunar surface when dust has built up on the equipment to the point that its effectiveness is compromised. Trade studies are needed to determine the efficiency of brushing away dust from surfaces as opposed to making all of the surfaces dust-repellant.

A.2.4   A "generic" sortie on a planetary surface

Table A.6 provides a list of high-level tasks that a robot would be required to perform during a planetary exploration traverse. In this table it is assumed that the functions are performed by a suited crew member as a baseline. The tasks could also be

Table A.6. A "generic" sortie on a planetary surface

Step 1	*Sortie preparation*
1A	Egress the airlock/start-of-task location
1B	Translate to rover
1C	Translate (with rover if needed) to tool locker
1D	Detach tools and equipment (if required)
1E	Attach tools to the rover
Step 2	*Acquire hardware, equipment, or science instruments*
2A	Translate to equipment storage area
2B	Detach equipment from storage area
2C	Transport equipment to rover
2D	Attach equipment to rover
2E	Translate with rover to worksite area
Step 3	*Set up worksite*
3A	Detach Tools
3B	Translate from rover to the worksite
3C	Set up support equipment (lights, cameras, tools)
3D	Translate back to the rover
3E	Deploy (open) doors, covers, or thermal blankets, MLI, etc.
Step 4	*Change out ORU/Set up equipment*
4A	Detach any equipment that is to be replaced from its installed location
4B	Detach replacement/new equipment from rover (or stow/dispose of old equipment)
4C	Set up replacement/new equipment
4D	Verify functionality of equipment installation (monitor or test)
Step 5	*Perform science activities*
5A1	Compare worksite with pre-existing maps and other data
5A2	Verbalize activities
5B1	Select areas of interest
5B2	Confirm areas of interest with base station crew who are monitoring the sortie
5C	Photograph areas of interest, moving on foot or via rover from one specific area to another, while being careful to avoid hazards due to tripping, steep gradients, etc.
5D	Select samples for further evaluation
5E	Photograph samples *in-situ*
5F	Collect samples:
5F1	acquire sampling tool
5F2	translate to sample,
5F3	operate sampling tool (acquire sample)
5F4	verify acquisition of sample
5F5	translate to rover
5F6	acquire sample container
5F7	place sample in container
5F8	label container
5F9	verbalize activities and sample description
5F10	stow sampling tool

5G	Collect non-sample data
5G1	acquire data acquisition tool
5G2	translate to area of interest
5G3	operate data acquisition tool
5G4	verify acquisition of data
5G5	verbalize activities and description of area of interest
5G6	translate to rover
5G7	stow data acquisition tool
	The substeps in Step 5 may be repeated numerous times
Step 6	*Tear down worksite*
6A	Retract thermal blankets, covers, doors, etc.
6B	Ingress left-over parts and trash items in trash bag
6C	Translate to rover
6D	Attach failed equipment and/or EVA support equipment to rover
6E	Translate/Transport items to next worksite or to stowage area
	(Steps 3–6 may be repeated at each worksite during the EVA sortie)
Step 7	*Stow equipment*
7A	Translate to vicinity of storage location
7B	Detach equipment/failed hardware
7C	Translate hardware/equipment to storage platform
7D	Attach equipment to storage platform
7E	Actuate equipment restraints
7F	Translate to rover
7G	Translate to EVA support equipment and tools storage area
7H	Detach support equipment from rover
7I	Attach EVA support equipment at storage location
7J	Translate to rover storage location
7K	Attach rover and related equipment to restraints/power-supply cables at storage site
Step 8	*Sortie closeout*
8A	Translate to airlock/end-of-task location
8B	Clean dust off of EMU
8C	Ingress airlock/end-of-task location

commanded tele-robotically, either from a lunar base or from Earth. The steps and substeps in this table provide an example of the level of detail that is required for understanding and planning robotic activities. Sixty-two individual tasks were identified. Each of the steps would include numerous motions and actuations, which could be pre-programmed so that higher-level commands, such as the ones listed in Table A.6, could be given to the robot.

A.3 CONCEPTUAL DESIGN FOR ROBOTRACTOR: A MULTI-PURPOSE EXCAVATING AND REGOLITH-MOVING MACHINE

A.3.1 Summary

This section provides a description of a semi-autonomous excavating/digging/soil-manipulating robot for use on the Moon. One of the "enabling technology" tasks identified earlier is the capability to robotically prepare a base campsite on the lunar surface. Robotic site preparation may entail leveling the ground for habitat emplacement, trenching to bury cables, and excavating to create emergency radiation-storm shelters.

The tasks required of a robotic "earth-moving machine"[5] (RoboTractor) are pervasive and frequent. Moreover, the class of tasks that could be performed by this machine would be highly beneficial on the Moon, because much of the work is too dangerous or too strenuous for humans.

The functions and requirements for RoboTractor may be satisfied by a combination of modified commercial off-the-shelf (COTS) technology and existing space flight equipment designs. The Marsokhod rover design is an appropriate size and mass for incorporation of the earth-work components that are commercially available.

A.3.2 Basic requirements and design elements

A.3.2.1 Basic requirements for RoboTractor

Safety: The RoboTractor will present a safety hazard for EVA astronauts since it will only be effective if it incorporates numerous sharp objects such as tool blades, auger tips, wheel spikes, and so on. Moreover, because of its mass and power, the consequences of uncontrolled collisions will be more serious than they would be for smaller machines. Thus, a primary focus for development should be methods of assuring the safety of humans who are working with (or who are in the vicinity of) the robot.

Communication/Navigation/Vision system: Autonomous navigation and operation will be required for lunar far-side robotic precursor missions. For lunar near-side missions, somewhat less autonomy is required. Research is underway at the Center for Intelligent Machines and Robotics (CIMAR), at the University of Florida, under the guidance of the Air Force Research Laboratory, to develop a series of autonomously-navigating vehicles. Other technologies are under development at Carnegie Mellon University, Omnitech Robotics (*www.omnitech.com*) and the

---

[5] Activities related to excavation and piling of rocks and soil, in connection with an engineering operation, are referred to as "earth work" or "dirt work". Rather than inventing a new name for this activity, it may be useful to continue using this term, in the same way that we use the term "geologist" to describe those who study the rocks, subsurface structure, and history of various planets.

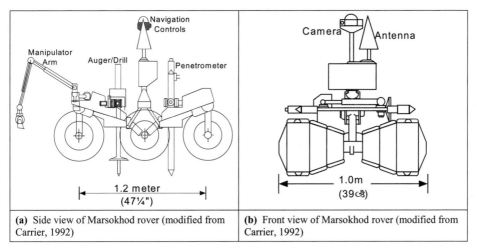

(a) Side view of Marsokhod rover (modified from Carrier, 1992)

(b) Front view of Marsokhod rover (modified from Carrier, 1992)

Figure A.10. Marsokhod rover.

D.O.E. University Research Project in Robotics (the Universities of Michigan, New Mexico, Tennessee, and Texas, to name only a few).

Power supply: Fuel cells hold promise as a source of power, and commercial groups are building fuel cells for the automotive industry. The power requirements for RoboTractor would be based on similarly-sized equipment that operates on Earth.

A.3.2.2   Design elements

Multi-purpose use, EVA compatibility, traction, and maneuverability are important design considerations. Departures from typical Earth-based designs will be needed to address the lunar gravity environment.

Rover body: The Marsokod rover (Figure A.10), has six wheels and moves by a combination of rolling and walking. The frame is articulated such that the vehicle can drive over an obstacle twice as high as the wheel diameter, or it can bridge over a crevice. The larger Marsokhod has a mass of 350 to 450 kg; the smaller one, developed for hard-to-reach locations, has a mass of 70 to 100 kg. The larger of these two vehicles, which has a 1.2-m wheel base and a 1-m axle width, is of similar size to the commercial lawn tractors upon which RoboTractor's capabilities are modeled.

A.3.2.3   Power sources

In addition to solar power (converted and stored in batteries), an alternative source of power to be considered is fuel cells. This alternative may be useful in situations where batteries are not efficient (e.g., during lunar night or in a permanently-shadowed region).

Typical Fuel Cell

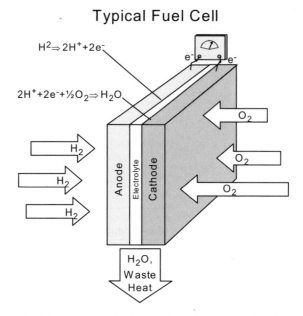

Figure A.11. Schematic of a typical fuel cell (Kartha and Grimes, 1994).

Fuel cells: Fuel cells have been used in NASA's spacecraft designs since the Gemini missions. On the space shuttles, three fuel cells, each weighing 260 pounds (118 kg) and rated at 10 kW, run everything except for a few battery-powered experiments.

Recent advances in fuel cell technology make them an attractive source of energy for planetary surface applications, and they are inherently efficient. On Earth, fuel cells have begun to appear as alternatives to internal combustion engines in buses and cars (e.g., *http://www.isecorp.com/*). Figure A.11 shows the basic design of a typical fuel cell. The key components are the anode, to which the fuel is supplied, a cathode, to which the oxidant is supplied, and an electrolyte, which permits the flow of ions (but not electrons or reactants) between the anode and the cathode. The net chemical reaction is exactly the same as if the fuel were burned.

The fuel cell intercepts the stream of electrons that flow spontaneously from the reducer (the fuel) to the oxidant, and it diverts the stream for use in an external circuit. The main distinction between a fuel cell and a conventional battery is that the fuel and oxidant are not integral parts of a fuel cell, but instead they are supplied as needed to provide power to an external load, while the waste products are continuously removed. In the case shown in Figure A.11, hydrogen is supplied to the anode, oxygen to the cathode, and water is produced.

One of the most appealing aspects of the fuel cell is its inherent modularity: both its efficiency and its cost per unit power are relatively insensitive to its size. This suggests that fuel cells would be effective in applications of wide-ranging scale, from large power plants to portable power generators and even laptop computer power supplies. The cost and technical risk involved in using a fuel cell vehicle should be lower than that of some competing power sources, due to greater fuel efficiency and ease of maintenance.

Fuel cells have efficiencies of approximately 50 percent, whereas advanced solar power cells typically have efficiencies of about 18 percent. The fuel cell is an attractive option for RoboTractors because they can benefit by having their electrical power generated onboard, on demand.

Fuel cells have lifetimes ranging from 1,000 to 50,000 hours. Rapid progress is being made as auxiliary equipment is developed and integrated with fuel cells, fuel-processing technology is improved, system designs are optimized, and mass-production processes are established.

A.3.2.4   COTS dirt-work equipment

High-end commercial lawn tractors have capabilities that are analogous to Robo-Tractor, and provide a baseline for the RoboTractor tool suite and vehicle size. Vermeer and DitchWitch brands are examples. Both these companies offer numerous attachments for their small tractors, which are analogous to the tools that would be needed on the Moon, and an example is shown in Figure A.12. The trencher and vibratory plow attachments are intended as examples only, and do not exclude the incorporation of a backhoe, portable drilling rig, or front-end loader. Rather than integrating all of the tools onto the rover as permanent fixtures, it may be more efficient to have a tool shed to contain the tools that are not in use. If properly designed, RoboTractor could drive up to the tool shed, open the door (or command

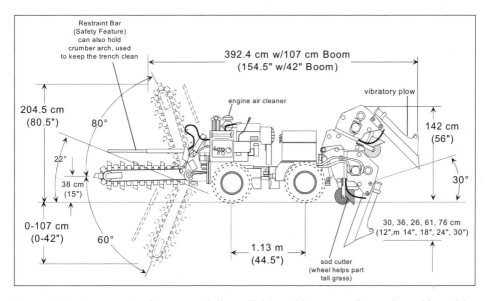

Figure A.12. An example of a commercially available multi-purpose dirt-work machine, with a body slightly larger than a riding lawnmower.

the door to open itself), then make direct contact with the capture latches or attachment mechanisms to secure the required tool to the body of the vehicle:[6]

- *Grader/Scraper/Blade*: Smoothing the surface will be useful in preparing a base camp or a platform for a telescope.
- *Vibratory plow*: As mentioned in Chapter 4, tripping hazards were noted during the Apollo surface EVAs, due to cables that do not lie flat on the ground (having been stowed in a coiled configuration, and not being constrained by gravity to the same extent as on Earth). A vibratory plow, with a cable reel, can be used to bury cables a few centimeters to a few tens of centimeters below the ground, depending on the subsurface structure (a trencher may be needed if the regolith is shallow or indurated.)
- *Cable reel/carrier*: The cable reel is used in conjunction with the vibratory plow to bury power and data cables below the surface.
- *Trencher*: In cases where the regolith is shallow and bedrock (or coarse rubble) is near the surface, a trencher may be needed to bury cables. It is also useful for geologic investigations.
- *Backhoe*: Used to transport regolith for use as radiation shielding.
- *Manipulator arm*: An arm may be required for picking up rocks in the 10–30-cm size range, for assisting in attaching and detaching other tools, and for other contingency uses.
- *Wagon*: The wagon may be used as an emergency transport for an injured crew member. Alternatively, if it is necessary to move regolith or boulders from one place to another, a wagon or "dump truck" attachment will allow RoboTractor to collect a larger amount of material before it becomes necessary to move to the off-load location.
- *Trailer-mounted portable drill*: Some COTS portable drills are small and light-weight enough to be pulled by a lawn tractor while being capable of drilling to approximately 60 meters (200 feet). Extensive automation would be required for this attachment, as its current embodiment requires human operation and continuous supervision (Figure A.13).

RoboTractor must also be smart. In cases where line-of-sight command is possible, the individual components can be operated tele-robotically, either from the lunar base or from Earth. When operations are performed on the lunar far side (to emplace telescopes, for example), these machines must have some degree of autonomy.

A.3.3   Preliminary design synthesis

The preliminary design concept is shown in Figure A.14. On Earth these machines rely on gravity to provide reactive force. With no alternative reactive force, the

---

[6] John Deere currently markets a "Quick-Tach" system which allows most implements to be hooked up in five minutes or less. This concept should be studied as a basis for an automated analogue.

(a)   (b)

Figure A.13. Portable drilling units currently require human operation. (a) Ardco-Traverse LLC man-portable drill, mounted on a wagon behind a lawn tractor. (b) Ditch Witch JT4020 is a self-contained directional-drilling unit, which can drill a horizontal bore up to 300 m in length.

digging capacities might decrease by as much as 84 percent on the Moon. To address this difficulty, dual augers could be mounted near RoboTractor's center of mass. These augers will dig into the ground below the vehicle, and feedback sensors will determine when the augers are firmly in place. Then, RoboTractor can operate (albeit slowly) against a mass that it seeks to manipulate in a vertical direction. However, the augers would not be useful in horizontal mode, when RoboTractor must push or pull

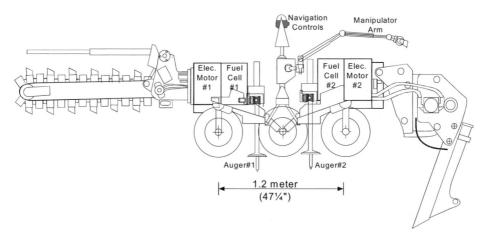

Figure A.14. Preliminary basis for design of the RoboTractor. The manipulator arm, mounted on the mast near the middle of the vehicle, should be capable of activating the capture mechanism for tool change-out, picking up rocks, and assisting in contingencies.

more than its own weight. Although RoboTractor will be able to manipulate some-what more than its own mass in a horizontal direction, wheel spikes should be considered for providing additional horizontal traction.

A.3.4   Example development cycle for RoboTractor

- Design pre-production prototype system. Select materials, power supply, wheel and body size, shape and performance characteristics, and software. Select subcontractors to provide COTS components such as guidance, navigation, and control software and hardware, radio and antenna, cameras, actuators, system I/O hardware, pitch, roll, and bearing sensors, power system control hardware, and so on.
- Research the gamut of designs for mobility systems and determine which of these can be efficiently incorporated into the RoboTractor. Do similar research for other systems that are specific to the space environment or to vehicles for other planetary surfaces.
- Fabricate the pre-production prototype system.
- Test the pre-production prototype system.
- Perform a field demonstration of navigation, obstacle avoidance, hazard detection, tool interchange, excavating operations, emergency shut-down, com-munication of stored data and supervision with a communication lag of up to 12 minutes.
- Complete technical feasibility testing of the full system in a field environment, including simulations of contingency operations.
- Perform reduced-gravity testing, simulating lunar-surface and Mars-surface gravity fields. For small items, this may be done in an aircraft that flies repeated parabolic cycles. Depending upon the angle of the downward part of the path, various levels of reduced gravity (including microgravity) can be simulated. However, this method only works for tests that can be conducted in a very short period of time. For tests that require a longer duration, other methods have been devised, such as using a device that applies a nearly-constant upward force to the object being tested. This is typically done with a pulley system. Other methods to simulate aspects of reduced gravity include air-bearing floors, or placing the test article in a pool of water.
- Perform endurance testing (destructive testing) to determine failure modes and weaknesses in the design. If the RoboTractor is designed to be capable of lifting a certain amount of mass for a certain number of repetitions, the test article is cycled far more often, or with a heaver load, than required. Repeating this testing until the item fails gives information about the real engineering limits of the design.
- Design the qualification unit.
- Select materials based on results of prototype testing and on the requirements for flight hardware.
- Fabricate the qualification unit.
- Test the qualification unit.

- Testing should include launch loads, response to thermal stress, electromagnetic interference (EMI), reduced-gravity testing and endurance testing.
- Design the flight unit.
- Fabricate the flight unit.

A.4 CONCLUSIONS

The first lunar explorers of the twenty-first century will be robots. They prepare the way for humans by establishing the availability and recoverability of consumables and other resources. They construct a solar power grid, and continue to add to it as humans begin arriving and more power is needed. They provide a safer way for us to learn how to "live off the land" on the Moon, so that we can be assured that resources such as rocket fuel are stored and ready for us to use. As technology matures, so will the capabilities of the robots, until they eventually begin autonomous exploratory journeys travelling towards the equatorial regions and the lunar far side.

A robotic task analysis includes a list of all possible tasks, and a metric applied to determine which tasks make sense to roboticize first, and which tasks are less useful. Tasks that are repetitive, tedious, strenuous, or dangerous are generally good candidates for robots, whereas tasks that are enjoyable and stimulating should probably be reserved for humans.

An example of near-term robotic capability is "RoboTractor", which represents a systems engineering and integration challenge, but none of its components are new. All the technologies exist that are needed to build the first RoboTractor. As we colonize the Moon, multiple copies of RoboTractor will be needed.

A.5 REFERENCES

Allen C.C.; Morris, R.V.; and McKay, D.S. (1994) Reduction of Lunar Mare Soil and Volcanic Glass. *Journal of Geophysical Research*, **99**, E11, 23173–23185.

Bates, J.R.; Lauderdale, W.W.; and Kernaghan, H. (1979) *ALSEP Termination Report*. NASA Reference Publication 1036.

Coombs, C.R.; Hawke, B.R.; and Wilson, I. (1987) Geologic and remote sensing studies of Rima Mozart. *Proc. Lunar Planet. Sci. Conf. 18th*, pp. 339–353.

Cooper, B.L. (1994) Reservoir Estimates for the Sulpicius Gallus Region. *Engineering, Construction, and Operations in Space IV*, American Society of Civil Engineers, New York.

Cooper, B.L. (2006) Craters and Channels on Malapert Mountain in the Lunar South Pole Region: Challenges Associated With High-Incidence-Angle Imagery. *Space Roundtable VIII, Golden, Colorado*.

Eppler, D. (1999) Silver Lake Test Suit Subject Report (Appendix A of Kosmo et al., 1999). Crew and Thermal Systems Division, NASA JSC, Houston, TX.

Feldman, W.C.; Maurice, S.; Lawrence, D.J.; Getenay, I.; Elphic, R.C.; Barraclough, B.L.; and Binder, A.B. (1999) Enhanced Hydrogen Abundances Near Both Lunar Poles. *Workshop on New Views of the Moon 2: Understanding the Moon through the Integration of Diverse Datasets*, Lunar and Planetary Institute, Houston, TX.

Hepp, A.; Linne, D.; Landis, G.; Wadel, M.; and Colvin, J. (1994) Production and Use of Metals and Oxygen for Lunar Propulsion. *Journal of Propulsion and Power*, **10**, No. 6, 834–840.

Hörz, F. (1985) Lava tubes – Potential shelters for habitats. *Lunar Bases and Space Activities of the 21st Century*, Lunar and Planetary Institute, Houston, TX, pp. 405–412.

Kaplan, D.; Baird, R.S.; Ratliff, J.E.; Baraona, C.R.; Jenkins, P.P.; Landis, G.A.; Scheiman, D.A.; Brinza, D.E.; Johnson, K.R.; Karlmann, P.B. *et al.* (2000) The 2001 Mars In-Situ Propellant Production Precursor (MIP) Flight Demonstration. *Proceedings of the Human Space Transportation and Exploration Workshop, February 29–March 1, 2000, Galveston, Texas.*

Kosmo, J.J.; Treviño, R.; and Ross, A. (1999) *Results and Findings of the Astronaut-Rover (ASRO) Remote Field Site Test, Silver Lake, CA*, Crew and Thermal Systems Division, NASA JSC, Houston, TX.

Landis, G.A.; Jenkins, P.P.; Baraona, C.; Wilt, D.; Krasowski, M.; and Greer, L. (1998) Dust Accumulation and Removal Technology (DART) Experiment on the Mars 2001 Surveyor Lander. *Proc. 2nd World Conference on Photovoltaic Energy Conversion, Vienna, Austria, July 1998.*

Margot, J.L.; Campbell, D.B.; Jurgens, R.F.M.; and Slade, A. (1999) Topography of the Lunar Poles from Radar Interferometry: A Survey of Cold Trap Locations. *Science*, 4 June 1999, **284**, No. 5420, 1658–1660.

McKay, D.S.; Fruland, R.M.; and Heiken, G.H. (1974) Grain size and evolution of lunar soils. *Proc. Lunar Sci. Conf. 5th*, pp. 887–906.

Olhoeft, G. (1984) Applications and limitations of ground penetrating radar, in: *Expanded Abstracts, 54th Ann. Int. Meeting and Expo. of the Soc. of Explor. Geophys., Atlanta, Georgia*, pp. 147–148.

Spudis, P.D. and Taylor, G.J. (1990) Rationale and Requirements for Lunar Exploration, in: *Proceedings of Space 90: Engineering, Construction and Operations in Space II.* American Society of Civil Engineers, New York, pp. 236–245.

Sullivan, T.A. (1994) *Catalog of Apollo Experiment Operations.* NASA-RP-1317, Johnson Space Center, Houston, TX.

Colour plates

Hubble Ultra Deep Field
Hubble Space Telescope • Advanced Camera for Surveys

NASA, ESA, S. Beckwith (STScI) and the HUDF Team   STScI-PRC04-07a

Plate 1. The Hubble Ultra Deep Field, or HUDF, is an image of a small region of space in the constellation Fornax, composited from Hubble Space Telescope data accumulated over a period of one million seconds. It is the deepest image of the universe ever taken in visible light, looking back in time more than 13 billion years.

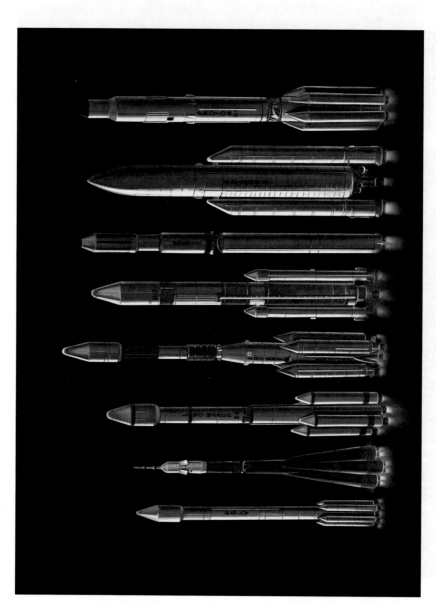

Plate 2. International fleet of heavy-lift launchers. Nations of the world are building ever-larger launchers that may be commissioned for a series of lunar missions. Depicted are the Indian PSLV, Russian Soyuz, Chinese Long March, ESA Ariane IV, Japanese H-2, Russian Zenit, ESA Ariane V, and Russian Proton Launch Vehicles (Paul DiMare).

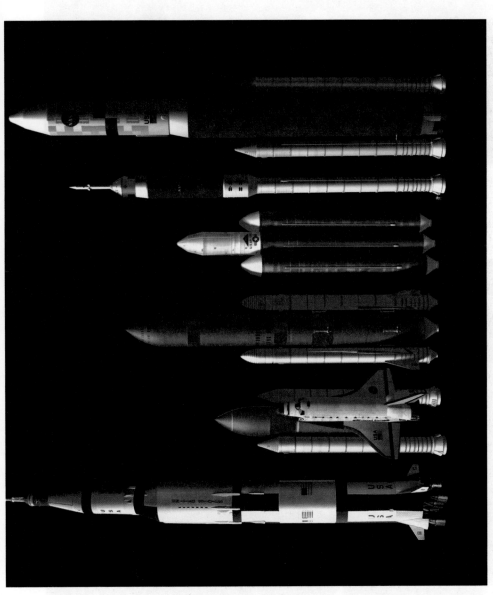

Plate 3. U.S. fleet of heavy-lift launchers capable of lofting 25–150 metric tonnes. Vehicles such as these are required to build a sustainable lunar settlement. Families of expendable and reusable launchers continue to be evolved in the U.S. The illustration depicts the decommissioned Saturn VB of the Apollo era, the Space Shuttle, a shuttle external tank-derived vehicle, the Delta IV Heavy, the Ares I and Ares V vehicles (the Atlas V Heavy is not shown) (Paul DiMare).

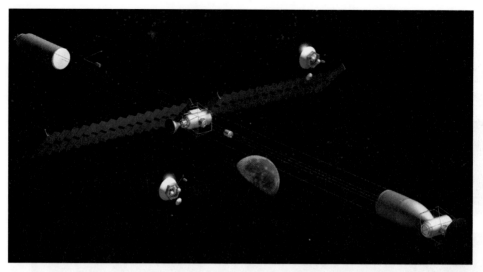

Plate 4. The Earth–Moon cycler in flight configuration. Cycling spaceships are large vehicles assembled in Earth orbit. Once commissioned, they will be set in continuous motion between the Earth and the Moon. The spinning configuration will provide partial gravity for the crew (Thangavelu/DiMare).

Plate 5. Cutaway section of cycler habitat. Cycling spaceships will offer large pressurized volumes and could evolve into interplanetary vehicles destined for multi-year voyages to Venus, Mars, Jupiter, and beyond (Thangavelu/DiMare).

Plate 6. Cargo cycler approaching the Earth. Cargo carriers containing trading goods arrive at Earth where landers descend into Earth's atmosphere using retrothrusters fueled on the Moon. Payloads are dropped off and picked up at the Earth and the Moon on each "figure-8" cycle between the Earth and Moon (Thangavelu/DiMare).

Plate 7. The International Lunar Observatory is a privately-sponsored effort to deploy and operate an observatory at the south polar region of the Moon. The Moon is an ideal platform for making astronomical observations in every segment of the electromagnetic spectrum (University of Southern California/Thangavelu)

Plate 8. Modular Assembly in Low Earth Orbit (MALEO) lander uses space-station-derived elements, assembled in low Earth orbit, to quickly create an interim lunar base. While the permanent base is being built, the crew will live and work from the MALEO base. The MALEO base will be landed using propulsion systems which will be jettisoned during descent (University of Southern California/Thangavelu).

Plate 9. The Nomad Explorer is a large pressurized multipurpose rover designed with versatility in mind. It is a long-traverse vehicle, capable of construction of lunar habitats, roads, landing pads, observatories, and other infrastructure. Using an "equipment lock" (such as that on the Japanese module of the International Space Station), delicate payloads can be transferred to and from the pressurized volume (Thangavelu/DiMare).

Plate 10. The lunar-orbiting hotel would accommodate 60 people and crew. Using shuttle-derived technology, the hotel would be serviced by a fleet of vehicles traveling between the Earth and the Moon. Vehicles dock at the bottom, and passengers would go up to their compartments. A ride on the elevator would take them further up to the observatory and lounge. The large solar array and radiator ring provide power and heat rejection for the facility (Thangavelu/DiMare).

Plate 11. Long-Range Rover with drill attachment will be used for deep drilling. Core samples retrieved from deep bore holes would help us better understand the origin of the Moon and also help locate useful minerals and volatiles that may lie deep in the lunar interior. Boreholes also offer one way to place science payloads like gamma-ray observatories and neutrino detectors deep in the lunar interior (Thangavelu/ DiMare).

Plates 12 and 13. SpaceDev, a private company based in Poway, California, is actively planning for lunar tourism. Collaborating with Bigelow Aerospace of Las Vegas, Nevada, the group intends to land inflatable habitats for tourists on the Moon. The Dreamchaser spacecraft would carry passengers and crew to Earth orbit and a separate Earth–Moon transfer vehicle would then carry them to the Moon. A lander would deliver the crew to the lunar surface where they would visit the inflatable habitats before returning to lunar orbit and back to Earth (Jim Benson/SpaceDev).

Plate 14. Lunar agriculture is an essential part of a sustainable lunar settlement. Food crops are cultivated using high-density crop growth methods including hydroponics, aeroponics, aquaponics, selective-wavelength lighting, and special nutrients. Plant growth, food production and aquaculture are part of a closed ecological life-support system (CELSS) where waste products are recycled using natural biological systems (Thangavelu/DiMare).

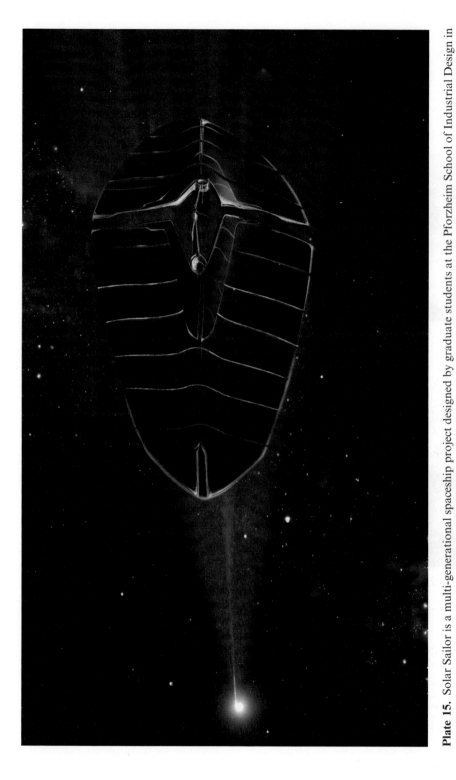

Plate 15. Solar Sailor is a multi-generational spaceship project designed by graduate students at the Pforzheim School of Industrial Design in Germany as a candidate architecture for interstellar flight. This design employs laser propulsion from the Moon to accelerate a spaceship built into a 30-km^2 light sail to velocities approaching ten percent the speed of light, needed to bridge interstellar distances in a reasonable timescale (Wallmeyer/Blasi/Thangavelu).

Plate 16. The interiors and cockpit of the Solar Sailor use advanced materials and systems including artificial intelligence and virtual reality to help the crew navigate the cosmos while feeling as close to home as possible (Wallmeyer/Blasi).

Appendix B

Lunar regolith properties

Bonnie Cooper, Ph.D.

The lunar regolith is different in important ways from terrestrial soils.[1] The lunar surface is devoid of moisture, and the weathering processes that transform landscapes on Earth do not occur. Lunar soils are susceptible to electrostatic charging to a degree that is not seen on Earth. The lack of atmosphere on the Moon permits meteorites and micrometeorites (dust-size particles) to strike the surface, creating breccias (the result of large impacts) and agglutinates (from smaller impacts), both of which are composed of smaller grains that have been welded together. The lack of free oxygen on the Moon results in the formation of minerals that are different from those found on Earth (although terrestrial mineral names are often used to describe lunar minerals). Because the weathering processes on the Moon are primarily micrometeorite and meteorite impacts, the lunar regolith also has different physical properties from terrestrial soils. In order to exploit the resources of the Moon, these differences must be understood, because chemical processes developed on the Earth, using terrestrial soils, may not work on the Moon.

The process of regolith creation has two phases. At first, a fresh rock surface is affected by every impact – both large and small – because it is directly exposed to the surface. Later on as a layer of rubble and soil slowly builds up, the rock below is shielded from the smaller impacts and is only affected by large impacts. At this point, the smaller impacts stir up ("garden") the regolith layer, so that it becomes well mixed. This again is different from terrestrial soils, in which distinct layers can be traced across distances of meters or kilometers.

The lunar regolith is approximately four to five meters thick in the mare areas, and about 10 to 15 meters thick in the (older) highlands regions. Beneath the regolith

---

[1] The term *regolith* is sometimes used for terrestrial materials, in which case it means the layer of loose rock, resting on the bedrock, that constitutes the surface of most land. It is most often used to describe ground-up rock particles that are devoid of organic life, such as fresh glacial till.

is a zone of rubble which is referred to as the megaregolith, with some blocks as much as a meter in diameter.

During the Apollo missions, we learned that the regolith can vary greatly between two locations that are only a few meters apart. Because of the complexity of the processes which create the regolith, understanding its history and development is a challenge. Numerous analytical methods are applied to this problem, including some methods that were developed specifically for lunar regolith studies.

It is likely that the top few centimeters of regolith will be the raw material that is used for lunar base construction, building roads, and resource extraction; thus, the economic importance of the lunar regolith is far greater than that of the underlying materials (Heiken *et al.*, 1991).

B.1 CHEMICAL AND MINERALOGICAL DIFFERENCES BETWEEN LU-NAR REGOLITH AND SOILS ON EARTH

Scientists often refer to the lunar regolith as "soil"; however, they are generally referring to the finer-grained portion of the unconsolidated material. Heiken *et al.* (1991) define "lunar soil" as the grains of regolith that are less than a centimeter in diameter. Regolith is a complex mixture of five basic particle types: crystalline rock fragments,[2] mineral fragments, agglutinates,[3] breccias,[4] and glasses.[5] The relative amount of each component varies from place to place.

One of the major differences between the lunar surface and the Earth's surface is that fewer than 100 different types of minerals have been found on the Moon, whereas several thousand have been found on Earth.[6]

B.1.1   Mineralogy and petrology[7]

Regolith samples from the lunar maria contain abundant basaltic[8] rock fragments; their minerals are largely pyroxene,[9] olivine,[10] and ilmenite[11] (Figure B.1).

---

[2] A rock is defined is an aggregate of one or more minerals.

[3] An agglutinate is a small (soil-sized) particle in which several other soil particles are "glued" together by a glassy groundmass.

[4] A breccia is a consolidated clod containing rock fragments. It may also contain glass.

[5] A glass is a non-crystalline solid.

[6] This probably has more to do with the lack of water on the Moon than with the limited sampling locations.

[7] Mineralogy: the study of minerals. Petrology: the study of the origin and structure of rocks.

[8] Basaltic: Basalt is a common gray to black volcanic rock. It is usually fine-grained due to rapid cooling of lava on the Earth's surface. Basaltic is used to describe rocks that have this character.

[9] Pyroxene: a mineral of the general chemical formula $ABSi_2O_6$, where A is chiefly magnesium, iron, calcium, or sodium, and B is magnesium, iron, or aluminum.

[10] Olivine: an iron/magnesium mineral series, ranging in composition from Mg_2SiO_4 to Fe_2SiO_4.

[11] Ilmenite: an iron/titanium oxide, $FeTiO_3$.

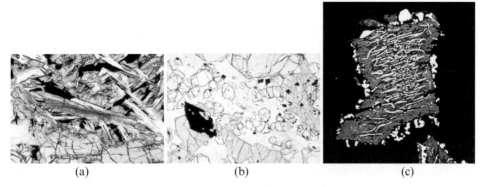

(a)   (b)   (c)

Figure B.1. Minerals found in mare regoliths. Other minerals also occur, but these are most common. (a) 100× magnification of lunar pyroxene crystals (elongated). (b) 40× magnification of lunar olivine crystals (width, length, and thickness are all similar). (c) A grain of ilmenite, after being processed to remove oxygen. White streaks are native iron that is left over from the process. Grain width is 0.1 mm.

Figure B.2: Lunar ferroan anorthosite[12] #60025 (plagioclase feldspar). Collected by Apollo 16 from the Lunar Highlands near Descartes Crater. This sample is currently on display at the National Museum of Natural History in Washington, DC.

Regolith samples from the lunar highlands areas (where Apollo 16 and Luna 20 landed) have a predominance of anorthositic[13] rock fragments and plagioclase feld-spar[14] (Figure B.2).

B.1.2 Maturity

Lunar soils change over time, due to exposure to solar wind, micrometeorite bombardment, and ionizing radiation. Mature soils have more evidence of long exposure

[12] Ferroan anorthosite refers to an anorthosite in which the pyroxene and olivine crystals are richer in iron compared to other anorthosites (that have more magnesium-rich compositions). Technically, a ferroan anorthosite is defined as one in which the atomic ratios of $Na/(Ca + Na)$ and $Fe/(Fe + Mg)$ are lower than those for the magnesian suite (or Mg-suite). However, the Mg-suite is not well-defined, and generally includes all highland rocks that are not ferroan.
[13] Anothositic: a type of rock that is composed almost completely of plagioclase feldspar.
[14] Plagioclase feldspar: a mineral that ranges in composition from $NaAlSi_3O_8$ to $CaAl_2Si_2O_8$.

times at the lunar surface, such as melting and re-agglomerating, ionizing radiation-damage, and even color change. Maturity is roughly equivalent to exposure age, although it has been challenging to develop a metric for maturity that directly correlates with age – the correlations are all indirect, and different ways of measuring maturity sometimes produce different results.

Mature (old) and immature (young) soils from the same area may have different mineralogical compositions, and mature soils are usually finer-grained than immature soils. Because different minerals have different cleavage[15] properties, some of them break apart much more readily than others, creating a fine-grained soil that has a relatively higher concentration of that mineral than did its parent rock.

The factors that contribute to differences between mature and immature soils are categorized generally as space weathering. This is a blanket term used the processes that act on any airless celestial body exposed to the space environment. Because it lacks an atmosphere, the Moon is exposed to galactic cosmic rays and solar cosmic rays; irradiation, implantation, and sputtering from solar wind particles; and bombardment by all sizes of meteorites and micrometeorites. These processes affect the physical and optical properties of the surface.

B.1.3   Agglutinates

Agglutinates consist of soil grains bonded by glass (Figure B.3). When micrometeorites strike the lunar soil, the kinetic energy melts some of the grains. The grains that are not melted, in contact with the melted ones, fuse together to form a particle that is an aggregate of the individual particles and the glass which was formed around them. They are usually less than 1 mm in diameter and contain tiny droplets of iron metal, which are sometimes so small that they cannot be seen with an optical microscope. Agglutinates contain solar wind gases, including helium and hydrogen. Agglutinates do not form on Earth, or on any other planet that has an atmosphere, because the micrometeorites that would have formed them are burned up in the atmosphere before reaching the ground.

On average, 25% to 30% of a lunar soil sample is in the form of agglutinates. In some mature soils, agglutinates are the major constituent, comprising up to 60 percent of the volume of a sample. Because the percentage of agglutinates increases over time, they are used as one measure of the maturity of a lunar soil sample.

B.1.4   Solar wind volatiles implantation

Because of the lack of magnetic field or atmosphere on the Moon, the solar wind[16] makes direct contact with the lunar surface. The ionized particles strike the grains of

---

[15] In mineralogy, cleavage refers to the way some minerals break along certain lines of weakness in their structure.

[16] Solar wind: atomic nuclei, mostly protons (H^+), approximately 4% alpha particles (He^{+2}), and trace amounts of noble gases (Kr, Xe, Ar) ejected radially outward from the solar corona. They strike the lunar surface with an energy of about 1 KeV/nucleon and penetrate to a depth of approximately 100 Å into the regolith grains.

Figure B.3. Image of a lunar agglutinate, taken with a scanning electron microscope (image courtesy of David S. McKay, NASA Johnson Space Center).

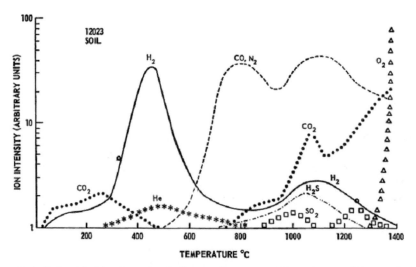

Figure B.4. Solar-wind-implanted volatiles can be extracted from lunar soil by gentle heating.

lunar regolith with enough energy to become implanted in the topmost layer of grains on the lunar surface. The predominant ions are hydrogen ($\sim$90%) and helium ($\sim$10%), with trace amounts of nitrogen, carbon, and noble gases. As we discuss in Appendix D, these implanted molecules are potential resources, because they consist of elements which are otherwise absent in lunar rocks and soils (at least the ones we have sampled so far). Moreover, they can readily be removed by gentle heating (Figure B.4).

Among the volatiles implanted by solar wind, the most abundant (and possibly the most important) is hydrogen. The amount of hydrogen varies by a factor of 600 from the highest to lowest concentration, and is correlated both with soil maturity (Fegley and Swindle, 1993) and with grain size (Carter, 1985). If

Table B.1. Hydrogen content of bulk lunar soils

Mission	# of soils	Hydrogen abundance (μg/g)			
		Minimum	Maximum	Median	Mean $\pm 1\sigma$
Apollo 11	1	38	66	46/54.2	52 ± 10
Apollo 12	5	1.9	46	30/38	31 ± 14
Apollo 14	5	35.9	106	61/67	67 ± 25
Apollo 15	14	28	120$^\beta$	52.6	57 ± 22
Apollo 16	15	19	79	36.3/39	46 ± 17
Apollo 17	14	0.2	53	22/25.7	26 ± 18
All soils$^\alpha$	54	0.2	120	45	46 ± 16

$^\alpha$ Excluding orange soil 74220, the minimum value becomes 1.9 μg/g (median and mean are unchanged).
$^\beta$ This value may include water contamination, in which case the maximum value for all lunar soils would be 106 μg/g.

water-ice is not found in the permanently-shadowed polar regions, the solar-wind-implanted elements are an alternative source for hydrogen, as well as other necessary volatiles such as nitrogen and carbon.[17]

B.1.5   Breccias

A breccia is a consolidated clod containing rock fragments, mineral fragments, glass, and, in some cases, soil grains. The surface of the Moon is continually bombarded by meteoroid impacts (although the evidence suggests that this bombardment was more intense in the distant past). Some of these impacts were large enough to heat and cement parts of the regolith back into rocks, which are referred to as *regolith breccias* – a subset of breccias in general. Radiometric age dating is used to determine the time at which these breccias were created; and some of them were formed as much as four billion years ago.

B.1.6   Glasses

Impact glasses may have been derived from either mare or highland compositions, because meteorites and micrometeorites are indiscriminate about what types of rocks they impact. Glasses of volcanic origin are typically associated with the mare areas. Three to five percent of all lunar soil consist of impact glasses that do not have any inclusions, and are not part of an agglutinate. One interesting group in this category is the high-alumina, silica-poor (HASP) glasses. These have been found on several areas of the Moon. HASP compositions are thought to be generated by impact-

---

[17] Nitrogen can be used to create an Earth-like atmosphere (Earth's atmosphere is predominantly nitrogen, with a lesser amount of oxygen). Carbon can be used to create methane, which would be useful as a propellant.

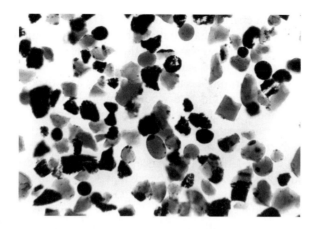

Figure B.5. Orange and black glass spheres from the Apollo 17 landing site. This type of lunar soil is more uniform in size than any other type that has been sampled to date.

induced volatilization processes on the Moon. The high Al/Si ratio of HASP is believed to be the result of preferential loss of SiO_2 relative to Al_2O_3. It has been estimated that 44% of the original SiO_2 and 35% of the FeO could be lost during the impact-induced volatilization which creates HASP glasses. It is unlikely that all of the SiO_2 and FeO would have escaped from lunar orbit; so it is expected that these volatile components became enriched in other lunar surface materials.

Some lunar glass particles are ropy in appearance. They contain fine dusty soil grains and have a wide variety of compositions. Also, some minerals are melted by impacts and form glasses that have the same composition as the mineral grain from which they formed. The resulting grains will have zones with varying degrees of alteration and are referred to as shocked minerals.

Volcanic glass particles are recognized by their uniform chemical composition, and the presence of a surface coating of condensed gases that were produced by the volcano. They also lack any evidence of chemical alteration that would have been caused by an impacting meteoroid. Typical of this group are the green and orange spheres which are thought to have come from volcanic fire fountain eruptions. These glasses form layers near the edges of the maria (Figure B.5).

B.1.7   KREEP (K, Rare Earth Elements, and P)

KREEP is an acronym that stands for potassium (atomic symbol K), rare earth elements (REE), and phosphorus (P), and is a component that is found in some lunar rocks. In addition to the potassium, KREEP contains other heat-producing elements – uranium and thorium. KREEP is probably a result of the Moon's origin, which is thought to be the result of a collision between the proto-Earth and a Mars-sized object that impacted the Earth 4.5 billion years ago. This impact put a large amount of material into Earth orbit that ultimately re-accreted to form the Moon.

Given the large amount of energy that was liberated in this event, it is assumed that a large portion of the Moon would have initially been molten, forming a near-global magma ocean. After crystallization was about three-quarters complete, anorthositic plagioclase would have begun to crystallize, and because of its low density, it would float to the top of the melt, forming a crust. Elements that are not compatible with the crystalline structure of this plagioclase would remain in the melt, which would become progressively more enriched in these incompatible elements. KREEP-rich material would have become sandwiched between the crust and mantle.

Before the Lunar Prospector mission, it was thought that KREEP-rich materials would have formed a near-global layer beneath the crust. However, results from Prospector's gamma ray spectrometer showed that KREEP-containing rocks are primarily concentrated within the region of Oceanus Procellarum and Mare Imbrium, a unique geological province that is now known as the Procellarum KREEP Terrane. This concentration of heat-producing elements is probably responsible for the longevity and intensity of volcanism on the nearside of the Moon.

B.2 PHYSICAL PROPERTIES OF THE LUNAR REGOLITH

The regolith covers almost the entire surface of the Moon, with the exception of very steep-sided features such as crater walls, scarps, and the sides of rilles. This again is different from the terrestrial environment, where outcrops of rock are frequently seen along highways, hiking trails, and rivers. On Earth, wind and water move soil particles from one place to another, exposing solid rock. On the Moon there is no similar process to expose fresh rock surfaces.

The meteorites that strike the lunar surface range in size from 1,000 km to less than 1 μm, but the effects of all these impacts are similar: a crater is excavated, and the target material is shattered, pulverized, melted, mixed and dispersed around the impact site. Figure B.6 shows examples of a medium-sized crater on the Moon, and also an example of one of the tiniest craters, referred to as a microcrater.[18] The average speed of impacting meteoroids is approximately 1.7 km/second. The kinetic energy imparted by these strikes breaks up surface layers, then creates new shapes and types of regolith particles. Secondary craters are also formed, as a result of fragments ejected from the impact site which follow ballistic trajectories and create additional, smaller craters when they land.

Information about the physical nature of the lunar regolith has been obtained from several sources, including direct measurements on the Moon itself during the Apollo missions (e.g., Costes *et al.*, 1971). For example, the penetration depth of the Lunar Modules' landing gear gave us information about the compressibility of the soil. Other Apollo experiments provided data on the electrical properties and heat flow of the regolith.

---

[18] Note that microcraters that are less than about 5 micrometers in diameter generally do not have raised rims or glassy pit linings. Microcraters were formerly referred to as zap pits, but this term has fallen out of usage.

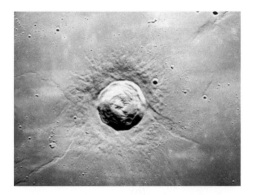 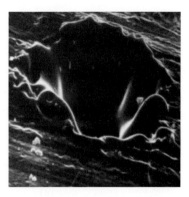

Figure B.6. Medium and small craters. Left: The crater Lambert is 32 kilometers (20 miles) in diameter, and is surrounded by a blanket of material blasted out by the impact that produced it. Right: A microcrater is a tiny crater on a grain of lunar soil, 50 μm (2/1000 inch) in diameter. It has a raised rim of glassy material caused by the impact.

Several experiments were performed specifically to study soil mechanics. These include use of penetrometers, which are rods that measure the force required to penetrate to various depths in the soil. Also, several small trenches were also excavated to study the soil properties along the trench walls. Finally, studies were performed on samples returned to Earth. For example, analysis of core tubes allows properties such as density, average grain size, strength, and compressibility to be measured as a function of depth.

Crew mobility, both on foot and in the Lunar Rover, was affected more by local topography such as craters and ridges than by soil properties. When astronauts inserted sampling tubes into the soil, they typically found that penetration was easy for the first 10 to 20 centimeters and increasingly difficult below that depth. The deepest penetration achieved on a hand-driven core tube was 70 centimeters, which required about fifty blows with a hammer. For sampling at greater depths, the Apollo 15, 16, and 17 crews used a battery-powered drill. This allowed sampling to depths of 1.5 to 3 meters, which was achieved easily on Apollo 16 but with much more difficulty on Apollo 15 and 17.

B.2.1   Geotechnical properties

Geotechnical properties are those physical properties of the lunar regolith and the lunar surface that affect mobility, construction, and general operations. These properties are important to understand if one wishes to extract resources. Even simple uses such as piling the regolith into a berm to provide a blast shield around a landing site, or grading a road, require an understanding of the physical properties of the soil. Grain-size properties are described in the terminology of statistics, and the shapes of the individual grains are expressed in terms of geometry.

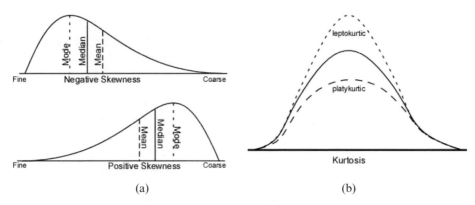

Figure B.7. (a) Illustrations of skewness, mean, and median. (b) Kurtosis.

Table B.2. Comparison of ϕ and micrometer measurements.

ϕ scale	Corresponding micrometer (micron) value
1	0.5
2	0.25
3	0.125
4	0.0625
5	0.03125
6	0.015625
7	0.0078125

B.2.2   Grain size distribution terminology

There are five measurements that are used by Graf (1993) when describing the grain size distribution of lunar soil samples. These are mean, median, mode, sorting, skewness, and kurtosis. Figure B.7a shows examples of mean, median, and skewness; Figure B.7b illustrates kurtosis. Table B.2 illustrates the relationship between phi and micrometer measurements.

Geologists sometimes use the symbol ϕ for measuring particle size, which is a logarithmic mean of the particle diameter for each size grouping. It is obtained by calculating, for each size category:

$$\phi = -\log_2 d,$$

where d is the diameter in millimeters. Figure B.9, which is a graphical description of the particle size distribution of lunar soil "78221,8", shows a ϕ scale along the bottom and its corresponding diameter scale along the top. A larger value for ϕ denotes a smaller grain size.

B.2.2.1   Mean

The mean is the sum of all the individual grain sizes in the sample, divided by the number of grains. It can be thought of as the center of gravity of the size distribution.

B.2.2.2   Median

The median is a number dividing the higher half of a sample from the lower half. The median can be found by arranging all the grain sizes from lowest value to highest value and picking the middle one.[19] The median is the value where 50 percent of the measured values are larger and 50 percent are smaller. The mean is primarily used for skewed distributions, which it represents differently than the arithmetic mean. Consider the set {1, 2, 2, 2, 3, 9}. The median is 2 in this case, as is the mode, and either of these might be seen as a better indication of central tendency than the arithmetic mean of 3.166.

B.2.2.3   Mode

The mode is the most frequent value of grain size occurring in the sample. It is the midpoint of the most abundant size class, or the number that occurs most frequently in a set of numbers. For example, in Figure B.11 the mode is the range between $\phi = 5$ (32 µm) and $\phi = 6$ (16 µm).

Of the three, the mode is the most variable and easily-manipulated term. Changing the increment of the step size will change the value of the mode. To keep internal consistency in his data set, Graf (1993) always uses a step size of 1ϕ for histogram graphs and for finding the mode. This limits the allowable values of the mode to 2.5ϕ, 3.5ϕ, and so on.

B.2.2.4   Sorting

Sorting means the degree of similarity of the sizes of the grains or particles in a specific sample. If there is a wide variation of sizes, the sample is said to be poorly-sorted; whereas if all the grains are nearly the same size, the sample is said to be well-sorted.

---

[19] The difference between the median and mean is illustrated in this example. Suppose 19 paupers and 1 billionaire are in a room. Everyone removes all money from their pockets and puts it on a table. Each pauper puts $5 on the table; the billionaire puts $1 billion (that is, 10^9) there. The total is then $1,000,000,095. If that money is divided equally among the 20 persons, each gets $50,000,004.75. That amount is the mean (or "average") amount of money that the 20 persons brought into the room. But the median amount is $5, since one may divide the group into two groups of 10 persons each, and say that everyone in the first group brought in no more than $5, and each person in the second group brought in no less than $5. In a sense, the median is the amount that the typical person brought in. By contrast, the mean (or "average") is not at all typical, since no one present – pauper or billionaire – brought in an amount approximating $50,000,004.75 (Wikipedia).

Sorting can be thought of as the standard deviation of the size distribution. In the strictest sense, we cannot use the term "standard deviation" if the size distribution has a ϕ scale (a logarithmic scale). Nevertheless, both sorting and standard deviation are measurements that describe the shape of a distribution. A nearly-perfect sorting with all grains of similar size would have a sorting parameter of less than 1ϕ. As the size distribution becomes more spread out, the sorting parameter becomes larger.

B.2.2.5   *Skewness*

Skewness is the degree of asymmetry of a distribution. It can be described with the diagram in Figure B.7. If the median is on the coarse side of the mean, the central trend is coarse, the extended tail is fine, and the distribution is said to be positively skewed. Similarly, if the median (central trend) is finer than the mean, the distribution is said to be negatively skewed.

B.2.2.6   *Kurtosis*

Kurtosis expresses the sharpness of the peak of a frequency distribution, and is usually described relative to a normal distribution. A distribution having a relatively-high peak is called "leptokurtic", while a flat-topped curve is called "platykurtic" (Figure B.7b).

B.2.3   **Particle shapes**

Particle shapes are described by terms such as elongation, aspect ratio, roundness, sphericity, volume coefficient, and specific surface area.

B.2.3.1   *Elongation*

Elongation is defined as the ratio of the major axis to the intermediate axis of the particle, or length to width. Particles with elongation values less than 1.3 are classified as equant, and those with values greater than 1.3 are classified as elongate.

B.2.3.2   *Aspect ratio*

Aspect ratio is inversely related to elongation. It is the ratio of the minor axis to the major axis of an ellipse fitted to the particle by a least-squares approximation.[20]

B.2.4.3   *Roundness*

Roundness is the ratio of the average of the radii of the corners of the particle image to the radius of the maximum inscribed circle (Figure B.8). A perfectly-rounded particle has a roundness value of 1.0. The term should not be confused with

---

[20] Least squares approximation: a statistical method of determining the trend of a group of data when the trend can be represented on a graph by a straight line.

Roundness

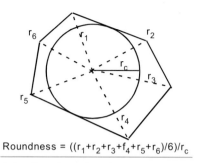

Roundness = $((r_1+r_2+r_3+f_4+r_5+r_6)/6)/r_c$

Figure B.8. Roundness is defined as the ratio of the average of the radii of the corners of the particle image, to the radius of the maximum inscribed circle.

sphericity: a nearly-spherical particle may have sharp corners and be angular, while a flat pebble may be well-rounded.

B.2.3.4 Sphericity

The sphericity, Ψ, of a particle is the ratio of the surface area of a sphere (with the same volume as the given particle) to the surface area of the particle:

$$\Psi = (\pi^{1/3}(6V_p)^{2/3})/A_p$$

where V_p is the volume of the particle and A_p is the surface area of the particle. A sphere has a sphericity of 1.0, a cube has a sphericity of 0.806, and a tetrahedron has a sphericity of 0.671. Referring to Figure B.3, one sees that the sphericity of this agglutinate particle would be a very small number; while the sphericity of a pyroclastic glass bead (Figure B-5) would be very close to 1.0.

Other important geotechnical properties include specific gravity, bulk density and porosity, relative density, compressibility, shear strength, permeability and diffusivity, bearing capacity, slope stability, specific surface area, and trafficability. The reader is advised to consult the previously-mentioned references, especially the *Lunar Sourcebook* (Heiken *et al.*, eds., 1991), for additional information on these.

B.2.4 Lunar soil grain shapes and sizes

Now that the terminology has been explained, we can discuss the lunar soil grains with less ambiguity. The shapes of individual lunar soil grains are highly variable, ranging from spherical to extremely angular. Because they are usually elongated, the grains tend to pack together with a preferred orientation of the long axes. Consequently, we can expect that some of the physical and geotechnical properties of the lunar surface will be anisotropic.[21] For example, the thermal conductivity in the

---

[21] Anisotropic: having some physical property that varies with direction.

horizontal direction may be different from the thermal conductivity in the vertical direction. Furthermore, many of the particles are not compact, but have irregular, often reentrant[22] surfaces. These surface irregularities affect the compressibility and shear strength of the soil.

At most locations on the Moon, continuous meteoroid impact has converted the surface into a poorly sorted[23] layer of grains and pebbles that has reached a "steady state" condition with regard to thickness, particle size, and other properties. The limited number of particle types, the absence of a lunar atmosphere, and the lack of water and organic material in the regolith constrain the physical properties of most of the lunar surface to a relatively-narrow range – smaller than the corresponding ranges for terrestrial materials. However, within a given sample of lunar soil, particle size and shape distributions vary much more than any single sample of terrestrial soil.

Particle size distribution indicates, to some extent, the strength and compressibility of any soil, and it also affects the soil's thermal properties. Particle size distribution is partially determined by the amount of meteorite bombardment and the evolution of the soil (these factors contribute to another type of measurement, soil maturity, described previously). The average size of lunar regolith grains is less than 250 μm. However, for any given site, both the grain size of the finest-grained portion and the mean grain size of all soils collected are different from those of other sites. This difference may be related to differences in the overall thickness of the regolith at each of these sites.

The majority of lunar soil samples fall in a fairly-narrow range of particle size distributions. They are generally poorly sorted, with an average particle size between 40 μm and 100 μm. About half of the soil by weight is finer than the human eye can resolve. Roughly 10% to 20% of the soil is finer than 20 μm, and a thin layer of dust adheres electrostatically to everything that comes in contact with the soil. The coarsest samples are the most poorly sorted, whereas soil samples with smaller average grain sizes are better sorted. The pyroclastic glasses (such as the sample shown in Figure B.5) are finer-grained and better sorted than any of the other lunar soils that have been sampled.

Soils are often described graphically by using a cumulative percent probability scale, which is the ordinate (*y*-axis) in Figure B.9. It shows, for example, that 10 percent of the sample is larger than 1.7ϕ (308 μm), 50 percent of the sample is larger than 4.5ϕ (44 μm), and 95 percent of the sample is larger than 7ϕ (8 μm).

In practice, a size class is really a size interval. Figures B.10 and B.11 show two other popular methods for describing soil size distributions. Figure B-10 shows a cumulative weight distribution curve for the same soil sample that was described by Figure B.9, and reading this graph shows the same result – 10 percent of the

---

[22] Reentrant: re-entering or directed inward; for example, a bay on a coastline or a bite out of a sandwich.

[23] Sorting means the degree of size-similarity in a sample of regolith or soil. Well-sorted means that the grains are fairly uniform in size; poorly-sorted means that there is a lot of variation in the sizes of individual grains.

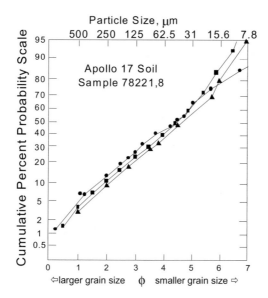

Figure B.9. An example of how grain size distributions are graphically displayed.

sample is larger than 1.7ϕ, and so on. The difference is that Figure B.10 has a linear ordinate[24] scale, rather than the semi-logarithmic[25] scale of Figure B.9, and Figure B.10 also shows more information about the extreme values for grain sizes in this sample. Figure B.10 includes information about sizes smaller than 7ϕ and larger than 0ϕ, whereas Figure B.9 does not.

Figure B.11 shows a size distribution histogram[26] for the same sample. It represents the calculated distribution of various grain sizes within the sample. From this we see that the largest percentage of grains fall in the size range from 6ϕ to 5ϕ (16 μm to 31 μm). Figures B.10 and B.11 are from Graf (1993). His book is a compilation of lunar soil size data, presented in a consistent format so that soil size distributions can be readily compared with each other. Each soil is described graphically with a cumulative weight distribution curve and a size distribution histogram, and information about soil maturity, sorting, and so on, is also given.

The most complete reference volumes for lunar soil data are the *Handbook of Lunar Soils*[27] (Morris *et al.*, 1983), and the *Lunar Soils Grain Size Catalog* (Graf, 1993). Both are in database format, so it is a straightforward task to compare

[24] Ordinate: the vertical coordinate of a point on a graph, or the vertical axis of a graph.
[25] Semi-logarithmic: having a logarithmic scale or coordinates on one axis and a standard scale on the other.
[26] Histogram: a graph of frequency distribution in which vertical rectangles or columns are constructed with the width of each rectangle being a class interval and the height corresponding to the frequency in that class interval.
[27] This two-volume set lists literature references for each soil as well as its chemical composition and some sample documentation.

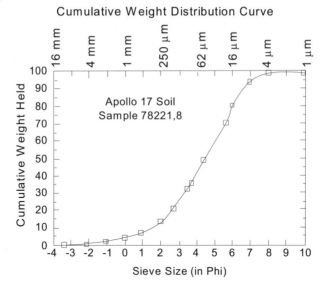

Figure B.10. Cumulative weight distribution curve of a lunar soil sample (redrawn from Graf, 1993).

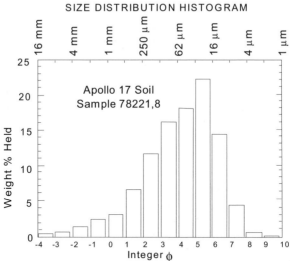

Figure B.11. Size distribution histogram of a lunar soil sample (redrawn from Graf, 1993).

one soil with another, as long as both soils have been examined. In some cases, the individual specimens have received only a cursory examination. Consequently, not all the entries give information about soil components, chemistry, or grain size.

The two books are more useful together. Whereas Morris's work gives comprehensive literature references for each soil that has been studied, Graf's book gives in-depth size distribution information, and the graphs are all plotted with the same scale. Furthermore, of the 203 soil samples listed in either of these works, there are 12 samples that were not included by Graf, but for which Morris gave some general grain size information. There were 34 samples for which grain size information was given by Graf, but no grain size information was given by Morris. Neither of these books provide coverage of all the lunar samples, because there are soil samples in the Apollo collection that have not yet been studied.

B.3   DUST

Lunar dust is the fraction of the lunar regolith that is less than 20 µm in diameter, and it makes up some 10–20 wt% of the soil (Liu *et al.*, 2006). Apollo 17 soil 78221,8 may contain up to 28 percent dust (Figure B.10). During the Apollo missions, we learned that lunar dust was a serious problem. It clings electrostatically to surfaces, and was difficult (impossible in some cases) to remove. The dust would cling to the astronauts' helmet visors, obscuring vision. Attempts to wipe the dust off the visors resulted in scratches that also obscured vision. The dark dust grains absorb sunlight, and equipment that became dust-coated sometimes became excessively hot. This resulted in a steady increase in temperature which sometimes resulted in false readings in instruments. The Lunar Rovers had less traction than was expected, and this was attributed to the fine lunar dust. The dust also affected seals on the spacesuits, and because of this the suits began to lose their ability to maintain pressure within safe limits. If the missions had lasted longer than they did, it is very likely that no additional excursions on the lunar surface would have been possible because of this safety issue. Finally, because the dust clings so tenaciously to the spacesuit fabric, it was impossible to reenter the lunar lander without bringing some of the dust inside. Astronauts reported that it smelled like gunpowder, and one astronaut had allergy-like symptoms during the return trip to Earth. Although simple measures were sufficient to mitigate some of the problems (such as loss of traction) they were ineffective in resolving many of the more serious problems (such as clogging, abrasion, and diminished heat rejection). The severity of the dust problems was consistently underestimated by ground tests, indicating a need to develop better simulation facilities and procedures (Gaier, 2005).

During the Apollo landings, the impact of rocket exhaust with the surface produced dust clouds. On some missions, dust became visible 30 to 50 meters above the surface, and during the final 10 to 20 meters of descent the surface was largely obscured by the dust cloud. On other missions, the dust cloud was not as dense and the surface remained clearly visible throughout the landing.

B.3.1   Electrostatic charging

The lunar environment causes dust to have some surprising qualities. The lunar regolith is completely devoid of moisture at all of the sites that have been sampled as of this writing (Apollo and Luna landing sites). And, because there is effectively no

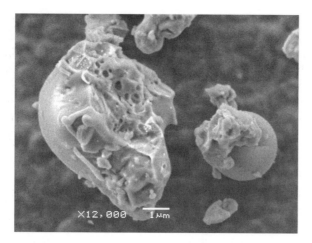

Figure B.12. Scanning electron microscope image of grains of lunar dust (courtesy of Y. Liu, University of Tennessee – Liu *et al.*, 2006).

atmosphere on the Moon, galactic cosmic rays, solar ultraviolet radiation, and solar wind particles reach the surface and interact with the regolith. The lunar regolith develops a negative electrical charge on the night side, and a positive electrical charge on the daylight side (Stubbs *et al.*, 2005). As the terminator (the line between day and night) moves across the lunar surface, dust is levitated and transported to higher altitudes.

Several of the Surveyor spacecraft returned photographs showing an unmistakable twilight glow low over the lunar horizon that persisted for about an hour after the Sun had set (Figure B.13a). An Apollo surface experiment, the Lunar Ejecta And Meteoroids experiment (LEAM) detected an order-of-magnitude increase in particles striking the detectors during local sunrise and sunset (Figure B.13b). Most of the events registered by this experiment were found to have been caused by electrostatically-charged and transported lunar dust. A pronounced increase in the dust impact rate occurred around the sunset and sunrise terminator crossings. The increase began about 150 hours before sunrise, while the Sun was still 70 degrees below the horizon.

Lunar horizon glow (LHG) was also observed by the Apollo crews. Apollo 17 astronauts orbiting the Moon drew sketches (Figure B.14) of what they referred to as "bands", "streamers", or "twilight rays" which were seen for about 10 seconds before lunar sunrise or after lunar sunset. This phenomenon was also reported (but not sketched) by the Apollo 8, 10, and 15 astronauts.

B.3.2   Dust mitigation

There are new technologies available that may reduce the problems associated with dust, and work is ongoing to adapt these technologies to the problems of lunar

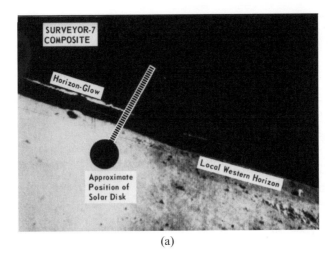

(a)

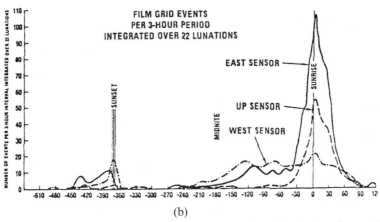

(b)

Figure B.13. (a) Composite image from Surveyor 7. The horizon glow is superimposed on an image of the sunlit surface of the Moon at the Surveyor site. (b) Data from the Apollo 17 Lunar Ejecta And Meteoroids (LEAM) Experiment, showing the increase in events at local sunrise and sunset.

exploration. Machines with moving parts will be designed with wider tolerances, so that a grain of dust that becomes lodged in a moving part will be less likely to result in breakdown. Wherever possible, rotating parts will be enclosed so that they are not directly exposed to dust. New seals and connectors are being developed that will be more robust in the lunar environment.

The innovative ideas that are being developed depend in part on using some of the characteristics of lunar dust in our favor. Because the dust is electrostatically charged, it can be moved (or removed) using devices that contain electromagnetic fields. This may be useful for creating an electromagnetic "curtain" inside an airlock

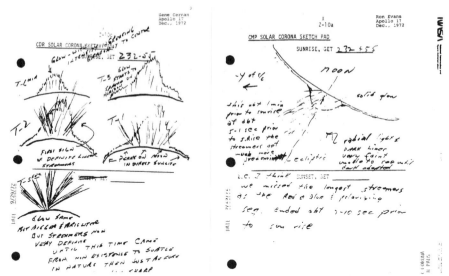

Figure B.14. Sketches made by astronaut Gene Cernan (*left*) and Ron Evans (*right*) while in lunar orbit as the space capsule (on the nighttime side) was approaching the sunlit side.

that cleans the spacesuits before they are removed by the crew. An electromagnetic screen could remove dust build-up from windows or helmet visors (Immer *et al.*, 2006). Lunar dust is also magnetic, and magnetic filters have been suggested as a means to remove dust that does find its way inside a habitation or vehicle.

Material selection will play a role in mitigating the problems caused by lunar dust. Parts can be made of Teflon or other materials that do not have a tendency to attract dust. Finally, operational controls will play a factor in reducing this problem. One example of this is having an airlock on the lunar-landing module (this was not done on the Apollo missions). An alternative to an airlock is a "suit lock". In this concept, the spacesuit would be attached to the outside of the habitat, and it would be possible to "enter" the back of the suit from inside the habitat, seal up the suit (and the hatch) and then detach the suit for the surface excursion.

There is also a new idea for how to make the area immediately surrounding the habitat less dusty: the "lunar lawnmower" (Figure B.15), which sinters the frequently-traversed areas near habitats so that the local dust is controlled. Sintering can be done with a microwave system, because tiny beads of native iron are abundant in lunar dust and soil, and they concentrate the microwave energy. Dr. Larry Taylor and Dr. Thomas Meek, both of the University of Tennessee, experimented with this idea by placing a small amount of lunar soil brought back by the Apollo astronauts into a microwave oven. They found that it melted within 30 seconds at only 250 watts (Taylor and Meek, 2003). This property of lunar soil could also be used to sinter roadbeds or areas where telescopes are to be emplaced.

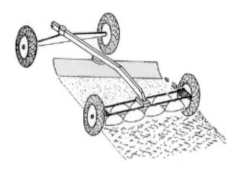

Figure B.15. The "lunar lawnmower" concept, proposed by Dr. L.A. Taylor of the University of Tennessee.

B.4   SUMMARY

The long history of micrometeorite bombardment on the airless Moon has resulted in the development of a layer of powdery lunar soil or regolith. Although the terms "lunar soil" and "regolith" are often used interchangeably, "soil" refers to the finer-grained portion (with grains less than 1 cm in diameter), and "dust" refers to the finest fraction – grains less than 20 μm in diameter.

Lunar regolith is different from terrestrial soils, not only in size and shape, but also in chemistry. Fewer than 100 different types of minerals have been found on the Moon, whereas several thousand have been found on Earth. Also, the minerals that comprise lunar rocks are also in a chemically-reduced state compared with many terrestrial rocks. Some of the properties of lunar soil, such as its magnetic susceptibility, have been impossible to duplicate so far. However, the need to test processes and operations before we return to the Moon has resulted in continuing research into how to create "imitation lunar soils", also known as simulants. Simulants will be discussed in Appendix C.

B.5   REFERENCES

Carrier, W.D. III (1994) *Trafficability of Lunar Microrovers* (Part 1). LGI-TR94-02, Lunar Geotechnical Institute, Lakeland, FL.

Carter, J.L. (1985) Lunar Regolith Fines: A Source of Hydrogen. In: W. Mendell, ed., *Lunar Bases and Space Activities of the 21st Century*, Lunar and Planetary Institute, Houston, TX.

Costes, N.C. *et al.*, (1971) Cone penetration resistance test – an approach to evaluating in-place strength and packing characteristics of lunar soils. *Proc. Second Lunar Sci. Conf.*, Vol. 3, MIT Press.

Dukhovskoy, Ye.A.; Motovilov, E.A.; Silin, A.A.; Smorodinov, M.I.; and Shvarev, V.V. (1992) *The Investigation of Frictional Properties of Lunar Soil and Its Analogs* (originally published in Russian in *Lunar Soil from the Sea of Fertility*, Nauka Science Publishing House, Moscow, 1974). LGI-TR92-03, Lunar Geotechnical Institute, Lakeland, FL.

Fegley, B. Jr. and Swindle, T. (1993) Lunar Volatiles: Implications for Lunar Resource Utilization. In: J. Lewis, M.S. Matthews, and M. L. Guerrieri, eds., *Resources of Near-Earth Space*, University of Arizona Press, pp. 367–426.

Gaier, J. (2005) *The Effects of Lunar Dust on EVA Systems during the Apollo Missions*. NASA Glenn Research Center Technical Memorandum TM-2005-213610, 72 pp.

Graf, J.C. (1993) *Lunar Soils Grain Size Catalogue*, NASA Reference Publication 1265, NASA Johnson Space Center, Houston, TX and Lunar Geotechnical Institute, Lakeland, FL.

Gromov, V.V. (1991) *Physical and Mechanical Properties of Lunar and Planetary Soils*. LGI-TR91-03, Lunar Geotechnical Institute, Lakeland, FL.

Heiken, G.H.; Vaniman, D.T.; French, B.M.; and Schmitt, H.H., eds. (1991) *Lunar Source-book: A User's Guide to the Moon*. Cambridge University Press and Lunar and Planetary Institute, Houston, TX, 736 pp.

Immer, C.; Starnes J.; Michalenko, M.; Calle, C.I.; and Mazumder, M.K. (2006) Electrostatic Screen for Transport of Martian and Lunar Regolith [abstract]. *37th Annual Lunar and Planetary Science Conference*, Lunar and Planetary Institute, Houston, TX.

Liu, Y.; Park, J.; Hill, E.; Kihm, K.D.; and Taylor L.A. (2006) Characterization of lunar dust for toxicological studies: Part 2 – morphology and physical characteristics. *Journal of Aerospace Engineering* (submitted).

Lunar Geotechnical Institute (1994) *Bibliography of Lunar Geotechnical Literature*, LGI-TR94-01.

Morris, R.V.; Score, R.; Dardano, C.; and Heiken, G. (1983a) *Handbook of Lunar Soils Part I: Apollo 11–15*. Planetary Materials Branch Publication 67, JSC 19069, NASA Johnson Space Center, Houston, TX.

Morris, R.V.; Score, R.; Dardano, C.; and Heiken, G. (1983b) *Handbook of Lunar Soils Part II: Apollo 16–17*. Planetary Materials Branch Publication 67, JSC 19069, NASA Johnson Space Center, Houston, TX.

Morrison, D.A. (1992) *Lunar Engineering Models: General and Site-Specific Data*, NASA Johnson Space Center, Houston, TX.

Stubbs, T.J.; Halekas, J.S.; Farrell, W.M.; and Vondrak, R.R. (2005) Lunar Surface Charging: A Global Perspective Using Lunar Prospector Data [abstract]. *Workshop on Dust in Planetary Systems 2005*, Lunar and Planetary Institute, Houston, TX.

Taylor, L.A., and Eimer, B. (2006) Lunar Regolith, Soil, and Dust Mover on the Moon [Abstract #1023]. Space Resources Roundtable VIII, Lunar and Planetary Institute, Houston.

Taylor, L.A. and James, J.T. (2006) Potential Toxicology of Lunar Dust [Abstract #1008]. *Space Resources Roundtable VIII*, Lunar and Planetary Institute, Houston, TX.

Taylor, L.A. and Meek, T.T. (2005) Microwave Sintering of Lunar Soil: Properties, Theory, and Practice. *Journal of Aerospace Engineering*, **18**, no. 3, 188–196.

Appendix C

Lunar soil simulants

Bonnie Cooper, Ph.D.

C.1 WHAT IS A SIMULANT?

A lunar simulant is a substance (usually granular) that mimics some property of the materials found on the lunar surface. It may be manufactured from natural or synthetic components. As described in Appendix B, lunar regolith is different from terrestrial rocks and soils in important ways. Simulants of the regolith are needed for testing rovers, airlocks, and other components that may come into contact with the lunar soil. They are also needed for testing various resource utilization concepts. The lunar simulants that were developed prior to the return of lunar samples were not entirely successful. Simulated lunar dust, for example, was used to test space suits so that they could be resistant to abrasion. However, mitigations that were developed to prevent damage to spacesuit seals, equipment, and experiments were found to be inadequate.

As we look forward to the next generation of robots and humans on the Moon, we are faced with budgetary concerns as well as the need to create a sustainable program of development. It has become more important than ever to test our systems and materials in a realistic way, to prevent expensive and time-consuming mistakes. Various types of lunar simulants will be needed that match the properties of the lunar material as closely as possible.

C.2 TYPES OF SIMULANTS

Simulants are needed for two general categories of research. One of these is engineering (designing robotic rovers, building roads, digging trenches, drilling holes, site preparation, and so on). Another need is for simulants than can be used to test resource extraction and processing schemes. Other proposals, such as building solar power cells using lunar regolith for both substrate and components, may require a simulant that meets both civil engineering and resource extraction requirements.

C.2.1   Simulants for consumables extraction experiments

Prior to 1985, lunar simulants were created by individual experimenters for specific purposes (e.g., Chang *et al.*, 1973, Pilcher *et al.*, 1972). The amounts of simulant produced were small – only enough for the use of the individual researcher. The first simulant that was produced for general use (for both NASA and other experimenters) was Minnesota Lunar Simulant-1 (MLS-1). Produced by Dr. Paul Weiblen of the University of Minnesota, it became available in 1988. MLS-1 mimics the chemistry and detailed grain properties of typical lunar soils, but does not adequately simulate the glass component. MLS-2, produced only in small quantities on an experimental basis, is a highlands simulant containing more silica and less titanium oxide than MLS-1.

To compare MLS-1 with lunar material, Allen *et al.* (1994) tested reduction of MLS-1, lunar orange soil, and lunar basalt samples (both from the Apollo 17 collection). The lunar basalt reaction was rapid, with major evolution of water occurring within minutes after the introduction of hydrogen. The reduction of MLS-1 was not quite as rapid as was the reaction with the lunar samples. However, the MLS-1 is similar enough to lunar material that we can have confidence in the previous experiments that used it. The main difference between lunar material and MLS-1 is that the MLS-1 has some magnetite in it, which results in a slightly higher overall oxygen yield when it is used for hydrogen reduction experiments (explained in Appendix E).

Some early work was done to produce glasses that match the chemistry of those found in lunar soils (McKay and Blacic, 1991). Figure C.1 shows the experiment that was used for these experiments. Ten to thirty percent of the input feedstock could be converted to glassy material. The glass that was produced contained iron blebs that are analogous to those found in lunar glasses.

In addition to simulating the chemistry and mineralogy of lunar soils, it is useful to simulate other properties. For some types of resource extraction experiments, one may also wish to simulate the occurrence of single-domain iron[1] blebs,[2] anhydrous minerals,[3] radiation damage, volatiles implantation,[4] and agglutinates.[5] These properties will be more challenging to duplicate.

---

[1] An occurrence of iron that has a single magnetic domain. In lunar regolith particles, single-domain iron is descriptive of the tiny metallic iron grains dispersed throughout the glassy material produced by micrometeorite impacts in the presence of solar wind protons.
[2] Bleb: borrowed perhaps from medical science; it refers to an object which is bubble- or blob-shaped.
[3] Anhydrous: devoid of water molecules in the mineral composition; devoid of moisture in any form.
[4] Because of the lack of magnetic field or atmosphere on the Moon, the solar wind has implanted ions of hydrogen, helium, carbon, nitrogen, and other elements into the outer surfaces of grains on the topmost layer of the regolith.
[5] Agglutinate: a fragile, irregularly-shaped particle composed of rock, mineral, and glass fragments welded together by glass splashes from micrometeorite impacts. Usually less than 1–2 mm diameter. They contain implanted solar wind gases such as hydrogen. See Appendix B for additional information.

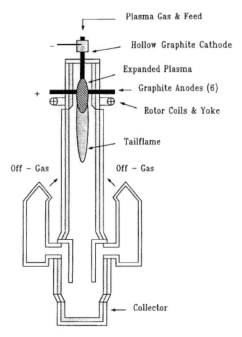

Figure C.1. In-flight sustained shockwave plasma.

C.2.2   Simulants for civil engineering experiments

JSC-1: Simulants that are to be used for engineering tests require a far greater volume of material than those required by resource extraction experiments. The first engineering simulant that was produced for general use was JSC-1. It was developed by Dr. David McKay of NASA's Johnson Space Center, with help from Dr. James Carter of the University of Texas at Dallas, and Dr. Carl Allen and Judy Allton, both of Lockheed. Texas A&M University was tasked with providing storage and preliminary characterization of the simulant. JSC-1 was produced specifically for large- and medium-scale engineering studies such as material handling, construction, excavation, and transportation.[6]

JSC-1 mimics the regolith's engineering properties and includes a glass component, but does not contain ilmenite, an important mineral in lunar soil grains. It has been ground to approximately the same grain size distribution as lunar regolith. The simulant was available in large quantities – up to 2,000 pounds (907 kg) – and was also intended to be useful for studies on dust control and spacesuit durability. However, JSC-1 could only be used to a limited extent as a mineralogical analog to some lunar soils.

JSC-1 was produced from a volcanic deposit located near Flagstaff, Arizona. The source material is usually black but sometimes red in color, and it is several meters

---

[6] While MLS-1 is used in sub-kilogram batches, JSC-1 is typically used by the tonne.

thick. Following coarse sieving, the simulant grains were processed in an impact mill, which breaks down the material by means of multiple impacts with other particles. This method results in less metal contamination than occurs in some other methods. After several grindings, the ash was dried to an average water content of 2.7 wt% and was then mixed. Finally, it was loaded into 50-pound bags that were heat-sealed.

The chemical composition of JSC-1 is listed in Table C.1, and the composition of Apollo 14 soil sample 14163 is shown for comparison. The normal convention for presenting rock chemistry data uses oxide formulae from an assumed oxidation state for each element (with the exception of Fe) and oxygen is calculated by stoichiome-try.[7] For example, silicon is analyzed as an element but presented as SiO_2. It is important to understand that these are only representations of the chemistry; they do not describe actual compounds or minerals in the simulant.

Scanning electron micrographs show that the simulant contains broken glass and crystal fragments with diameters up to 300 μm. The glass particles always display broken vesicles and sharp edges. The crystal fragments, on the other hand, are angular to sub-rounded, and many of them display impact scars from the milling process. About 25 tons of JSC-1 were produced in 1994, but all of this material has since been used up. A new batch was requisitioned in 2005, and in order to differentiate this material from the original simulant it was proposed that the new simulant be named JSC-2 (Carter *et al.*, 2005). Ultimately, however, the name JSC-1A was chosen for the new simulant.

JSC-1A: JSC-1A matches the composition and particle size distribution of the original JSC-1 simulant, and is intended to be as similar to JSC-1 as possible. However, the two simulants cannot be said to be identical because the material was acquired in a different batch (albeit from the same rock formation). JSC-1A was mined from the same location at which JSC-1 was obtained.

One noteworthy improvement has been made: JSC-1A may be obtained in various size fractions, which include JSC-1AC (coarse, 1 mm to 5 mm in diameter), JSC-1AF (fine, particles less than or equal to 50 μm in diameter); and JSC-1AVF (very fine, particles less than or equal to 20 μm in diameter). The fine size fractions are useful for tests related to dust mitigation and for experiments in toxicology. The results of analysis completed on six samples of JSC-1AF to determine major element compositions are presented in Table C.2.

JSC-1A is available to NASA researchers and projects as the supply remains. Other researchers, educators, students, and commercial organizations may also purchase this material from an additional 15 tonnes that are being made available by Orbital Tehcnologies Corporation (ORBITEC). Each one-tonne batch of simulant will be characterized by the United States Geological Survey.

OB-1: OB-1 is an engineering simulant that was developed by EVC/NORCAT[8] and the University of New Brunswick. It was designed to mimic the physical

---

[7] Stoichiometry is the accounting, or math, behind chemistry. Given enough information, one can use stoichiometry to calculate masses, moles, and percents within a chemical equation.

[8] EVC/NORCAT: a consortium which includes the Electric Vehicle Controllers Ltd. and the Northern Centre for Advanced Technology Inc., both of Ontario, Canada.

Table C.1. Chemical compositions of JSC-1 and lunar soil 14163.

Oxide	JSC-1 (ash) (wt%)	Lunar soil 14163 (wt%)
SiO_2	48.77	47.3
TiO_2	1.49	1.6
Al_2O_3	15.65	17.8
Fe_2O_3	1.71	0.0
FeO	8.88	10.5
MgO	8.48	9.6
CaO	10.44	11.4
Na_2O	2.93	0.7
K_2O	0.81	0.6
MnO	0.19	0.1
Cr_2O_3	–	0.2
P_2O_5	0.66	–
Total	100.01	99.8

Note the complete absence of Fe_2O_3 in the lunar soil, whereas JSC-1 contains a percentage of this compound.

Table C.2. Major element composition of JSC-1AF

Oxide	Weight % (average)	Standard deviation	% Relative standard deviation
SiO_2	47.10	0.24	0.51
TiO_2	1.87	0.01	0.74
Al_2O_3	17.10	0.08	0.44
Fe_2O_3	3.41	0.05	0.44
FeO	7.57	0.01	0.16
MnO	0.18	0.00	–
MgO	6.90	0.10	1.51
CaO	10.30	0.05	0.5
Na_2O	3.30	0.04	1.27
K_2O	0.86	0.01	0.87
P_2O_5	0.76	0.01	1.39
Total	100.20	–	–

Note: For comparison with other data sets, total Fe as FeO is 10.64%, and total Fe as Fe_2O_3 is 11.83%.

properties of lunar anorthositic (highlands) rocks. There has been 800 kg of this simulant produced, packaged in 10-kg quantities. Proprietary crushing techniques were developed with the aim of producing a simulant with correct particle size distribution while minimizing the impact of the crushing process on the desired particle shape.

In recognition of the fact that the lunar surface also contains abrasive glass particles and agglutinates, a glassy material was identified that serves to simulate these components. As a result, the simulant contains 40 percent of this crushed glassy material and 60 percent crushed anorthosite. After it is crushed and sized, compaction of the material, moisture content regulation, freezing, and the application of a vacuum are also required in order to produce a simulant that approximates the characteristics of the lunar materials that we expect to find in the polar regions.

FJS-1: FJS-1 is a simulant that was developed in Japan in 1995, by crushing basaltic rock from Mount Fuji. It is a good simulant for the bulk mechanical properties of the lunar soil, and also simulates the chemistry of lunar basaltic material to some degree. However, it has only a small percentage of glass, which may be a detriment if one wishes to use the simulant for micro-engineering studies (such as soil handling in an oxygen production plant).

Because supplies of FJS-1 are nearly depleted, Japan has begun the development of new simulants. They will include variations that represent highlands material in addition to basaltic mare material. The Japanese plan incorporates recommendations made for standardized simulants, discussed below.

C.3   FUTURE DEVELOPMENT OF SIMULANTS

C.3.1   Ongoing research

Several properties of lunar dust and soil have not been incorporated into the existing simulants. These include electrostatic and magnetic properties, the friability[9] of agglutinates,[5] single-domain iron,[10] and the presence of implanted solar wind gases. Of these characteristics, agglutinates and single-domain iron seem most immediately useful, because agglutinates comprise an average of 50 percent by weight of lunar soils, and the iron droplets on their surfaces give the particles unique properties with respect to microwave and magnetic susceptibility. ORBITEC has begun developing simulated agglutinates with single-domain iron under a NASA contract (Gustafson *et al.*, 2006). If successful, this research will enhance the fidelity of future regolith simulants. Preliminary results are promising, with the creation of grains that mimic the physical appearance of lunar agglutinates (Figure C.2).

The iron droplets on lunar agglutinate particles are reasonably simulated by the ORBITEC process. Approximately 99 percent of the Fe^0 globules in lunar agglutinates have a diameter of 1 μm or less. Figure C.3 shows a comparison between the nanophase Fe^0 globules on a lunar agglutinate and similar blebs on a grain of ORBITEC agglutinate simulant.

---

[9] Friability: the quality of being easily crumbled or reduced to powder.

[10] Single-domain iron: a grain of iron within which all of the atoms display uniformly oriented magnetic momenta. Pierre-Ernest Weiss discovered in 1907 that the magnetic moment of atoms of ferromagnetic materials become oriented, even without an external magnetic field. The size of these oriented domains is in the range of 10^{-3} to 10^{-5} mm including a volume of about 10^6 to 10^9 atoms.

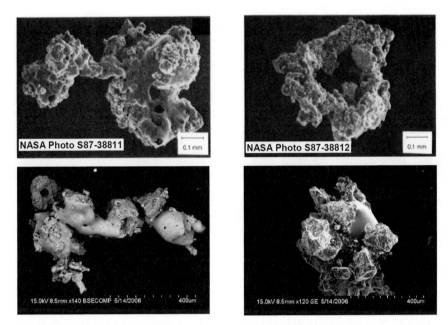

Figure C.2. (*top*) Lunar agglutinate particles. (*bottom*) Simulated agglutinate particles (from Gustafson *et al.*, 2006).

C.3.2   Standardized simulants

In the 1960s only two organizations – NASA and the Soviet Space Agency – were involved in missions to the Moon. The "space race" was a contest between two superpowers, and when one group determined that they had "won"[11] the race, funding was cut off and exploration ended, except for telescopic observations.[12]

The politics of space is now different. International cooperation is a goal, and commercial entities have also become major players. These developments offer opportunities to create an affordable and sustainable infrastructure on the Moon, but they also present challenges.

---

[11] History books are inconsistent in their judgement of who "won" the space race. In the United States it is commonly believed that the U.S. won because they were the first to send humans to the Moon and return them safely. However, the Soviet Union (and the reporters for the BBC) maintained that the Soviets had "won", because they were the first to sucessfully place a human in orbit and return him safely. It is our opinion, however, that the entire world "won" because everyone has benefited from the technological leaps that were made during the "space race".

[12] From 1972 until 1990, no spacecraft went to the Moon – or even near it. After the Apollo program, the first robotic spacecraft to image the Moon was Galileo. On its way to Jupiter, it acquired several multispectral images of the lunar surface in order to calibrate its imaging system.

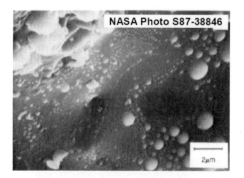 

Figure C.3. Left: Nanophase Fe^0 globules on a lunar agglutinate. Right: Similar blebs formed on the surface of ORBITEC's agglutinate simulant (from Gustafson *et al.*, 2006). Note the scale difference between the two images; which shows that the ORBITEC simulant's blebs are occasionally larger than those found on actual lunar particles.

Universities, private companies, and space agencies from all over the world are involved in lunar development. Each group is conducting experiments on resource extraction as well as testing robotic rovers and other types of lunar-surface equipment. In order for these tests and experiments to be useful to everyone, it must be possible to compare the results of one group against the results of another.

For example, a university in Japan may be building a robotic rover whose wheels have deep treads in a chevron pattern, and a company in the United States may be building a rover that uses tracks instead of wheels. Each company must test their vehicle to determine how well it performs. They want to compare their results in a meaningful way, so that they can determine which methods are most likely to be successful on the Moon. Like any scientific experiment, a meaningful comparison can only be made when all of the variables are understood and controlled. For example, if the Japanese university uses beach sand for traction testing, and the U.S. company uses gravel, no meaningful comparison can be made.

For this reason, the idea of "certified" simulants is gaining popularity. It requires that some organization which is not competing with the experimenters provide a service to all parties – namely, an analysis and description of a simulant, so that the users can confirm that their tests are performed in equivalent settings.

Carter *et al.* (2004) proposed that the concept of a standardized simulant be followed by the community, in which large quantities (more than 100 tonnes) of simulant are produced in a manner that homogenizes it, so that all sub-samples are equivalent. From this "root" simulant, it would be possible to produce other, more specialized simulants. For example, implanting solar wind, adding ice in various proportions, or adding specific components, such as metallic iron, carbon, organics, or halogens, will simulate special properties of lunar regolith needed for specific kinds of tests and experiments.

The Japanese simulant-manufacturing plan includes the development of simulants that can mimic the chemical and mineralogical characteristics of lunar soils in

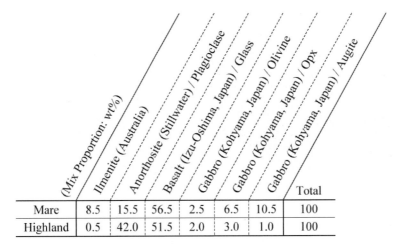

	(Mix Proportion: wt%)	Ilmenite (Australia)	Anorthosite (Stillwater) / Plagioclase	Basalt (Izu-Oshima, Japan) / Glass	Gabbro (Kohyama, Japan) / Olivine	Gabbro (Kohyama, Japan) / Opx	Gabbro (Kohyama, Japan) / Augite	Total
Mare	8.5	15.5	56.5	2.5	6.5	10.5		100
Highland	0.5	42.0	51.5	2.0	3.0	1.0		100

Figure C.4. Recommended modal abundances to simulate the characteristics of mare and highland soils. (from Kanamori *et al.*, 2006).

addition to their physical and engineering properties. These simulants will be available to researchers all over the world, and will serve as one of several new standards.

The plan for new Japanese simulants is based on the concept of "root simulants" and additives, following the proposal of Carter *et al.* (2004) discussed above. "Generic" mare and highland soils will be created by mixing appropriate percentages of components from various sources, as shown in Figure C.4.

Kanamori *et al.* (1998) have also developed a formula that allows one to determine the deviation of a material's composition from that of lunar samples. In this way, if a researcher chooses not to use a standard simulant, they can calculate the difference between their chosen simulant and a lunar sample. The similarity, R, is evaluated by comparing percentages of chemical elements in their representation as oxides – SiO_2, TiO_2, Al_2O_3, MgO, CaO, Na_2O, and K_2O – with those in the Apollo sample of interest. A smaller value for R indicates a better simulation. Using the various components and the calculation for R, a simulant may also be produced that closely matches the characteristics of soil from a specific Apollo landing site. This is done by mixing raw materials in differing proportions, as shown in Figure C.5.

C.4 SUMMARY

Simulants of lunar soils are needed to develop machinery and processes that will perform as expected on the Moon. This is exemplified by our Apollo experience, in which methods that were devised (prior to the first sample return mission) to protect spacesuits and equipment from lunar dust were found to be inadequate. The nature of lunar dust was not well known, and the dust simulants that were developed for testing did not reflect its unique properties.

Mix Proportion: wt%	Bytownite (Minnesota)	Forsterite (S. California)	Forsterite (Pakistan)	Ilmenite (Australia)	Anorthosite (Stillwater)	Gabbro (Kohyama, Japan)	R
Apollo-11		1		18	81		1.336
Apollo-12				10	81		1.591
Apollo-14				5	76	19	0.737
Apollo-15		9		6	85		1.220
Apollo-16	87		8	5			2.728
Apollo-17			4	12	84		1.616

Figure C.5. Mixtures of raw materials to simulate soils from specific Apollo landing sites.

In addition to mineralogical and engineering simulants, there are other disciplines (such as agriculture and medicine) that will also require simulated lunar soil or dust materials. To address these needs, standard simulants will be valuable. They will allow a common basis for comparison of experimental results, and help us to gain confidence that the tools and techniques we are developing will be useful and robust on the Moon.

C.5   REFERENCES

Basu, A.; Wentworth, S.; and McKay, D. (2001) Occurrence and Distribution of Fe^0-Globules in Lunar Agglutinates [extended abstract]. *Lunar and Planetary Science*, **XXXII**.

Carter, J.L.; McKay, D.S.; Taylor, L.A.; and Carrier, W.D. III (2004) Lunar Simulants: JSC-1 is Gone; the Need for New Standardized Root Simulants [extended abstract]. *Space Resources Roundtable VI*, Lunar and Planetary Institute, Houston, TX.

Carter, J.L.; McKay, D.S.; Allen, C.C.; and Taylor, L.A. (2005) New Lunar Root Simulants: JSC-2 (JSC-1 Clone) and JSC-3 [extended abstract]. In: *The Lunar Regolith Simulant Materials Workshop*, Marshall Space Flight Center, Huntsville, AL.

Chang, S.; Mack, R.; Gibson, E.K. Jr.; and Moore, G.W. (1973) Simulated solar wind implantation of carbon and nitrogen ions into terrestrial basalt and lunar fines. *Proceedings of the Fourth Lunar Science Conference* (Supplement 4, *Geochim. et Cosmochim Acta*, **2**), pp. 1509–1522.

Kanamori, H.; Matsui, K.; Miyahara, A.; and Aoki, S. (2006) Development of New Lunar Soil Simulants In Japan [extended abstract]. *Space Resources Roundtable VIII*, Lunar and Planetary Institute, Houston, TX.

Gustafson, R.; White, B.; Gustafson, M.; and Fournelle, J. (2006) Development of New Lunar Soil Simulants In Japan [extended abstract]. *Space Resources Roundtable VIII*, Lunar and Planetary Institute, Houston, TX.

Klosky, J. (1996) Mechanical Properties of JSC-1 Lunar Regolith Simulant. *Engineering, Construction, and Operations in Space V*, American Society of Civil Engineers, New York, pp. 680–688.

McKay, D.S. and Blacic, J.D. (1991) *Workshop on Production and Uses of Simulated Lunar Materials*. LPI Technical Report No. 91-01, Lunar and Planetary Institute, Houston, TX.

McKay, D.S.; Carter, J.L.; Boles, W.W.; Allen, C.C.; and Allton, J.H. (1994) JSC-1: A New Lunar Soil Simulant. *Engineering, Construction, and Operations in Space IV*, American Society of Civil Engineers, New York, pp. 857–866.

Pillinger, C.T.; Cadogan, P.H.; Eglinton, G.; Maxwell, J.R.; Mays, B.T.; Grant, W.A.; and Nobes, M.J. (1972) Simulation study of lunar carbon chemistry. *Nature, Phys. Sci.*, **235**, 108–109.

Appendix D

In-situ resource utilization (ISRU)

"Living off the land" is the principle of *in-situ* resource utilization (ISRU). On the Moon, this means using lunar resources for the production of everything that is needed for human settlement. The history of human migration shows that the settlers of a new land brought their essential tools and "seed corn" with them, but they otherwise relied upon the natural resources of the new land for their survival and growth. The migration of humanity to the Moon will be no different: the first human settlers on the Moon will bring their tools and survival equipment with them, but they will use lunar resources and solar energy for long-term development and colonization (see Figure D-1).

"Living off the land" will be impossible for humans during the first stages of lunar development. Not only are "the comforts of home" missing from the Moon, there is also no atmosphere, vegetation, wildlife, streams, or habitats that allow life to flourish. Here, robots will assist us by paving the way for human settlement. As robotic technology continues to develop, the robots and automated systems on the Moon will be increasingly capable of producing complex items, and fewer things will be imported from Earth.

As explained in Chapter 3, the Moon has all of the elements that are found on the Earth, including an abundance of iron, oxygen, aluminum, titanium, and silicon. These elements will be feedstock for lunar manufacturing processes. For example, iron, titanium, and aluminum will be used for structures; aluminum will also be used for electrical cable and (possibly) rocket propellant; and silicon will be used for solar cells, computer chips, and telecommunication (fiber-optic) cable. When the mining and manufacturing equipment for ISRU are in place on the Moon, a permanent utilities infrastructure can be constructed (see Chapter 7), and global human settlement will then be possible.

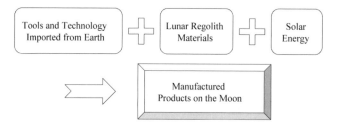

Figure D-1. Manufacturing on the Moon.

D.1   POWER

In planning the development of a lunar outpost, the first consideration is power – for operating machinery and for transportation. Getting to and from the lunar surface is currently expensive, and the expense could be reduced if the vehicles that go there did not have to carry enough fuel for the return trip. There is also a need for power for robots and machinery on the Moon. Again, if their power must be completely supplied by batteries that are sent with them, the cost of transportation is much greater than if they can generate some or all of their power from sunlight (Chapter 4 describes how this is done for the Mars rovers).

Electrical, thermal, and chemical power have advantages over nuclear and hydraulic power on the Moon. Historically, it has been possible to power small spacecraft with RTGs (radioisotope thermoelectric generators); however, RTGs are no longer being made because there is no plutonium available from which to make them. Other types of nuclear power have been discussed, but it is difficult to gain public support for nuclear energy, and in the case of the Moon it is also expensive to transport the equipment to the lunar surface. Hydraulic power requires a working fluid, and the Moon does not have any known source for the fluids that are typically used in these systems. For these reasons, mission planners are focusing on electrical power from solar cells, thermal power from solar collectors, and the production of rocket fuel from indigenous materials on the Moon.

D.1.1   Solar power cells for electricity

As discussed in Appendix A, concepts for "paving" the Moon with solar power cells are being developed, because most alternative sources of power are either more expensive, more complex, or both. Fuel cells are another emerging possibility, especially for rovers that must operate during the lunar night, but they require replenishment of their constituents – converting the output water back into oxygen and hydrogen. Solar power cells are likely to be the primary method of obtaining electrical power on the Moon. Electrical power can be used directly for many operations, and can be the power source for converting the output water of fuel cells back into hydrogen and oxygen.

A solar cell (also referred to as a photovoltaic cell) is a device that converts photons from the Sun (solar radiation) into electricity. When they first became available, they were used in solar-powered calculators and other low-power applications. As their efficiency improved, they began to be used in remote areas where electrical power from transmission lines (the "grid") is unavailable. They are now becoming increasingly popular as a method of reducing our dependence on non-renewable energy sources, such as coal and petroleum products. Parking lot lights and traffic signs powered by solar cells are becoming more common. Because photovoltaic systems are modular, their electrical power output can be engineered for almost any application.

Large arrays of photovoltaic cells are also used to power spacecraft. The International Space Station, as well as the Russian Mir space station that preceded it, were both designed to use solar-generated electricity as their main source for operational power. Their solar arrays must include gimbals or other methods of rotating the solar panels so that they are in direct sunlight for as much of the time as possible, and this technology is also applicable to the devices that will be sent to the Moon in the first robotic missions.

Solar cells are a type of semiconducting device. Semiconducting devices have many applications besides photovoltaic cells, such as small semiconductor chips containing tens of millions of transistors. These chips have made possible the miniaturization of electronic devices. In addition, extremely-small devices are being made using the technique of molecular beam epitaxy.[1] A laser beam is used to vaporize a substance, which is then deposited on a substrate (such as a silicon chip) in a single molecular layer.

Solar cells use the photovoltaic[2] principle for converting the energy of the Sun's photons into electrical energy. The electricity is direct current and can be used that way, or it can be converted to alternating current, or it can be stored for later use. There are no moving parts, and if the device is correctly encapsulated against the environment, nothing will wear out.

Figure D.2 illustrates a common photovoltaic structure. It is a semiconductor material into which a diode,[3] or p–n junction,[4] has been formed. Electrical current is taken from the device through a grid structure on the front (which makes a good contact at the same time as simultaneously allowing sunlight to enter the cell) while a

---

[1] Molecular beam epitaxy is one of several methods of thin-film deposition. Elements such as gallium and arsenic are heated until they each begin to evaporate. The evaporated elements then condense on a substrate. The process takes place in high vacuum, thus the Moon is a natural place to produce them. The term "beam" simply means that evaporated atoms do not interact with each other or any other gases until they reach the substrate, due to the large mean-free path lengths in vacuum.

[2] Photovoltaic: generating an electric current when acted on by light or a similar form of radiant energy.

[3] Diode: a two-terminal electronic device that has a high resistance to current in one direction but a low resistance in the other direction [from di (two) + (electr)ode].

[4] p–n junction: two layers of semiconductor, one of which contains excess electrons, and the other of which lacks electrons (it is said to possess holes) and consequently has a positive charge. "p–n" refers to the positive and negative "doped" layers, which are metallurgically joined.

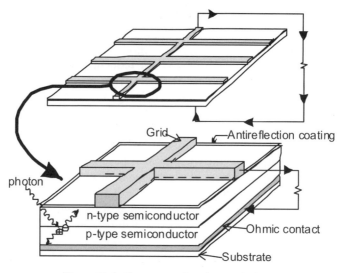

Figure D.2. Structure of a photovoltaic cell.

second contact on the back completes the circuit. An anti-reflection coating mini-mizes the amount of sunlight that reflects away from the device.

Silicon is the most commonly-used semiconductor at present. It is a Group IV element on the Periodic Table, which means it has four electrons in its outermost orbit. In the silicon tetrahedral crystal structure, each atom shares its four valence electrons[5] with its four nearest neighbors, forming four covalent bonds.[6] The valence energy band, where these electrons generally reside, has a lower energy level than the conduction band, where electrons can move about freely to carry charge. The amount of energy needed to push an electron from the valence band into the conduction band (in other words, to release a bound electron) is called the band-gap energy, which is about 1.12 eV for silicon. Even at room temperature, the amount of conductivity in pure silicon is very small: 1.6×10^{10} conductors (carriers) per cubic centimeter, as compared with 10^{22} carriers per cubic centimeter in a typical metal.

To modify the conductivity, small amounts of impurities are introduced into the crystal lattice. This is called "doping". If you dope silicon with a Group III element such as boron (three electrons in the outermost orbit), then that part of the crystal lattice will have a covalent bond that is deficient by one electron. In other words, it will have a "hole". Electrons and "holes" are both charge carriers. Similarly, if you dope the silicon with a Group V element, you will have an extra, loosely-bound electron that can easily jump up into the conduction band. Doping creates another

---

[5] Valence electron: an electron in the outer shell of an atom. In a chemical reaction, the atom gains, loses, or shares these valence electrons, so that they combine with other atoms to form molecules, or ions.

[6] Covalent bond: a bond formed when electrons are shared by two atoms.

Table D.1. Properties of various semiconductors

	Ilmenite	Silicon	Silicon carbide	Gallium arsenide
Resistivity	$1.11\,\Omega m$	$0.23\,\Omega m$	$0.1\,\Omega m$	$0.004\,\Omega m$
Band gap	$2.54\,eV$	$1.12\,eV$	$2.9\,eV$	$1.43\,eV$
Melting point	$1683\,K$	$1385\,K$	$3070\,K$	$1784\,K$
p-type semiconductor	Pure ilmenite	On doping	On doping	On doping
n-type semiconductor	In solution with $\alpha\text{-}Fe_2O_3$	On doping	On doping	On doping

energy level close to the valence band or conduction band, resulting in lower activation energies.

In order for electrons to jump up into the conduction band (where they can carry charge), energy must be added to the system. In the case of a photovoltaic cell, the energy is provided by light whose wavelength is short enough (and whose energy per photon is thus high enough) to excite individual electrons.

Other materials besides silicon have been used to make solar cells. Gallium arsenide is a well-known example. It can achieve a higher energy conversion efficiency than silicon because it has a larger band gap than silicon – 1.43 eV rather than 1.12 eV. Materials with larger band gaps can withstand higher power, higher frequency, and higher temperature environments, and they are more resistant to radiation damage. Lunar ilmenite has good potential to be a large-band-gap semiconductor of considerable usefulness. Its band gap is 2.54 eV – over twice as large as the silicon band gap.

Table D.1 shows some of the relevant properties of lunar ilmenite compared with better-known semiconductors.

D.1.1.1 *Lunar ilmenite for solar panels*

As discussed above, most photovoltaic cells are made with silicon or gallium arsenide. However, ilmenite – particularly ilmenite which is formed in a reducing atmosphere such as the Moon's – has excellent potential as a material for photovoltaic cells. Ilmenite comprises up to 20 percent of the lunar regolith in its richest locations. It has been studied for use as a feedstock for the hydrogen reduction process to manufacture oxygen. However, its use as a semiconductor and for solar cell manufacture is a relatively new idea.

There is an important difference between lunar ilmenite and terrestrial ilmenite. Lunar ilmenite is formed under reducing conditions, and consequently its iron is in the 2+ valence state.[7] On Earth, the iron in ilmenite is only partially in the 2+ state, because there was more oxygen available during its formation. The iron in the 2+

---

[7] In the 2+ valence state, the atom has lost two of its (negatively charged) valence electrons, and thus has a net positive charge of 2.

state is responsible for the good semiconducting qualities of ilmenite, thus lunar ilmenite would be preferred for semiconductors.

Engineers at Texas A&M University have grown monocrystalline ilmenite in the lab under reducing conditions to learn more about the benefits of lunar ilmenite as a semiconductor (Sankara, 1995). They use the Czochralski technique, which is commonly used in the semiconductor industry to grow bulk silicon monocrystals. It utilizes the chemical reaction

$$Fe + Fe_2O_3 + 3TiO_2 \Rightarrow 3FeTiO_3.$$

The procedure involves first grinding the material to mix it, then pre-heating it to 1,000°C in a reducing atmosphere, with about 258 torr (34,474 Pa) of nitrogen. This is not a perfect duplication of the lunar environment, but is one that can be maintained for long periods with the equipment that is available. X-ray diffraction plots are made of the sintered material. If it has not reacted into ilmenite, then the pre-heating cycle is repeated until there are good ilmenite lines in the X-ray diffraction plot.

Inductive[8] heating is used because of the very high temperatures (1,405–1,410°C) required. In inductive heating, radio frequency is fed into a coil. The radio frequency couples with the crucible and causes eddy currents that heat it to very high temperatures. The temperature can be controlled by changing the power fed into the coils.

For this process to work, the right material must be chosen for the crucible. Platinum does not work well in a reducing atmosphere, so iridium is used instead. A graphite crucible, which is resistant to the high temperature and reducing atmosphere, surrounds the iridium crucible. The graphite is heated, and it heats the iridium which in turn heats the ilmenite inside.

In order to start crystal growth, a crystal "seed" must be immersed in the melt. They use a material that has a close lattice match to ilmenite to dip into the liquid, in order to grow a boule (a cylinder) with several single-crystal sections, one of which can be isolated and used as a seed for the next phase. The seed is dipped into the liquid at the correct temperature using sophisticated control systems. The temperature and the pull rate (how fast the seed is pulled out of the liquid) have to be monitored and continuously adjusted to achieve good crystal growth.

By controlling the operating parameters, the crystal can be grown to any diameter needed. A photograph of a completed boule is shown in Figure D.3. Wafers are cut from the boule and polished, and these become the substrate material for semiconductor devices. Very accurate control is required to avoid defects in the crystal structure. The crystal grows at a rate of about 2 mm per hour, so it takes about 20 hours to grow a few centimeters of material.

The researchers have also experimented with neural networks for control systems so that the process of creating ilmenite semiconductors can be automated. (This would be a good candidate for automation, as explained in Appendix A.) The goal is to eliminate human involvement in the process and make it fully automatic. The

---

[8] Induction is the process by which an object having electrical or magnetic properties produces similar properties in a nearby object, without direct contact.

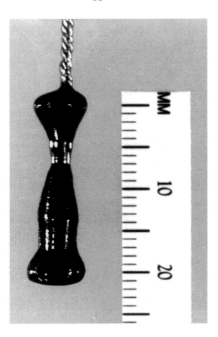

Figure D.3. A completed boule of synthetic ilmenite.

team has modeled zero-G growth on the computer, in preparation for future zero-G experiments.

Because large-band-gap semiconductors such as ilmenite may display electrical resistivities that are two to three orders of magnitude greater than the resistivity of silicon, they can reduce leakage currents and hence reduce the power that is lost to heat dissipation. Extensive cooling equipment is required for silicon-based electronics on spacecraft. Much of the cooling equipment could be eliminated by using a lunar ilmenite semiconductor.

Another problem with silicon and gallium arsenide is that their conversion efficiency is limited at the shorter light wavelengths of the space environment. Ilmenite, with its higher band gap, will have a higher response to this full spectral range and thus may be more suitable for space-based solar cells.

Radiation damage is the major life-limiting factor for a photovoltaic array, especially in space. A typical silicon solar array may be oversized by as much as 40 percent to assure sufficient power over the life of the mission, whereas the large band gap of ilmenite makes it naturally radiation-resistant.

Lunar ilmenite would also be an excellent material for solid-state lasers for optical data storage systems. Optical data systems are limited by the wavelength of the laser light used in read and write operations. Because of its higher band gap, ilmenite can be used for shorter-wavelength lasers, which will enable higher data densities.

The initial work on synthesizing and characterizing lunar ilmenite for its semi-conducting properties has been completed, but much work remains. For example, we need to know more about its structure and the nature of the oxidation states of the iron. Then modifications can be made in the system to suit the needs of technology. Another area of research is the annealing of the material to stabilize it and enhance atomic ordering.

D.1.1.2   Thin films

Thin-film technologies are being developed to reduce the mass of light-absorbing material required in a solar cell. This can lead to reduced processing costs compared with bulk materials, but also tends to reduce energy conversion efficiency. To address this challenge, multi-layer thin films have been used, which demonstrate efficiencies above those of bulk silicon wafers.

Thin-film solar cells developed directly on the lunar surface, using ilmenite-based semiconductors and robotic manufacturing equipment, are one of the enabling technologies for a lunar electric grid (Chapter 7) to be constructed.

D.1.2   Mirrors and solar concentrators

D.1.2.1   Mirrors

Because there is no wind (other than the solar wind) on the lunar surface, mirrors and solar concentrators can be lightweight and made of Mylar film or other materials that can be compressed for stowage, then unfolded for use. At first, these items will come from Earth and be deployed automatically when a suitable flat surface is found (or prepared). Lightweight mirrors on masts, located at the peak of Mons Malapert, can be installed on masts to provide sunlight in areas farther south. They can also be used to provide sunlight from above the peak, during periods when Mons Malapert itself would otherwise be in shadow.

D.1.2.2   Solar concentrators

Solar concentrators gained popularity during the 1980s when petroleum was in short supply. They produce electric power by converting the Sun's energy into high-temperature heat using various mirror configurations. The heat can then be channeled through a conventional generator to produce electricity, or it can be used as a heat source for resource processing.

There are three kinds of concentrating solar power systems – troughs, dish/engines, and power towers.

Troughs: In trough systems, the Sun's energy is concentrated by parabolically-curved, trough-shaped reflectors onto a receiver pipe running along the inside of the curved surface. This energy heats a fluid that flows through the pipe, and the heat energy is then used to generate electricity in a conventional steam generator.

A collector field comprises many troughs in parallel rows. The troughs track the Sun to ensure that it is continuously focused on the receiver pipes. Trough designs can incorporate thermal storage – setting aside the heat transfer fluid in its hot phase – allowing for electricity generation several hours into the lunar night.

Dish/engines: A solar dish/engine system is an electric generator that "burns" sunlight instead of gas or coal to produce electricity. The major components are the solar concentrator and the power conversion unit.

The dish (concentrator) is the primary solar component of the system. It collects the solar energy and focuses it on a small area. The resultant solar beam has all of the power of the sunlight hitting the dish, but is concentrated in a small area so that it can be more efficiently used. The dish structure must track the Sun continuously to reflect the beam into the thermal receiver.

The power conversion unit includes the thermal receiver and the engine/generator. The thermal receiver absorbs the concentrated beam of solar energy, converts it to heat, and transfers the heat to the engine/generator. A thermal receiver can be a bank of tubes with a cooling fluid (usually hydrogen or helium), which is the heat transfer medium and also the working fluid for an engine. Alternate thermal receivers are heat pipes, in which the boiling and condensing of an intermediate fluid is used to transfer the heat to the engine.

The engine/generator system is the subsystem that takes the heat from the thermal receiver and uses it to produce electricity. The most common type of heat engine used in dish/engine systems is the Stirling engine. A Stirling engine uses heat to move pistons and make mechanical power, similar to the internal combustion engine in a car. The mechanical work, in the form of the rotation of the engine's crankshaft, is used to drive a generator and produce electrical power.

Power towers: A power tower converts sunshine into clean electricity for the world's electricity grids. The technology utilizes many large, sun-tracking mirrors (heliostats) to focus sunlight on a receiver at the top of a tower. A heat transfer fluid is converted to steam, which in turn is used in a conventional turbine generator to produce electricity. Early power towers used water as the heat transfer fluid, but more recent designs use molten nitrate salt because of its superior heat transfer and energy storage capabilities. Power towers are unique among solar electric technologies in their ability to store solar energy and dispatch electricity when needed.

D.1.3   Rocket fuels and oxidizers

Chemical rockets require both a fuel and an oxidizer to function. Liquid-fueled rockets have better specific impulse[9] than solid rockets and are capable of being throttled, shut down, and restarted. Liquid oxygen (LOX) and liquid hydrogen (LH$_2$) are used in the Space Shuttle, the Centaur upper stage, the newer Delta IV rocket, and most stages of the European Ariane rockets. It is a well-understood technology that can be used for lunar propulsion systems. Methane has also been

---

[9] Specific impulse is the impulse (change in momentum) per unit mass of propellant. It is used as a measure of the efficiency of a propulsion system.

Table D.2. Propellant mass and volume for a proposed Mars orbital transfer vehicle (from de Weck *et al.*, 2005).

Propellant component	Mass (kg)	Density (kg/m^3)	Volume (m^3)
Liquid hydrogen (LH$_2$)	21,430	70.8	302.7
Liquid oxygen (LOX)	128,570	1,141	112.7

proposed for lunar-based transportation systems, because the technology for using it could be applied to Mars exploration. In either case, oxygen is the most likely oxidizer to be produced on the Moon, and many processes have been proposed for extracting it from lunar regolith. Appendix E discusses these processes in detail.

In all chemical rockets, the mass of oxidizer exceeds the amount of fuel. For example, in a LOX–LH$_2$ propulsion system, the oxidizer-to-fuel ratio is typically 4 : 1. Some proposed systems (Table D.2) have oxidizer-to-fuel mass ratios as high as 6 : 1 (de Weck *et al.*, 2005).

If the oxidizer can be produced from local materials, the amount of propellant mass that must be launched from Earth can be reduced by a factor of 4 to 6. For this and other reasons, the production of oxygen on the Moon is the second-highest priority (following power).

D.1.4   Other consumables

As discussed in Chapter 1, there is some evidence that water ice may be found in the permanently-shadowed polar regions of the Moon. Although water can be created at any lunar location by combining the hydrogen and oxygen that are present in the soil, it may be easier to recover water in the form of ice or hydrated minerals (if they are indeed present in the north and south polar regions).

D.1.4.1   Ice at the lunar poles

Calculations by Allen and Zubrin (1999) have shown that even if water ice exists in the permanently-shadowed craters at the lunar poles, extracting it may not be more energy-efficient than some of the proposed processes for obtaining oxygen from lunar rock (magma electrolysis actually requires less power). This surprising result stems from the fact that frozen water-ice at 40 K requires a substantial amount of heat in order to bring it to its melting point. Moreover, transporting the feedstock from a crater floor to an area where it can be processed will require additional energy as well as equipment. Finally, operating machinery at 40 K will be a challenge that is unique to this process.[10]

In spite of these challenges, there are several reasons why using water ice in permanently-shadowed regions may still be the method of choice for supplying water and oxygen to the lunar base. First, the technology for warming something to its

---

[10] While it is true that liquid-hydrogen-handling systems on Earth operate at 20 K, they do not have moving parts. Thus, we have no terrestrial analogue for operating machinery at these temperatures.

melting temperature is well-understood; little research on this process is required. Second, one avoids the hazard of hydrogen sulfide gas (H_2S) which would be created by heating regolith above $\sim 700°C$ (973.15 K). To extract water, one only needs to heat the regolith above the freezing point of water ($0°C$ at atmospheric pressure). Third, ice mining at the lunar poles would allow the use of solar power for much of the work, whereas mining operations at the lower latitudes will require either an alternative power source for lunar nighttime operations, or a "sleep" mode in which production only occurs during daylight hours.

Solar wind volatiles: Because solar wind gases (described in Appendix B) are weakly bound in the lunar regolith, it will be relatively easy to extract them by heating the regolith to about $600°C$. Because there is a higher concentration of solar wind gases in the smaller particles (Carter, 1985), it may be useful to size-sort the regolith, retaining only the smaller particles.[11] The feedstock could then be contained and heated to extract these gases. Solar wind volatiles include ^3He, H_2, ^4He, various carbon compounds, N_2, and the noble gases.

The most abundant of these volatiles is hydrogen, and it is also the most immediately useful. Oxygen is needed as the oxidant for any combustion process, which includes liquid-fuel rockets. However, in addition to the oxidant, the fuel itself (hydrogen or methane) is also needed. By obtaining hydrogen from the lunar soil, all of the components for rocket propulsion can be re-supplied on the Moon.

In addition to its usefulness as a fuel, hydrogen is also needed for fuel cells and to produce water for agriculture, human consumption, and hygiene. Fortunately, most of the processes that have been proposed for oxygen extraction (detailed in Appendix E) involve heating the lunar soil, and the pre-heating step of these processes could include a system to collect, sort, and store the solar wind volatiles.

In addition to hydrogen, nitrogen and carbon will both be useful. Nitrogen can be used to make up for losses of the "air" inside habitats,[12] which would be comprised of a mixture of nitrogen and oxygen that is similar to Earth's atmosphere. Carbon and nitrogen are both useful for developing a plant growth medium that provides proper nutrients for growing food crops.

Another solar-wind volatile that has gained attention in recent years is Helium-3 (^3He), because of its potential use as a feedstock for clean nuclear fusion. Appendix H describes the importance of Helium-3 and how it can be extracted from lunar soil.

D.1.4.2   Leftover fuel

For the first humans who return to the Moon, an additional consumables resource exists in the form of leftover fuel on the lunar descent vehicle. When it arrives at the

---

[11] Size-sorting the lunar soil grains may not be preferred, because some of the volatiles are so loosely bound to the grains that they would be lost by that process.

[12] Small losses of atmosphere are almost inevitable during operations where an airlock is cycled. Over time, this lost air must be re-supplied to maintain an Earth-normal atmosphere inside the habitat. Use of an oxygen atmosphere (at lower pressure) is possible, but is also dangerous, because many substances become flammable in an oxygen atmosphere.

Moon, the descent vehicle is required to have enough fuel to enable an "abort to orbit" should a contingency arise in which this was needed. Consequently, each lunar descent module that lands safety on the lunar surface will have about three tonnes of propellant – pure hydrogen and pure oxygen – which can be mixed together for water, used for fuel cells, or used for supplementary breathing oxygen.

D.2   ROADS, HABITATS AND FACILITIES

D.2.1   Excavation and transport

While producing electrical power and propellants are a necessary first use of lunar resources, there are also significant cost savings to be gained by producing any high-mass object on the Moon – especially objects that are technologically mature. "Dirt work" and construction are two areas that will provide significant cost savings for future lunar development.

The longevity of machines in lunar conditions is unknown. The abrasive regolith is hard on moving parts, and even during the few days duration of each Apollo mission, mechanisms failed because of the abrasive dust. Some lubricants exist which can mitigate this problem, but a complete solution has not yet been found. During the early phases of lunar bases, it would be interesting to revisit the Apollo 17 site. Here we could determine the mechanical condition of the lunar rover, which has been sitting idle on the lunar surface for over 30 years. Would it fall apart as soon as someone touched it, or could we re-charge the battery and drive it away? Perhaps an experiment could be designed in which a robot would approach the rover and find out something about its condition.

D.2.1.1   *Regolith excavation and dust control*

Wire brush excavator: Instead of using conventional shovels or other excavating equipment, a rotary wire brush with stiff bristles has also been suggested (Boles, 1992), and initial testing achieved good results. A rotary brush requires less energy to move the regolith than does a scraper, and it is tolerant of cobbles and boulders. The rotary brush takes advantage of the de-compaction ("reverse rutting") of lunar soil so that as the operation continues, the regolith loosens without substantial effort. Furthermore, a rotary brush should be easier to automate than a shovel or front-end loader, and it would be easier to maintain. It is still unknown, however, if this method of excavation will release valuable solar wind gases that are trapped in the regolith, how much power would be required, and how long it would endure the abrasive dust that would be churned up by the operation.

Pneumatic conveying (Sullivan *et al.*, 1992): Pneumatic conveying of sand-size particles is a technique that is used extensively on Earth for transporting materials in factories. The sand- or smaller-size particles are placed in a closed pipe with a moving stream of air or other gas. Pneumatic conveying might be a useful way to move

Figure D.4. A technician checks the control mechanism of a pneumatic conveying system (Pat Rawlings and NASA).

regolith within an ISRU plant as an alternative to mechanical conveyors. Bucket elevators, screw feeders, conveyor belts, and other mechanical conveyors are difficult to operate on the lunar surface because of the abrasive nature of the regolith. Furthermore, the lubrication of such machines is difficult in a vacuum.

Pneumatic conveying would also be consistent with a fluidized bed system (see Appendix E) for solar wind volatile release or oxygen production. Since there is a need to go from vacuum to pressure at some point in the system, it may be sensible to do so early in the process and then keep the system pressurized until the material is ready to leave. A few pounds of air will go a long way in a pneumatic conveyor system – 3.4 kg of air at 760 torr (101,353 Pa) would fill 1,397 meters of 5-centimeter-diameter pipe. Gas lost to leaks can be replenished with little difficulty. Oxygen, hydrogen, helium-4, or other gases (that have been recovered from the regolith) could be used as the conveying medium (see Figure D-4).

Pneumatic conveying can be used to move soil from a mine site to the processing plant, and is thus an alternative to lunar dump trucks, which suffer the problems of abrasion and difficulty of lubrication. There are comparatively few moving parts in a pneumatic system; furthermore, pneumatic conveying would eliminate the need for mobile thermal control systems and power sources, or for recharging battery-operated vehicles.

Pneumatic conveying can be coupled with a gas classifier as an alternative to vibrating screens, which do not work well even on Earth. A gas classifier system works by forcing air through a screen located below the incoming feedstock. Oversize particles will not be carried over a barrier as well as smaller particles, thus the fines can be moved to a different bin. This would be preferred over vibrating screens because the screens used in gas classifying do not move; therefore, much less screen maintenance will be needed.

As an alternative to bagging sand for radiation shielding, pneumatic conveying can be used to fill the exterior shells of inflatable structures. Pneumatic conveying would simplify the process of bagging regolith for radiation shielding, because one large bladder could be used, rather than many small ones. This would eliminate the problems associated with using loose, uncontained regolith, and would eliminate the need for a conveying system to emplace small bags after they were filled.

Sullivan *et al.* (1992) built an experiment to test pneumatic conveying at $\frac{1}{6}$ G that was flown on NASA's KC-135 airplane. Vertical "air lift" conveying requires about half the pressure at $\frac{1}{6}$ (lunar) gravity as compared with full Earth-normal gravity. "Choking" velocities are two to three times lower at $\frac{1}{6}$ gravity, and horizontal salt-ation[13] velocities are similarly lower. This means that lower flow rates will be required and consequently less abrasion will occur. Overall, at $\frac{1}{6}$ G, twice the mass can be moved at less than half the flow rate as compared to a similar system: at 1 G. The KC-135 flight also provided other important information about the system. The cyclones that were used in the system did not empty very well. Some experimentation needs to be done in changing the geometry of the cyclone, increasing the diameter of the downcomer,[14] and perhaps adding a secondary separator. The behavior of the feeder and standpipe also presented some new challenges.

Other experimental work can now be defined more clearly. Size classification using air also needs experimental verification, and further work should be done to improve the basic design of the horizontal conveying system. Experiments are also envisioned to test potential improvements in the air lift design to increase capacity.

Electromagnetic conveying (Taylor and Eimer, 2006): Professor Larry Taylor of The University of Tennessee has devised a potential method to mitigate the dust problem by utilizing its ferromagnetic properties. The ferromagnetism is due to the presence of nanophase metallic iron in the lunar soil. The nanophase iron grains are present in the impact glasses (see Appendix B) which make up ~40–50 percent of the lunar soil. Moreover, 80–90 percent of the smallest grains (less than 20 µm in diameter) are glass, therefore this portion of the soil can be readily attracted by a magnet.

It should be possible to effectively suck up the regolith using this magnetic potential. It can be done in a similar fashion to the way maglev trains and coil guns

---

[13] Saltation occurs when the lift and moment exerted by a fluid on a particle is enough to pull the particle away from the surface and into the flow. Initially the particle moves quite rapidly compared with the flow and so has high lift, moving it away from the surface. As the particle moves into the faster flow away from the bed, the velocity difference between particle and flow decreases, and so lift decreases. When the particle weight is greater than the lift force, the particle sinks back towards the surface. During its descent, the particle keeps some of the speed it picked up in the faster-moving flow, and so returns to the surface at higher speed than the fluid near the surface. This gives the particle a parabolic trajectory through the fluid, which is the defining characteristic of saltation.

[14] Downcomer: a pipe to convey something downwards.

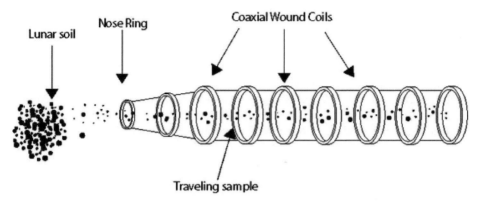

Figure D.5. Lunar electromagnetic conveying device (Taylor, 2006). Coaxial rings are sequentially powered up and down in a rippling effect to move ferromagnetic lunar soil to the right.

(or mass drivers) work. These two technologies use consecutive electromagnets to pull an object along. The biggest advantage is that there are no moving parts in the device. Most importantly, this method would not only pull the soil along, but would mitigate the dust as well (Appendix B).

The operation of this "coil vacuum" is shown in Figure D.5. It consists of a series of wound coils that are individually powered to generate magnetic fields. Soil is picked up by a "nose coil" and pulled into the center of the tube where it is suspended by the electromagnetic force of the coil. As the material approaches this first coil, the coil is powered down, and the next coil in the sequence is powered up, which attracts the particles further into the tube. This process of turning coils on and off continues in a "caterpillar/millipede effect", moving the soil particles along this electronic conveyor belt.

Conceptually, a lunar surface-mining operation might use this device to gather and transport soil (and dust) across great distances to processing plants. One possibility is to have a "trunk line" that is capable of generating large magnetic fields and moving large amounts of material, with several feeder lines into it. The feeder lines would branch off the trunk line, pulling in material from the surrounding area. This allows several areas to be excavated simultaneously, and as the regolith is exhausted in one area, the trunk line can be extended to new ones.

Several issues remain to be overcome. First, the magnetic fields must be strong enough to attract the soil from a reasonable distance and accelerate it to a speed sufficient to carry it to the next coil through momentum. In the case of the Moon, this is made somewhat easier by both the absence of atmosphere and its $\frac{1}{6}$ gravity (lighter to pick up vertically, and less drop in horizontal transport). Second, it will be necessary to determine the on–off timing needed to energize and relax consecutive rings, in order to keep a continuous flow of soil through the tube.

Boring machines: Other terrestrial technologies could be refined for use on the lunar surface. For example, automatic mining machines already exist. The Subselene, conceived by Rowley and Neudecker (1985), could core holes in regolith and sinter the walls as it cores.

However, Apollo data on lunar drilling and coring have shown that removal of cuttings from boreholes is difficult, and that the lunar dust is as abrasive as sandpaper. While shoveling is easy for the first few centimeters, soil compaction occurs at a depth of approximately 15 centimeters. Below that point, more energy is required for excavation. The challenges that confront those who wish to excavate the lunar regolith have been studied in some detail, and are referenced in Appendix B.

Mining techniques that are used on Earth will typically need to be modified for lunar applications. For instance, the Moon's $\frac{1}{6}$ G will cause bulldozers and other excavating machines to exert much less downward force, meaning that shallower excavations must be attempted initially (although this may not be a problem if time is not an important constraint). To counter the effects of lesser gravity, scrapers pulled by draglines have been suggested, and this approach seems feasible. Augers placed in the subsurface soil can anchor the pulleys and lines. Auger systems could also be used to improve the load-bearing capability of mobile excavation equipment. The auger can be extended in the ground as an anchor, and retracted when it is time for the machine to move to another location.

Railroad: The lunar railroad will be a useful means of long-distance transportation of raw materials on the Moon. Railroads have the advantage of being able to transport large quantities of raw materials and other cargoes over long distances at high speeds and at lower cost, with a minimum of tele-operation control. If the railway is designed so that it does not stir up dust, it will be a reliable system for transporting materials across the Moon.

The construction of the lunar railroad and other lunar utilities will require grading of the regolith along the railroad right of way. The railroad can transport construction materials and equipment to work sites, and the regolith that is removed can be transported by the railroad to industrial parks where useful materials will be recovered from it.

Mechanical ballistic transport – slingshots and "the advanced Roman catapult": Slingshot systems, featuring high-strength tethers or rigid spinning arms with cargo buckets, could be deployed for sending raw materials or small cargo packages between mining and manufacturing facilities a few kilometers apart.

Using the same technology that was used by the Romans to hurl stones and flaming missiles over enemy fortifications, it is possible to throw large shock-tolerant payloads on ballistic trajectories over the lunar surface. It may be possible to throw chunks of icy regolith that are mined in the permanently-shadowed craters of the south pole[15] to sites at higher elevation near the lunar railroad (see Figure D.6).

---

[15] The area from which icy regolith is extracted will be below the level of the rail line that courses along the adjacent high elevations. In its operation, the catapult would throw the payload with the precise force that would cause the payload to land at its predetermined destination with close to zero kinetic energy (at the top of its trajectory).

Figure D.6. Catapult and slingshot transportation of materials (Thangavelu/DiMare).

D.2.2   Construction

D.2.2.1   *Sintering*

Sintering of compressed regolith can create "bricks" for use in buildings, roads, landing pads, blast shields, and radiation shields. The concept of the "lunar lawn-mower," which sinters the upper surface of the area around habitats, landing sites, and other facilities, was described in Appendix B.

Allen *et al.* (1994) ran tests at varying heating rates and time spans, using different size fractions of lunar soil simulant. Figure D.7 shows some of the results of tests run on different sizes of basalt grains. When all other variables were held constant, only grain sizes smaller than 400 microns would produce a sintered brick with strength comparable with that of standard concrete.

In a second series of tests (Figure D.8), it was learned that a temperature of at least 1,100°C was required to produce concrete-strength sintered blocks from basalt particles the size of typical lunar soil. Because these experiments were performed using convection heating rather than microwave heating, further work is needed to understand how microwave heating might affect these results.

Microwave sintering: Until 2005, microwave sintering of the lunar regolith was considered to be only marginally useful, because experiments performed with lunar simulants had indicated that the process was inefficient, and that even distribution of heat would be difficult to achieve. However, in 2005 Dr. Larry Taylor and Dr. Thomas Meek, both of the University of Tennessee, had the opportunity to test microwave heating on an actual lunar sample from Apollo. The results of that experiment confirmed Meek's (1992) prediction that the results would be different

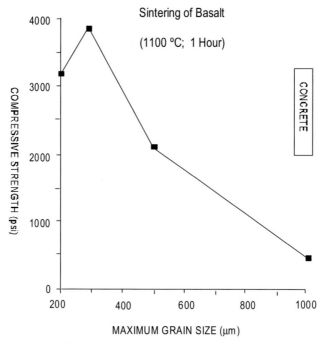

Figure D.7. Grain-size effects on the strength of sintered lunar materials (Allen *et al.*, 1992).

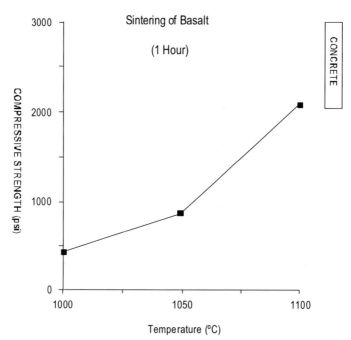

Figure D.8. Temperature effects on the strength of sintered basalt grains.

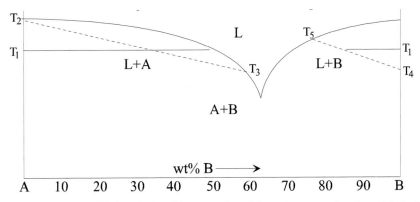

Figure D.9. Phase equilibria relationships are altered by microwave heating (Meek, 1992). *Note.* A and B represent two distinct minerals that are mixed in different proportions, represented by a point along the X axis. L = Liquid, L + A = Liquid + Mineral A, L + B = Liquid + Mineral B, and A + B indicates that the mixture of the two components is completely solid. See text for additional discussion.

from what would be expected with conventional heating. The kitchen microwave oven that they used was able to melt the lunar soil within a very short period of time. As mentioned in Appendix B, this characteristic can be used to control dust.

Taylor and Meek's work represents significant progress in understanding the usefulness of microwave heating for lunar construction. Figure D.9 shows a phase equilibrium[16] diagram for a two-component "regolith" composed of titania[17] and silica[18] (Meek, 1992).

"A" and "B" represent two mineral phases which are combined in a rock. The lower horizontal line is the temperature at which all of the material becomes completely solid (known as the solidus). T1, the next horizontal line, represents a hypothetical temperature at which various parts of the mixture will be solid and others will be liquid. The composition of the mixture at this temperature determines which mineral phases begin to solidify first, and how much of the liquid will be left as those mineral phases crystalize more and more. "L + A" indicates that at this temperature (and with the specific composition represented by the horizontal distance from "A"), crystals of phase "A" and a liquid will exist, but none of phase "B" will have crystalized. Similarly, "B + L" indicates that with a higher percentage of mineral "B" in the starting mixture, there will be certain temperatures where crystals of B exist, and a liquid, but none of mineral "A" will have solidified. The arcuate lines represent the temperature at which all of the material will be liquid (the "liquids"); and again the temperature where this occurs will vary depending upon the percentage of each phase that co-exists. The "L" in the field above the solidus shows that at any

[16] Phase equilibria: the study of the chemical balance between liquids and solids in a closed system, generally including mixtures of two or more compounds.
[17] Titania: common name for titanium dioxide, TiO_2.
[18] Silica: common name for silicon dioxide, SiO_2.

composition above the temperature of the solidus, all of the mixture will be 100% liquid.

What we have described so far are the components of the traditional "phase diagram"; but in this diagram there are two additional lines. The slanted lines (T2–T3 and T5–T4) show how these mixture relationships change when microwave heating is used instead of a more traditional heating source. For example, at a point along the "X" axis at which the mixture contains 40% of mineral "B", there will be more liquid in the mixture if microwave heating is used compared to the amount of liquid at that same temperature and composition if conventional heating were used. Conversely, with a composition that is 100% mineral "A" (at $X = 0$), there would be more solid material with microwave heating than there would be with conventional heating.

Microwave sintering can create a "pavement" for roads or landing pads. However, part of the reason that a microwave oven works so well is that the energy waves are reflected by the oven walls, and they do not escape the system until they are absorbed by the material being heated. Outdoors, this would not be the case. In fact, the radiation would only have one chance to be absorbed by the surface or subsurface, after which it would dissipate. Experiments are needed on the lunar surface to determine the most effective design for a system to form a microwave-sintered road bed.

Microwave production of aerobraking heat shields: One potential application of microwave sintering is the production of aerobraking heat shields from titanium dioxide (TiO_2). Aerobraking heat shields are used to reduce the speed of spacecraft upon entry into the atmosphere. The heat shield is eroded (and thus consumed) by friction with the atmosphere, but, using proper entry techniques, the next stages of the braking maneuver will allow the spacecraft to proceed safely to its destination. The ideal materials for heat shields are refractories[19] which consist of the oxides of aluminum, calcium, or titanium. Because TiO_2 is the by-product of the production of iron and oxygen from ilmenite ($FeTiO_3$), it will be available for the production of heat shields.

D.2.2.2   "Smart bricks"

The idea of sintering bricks has given rise to the more advanced concept of "smart bricks" (Allen *et al.*, 1994b). The name may have come from an old SDIO[20] project, but it is nevertheless very descriptive of the concept. The primary features of smart bricks are that they are interlocking and magnetic, and they can be made as part of the oxygen production process. As described in Chapter 7, this technology can also be

[19] Refractory: material, such as brick, that has a very high melting point.
[20] SDIO: Strategic Defense Initiative Organization. The Strategic Defense Initiative (SDI) was proposed by U.S. President Ronald Reagan in 1983. The initiative focused on strategic *defense* rather than the previous strategic *offense* doctrine of Mutual Assured Destruction (MAD). Under the administration of President Bill Clinton in 1993, its name was changed to the Ballistic Missile Defense Organization (BMDO) and its emphasis was shifted from national missile defense to theater missile defense (i.e., from global to regional coverage).

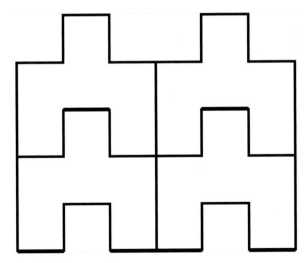

Figure D.10. Interlocking designs will facilitate robotic construction.

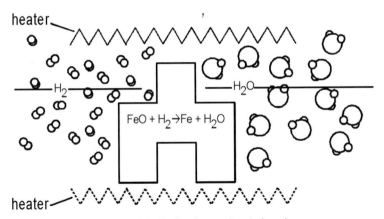

Figure D.11. Reduction-assisted sintering.

used in creating a lunar railway system, eliminating the need for connectors between rail sections.

Interlocking design allows the bricks to be fitted together more accurately. This will allow automation of construction operations, which will in turn reduce costs. Figure D.10 shows a simple interlocking design. Many other interlocking designs may be envisioned which would make robotic construction easier, such as conical connecting shapes.

Reduction-assisted sintering (Figure D.11) is the process by which smart bricks can be made. This allows oxygen to be extracted from the regolith material that is being used to make the bricks, concurrent with the brick-making process. This is an economical approach which allows utilization of every part of the regolith that has been excavated and hauled to the processing location. Either high-titanium soil or

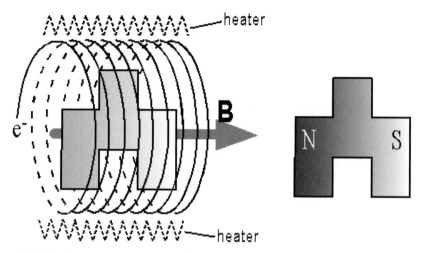

Figure D.12. Induction creates a magnetic polarity on each brick or other component, facilitating assembly.

iron-rich glass could be used in the process, in which hot hydrogen gas flows through the system to reduce the oxides. The reactions for high-titanium soil and for iron-rich volcanic glasses are described in Appendix E.

Volcanic glass would be the optimum feedstock for this process, because it is uniformly fine-grained, it reacts rapidly, and it can be used with little or no preprocessing. Additionally, iron-rich glass produces a higher oxygen yield than does high-Ti regolith.

The production of free iron in either of the above processes makes it possible to induce magnetism in the bricks. If the reduction process takes place in a magnetic field, each brick becomes magnetized. Its north pole will be repelled by the north poles and attracted by the south poles of bricks around it. As a consequence, the bricks will tend to naturally align themselves in the correct position. Figure D.12 shows the concept of induced magnetism (the reduction-assisted sintering would happen concurrently with this magnetization, but is left out of Figure D.12 for simplicity).

Magnetized bricks would have an advantage over non-magnetic ones for robotic construction operations. A robotic operation could be devised where an electromagnetic device would be placed on the end of the robot's arm, instead of a claw or other grasping device. When the electricity is turned on, the electromagnetic disk can pick up a magnetic brick by the brick's opposite pole. Then the arm would move the brick to a location close to where it is to be placed. When the electricity is turned off, the brick will drop into place, and its magnetic polarity will cause it to orient itself properly in relation to the other bricks or rails that have already been placed. Figure D.13 shows the concept of a lunar construction robot carrying a load of bricks and building a wall.

Magnetic data imprinting is also possible if the bricks are magnetized. This will

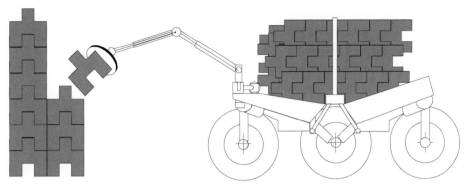

Figure D.13. Construction using "smart" bricks and an electromagnetic end effector.

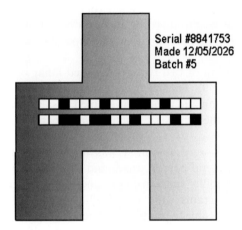

Serial #8841753
Made 12/05/2026
Batch #5

Figure D.14. Lot and part traceability can be incorporated into construction components via magnetic data imprinting.

allow information to be stored on each brick – for example, giving its manufacture date and other data which could be useful for quality control. This is illustrated in Figure D.14.

Magnetized bricks such as those described here have been produced in a hydrogen-reduction furnace, to demonstrate proof of principle. The bricks did indeed contain enough iron metal to permit lifting by a relatively-small magnet (Allen *et al.*, 1994b). Moreover, the electromagnetic force required for lifting would be reduced in lunar gravity.

D.2.2.3 Cast basalt

Some of the oxygen-extraction processes discussed in Appendix E require that the regolith be fully melted, after which various components can be extracted by electrolysis or other chemical reactions. If such processes are used on the Moon, the melted basalt which is waste material from these processes will become the feedstock for another process – the production of cast basalt bricks, paving stones, pipes, mineral wool, and other items (e.g., Jakeš, 1998). The first lunar manufacturing processes will use sintering and melting techniques that do not require highly-refined feedstock. Simple things such as chairs and tables might be early products.

D.3 MANUFACTURING OF OTHER ITEMS

D.3.1 Beneficiation

There may be some manufactured items that are more easily produced from pure minerals than from lunar soil. As explained in Appendix B, lunar soil contains agglutinates, and it is challenging to separate these into their constituent parts. Lunar soil is a better feedstock material than lunar rock for most purposes, because using soil would eliminate the need for blasting and grinding. However, if pure minerals are needed, it will be more practical to crush and separate rocks, which are devoid of agglutinates.

Magnetic beneficiation of two types of crushed lunar rock has been studied by Vaniman and Heiken (1990). Tests were conducted on a fine-grained basalt from the Apollo 17 sample collection and a medium-grained basalt from Apollo 11. The rocks were crushed and sized, then each size fraction was separated into 20 fractions based on magnetic susceptibility.

Because agglutinates were absent, it was much easier to separate the grains according to size and mineral type. It was found that ilmenite was concentrated into several magnetic susceptibility fractions, as a function of the amount of troilite (FeS) that was attached to the grains – the more troilite, the higher the magnetic suscept-ibility. It was therefore possible to produce an ilmenite fraction that was free of troilite. This is an important finding, because the presence of sulfur in the feedstock could cause sulfuric acid to be formed in the product water, or could result in the release of toxic H_2S gas (see Appendix B). With the 44–150 µm size fraction of the medium-grained basalt, concentrates with 64 wt%[21] free ilmenite grains were obtained. It was also possible to produce fractions which had about 90 vol%[22] plagioclase,[23] and others with greater than 90 vol% clinopyroxene.[24]

The medium-grained basalt, which contained only 13 wt% TiO_2, produced better grades of free ilmenite than did the finer-grained basalt, which had 20 wt% TiO_2. A coarse-grained basalt may be best for production of high-grade mineral separates.

---

[21] Wt% means percent by weight.

[22] Vol% means percent by volume.

[23] Plagioclase refers to any member of the solid-solution series of minerals that range in composition from $NaAlSi_3O_8$ to $CaAl_2Si_2O_8$.

[24] Clinopyroxene and orthopyroxene are the two general types of pyroxene, which is a class of silicate mineral found in many igneous and metamorphic rocks. Pyroxenes have the general formula $XY(Si,Al)_2O_6$ (where X usually represents calcium, sodium, iron^{+2} and magnesium and Y represents ions of smaller size, such as chromium, aluminum, iron^{+3}, magnesium, and manganese).

However, the additional cost in equipment mass and energy for mining (with possible blasting), crushing, transportation, and processing of these rocks means that this type of manufacturing will not be the first priority as the lunar base develops. Instead, it is more likely that this type of beneficiation may not occur until the more basic needs of water, oxygen, and propellant have been met.

D.3.2   Processing of beneficiated materials

After the regolith has been mined, beneficiated, and transported to the manufacturing area, further processing of the feedstock creates the raw materials for industry wire, ceramics, sheet metal, glass,[25] and so on.

The processes which are used to extract oxygen from the lunar regolith (Appendix E) will also yield other valuable elements such as silicon, aluminum, iron, and titanium. In addition, gentle heating releases the volatile elements that have been deposited into the regolith by the solar wind, including carbon, helium, and nitrogen.

Ion sputtering (see Appendix E) is an example of a higher-temperature, one-step refining process that has the potential of converting raw regolith materials into pure elements[26]. Both regolith and waste materials[27] from other processes could be converted into their component elements. Each of the recovered elements, such as silicon, aluminum, iron, hydrogen, and carbon, would then be used for producing computers, electric cable, structural components, plastics, and other items.

Ion sputtering is not likely to be used in the early phases of lunar development, because it is the most power-intensive of all of the proposed resource extraction processes. However, as power levels at the lunar base increase by the expansion of the lunar power system, experiments can be performed with processes that require larger amounts of power. Chapter 7 describes how the lunar electric power system will eventually generate abundant megawatts of electricity on the Moon. This wealth of energy will accommodate technological processes that allow all of the resources of the lunar regolith to be made available for industrial uses.

D.3.3   Chemical vapor deposition process

In the chemical vapor deposition (CVD) process, metals such as nickel, iron, or aluminum are chemically converted into a vaporizable compound, from which metal is deposited, molecule by molecule, onto a heated form. It creates full-strength metal

---

[25] Experiments on forming glass in a vacuum (Carsley *et al.*, 1992) suggest that lunar silicate glasses may possess superior mechanical properties compared with terrestrial glasses because the anhydrous lunar environment should prevent hydrolytic weakening of the strong Si–O bonds.

[26] Ion sputtering can also separate the isotopes of an element from one another, such as Helium-3 from Helium-4.

[27] If it can be proven, ion sputtering is the ultimate method of waste disposal and recycling – all lunar wastes will be separated into their constituent elements, which then become the feedstock for other manufacturing processes.

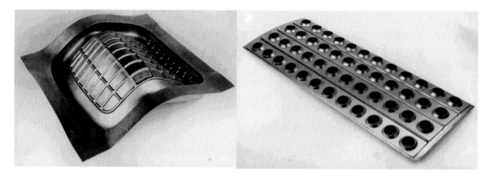

Figure D.15. Examples of metal objects formed using nickel carbonyl chemical vapor deposition (CVD). Left: tractor seat mold, 60 cm × 60 cm. Right: 48-cavity mold, used to form plastic cups.

objects of almost any desired shape, which are generally stronger than objects made by casting.

In order to achieve a high-quality product, the rate of deposition should not exceed ~0.25 mm of metal per hour. Deposits over large areas are uniform, and internal corners are filled better than with other deposition technologies. Also, projections of any parts of the mandrel (an object used to shape machined work) do not acquire excessive buildup as is typical with other deposition technologies. Molds can be made using this process, and they are typically utilized in reaction injection molding operations.

Figure D.15 shows two examples of objects formed using nickel carbonyl CVD. This is the starting point for developing iron carbonyl CVD for use on the Moon.

CVD can also be used as a method to extract metals from ores. The ore concentrate is reacted with carbon monoxide to make a metal carbonyl,[28] which is then decomposed to make metal powder or pellets. The resulting material will not be elementally pure, but the process can be tailored to minimize impurities. Some additional post processing or reprocessing may be required to reach acceptable contaminant levels.

It should also be noted that metal carbonyls are toxic. The OSHA ceiling for nickel carbonyl exposure is on the order of 1 part per billion (ppb) and the ceiling for iron carbonyl is 0.1 part per million (ppm). These are orders of magnitude smaller than the OSHA ceiling for hydrogen sulfide (H_2S) of 20 ppm. Moreover, iron carbonyl may require more complex equipment and processes than does nickel carbonyl.

Figure D.16 is a cross section of one type of CVD manufacturing chamber. It is divided into two sections by the mandrel in the middle and the cast silicon rubber separators. Vapors are injected into the upper half of the chamber, and the bottom half of the chamber is used to heat the mandrel. The heating section consists of a

---

[28] In inorganic chemistry, carbonyl refers to carbon monoxide. Carbon is triple-bonded to oxygen: C≡O.

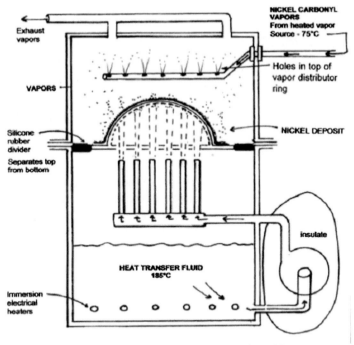

Figure D.16. Example of a chemical vapor deposition system.

series of jets of a heat exchange fluid, typically at 185°C, directed at the back side of the mandrel.

The mandrel is a cast high-temperature epoxy resin mixed with aluminum powder and granules to make it thermally conductive. With the heat transfer fluid at 190°C on the bottom, the top of the mandrel (where deposition occurs) reaches a temperature of 180°C. (The mandrel can also be made of cast tin or lead metal.) The parts that are not to be coated are kept below 125°C. Other methods of heating the mandrel (not shown) include using electrical heating elements or infrared heat supplied via transparent windows.

While nickel CVD is a well-understood process, iron CVD is still in the experimental stages. It is expected that a temperature of 200°C will be needed for iron CVD, and native iron can be supplied via one of the processes described in Appendix E. The iron is then reacted with CO to form a metal–carbonyl complex at 100°C. Although the temperature for this reaction is easily achievable, the pressures used in terrestrial operations are on the order of ~13,800 KPa. Research is ongoing to determine if an alternate method can be found in which these high pressures are not required.

D.3.4   Manufacturing of high-tech materials

As experience with manufacturing is gained, and as more specialized robotic manufacturing and other tools are delivered from the Earth to the Moon, it will

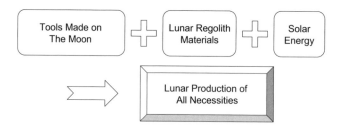

Figure D.17. The goal of lunar manufacturing: autonomy.

be possible to manufacture a wider variety of products. Eventually, it will be possible to manufacture high-technology items such as metal lathes, drills, electric motors, and electronic components from regolith materials.

When manufacturing tools are produced on the Moon, those tools can then be used to create other tools, including replicas of themselves. After many iterations, it will be possible to manufacture anything on the Moon that can be manufactured on Earth.

At this stage of development, lunar manufacturing will be self-sufficient, without the need to import any additional equipment from Earth. Automated machinery will manufacture robots, computers, and construction materials, and the pace of lunar development will then be a function of lunar-based industrial capability. Independent of the Earth, lunar manufacturing autonomy will be a reality (see Figure D.17).

D.4   CHALLENGES TO BE OVERCOME

D.4.1   Mineral rights and commercial use rights

Legal and political challenges, as well as habitual ways of thinking, must be overcome in order for Planet Moon to become a reality. The right to use lunar resources for profit needs to be clearly established and understood by all stakeholders. Without a clear path to profit from their efforts, private industry will continue in the role of government contractors. However, if industry has a way to profit from its efforts, much more private investment in lunar development will occur (see Chapter 9).

D.4.2   Political issues

International collaboration in the field of high-technology space-based operations is perceived by some nuclear proliferation authorities to be in conflict with the need for global security. Clear guidelines, agreed upon by all international partners, are needed to permit increased collaboration between the governments of the world.

D.4.3 Breaking old habits

Almost every country in the world now uses the metric system. In the U.S., high-technology machines that are used for fabricating metal parts are capable of working with either English or metric units. Moreover, scientists worldwide (including U.S. scientists) use the metric system exclusively. Not using the metric system for engineering efforts reduces the U.S. ability to work with international partners in the exploration of the Moon. The lunar government will almost certainly adopt the metric system of measurements (see Chapter 9).

D.5 SUMMARY

The resources of the Moon can be used from the beginning phases of lunar exploration and development, and will become increasingly useful over time as new technologies are fielded. *In-situ* resource utilization (ISRU) will begin by developing the most critical systems and capabilities: electrical power, rocket propellants, oxygen, and other consumables. Of equal importance will be the capability to make roads, habitats, and other facilities using local resources, instead of bringing everything from Earth. The waste products from oxygen-extraction processes will provide feed materials for construction.

The initial elements of the industrial base that are delivered to the Moon must be able to withstand the harsh conditions (thermal stress, vacuum, micrometeorite bombardment, and cosmic radiation) that exist on the lunar surface. Their task will be to convert unprocessed regolith into commodities that will be used for the initial build-up of the base. To that end, the first manufacturing elements to be sent to the Moon are likely to include solar cell fabrication equipment and regolith-sintering machines.

Solar ovens and sintering machines will be used to manufacture bricks, glass, and other construction materials. Vertical walls that shield equipment from the Sun, level pathways that are relatively free from lunar dust, and berms to protect buildings from rocket blast debris will be among the first uses of lunar materials. From that beginning, semi-closed shelters will be constructed so that the industrial base will move underground, and all equipment that is not needed on the lunar surface will be moved into the shelters for protection from thermal stress, meteorites, and radiation.

For the final stages of industrial base development, continually-growing electric power levels and the delivery of additional manufacturing equipment from the Earth will allow the expansion of the base to include completely-closed underground chambers. The chambers can be lined with glass or other sealant material, then pressurized to create manufacturing areas, plant-growth areas, or human-habitation areas. Eventually, it will be possible to replicate virtually any Earth-based technology at the industrial base on the Moon, and lunar manufacturing autonomy will be assured.

The harsh conditions of the lunar environment and $\frac{1}{6}$ gravity will present challenges for the establishment and operation of the industrial base, but no insurmountable technological barriers are apparent. When it becomes possible to construct sealed and pressurized underground chambers on the Moon, optimum conditions can be created for every manufacturing, agricultural, scientific, and human-habitation purpose.

D.6   REFERENCES

Allen, C.C. (1998) Bricks and Ceramics. In: M.B. Duke, ed., *Workshop on Using In Situ Resources for Construction of Planetary Outposts.* LPI Technical Report 98-01, Lunar and Planetary Institute, Houston, TX.

Allen, C.C. and Zubrin, R. (1999) In-situ resources. In: W.J. Larson and L.K. Pranke, eds., *Human Spaceflight: Mission Analysis and Design* (Space Technology Series), McGraw-Hill, New York, pp. 477–512.

Allen, C.C.; Hines, J.A.; McKay, D.S.; and Morris, R.V. (1992) Sintering of Lunar Glass and Basalt. In: *Engineering, Construction and Operations in Space III, Space '92, Proceedings of the Third International Conference,* American Society of Civil Engineers, New York, pp. 1209–1218.

Allen, C.C.; Bond, G.G.; and McKay, D.S. (1994a) Lunar oxygen production – a maturing technology. In: *Engineering, Construction and Operations in Space IV, Space '94, Proceedings of the Fourth International Conference,* American Society of Civil Engineers, New York, pp. 1157–1166.

Allen, C.C.; Graf, J.C.; and McKay, D.S. (1994b) Sintering Bricks on the Moon. In: *Engineering, Construction and Operations in Space IV, Space '94, Proceedings of the Fourth International Conference,* American Society of Civil Engineers, New York, pp. 1220–1229.

Altenberg, Barbara H. (1992) Oxygen Production on the Lunar Materials Processing Front. *Proceedings of Lunar Materials Technology Symposium,* NASA Space Engineering Research Center, Arizona University.

Anthony, D.L.; Cochran, W.C.; Haupin, W.E.; Keller, W.E.; and Larimer, K.T. (1988) Dry extraction of silicon and aluminum from lunar ores [abstract]. In: W.W. Mendell, ed., *Second Symposium on Lunar Bases and Space Activities of the 21st Century,* NASA Conference Publication 3166, paper no. LBS-88-066.

Beall, George H. (1992) Glasses, Ceramics and Composites from Lunar Materials. *Proceedings of Lunar Materials Technology Symposium,* NASA Space Engineering Research Center, Arizona University.

Beck, Theodore R. (1992) Metals Production. *Proceedings of Lunar Materials Technology Symposium,* NASA Space Engineering Research Center, Arizona University.

Bents, D.J. (1987) High-temperature solid oxide regenerative fuel cell for solar photovoltaic energy storage. *IECEC '87; Proceedings of the Twenty-second Intersociety Energy Conversion Engineering Conference, Philadelphia, PA, August 10–14, 1987,* Vol. 2 (A88-11776 02-20). American Institute of Aeronautics and Astronautics, New York, pp. 808–817.

Bhogeswara, R.; Choudaray, U.; Erstfield, T.; Williams, R.; and Chang, Y. (1979) Extraction processes for the production of aluminum, titanium, iron, magnesium and oxygen from nonterrestrial sources. In: J. Billingham and W. Gilbreath, eds., *Space Resources and Space Settlements*, NASA SP-428, pp. 257–274.

Binder, A.B.; Culp, M.A.; and Toups, L.D. (1990) Lunar derived construction materials: cast basalt. In: S.W. Johnson and J.P. Wetzel, eds., *Engineering, Construction and Operations in Space II, Space '90, Proceedings of the Second International Conference*. American Society of Civil Engineers, New York.

Boles, W. (1992) Drilling and digging techniques for the early lunar outpost [abstract]. In: *Joint Workshop on New Technologies for Lunar Resource Assessment*. Lunar and Planetary Institute, Houston, TX, pp. 12–14.

Boundy, R.A. (1983) Executive summary. In: W.F. Carroll, ed., *Research on the Use of Space Resources*, NASA Jet Propulsion Laboratory, Pasadena, CA.

Briggs, R.A. and A. Sacco (1988) Oxidation and reduction of ilmenite: Application to oxygen production on the Moon. In: W.W. Mendell, ed., *Second Symposium on Lunar Bases and Space Activities of the 21st Century*, NASA Conference Publication 3166, paper no. LBS-88-170.

Burt, D.M. (1988a) Lunar production of oxygen and metals using fluorine: Concepts involving fluorite, lithium, and acid-base theory [abstract]. *Lunar Planet. Sci. Conf. 19th*, Lunar and Planetary Institute, Houston, TX.

Bustin, R. and Gibson, E.K. Jr. (1988) Availability of hydrogen for lunar base activities. In: W.W. Mendell, ed., *Proceedings, Second Conference on Lunar Bases and Space Activities of the 21st Century*, NASA Conference Publication 3166.

Cammarota, V. Anthony Jr. (1984) America's Dependence of Strategic Minerals. In: Gerard J. Mangone, ed., *American Strategic Minerals*, Crane Russak & Co., New York, pp. 29–58.

Capps, S. and Wise, T. (1990) Lunar basalt construction materials. In: S.W. Johnson and J.P. Wetzel, eds., *Engineering, Construction and Operations in Space II, Space '90, Proceedings of the Second International Conference*, American Society of Civil Engineers, New York.

Carroll, W.F., ed. (1983) *Research on the Use of Space Resources*, NASA Jet Propulsion Laboratory, Pasadena, CA.

Carsley, J. E.; Blacic, J. D.; and Pletka, B. J. (1992) Vacuum melting and mechanical testing of simulated lunar glasses. In: S.W. Johnson and J.P. Wetzel, eds., *Engineering, Construction and Operations in Space III, Space '92, Proceedings of the Third International Conference*, American Society of Civil Engineers, New York.

Carter, J.L. (1975) Surface morphology and chemistry of rusty particle 60002,108. *Proc. Lunar Sci. Conf. 6th*, Lunar Science Institute, Houston, TX, pp. 711–718.

Carter, J.L. (1985) Lunar regolith fines: A source of hydrogen. In: W.W. Mendell, ed., *Lunar Bases and Space Activities of the 21st Century*, Lunar and Planetary Institute, Houston, TX, pp. 571–581.

Carter, J.L. and McKay, D.S. (1972) Metallic mounds produced by reduction of material of simulated lunar composition and implications on the origin of metallic mounds on lunar glasses. *Proc. 3rd Lunar Sci Conf.* (Supplement 3, *Geochim Cosmochim. Acta*, **1**), MIT Press, Cambridge, MA, pp. 953–970.

Chang, M.C.S. (1959) *Process for treating materials containing titanium and iron*. U.S. Patent No. 2,912,320, United States Patent Office, Washington, D.C.

Christiansen, E.L.; Euker, H.; Maples, K.; Simonds, C.H.; Zimprich, S.; Dowman, M.W.; and Stovall, M. (1988) *Conceptual Design of a Lunar Oxygen Pilot Plant*. NAS9-17878, EEI Rpt. No. 88-182, Eagle Engineering Inc., Houston, TX.

Cooper, B.L. (1990) Sources and Subsurface Reservoirs of Lunar Volatiles. In: *Proceedings of the 20th Lunar and Planetary Science Conference*, Lunar and Planetary Institute, Houston, TX.

Cooper, B.L. (1995) Lunar ilmenite for solar power cells. *Space Resource News*, Vol. 4, No. 6, League City, TX.

Criswell, D.R. (1972) Lunar Dust Motion. In: *Proc. 3rd Lunar Sci. Conf.*, MIT Press, Cambridge, MA, pp. 2671–2680.

Criswell, D.R. (1973) Horizon-Glow and the Motion of Lunar Dust. In: *Photon and Particle Interactions with Surfaces in Space*, D. Reidel Publishing Company, Dordrecht, Holland, pp. 545–556.

Criswell, D.R. (1983) A transportation and supply system between low Earth orbit and the Moon which utilizes lunar derived propellants [abstract]. *Lunar Planet. Sci.*, **14**, Special Session Abstracts, Lunar and Planetary Institute, Houston, TX.

Curreri, P.A. (1990) *Ore nonspecific process for differentiation and the restructuring of minerals in vacuum*. Patent application, NASA in-house memorandum.

Cutler, A.H. and Krag, P. (1985) A carbothermal scheme for lunar oxygen production. In: W.W. Mendell, ed., *Lunar Bases and Space Activities of the 21st Century*, Lunar and Planetary Institute, Houston, TX, pp. 559–569.

Cutler, A.H. and Wolff, A.R. (1990) Impact of silane fuel on lunar oxygen supply to low Earth orbit. *Proc. AIAA Space Prog. and Technologies Conf, 1990*. American Institute of Aeronautics and Astronautics, Washington, D.C.

Dalton, C. and Hohmann, E. (1972) *Conceptual Design of a Lunar Colony*, NASA/ASEE Systems Design Institute, NGT 44-005-114.

Davis, H.P. (1983) Lunar oxygen impact upon STS effectiveness. *Lunar Planet. Sci.*, **14**, Special Session Abstracts, Lunar and Planetary Institute, Houston, TX.

de Weck, O.L.; Nadir, W.D.; Wong, J.G.; Bounova, G.; and Coffee, T.M. (2005). Modular Structures for Manned Space Exploration: The Truncated Octahedron as a Building Block. In: *1st Space Exploration Conference: Continuing the Voyage of Discovery*. American Institute of Aeronautics and Astronautics, New York, paper no. AIAA 2005-2764.

du Fresne, E. and J.E. Schroeder (1983) Magma electrolysis. In: W.F. Carroll, ed., *Research on the Use of Space Resources*, NASA Jet Propulsion Laboratory, Pasadena, CA.

Franklin, H.A. (1998) Materials Transportation. In: Duke, M.B., ed., *Workshop on Using In Situ Resources for Construction of Planetary Outposts*, LPI Technical Report 98-01, Lunar and Planetary Institute, Houston, TX.

Gibson, E.K. Jr. and Moore, G.W. (1973) Variable carbon contents of lunar soil 74220. *Earth and Planet. Sci. Lett.*, **20**, no. 3, 404–408.

Gibson, E.K. Jr.; Moore, G.W.; and Johnson, S.M. (1974) *Summary of analytical data from gas release investigations, volatilization experiments, elemental abundance measurements of lunar samples, meteorites, minerals, volcanic ashes and basalts*. NASA Johnson Space Center, Houston, TX.

Gibson, M.A. and Knudsen, C.W. (1985) Lunar oxygen production from ilmenite. In: W.W. Mendell, ed., *Lunar Bases and Space Activities of the 21st Century*, Lunar and Planetary Institute, Houston, TX, pp. 543–550.

Gibson, M.A. and Knudsen, C.W. (1988) Lunar oxygen production from ilmenite. In: W.W. Mendell, ed., *Second Symposium on Lunar Bases and Space Activities of the 21st Century*, paper no. LBS-88-056.

Gibson, M.A.; Knudsen, C.W.; and Roeger, A. III (1990) Development of the Carbotek process (TM) for lunar oxygen production. In: S.W. Johnson and J.P. Wetzel, eds.,

Engineering, Construction and Operations in Space II, Space '90, Proceedings of the Second International Conference, American Society of Civil Engineers, New York.

Heiken, G.H.; Vaniman, D.T.; and French, B.M., eds. (1991) *Lunar Sourcebook*, Cambridge University Press and Lunar and Planetary Institute, Houston, TX.

Henderson, B. (1990) Sandia researchers test "coil gun" for use in orbiting small payloads, *Av. Wk. & Space Tech.*, May 7.

Henderson, P. (1984) General geochemical properties and abundances of the Rare Earth Elements. In: P. Henderson, ed., *Rare Earth Element Geochemistry*, Elsevier Science Publishers, Amsterdam, pp. 1–29.

Hörz, F (1985) Lava Tubes: Potential Shelters for Habitats. In: W. Mendell, ed., *Lunar Basins and Space Activities of the 21st Century*, Lunar and Planetary Institute, Houston, TX.

Jakeš, P. (1998) Cast Basalt, Mineral Wool, and Oxygen Production: Early Industries for Planetary (Lunar) Outposts. In: *Workshop on Using in situ Resources for Construction of Planetary Outposts*, LPI Technical Report 98-01, Lunar and Planetary Institute, Houston, TX.

Keller, R. (1986) *Dry extraction of silicon and aluminum from lunar ores*. Final report, SBIR contract NAS9-17575, EMEC Consultants.

Kulcinski, G.L.; Cameron, E.N.; Santarius, J.F.; Sviatoslavsky, I.N.; Wittenberg, L.J.; and Schmitt, H.H. (1988) Fusion energy from the Moon for the twenty-first century. In: W.W. Mendell, ed., *The Second Conference on Lunar Bases and Space Activities of the 21st Century*, NASA Conference Publication 3166, NASA Johnson Space Center, Houston, TX.

Landis, G.A. (1998) Materials refining for structural elements from lunar resources. In: M.B. Duke, ed., *Workshop on Using In Situ Resources for Construction of Planetary Outposts*, LPI Technical Report 98-01, Lunar and Planetary Institute, Houston, TX.

Lewis, J.S.; Jones, T.D.; and Farrand, W.H. (1988) Carbonyl extraction of lunar and asteroidal materials. In: S.W. Johnson and J.P. Wetzel, eds., *Engineering, Construction and Operations in Space I, Space '88, Proceedings of the First International Conference*, American Society of Civil Engineers, New York.

Lin, T.D *et al.* (1998) Lunar and Martian Resource Utilization – Cement and Concrete. In: M.B. Duke, ed., *Workshop on Using In Situ Resources for Construction of Planetary Outposts*, LPI Technical Report 98-01, Lunar and Planetary Institute, Houston, TX.

Lynch, D.C. (1989) Chlorination processing of local planetary ores for oxygen and metallurgically important metals. In: *Space Engineering Research Center for Utilization of Local Planetary Resources Annual Progress Report 1988–89*, University of Arizona, Tucson.

Maskalick, N.J. (1984) *High-Temperature Electrolysis Cell Performance Characterization*. T.N. Veziroglu and J.B. Taylor, eds., Pergamon Press, New York, pp. 801–812.

McCullough, E. and Mariz, C. (1990) Lunar oxygen production via magma electrolysis. In: S.W. Johnson and J.P. Wetzel, eds., *Engineering, Construction and Operations in Space II, Space '90, Proceedings of the Second International Conference*, American Society of Civil Engineers, New York.

McGannon, H., ed. (1964) *The Making, Shaping, and Treating of Steel*, 8th edn., United States Steel Corporation, Pittsburgh, PA.

Meek, T.T. (1992) Interaction of Microwave Radiation with Matter: A Thermodynamic Approach. In: S.W. Johnson and J.P. Wetzel, eds., *Engineering, Construction and Operations in Space II, Space '90, Proceedings of the Second International Conference*, American Society of Civil Engineers, New York.

Morris, R.V.; Score, R.; Dardano, C.; and Heiken, G. (1983) *Handbook of Lunar Soils*, NASA Planetary Materials Branch Publication 67, JSC 19069, NASA Johnson Space Center, Houston, TX.

Morrison, D.A. (1992) *Lunar Engineering Models: General and Site Specific Data*, Review Draft, NASA Johnson Space Center Exploration Programs Office, Houston, TX.

NASA Space Engineering Research Center (1989) *Space Mining and Manufacturing*, University of Arizona, Tucson.

Neary, C.R. and Highley, D.E. (1984) The economic importance of the Rare Earth Elements. In: P. Henderson, ed., *Rare Earth Element Geochemistry*, Elsevier Science Publishers, Amsterdam, pp. 423–461.

Ness, R.O. Jr.; Runge, B.D.; and Sharp, L.L. (1990) Process design options for lunar oxygen production. In: S.W. Johnson and J.P. Wetzel, eds., *Engineering, Construction and Operations in Space II, Space '90, Proceedings of the Second International Conference*, American Society of Civil Engineers, New York.

Oberbeck, V.R.; Greely, R.; Morgan, R.B.; and Lovas, M.J. (1969) On the origin of lunar sinuous rilles, *Modern Geology*, **1**, 75–80.

Oder, R.R. and Taylor, L.A. (1990) Magnetic beneficiation of highland and hi-Ti mare soils: magnet requirements. In: S.W. Johnson and J.P. Wetzel, eds., *Engineering, Construction and Operations in Space II, Space '90, Proceedings of the Second International Conference*, American Society of Civil Engineers, New York.

Phinney, W.C.; Criswell, D.; Drexler, E.; and Garmirian, J. (1977) Lunar resources and their utilization. In: J. Grey, ed., *Space Manufacturing Facilities II, Proceedings of the Third Princeton/AIAA Conference*, pp. 171–182.

Register, B.M. (1989) *Proceedings of the Second Workshop on In Situ Resource Utilization*, NASA OEXP Annual Report, FY 1989, NASA TM 4170, Vol. VI.

Rosenberg, S.D.; Guter, G.A.; and Miller, F.E. (1966) The on-site manufacture of propellant oxygen from lunar resources. *Aerospace Chemical Engineering, AIChE*, **62**, no. 61, 228–234.

Rowley, J.C. and Neudecker, J.W. (1985) In situ rock melting applied to lunar base construction and for exploration drilling and coring on the Moon. In: W.W. Mendell, ed., *Lunar Bases and Space Activities of the 21st Century*, Lunar and Planetary Institute, Houston, TX.

Sammells, A.F. and Semkow, K.W. (1988) Electrolytic cell for lunar ore refining and electric energy storage [abstract]. *Proc. Second Symposium on Lunar Bases and Space Activities of the 21st Century, Houston*.

Semkow, K.W. and Sammells, A.F. (1987) The indirect electrochemical refining of lunar ores. *J. Electrochem. Soc.*, **134**, no. 8, 2088–2089.

Sherwood, B. (1992) Lunar Materials Processing System Integration. In: *Proceedings of Lunar Materials Technology Symposium*, NASA Space Engineering Research Center, Boeing Defense and Space Group, 1992.

Simon, M.C. (1985) A parametric analysis of lunar oxygen production. In: W.W. Mendell, ed., *Lunar Bases and Space Activities of the 21st Century*, Lunar and Planetary Institute, Houston, TX, pp. 531–541.

Steurer, W.H. and Nerad, B.A. (1983) Vapor phase reduction. In: W.F. Carroll, ed., *Research on the Use of Space Resources*, NASA Jet Propulsion Laboratory, Pasadena, CA.

Stone, J.L. (1993) Photovoltaics: Unlimited electrical energy from the Sun. *Physics Today*, September 1993, pp. 22–29.

Sullivan, T.A. (1990) Process engineering concerns in the lunar environment. *Proc. AIAA Space Prog. and Technologies Conf.* American Institute of Aeronautics and Astronautics, Washington, D.C. (in press).

Sullivan, T.A.; Koenig, E.; Knudsen, C.W.; and Gibson, M.A. (1992) Pneumatic Conveying in Partial Gravity. AIAA Paper #92-1667. *Proc. AIAA Space Prog. and Technologies Conf, 1992.*

Sunkara, S.S. (1995) *Growth and evaluation of ilmenite wide band gap semiconductor for high temperature electronic applications.* Ph.D. dissertation, Texas A&M University.

Taylor, L.A., and Eimer, B. (2006) Lunar Regolith, Soil, and Dust Mover on the Moon [abstract]. In: *Space Resources Roundtable VIII*, Lunar and Planetary Institute, Houston, TX.

Taylor, L.A. and Meek, T.T. (2005) Microwave Sintering of Lunar Soil: Properties, Theory, and Practice. *Journal of Aerospace Engineering*, **18**, no. 3, 188–196.

Taylor, L.A and Oder, R.R. (1990) Magnetic beneficiation and hi-Ti mare soils: Rock, mineral, and glassy components. In: S.W. Johnson and J.P. Wetzel, eds., *Engineering, Construction and Operations in Space II, Space '90, Proceedings of the Second International Conference.* American Society of Civil Engineers, New York.

Turkevich, A.L. (1973) The average chemical composition of the lunar surface. *Proc. 4th Lunar Sci. Conf.* (Supplement 4, *Geochim. et Cosmochim. Acta*), pp. 1159–1168.

Vaniman, D.T. and Heiken, G.H. (1990) Getting lunar ilmenite: From soils or rocks? In: S.W. Johnson and J.P. Wetzel, eds., *Engineering, Construction and Operations in Space II, Space '90, Proceedings of the Second International Conference.* American Society of Civil Engineers, New York.

Vaniman, D.; Pettit, D.; and Heiken, G. (1988) Uses of lunar sulfur [abstract]. In: *Proc. second symposium on Lunar Bases and Space Activities of the 21st Century*, Lunar and Planetary Institute, Houston, TX.

Volk, W. and Stotler, H.H. (1970) Hydrogen reduction of ilmenite ores in a fluid bed. *Journal of Metals*, **22**, no. 11, 50–53.

Waldron, R.D. (1985) Total separation and refinement of lunar soils by the HF acid leach process. *Proc. 7th Princeton Space Manufacturing Conf.*, American Institute of Aeronautics and Astronautics, New York.

Waldron, R.D. (1988) Lunar Manufacturing: A Survey of Products and Processes. *Acta Astronautica*, **17**, no. 7, 691–708 (Pergamon Press).

Waldron, R.D. (1989) Magma partial oxidation: A new method for oxygen recovery from lunar soil. *Space Manufacturing 7: Space Resources to Improve Life on Earth*, American Institute of Aeronautics and Astronautics, Washington, D.C.

Waldron, R.D. and Criswell, D.R. (1979) Overview of methods for extraterrestrial materials processing. In: *Fourth Princeton/AIAA Conference on Space Manufacturing Facilities*, American Institute of Aeronautics and Astronautics, New York.

Waldron, R.D. and Criswell, D.R. (1982) Materials processing in space. In: B. O'Leary, ed., *Space Industrialization*, Vol. 1, CRC Press, Boca Raton, FL, pp. 97–130.

Williams, R.J. (1985) Oxygen extraction from lunar materials: An experimental test of an ilmenite reduction process. In: W.W. Mendell, ed., *Lunar Bases and Space Activities of the 21st Century*, Lunar and Planetary Institute, Houston, TX.

Williams, R.J. and Mullins, O. (1983) Enhanced production of water from ilmenite: An experimental test of a concept for producing lunar oxygen [abstract]. *Lunar Planet Sci.*, **14**, Special Session Abstracts, pp. 34–45.

Woodcock, G.R. (1986) Economic potentials for extraterrestrial resources utilization. *37th Congress of the International Astronautical Federation, Innsbruck, Austria*, IAA-86-451, American Institute of Aeronautics and Astronautics, New York.

Woodcock, G.R. (1989) Parametric analysis of lunar resources for space energy systems. *Space Manufacturing 7: Space Resources to Improve Life on Earth*, American Institute of Aeronautics and Astronautics, Washington, D.C.

Appendix E

Proposed processes for lunar oxygen extraction

E.1 INTRODUCTION

As explained in Appendix D, the resources of the Moon can be used to create solar cells, produce rocket propellants, and provide building materials. As these technologies are developed further, they will become increasingly useful. A number of trade studies have addressed the economic viability of producing oxygen from lunar resources. In spite of widely-varying initial assumptions, almost all of these studies conclude that oxygen production would be economically beneficial for lunar base development. These studies are described in the first part of this appendix. The second part is a review of some of the proposed processes for lunar oxygen production, and some experiments that have been performed to validate the concepts.

E.2 TRADE STUDIES

Studies have shown that large quantities of propellant will be required for an expanded human presence in space. It costs at least \$8,800 per kilogram (\$4,000.00 per pound) to lift anything, including propellant, from Earth's surface into low Earth orbit (e.g., Adler, 1985). For transportation aboard the shuttle, the cost is estimated at \$19,000 per kilogram (\$8,700 per pound). It would seem reasonable that a saving could be obtained by producing as much of the raw materials as possible on the Moon, rather than lifting all of the propellant for the round trip from Earth.

Several trade studies have been done in recent years that address the pros and cons of producing liquid oxygen (LOX) on the Moon as opposed to exporting it from Earth. These will be presented in chronological order, along with the assumptions that were used and conclusions reached. None of the trade studies assumed the existence of ice in the lunar polar regions, because at the time the studies were done there was no evidence to support the hypothesis.

E.2.1   Criswell, 1983; and Davis, 1983

Criswell (1983) suggested that the successful operation of a lunar supply system could almost eliminate the need for the Space Shuttle to carry propellant to low Earth orbit (LEO).[1] This would increase its useful cargo capacity. The savings created by a lunar-based transportation system could quickly exceed the costs involved in developing it. These results were based on the assumption that hydrogen in a LOX–hydrogen rocket system is provided from Earth, and only the oxygen is produced on the Moon. Other workers have asserted that lunar oxygen production would be cost-effective to the point of "fully supporting" a lunar surface outpost (Davis, 1983).

E.2.2   Simon, 1985

Simon (1985) found that four major systems would be necessary for oxygen production on the Moon: a processing and storage facility, habitation modules, a power system, and a transportation and logistics system. He assumed an oxygen production rate of 1,000 tonnes per year, and chose the most mature processes available for each system. It was assumed that an initial lunar base would be in place prior to the emplacement of the oxygen production facility. Thus, the capital costs included in Simon's (1985) study were only the marginal costs of expanding the lunar base to support oxygen production. The cost of the lunar base oxygen production facility was predicted to be \$3.1 billion. This would result in a cost of \$2,340 per kg (\$1,100 per pound) for oxygen delivered to low Earth orbit from the Moon, an order of magnitude lower than the cost of transportation from Earth. Best-case and worst-case values were also computed, and the range was \$1,450 to \$3,410 per kg (\$656 to \$1,543 per pound).

The principal driver for the capital costs is the cost of power, while the principal driver for the operations cost is the cost of transportation. Because capital costs were assumed to be amortized over a ten-year period, the cost of transporting equipment from Earth to the Moon has a much greater overall impact on the cost of lunar oxygen than the cost of the equipment itself. Three major cost-reducing strategies are (1) reduce or eliminate the need to transport hydrogen from Earth to the Moon; (2) reduce space transportation costs, particularly the cost of Earth-to-Moon transportation; (3) reduce lunar base re-supply requirements.

E.2.3   Eagle Engineering, 1985

A study done by Eagle Engineering in 1985 presented more pessimistic results (Stump *et al.*, 1985). This study assumes the sale of lunar liquid oxygen (LLOX, or LUNOX) at LEO after it has been produced on the Moon. A market of 4,316 tonnes/year of

---

[1] Low Earth orbit (LEO) is commonly defined as an orbit at an altitude between 200 and 2,000 km (124 to 1,240 miles) above the Earth's surface.

LLOX was found to be necessary for breakeven, with minimum development/transportation/infrastructure costs of $30 to $60 billion. Only if the government were to absorb this initial cost could the operation become profitable for a commercial enterprise. This study ignores all possible markets for LLOX other than LEO, because a local lunar market does not in itself provide a rationale for a return to the Moon.

This study places the entire burden of the lunar base on the sale of oxygen at LEO, rather than assessing a marginal cost as was done in the other analyses. No science, national goals, or political reasons are considered as possible justifications for any part of the operation. In effect, a different question is asked: "Should we go back to the Moon just to produce LLOX for LEO?" The answer was "no". In other trade studies, the question being asked is: "If we go back to the Moon, would it be worthwhile to produce oxygen there while we are at it?" In those cases, the answer was "yes".

E.2.4   Woodcock, 1986

Woodcock (1986) studied three economic models, each driven by a different motivation for using ISRU: (1) using lunar resources to enhance a lunar base; (2) using lunar resources for a space transportation operation; and (3) using ISRU to achieve self-sufficiency of a space settlement.

E.2.4.1   Woodcock Case Study #1

The use of lunar oxygen in a simple refueling mode for ascent from the lunar surface would reduce Earth-launch requirements by about half. This would result in a savings of approximately $200 million (1986 dollars) annually. This level of savings makes the investment of marginal value; however, the economics would improve as the facility expanded. More important for lunar base build-up is the use of local construction materials. If all the required construction materials were delivered from Earth, Woodcock (1986) projected that 60 years would be required to deliver the material to the Moon.

Woodcock concluded that construction materials may be more important than oxygen, due to the greater mass required and the correspondingly-greater launch costs. However, he determined that using ISRU for refuelling was not a viable option.

E.2.4.2   Woodcock Case Study #2

Here, a lunar base is established for the primary purpose of delivering lunar materials to another location in space for use. The comparison is then made with delivering those same materials from Earth. This scenario assumes the use of a mass driver to launch products from the Moon to LEO. Woodcock found a surprising result: the potential advantage of lunar oxygen is not dependent on a series of highly-optimistic assumptions – it only depends on an efficient mass driver system.

Since 1986, considerable advancement in mass driver technology has occurred. Researchers at Sandia National Laboratories have demonstrated a mass driver which was smaller than a refrigerator and capable of 5,500 gravities of acceleration (Henderson, 1990). They calculated that a full-size launcher, with a booster rocket included with each projectile, would be price-competitive with heavy-lift launchers over a 2,000-launch amortization. The mass driver has become feasible only recently, as a result of the development of smaller, lighter capacitors, and the development of a non-contact flight path (Henderson, 1990).

If lunar oxygen is only used in LEO, Woodcock (1986) finds the economics to be favorable. If, however, there was an opportunity to use lunar oxygen at space locations more distant than LEO, the economic advantage would increase.

E.2.4.3   Woodcock Case Study #3

Case Study #3 deals with using lunar oxygen to promote the self-sufficiency of space colonies through export. This case study required many assumptions. The preliminary indication is that unless the colony is almost fully self-sufficient, it will probably present a re-supply burden which is quite similar to that of the "company town" scenario.

Woodcock (1986) concludes that a rational lunar development scenario begins with the establishment of a small base, with early and continuing emphasis on developing local resources. Through expansion of the base – and lower cost space operations using lunar oxygen – the cost of a trip to Mars would be substantially reduced. The continual development of a lunar base into a permanent and self-sufficient colony would not be a result of special development but rather a natural outcome of the economic development of space.

E.2.5   Astronautics Corporation of America, 1987

In a study prepared by Astronautics Corporation of America (1987), only chemical propulsion systems were considered. The conclusions were less optimistic than those of the Woodcock (1986) study, in part because the potential advantages of mass drivers were ignored. A second key assumption implicit in the report is that a lunar science outpost will be built without consideration for human planetary exploration or eventual colonization. In other words, lunar resources are only considered as a means of returning personnel and scientific samples to Earth.

In the context of the Astronautics (1987) report, round trips from Earth's surface to the Moon to establish and maintain an outpost are the major expense. In this particular case, an aerobrake can reduce the Earth launch mass by approximately 25 percent when produced on the Moon, even if no lunar oxygen is available. Alternatively, using lunar oxygen re-supply can reduce Earth launch mass by 15 percent when the aerobrake is carried up from Earth, to the Moon, and back to low Earth orbit where it is used in the process of slowing the spacecraft to orbital speeds.

To generalize the conclusions of this study, we can say that lunar oxygen production represents cost savings for any lunar base scenario. Even if oxygen is not being delivered to LEO and the only use of ISRU is to build heat shields for return to Earth from the Moon, there are cost savings.

E.2.6   Weaver, 1989

A report by David Weaver (Office of Exploration, 1989) used the actual dollar cost of the launched mass, plus all the other intangibles which are associated with it, to determine the activity level at which oxygen at a lunar base becomes economically attractive. The results given in Weaver's analysis are less optimistic than others.

One of the conclusions of his study was that the potential benefit of lunar oxygen is strongly dependent upon the location at which the demand occurs. The potential return on investment is higher at a location which is closer to the production site itself. According to Weaver, most trade studies have assumed that the oxygen market would be at LEO. This is referred to as the "LLOX-to-LEO" scenario, where "LLOX" is an acronym for Lunar Liquid Oxygen. In order for the LLOX-to-LEO scenario to be economically viable, over 1,000 tonnes of oxygen per year would need to be shipped from the Moon to LEO. Instead of using the LLOX-to-LEO assumption, Weaver focused on using LLOX for operations between low lunar orbit (LLO) and the lunar surface.

Various approaches can be used for modeling a lunar base to determine the effects of LLOX on the system. In one of these approaches, the model tracks the detailed evolution of a lunar base, with and without LLOX production, starting with the first flight and ending at some arbitrary point. This technique is highly accurate for the specific development scenario chosen, but it is not robust for describing other development scenarios with different assumptions.

Another approach tracks only the differential effects of adding LLOX production to a base that has already reached a steady-state level of activity. Several different steady-state levels are examined to determine the level at which a reasonable return on investment can be achieved. This method is somewhat less precise than the previous one, but it has the advantage of being much less time-consuming. The second method was chosen for the Office of Exploration (1989) study.

Three different levels of development of the lunar base were chosen, based on the number of inhabitants. Best-case, average-case, and worst-case assumptions were made for the three different categories. The first category of development was a four-person base with an SP-100-class nuclear power supply, space-station-derived habitation modules, and 90 percent closure of the air and water loops for life support. It was found that a return on investment occurs within 5 to 10 years of start-up in all but the worst combination of assumptions. Furthermore, the reduction in total base program costs after a 20-year period could be as high as 15 to 25 percent. The most important variables are the power and mass requirements of the oxygen-production plant, including beneficiation equipment.

In all but the most pessimistic cases considered, a return on investment was achieved. If Earth-to-orbit (ETO) transportation costs are high relative to the development/production costs of LLOX hardware, then LLOX has a very early payback. On the other hand, if the LLOX development/production costs are greater than the ETO transportation costs, then the positive impact of LLOX on the base is reduced.

To summarize the results of this trade study: even if LLOX is used only for operations from the lunar surface to low lunar orbit (LLO) and back (not for sale to LEO), and even with a crew as small as four people, significant savings are seen within ten years. Only in the worst case does LLOX show a negative impact on the base costs.

E.2.7   Woodcock, 1989

Woodcock did another trade study with the view of deriving energy from space for use on Earth. This study focused on the ways that energy might be produced in space: by space-based solar-power cells, by lunar-based solar-power cells, and by collecting and using Helium-3. In order for any of these three methods to be economically advantageous, the cost of space transportation must be brought down to less than $2,200 per kg ($1,000 per pound) to launch.

Woodcock (1989) assumes in this scenario that hydrogen and xenon (electric propulsion reaction mass) are obtained from Earth and that oxygen is obtained from the Moon, except for the oxygen used in low Earth orbit. For the solar-power satellite, 96.5 percent of the required materials could be produced on the Moon. This would be helpful because of the lower launch cost and the lack of lunar atmosphere, which opens the possibility of using a mass driver to launch the materials into space. The concern is the amount of human labor required for any of these options. However, with continuing progress in automation and robotics, a re-examination of these seems warranted.

For the production of lunar Helium-3, the real transportation problem is the delivery of the mining and extraction equipment to the lunar surface. Returning the Helium-3 to Earth is trivial by comparison. Hydrogen and oxygen would also be obtained as by-products of the Helium-3 extraction process, and this adds to its potential efficiency and cost-effectiveness. In all the cases discussed by Woodcock (1989), lunar oxygen supplied to low lunar orbit was found to be the highest-leverage use of lunar resources.

E.2.8   Understanding the assumptions

The question of when lunar oxygen production becomes economically viable raises a series of related questions. In any trade study or economic analysis, assumptions must be made about where the oxygen is to be used, how it will be extracted from the regolith, whether there are useful co-products, and how much overhead will exist in equipment, personnel, and energy requirements. The assumptions about the form of transportation throughout the system (e.g., whether there is a mass driver on the Moon or aerobraking in LEO) also have an impact on the results. Finally, the development and production costs for the systems should be considered.

E.3 SUMMARY OF TRADE STUDIES

The high transportation cost of propellant to LEO is the motivation for considering lunar oxygen production; and yet it is likely that this cost would be lowered by using LLOX. Surprisingly, if the transportation cost from Earth to the Moon is lowered as a result of a decrease in Earth-to-LEO launch costs, the economic incentive for mining the Moon gets better instead of worse. This is because no matter how inexpensive it becomes to lift materials, supplies, and propellants from Earth, it will always be proportionately cheaper to set up a lunar production facility and get the needed supplies from the Moon. The fact that the lunar base propellant plant would be profitable under current conditions implies that it would be even more profitable as space transportation costs decrease.

The trade studies done to date have covered many possible situations and economic conditions which could exist for a LLOX plant. Although many other studies could be done, almost all of these studies agree on the economic viability of lunar oxygen production at some level. There can be little doubt that production of oxygen on the Moon makes sense.

E.3.1 Subsystems

All of the proposed LLOX plants can be broken down into subsystems. When this is done, it is seen that many of the processes discussed employ the same or similar subsystems. For example, all of the processes eventually require either liquefaction or high-pressure compression of the product oxygen. Almost all of them require an electrolysis or thermochemical step to split water (or some other intermediate chemical) to release its oxygen. Confusion has resulted from the fact that some studies include these steps in their total cost/mass/volume estimates, while others do not. We have attempted to include all relevant subsystems in the descriptions that follow.

E.4 PROPOSED OXYGEN-EXTRACTION PROCESSES

On Earth, the most efficient method of transforming an ore into a useful product often includes extensive chemical processing with many intermediate steps. On the Moon, this will not be the case. Simon (1985) argues that for a lunar base, the transportation cost of materials from Earth drives the system design, rather than efficiency. On the Moon, a method that is less effective or more energy intensive may be more economical if it required less mass launched from Earth. On the other hand, Calkins (2006) concludes that the recurring expense of maintenance and repair may be the largest cost driver. Both of these considerations represent a shift in thinking with respect to the economics of mining and materials production.

Over the past 20 years, many processes have been suggested for use in the production of lunar oxygen. A few of these processes have been tested in the laboratory. What follows is a description of the processes that have been proposed.

E.5   GAS/SOLID SYSTEMS

The first group of processes that we consider require interactions between gases and solids. All of these systems would share a dust containment problem. Also, at least one beneficiation[2] step (to eliminate over-size particles) will be required.

E.5.1   Ilmenite reduction by hydrogen (mare only)

Product: oxygen (only) from ilmenite ($FeTiO_3$, also written as $FeO \cdot TiO_2$)

The chemical reaction for this process is $FeTiO_3(s) + H_2(g) \Rightarrow Fe(s) + TiO_2(s) + H_2O(g)$. This reaction is endothermic (heat is absorbed), thus heat must be added to drive the reaction towards completion. Reaction temperatures of 900°C or higher are needed in order to obtain acceptable rates and conversion efficiencies (Christiansen *et al.*, 1988; Allen *et al.*, 1994). The water that is produced is separated by electrolysis into hydrogen and oxygen, and the hydrogen is recycled.

A schematic of the hydrogen-reduction process is shown in Figure E.1. A fluidized-bed reactor has been proposed for the reduction step (Dalton and Hohmann, 1972; Gibson and Knudsen, 1985; Gibson *et al.*, 1990). According to this early concept, a beneficiation step (not shown) separates the small fraction of ilmenite from the rest of the soil.[3] A system of storage hoppers then feeds the ilmenite into the top of a three-stage fluidized-bed reactor chamber. As the ilmenite flows from the first to the second level, it is heated to about 1,000°C. Hydrogen gas flows up through the beds and reacts with the ilmenite to form water vapor. An electrolysis cell separates the water vapor into hydrogen and oxygen. The oxygen is liquefied and stored, and the hydrogen flows back into the bottom of the reactor. A solids settling tank removes the reactor residuals from the bottom of the reactor (description from Stump *et al.*, 1989).

If the electrolysis phase is carried out at the same temperature as the reduction reaction, an energy saving would be achieved. Substantial progress has been made in developing the electrolytic cells that could carry out this high-temperature electrolysis (Weissbart *et al.*, 1969; Maskalick, 1984; Lawton, 1986; Bents, 1987). The chemistry of this process is fairly simple, and it has been demonstrated in the laboratory (Volk and Stotler, 1970; Williams and Mullins, 1983; Williams, 1985; Briggs and Sacco, 1988; Gibson and Knudsen, 1988). The reactions are known to work, and further refinement will reduce plant mass and energy requirements. As with any of the processes that produce water in an intermediate step, oxygen production and hydrogen recovery are simultaneous events. This decreases complexity, which increases reliability.

---

[2] Beneficiation: any sorting operation which improves the properties of a feedstock prior to smelting or chemical processing. Size-sorting is included in this definition. Magnetic separation, flotation, or gravity separation would also be considered beneficiation processes.

[3] Later research demonstrated that separation of ilmenite may not be so useful, because the glass (agglutinate) phase of the lunar regolith also contains a significant amount of extractable oxygen.

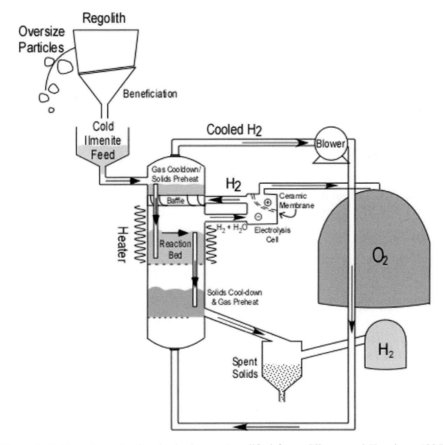

Figure E.1. Ilmenite reduction by hydrogen (modified from Gibson and Knudsen, 1985).

Many Earth analogues exist for this kind or process (McGannon, 1964; Volk and Stotler, 1970). As long as the reaction temperature remains below the melting point of the ilmenite feed, reactor materials problems are minimal. It also appears that some aspects of the fluidized-bed process will work better on the Moon than on Earth. For example, the low lunar gravity allows wider particle size distribution in the fluidized bed (e.g., Ness et al., 1990). This means that less material needs to be removed at the high and low ends of the size distribution before the feedstock enters the reactor, when compared to equivalent terrestrial processes. However, reduced-gravity modeling and experiments have also shown that gravity-fed systems behave in unexpected ways in simulated lunar gravity

A disadvantage of a process that requires hot hydrogen is that the gas will be difficult to retain. Hydrogen is a very small molecule and it is also quite reactive. Another issue with ilmenite reduction processes is that only the ilmenite is utilized. Although early estimates suggested that as much as 14 percent of the lunar mare regolith was ilmenite, more recent calculations show that the actual portion of separable ilmenite crystals is closer to 2 percent (Vaniman and Heiken, 1990). If this is the case, then 475 tonnes of mare soil would have to be mined in order to obtain 9.5 tonnes of ilmenite, to subsequently obtain one tonne of oxygen. An alternative method would be to collect and crush rocks and boulders of high-titanium basalt (Vaniman and Heiken, 1990). The problem in that case would be to separate basalt boulders from regolith breccia boulders, and that technique has not yet been tested. A process that was able to use unbeneficiated soil and could consume all of the ilmenite within this feed would have an advantage in mining rates; and Allen *et al.* (1992, 1994) have now shown that hydrogen reduction has approximately the same efficiency on unbeneficiated lunar mare soil as it does on beneficiated ilmenite.

Another problem is that the kinetics for the hydrogen-reduction reaction are very slow. It takes about an hour, at a temperature of 1,000°C, to remove approximately 70 percent of the available oxygen (Dalton and Hohmann, 1972; Gibson and Knudsen, 1985). It seems that by raising the temperature to 1,073°C, the rate is sped up to 0.25 hours for equal conversion (Briggs and Sacco, 1988). However, caution is advised when comparing the results of one experiment with the results obtained from a different worker in a different laboratory. There are many unreported variables in such a case and the comparison is therefore ambiguous.

At 1,000°C, approximately 10.5 percent of the available hydrogen is converted to water during each pass, and at 1,200°C, about 19 percent of the hydrogen is converted (Gibson and Knudsen, 1985).

The energy cost of this process has been calculated at 5,000 calories per gram of oxygen produced (Lewis *et al.*, 1988).

E.5.1.1   Experiments

Chang, 1959: Experimental success with the hydrogen reduction of ilmenite concentrate was reported in a patent by Chang (1959). In another early experiment, simulated lunar fines were heated in a closed vessel with a mixture of helium and hydrogen. The hydrogen concentration was less than 5.0 percent by volume, and the total pressure inside the vessel was 817 torr (15.8 psia). The simulant was heated to between 1,000 and 1,200°C. The measured weight loss of two separate samples after heating was 0.96 and 0.75 percent. In these experiments, water was produced in amounts sufficient to account for 0.83 and 0.72 percent, respectively, of the starting weight. There was a six-fold reduction in the amount of hydrogen added at the beginning of the second test. However, we cannot assume that this weight change is due to oxygen loss alone, because a certain amount of atmospheric water contamination or hydrated minerals exist in any terrestrial simulant.

Williams and Mullins, 1983; and Williams, 1985: In 1983, an apparatus was constructed to test the premise that removing the water vapor as it was produced could improve the per-pass yield (Williams and Mullins, 1983). Theoretically, this could improve the per-pass conversion of hydrogen to water by roughly a factor of 15.

The experimental setup consisted of a stainless steel reaction vessel with appropriate valves and accessories so that it could be isolated and pressure-monitored. One outlet of the vessel was connected to a cold trap. The vessel was first filled with powdered ilmenite, then the atmosphere was evacuated from it, after which it was flushed with argon several times. Next, the ilmenite-filled vessel was heated to the desired process temperature. Then, hydrogen was introduced into the vessel, at which point the vessel was sealed and its pressure was monitored to check for leaks.

Finally, the valve to the cold trap was opened and the pressure was then recorded as a function of time. A reaction temperature of 720°C and a hydrogen pressure of 20 psi (1,094 torr) was reported. When the system was then connected to a cold trap, the pressure was seen to drop smoothly from 18 psi (930.8 torr) to 9.2 psi (475.78 torr). Using this system, a water production rate of 0.003 moles/hour was measured.

Later, the cold-trap enhancement method was used in a continuous process (Williams, 1985). Yield was increased by 4.33 times when using a cold trap in the continuous process, as opposed to a factor of 15 times in the batch process. Unfortunately, the simulant used was found to contain traces of FeS, which released H_2S during processing, which in turn damaged several components of the test apparatus.

Williams (1985) also reported the masses of the test equipment and the power consumption used by each part of the process. These tests resulted in a yield of 0.80 mg/minute of oxygen, with a power consumption of 2,404 watts operating on a total system of mass = 18.5 kg and volume = 202 cm.[3] These values can also be expressed as 21.34 mg/kW-hour or as 0.043 mg/kg-minute (Williams, 1985).

Gibson and Knudsen, 1985–1990: Hydrogen reduction of ilmenite has been studied extensively by Gibson and Knudsen (1985, 1988) and by Gibson et al. (1990). Laboratory tests of ilmenite reactivity confirmed the initial design assumptions. Also, "two-dimensional" fluidized-bed experiments at Earth gravity and at lunar gravity were conducted aboard NASA KC-135 flights (Gibson et al., 1990). Other parts of the proposed design, such as standpipes, have been tested in the laboratory at Earth-normal gravity. These test results can be scaled to give an indication of lunar parameters. Lunar-gravity bubble size data and the 1-G standpipe flow data both indicate that the proposed three-stage, stacked fluidized-bed design is operable on the Moon (Gibson et al., 1990).

Allen et al., 1992–1994: Allen et al. (1994b) report on experimental reduction of lunar soils with high-ilmenite content as well as lunar basalts and lunar simulants. They used Mossbauer spectroscopy,[4] petrographic microscopy, SEM (scanning electron microscope), TEM (transmission electron microscope), and XRD (X-ray diffraction) to determine their effectiveness. Mossbauer analysis of the lunar basalt sample reduced at the lowest temperature, 900°C, showed complete reduction of the Fe^{2+} in ilmenite to iron metal. All the reduced ilmenite examined by SEM displayed phase separation and reduction throughout. Samples reduced at 1,000 and 1,050°C showed

---

[4] Mossbauer spectroscopy depends on the Mossbauer effect, which is the emission of gamma rays by radioactive nuclei of crystalline solids, and the subsequent absorption of the emitted rays by other nuclei.

evidence of iron migration to grain surfaces, while the single 900°C sample did not show such evidence. High-resolution TEM photographs of ilmenite from the 1,050°C experiment showed additional reduction of the TiO_2 to Ti_6O_{11}.

These tests allow us to estimate the output from a lunar oxygen plant. We can expect an oxygen yield of 10.5 percent of the ilmenite mass, or 20.0 percent of the mass of TiO_2 in the mineral. The TiO_2 formed by ilmenite reduction can itself be reduced to Ti_6O_{11} at 1,050°C and to Ti_4O_7 at 1,100°C. This process yields a maximum of 5.3 percent of the TiO_2 mass as oxygen.

E.5.2 Variants on the hydrogen-reduction-of-ilmenite process

A slight modification of this process includes a pre-oxidation step. Even though it would be more complex, the hope is that the faster processing would allow for a smaller plant. Briggs and Sacco (1988) have reported some preliminary experimental results.

E.5.2.1 Carbon monoxide reduction

In another variant of the ilmenite-reduction process, Chang (1959) suggested that either H_2 or CO, or a combination of both, could be used as the reducing gas. The relevant reactions are:

$$FeTiO_3 + H_2 \Rightarrow Fe + H_2O + TiO_2$$

$$FeTiO_3 + CO \Rightarrow Fe + CO_2 + TiO_2$$

After this, the reduced ore is cooled and transferred to a closed vessel containing about 40 percent HNO_3, with an overpressure of air or oxygen. During this step, iron is removed from the reduced ore by the reaction:

$$Fe + 4HNO_3 \Rightarrow Fe(NO_3)_2 + 2H_2O + NO$$

In the presence of oxygen another reaction occurs:

$$2NO + O_2 \Rightarrow 2NO_2$$

Now, the NO_2 dissolves in the leaching medium (water) to create more nitric acid:

$$3NO_2 + H_2O \Rightarrow 2HNO_3 + NO$$

The leached residue is then filtered. The solid residue contains 70 percent or more of TiO_2, while the filtrate is composed of ferric nitrate and any excess nitric acid. Next the filtrate is hydrolyzed under oxygen overpressure in the reaction:

$$2Fe(NO_3)_3 + 3H_2O \Rightarrow Fe_2O_3 + 6HNO_3 \qquad (400–500°F; \; 204–260°C)$$

The ferric oxide forms as a powder and can be filtered out of the nitric acid. Thus, the bulk of the nitric acid is recovered. The ferric oxide in turn is very pure and can easily be converted into iron powder – for example, in a fluidized bed by reaction with hydrogen. This process seems suited to the lunar environment because most of the reagent can be recycled. We are not aware of any experimental tests of this process.

E.5.2.2 Ilmenite reduction by methane (mare only)

This concept was first proposed by Chang (1959) in the patent discussed above. When CH_4 is used as the reactant with the feedstock, the chemical reactions are:

$$FeTiO_3 + CH_4 \Rightarrow Fe + CO + 2H_2O + TiO_2$$

$$CO + 3H_2 \Rightarrow CH_4 + H_2O$$

$$H_2O \Rightarrow H_2 + \tfrac{1}{2}O_2$$

$$NET: \quad FeTiO_3 \Rightarrow Fe + \tfrac{1}{2}O_2 + TiO_2$$

E.5.3 Hydrogen reduction of glass

Products: oxygen, solar wind volatiles (secondary)

The reduction of lunar glass is also an effective way to produce oxygen:

$$FeO(glass) + H_2O \Rightarrow Fe + H_2O$$

or more generally

$$MO(glass) + H_2 \Rightarrow M + H_2O$$

because this process can also be used with other metal oxides in the glassy component of lunar soil. However, it is easier to break the FeO bond than any other bond in lunar soil chemistry.

Measurements were made of the amount of iron reduction and the sample weight loss versus temperature, time and hydrogen flow rate. Figure E.2 shows the results of heating samples at varying temperatures for one hour. The highest percentage of oxygen was released at 1,100°C. At higher temperatures, less oxygen was released because the sample began to melt, which destroyed its permeability to the hot hydrogen gas. Figure E.3 shows the results of heating samples at 1,000°C for varying lengths of time. Very little change in the percentage of oxygen released was seen when heating times were extended beyond 3 hours.

Allen *et al.* (1994) showed that iron oxide blebs[5] were produced on lunar simulant glass grains by passing hot hydrogen gas over the grains in a closed system. Thirty-nine tests were run on simulants of Apollo 11 glass. The samples were reduced with hydrogen at 1,000–1,100°C. Maximum weight losses of 4.5 percent were obtained at 1,100°C after 3 hours. The glass was partially to totally devitrified[6] during this process, so the kinetics of "glass" reduction were partially controlled by the reduction of ilmenite and pryoxene.

Lunar glass is a viable source of oxygen when heated in flowing hydrogen at 1,100°C. Yields approaching 5 wt% from iron-rich material such as the Apollo 17 orange glass should be possible. Furthermore, the Apollo 17 orange glass may be a source for other volatiles, since the glass was probably produced by lunar fire-fountaining and may be coated with magmatic volatiles.

[5] Bleb: a small, usually rounded inclusion of one material in another.
[6] Devitrification: conversion of glass into crystalline material.

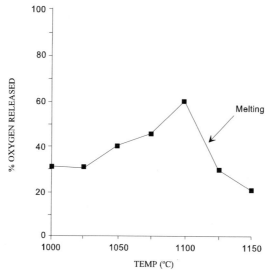

Figure E.2. Hydrogen reduction of glass. Percent oxygen produced as a function of temperature.

The Apollo 17 orange glass is also very fine-grained, with an average grain size of 40 microns. This is another advantage because experiments have shown that smaller grains (less than 74 microns) reduce more efficiently than larger grains. The overall conclusion is that the Apollo 17 orange glass would be an excellent feedstock material. Furthermore, it would require no beneficiation.

E.5.4   Hydrogen reduction of other lunar materials and lunar simulants

Allen *et al.* (1994b) also tested the reduction of a lunar basalt sample from the Apollo 17 collection. The reaction was rapid, with major evolution of water occurring within minutes after the introduction of hydrogen. The reduction of Minnesota Lunar Simulant-1 (MLS-1) basalt by hydrogen at 1,100°C was not quite as rapid as with the actual lunar samples. In this case, about 90 percent of the weight loss occurred in the first ten minutes. However, the MLS-1 behaves almost exactly the same as actual lunar material. The main difference between lunar material and MLS-1 is that MLS-1 has some magnetite in it, which results in a slightly-higher overall oxygen yield. Because lunar regolith was formed in a highly-reducing[7] environment, reducing it further to extract oxygen can be challenging. MLS-1 is easier to reduce because it was made in a more oxidizing[8] environment.

---

[7] Reducing: causing oxygen to be removed from a compound, or causing electrons to be added to a compound.
[8] Oxidizing: to cause oxygen to be added to a compound, or to cause electrons to be removed from a compound.

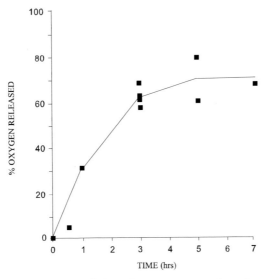

Figure E.3. Hydrogen reduction of glass. Percent oxygen released as a function of time.

The predicted output from a lunar oxygen plant depends on the ilmenite contents of the soil (unless the feedstock is a pyroclastic glass, as discussed above). In most lunar soils, almost all the TiO_2 is incorporated in ilmenite. The maximum oxygen yield therefore will be equal to 20 percent of the TiO_2 content if only ilmenite is reduced, and 25 percent if further conversion from TiO_2 to Ti_4O_7 occurs. Lunar soil 78221 contains 3.84 wt% TiO_2. The maximum predicted oxygen output from a plant using this feedstock is just under 1 per cent of the total input mass. The output from a high-Ti soil such as 75061, with 18.02 wt% FeO and 10.38 wt% TiO_2, is 2.6 percent. Such an output rate means that 38 tons of lunar soil would be required to produce one tonne of oxygen.[9] Similar to the ilmenite feedstock process, pre-oxidation of glass, followed by reduction, may improve the yield; however, no experimental data are available to validate this assumption. Moreover, as discussed previously, using the agglutinate in the soil may improve the yield significantly.

E.5.5   Fluorine extraction (feedstock-independent)

Products: oxygen and metals

A summary of the fluorine-extraction process in included in Christiansen *et al.* (1988). Because fluorine gas, F_2, is highly reactive, it is able to liberate metal from all oxides, creating metal fluorides. This concept has been studied by Burt (1987, 1988a, b). The general reactions are:

[9] By terrestrial standards this is a small amount of feedstock. A front-end loader takes less than ten minutes to load a single medium-sized dump truck, which can hold 40 tons.

metal oxides $+ F_2$(gas) $\Rightarrow$ metal flouride $+ O_2$(flourine exchange at 500°C)

metal flourides $+ K \Rightarrow$ metals $+$ KF(reduction of flourides with K vapor)

KF $+$ electricity $\Rightarrow K + \frac{1}{2}F_2$(electrolysis of KF at 846°C, the melting point, to recover the reagents)

The metals may include calcium, aluminum, iron, silicon, magnesium, or titanium. Anorthite ($CaAl_2Si_2O_8$) and ilmenite ($FeTiO_3$) are possible feedstocks. Fluorination proceeds fairly rapidly and can be safely done in nickel reaction vessels. The process could be done in a two-stage fluidized-bed reactor, as described by Burt (1988b), using anorthite feed. The reaction in the first stage is:

$$CaAl_2Si_2O_8 + 2F_2 \Rightarrow CaF_2 + Al_2SiO_4F_2 + SiO_2 + O_2(g)$$

Then, fresh fluorine is fed into the bottom of the second stage:

$$CaF_2 + Al_2SiO_4F_2 + SiO_2 + 6F_2 \Rightarrow CaAlF_5 + AlF_3 + 2SiF_4(g) + 3O_2(g)$$

The product gas leaves the fluidized bed and passes through a bed of NaF which scrubs out the SiF_4 gas:

$$2SiF_4(g) + 4NaF(s) \Rightarrow 2Na_2SiF_6$$

The sodium silicofluoride is separated out, then gaseous sodium is added to it to create silicon and sodium fluoride at temperatures above 992°C (the NaF melting point):

$$2Na_2SiF_6 + 8Na(g) \Rightarrow 2Si(s) + 12NaF(l)$$

The liquid NaF is then separated out. A third of the amount obtained is recycled to the NaF bed discussed above, and the remainder is used in a later step, in the electrolysis cell. The other residual solids and liquids are also reduced to sodium metal at 992°C:

$$CaAlF_5 + AlF_3(s) + 6Na(g) \Rightarrow CaF_2(s) + 2Al(l) + 6NaF(l)$$

The sodium fluoride is separated and used later in the electrolysis cell. The fluorite (CaF_2) is removed and mixed with sodium monoxide:

$$CaF_2 + Na_2O \Rightarrow CaO + 2NaF$$

This reaction is expected to be slow, and temperatures above 1,275°C may be needed to effect a reasonable reaction rate. The sodium fluoride is transferred to the electrolysis cell. The CaO may be used to scrub the final traces of F_2 from the product oxygen:

$$CaO + F_2 \Rightarrow CaF_2 + O_2$$

The necessary sodium monoxide is produced by the oxidation of sodium:

$$2Na + \frac{1}{2}O_2 \Rightarrow Na_2O$$

The sodium fluoride generated in the various steps above can be electrolyzed to yield Na and F_2 for recycling:

$$16NaF(l) + \text{electricity} \Rightarrow 10Na(l \text{ or } g) + 8F_2(g)$$

An advantage of the fluorine-extraction process is that fluorine reacts rapidly with any of the available metal oxides. The only beneficiation that is required is size-sorting. Oxygen is also liberated directly, unlike some other processes which require the splitting of water, carbon monoxide, or carbon dioxide.

A disadvantage is that the recovery of the fluorine for recycling is complicated and several processing steps are required. Because 2.375 tonnes of fluorine are required to produce 1 tonne of oxygen, recycling of this fluorine is essential. Furthermore, the overall process is complicated, requiring eight reaction vessels (Burt, 1988b). This complexity is due to the difficulty of recovering the fluorine for recycling. Some of the steps are also likely to be energy-intensive since they require temperatures of up to 1,200°C. The process can be simplified by using pure anorthite, but this then adds the complexity of additional beneficiation to obtain it from lunar regolith. Moreover, considerable technology development would be required to pursue this process. Finally, the NaF electrolysis step is unproven technology (Burt, 1988b).

E.5.6   Hydrogen sulfide (H_2S) reduction

Products: oxygen and metals
The use of hydrogen sulfide gas to reduce iron, calcium, and magnesium oxides has been under consideration for some time (e.g. Dalton and Hohmann, 1972). It is practical to use bulk, size-sorted lunar soil, without further beneficiation, for this process. The general reaction sequence is:

$$(Fe, Ca, Mg) + H_2S \Rightarrow (Fe, Ca, Mg)S + H_2O \text{ (reduction)}$$

$$(Fe, Ca, Mg)S + heat \Rightarrow (Fe, Ca, Mg) + S \text{ (thermal decomposition)}$$

$$H_2O + electricity \Rightarrow H_2 + \tfrac{1}{2}O_2 \text{ (electrolysis)}$$

$$H_2 + S \Rightarrow H_2S \text{ (H_2S regeneration)}$$

This process should yield more oxygen per unit soil mass than hydrogen reduction of ilmenite, because all of the feedstock is used, not just the ilmenite. With additional processing steps, it would also be possible to recover the metals from the oxides. However, the thermal decomposition step requires very high temperatures, and the process yield is uncertain. To our knowledge, no experiments have been done to verify the chemical reactions proposed in this process.

If the oxygen which is produced is to be used in environmental systems, oxygen purification steps will be necessary due to the toxic nature of H_2S (Christiansen *et al.*, 1988). However, as will be explained later, the H_2S problem will occur to some extent in almost all of these processes.

E.5.7   Carbochlorination

Products: oxygen, silicon, iron, and aluminum
One advantage of using chlorine in a chemical process is that almost all the elements in the periodic table will react with chlorine to form chlorides, except for a few noble

gases. However, metal chlorides are difficult to form from metal oxides. This difficulty can be overcome by using a CO–Cl_2 gas mixture or by reacting the Cl_2 in the presence of solid carbon. In carbochlorination, carbon monoxide acts as a reducing agent, forming CO_2, while the chlorine oxidizes the metal, forming a volatile chloride. In this way, a new surface is continually exposed for reaction (Lynch, 1989).

As described by Christiansen *et al.* (1988), the carbochlorination process would operate in a fluidized bed, with solid carbon and chlorine gas reacting with anorthite and ilmenite. The reaction sequence is:

$$CaAl_2Si_2O_8(s) + 8C(s) + 8Cl_2(g) \Rightarrow CaCl_2(s) + 2AlCl_3(g) + 2SiCl_4(g) + 8CO(g)$$

$$FeTiO_3(s) + C(s) + \tfrac{3}{2}Cl_2(g) \Rightarrow FeCl_3(g) + TiO_2(s) + CO(g)$$

$$2SiCl_4 + 4CO(g) \Rightarrow 2SiO_2(s) + 4C(s) + 4Cl_2(g)$$

Staged condensation steps are used to separate the gas components. A condenser at 225°C removes $FeCl_3$ as a liquid. Another condenser at 90°C is used to liquefy and separate the $AlCl_3$. A third condenser operates at −30°C to remove $SiCl_4$.

The $SiCl_4$ is recycled back into the reactor. There, its concentration builds up until it reaches equilibrium with the CO. From then on, the amount of silica produced offsets the amount of new $SiCl_4$ being added. The residual solids from the reactor are SiO_2 and $CaCl_2$. After removal from the fluidized bed, these are heated to 800°C to melt the $CaCl_2$, then the two are separated in a centrifuge. The silica will be useful for producing solar-power cells. The chlorine is needed to continue the carbochlorination process, and must be recovered. The reactions involved are:

$$CaCl_2(s) + 2H_2O(g) \Rightarrow Ca(OH)_2(s) + H_2(g) + Cl_2(g) \text{ hydrolysis at 400°C}$$

$$Ca(OH)_2(s) \Rightarrow CaO(s) + 2H_2O(g) \text{ calcination}^{10} \text{ at 600° C}$$

The iron chloride, $FeCl_3$, can be reduced directly by hydrogen gas at 700°C to produce metallic iron and hydrochloric acid, or it can be oxidized to hematite, Fe_2O_3, which is then reduced by hydrogen or carbon below 1,000°C to obtain low-carbon iron via the following reactions:

$$FeCl_3(g) + \tfrac{3}{4}O_2(g) \Rightarrow Fe_2O_3(s) + \tfrac{3}{2}C_2(g) \text{ oxidation at 300°C}$$

$$\tfrac{1}{2}Fe_2O_3(s) + \tfrac{3}{2}H_2(g) \Rightarrow Fe(s) + \tfrac{3}{2}H_2O(g) \text{ reduction at 1,000°C}$$

Carbon monoxide and water from the above reactions can be reduced to solid carbon, hydrogen, and oxygen, and thus all may be recovered.

As with H_2S reduction, this process has the advantage of being able to use many materials that occur in the lunar soil. Beneficiation is not required beyond size-sorting, which will reduce the amount of equipment that must be transported to the Moon to begin operations. Furthermore, production of aluminum and low-

---

[10] Calcination (also referred to as Calcining) is a thermal treatment process applied to solid materials in order to bring about a Thermal decomposition or removal of a volatile fraction. The calcination process normally takes place at temperatures below the melting point of the product materials.

carbon iron or steel is a necessary by-product of the reaction, and these products will be useful.

A disadvantage of the carbochlorination process is the large number of processing steps involved. Increased process complexity decreases reliability. There are also reactant-recovery-efficiency and materials-corrosion considerations (Christiansen *et al.*, 1988). Lynch (1989) also pointed out that 136 identifiable C–Cl–O by-product compounds may be produced during carbochlorination between 600 and 1,000°C, by oxides of Ti, Zr, and Al. The loss of scarce reactants and oxygen is a serious concern. Finally, although the carbochlorination process has terrestrial analogues, we are unaware of any experiments in which it has been tested on lunar simulants.

E.5.8 Chlorine plasma extraction

Products: oxygen, metals

As described above in the carbochlorination process, cold chlorine will only react with iron oxide (such as in ilmenite), and thus cannot be used without carbon. Another process was described by Lynch (1989), which uses a "cold" plasma reactor to create a chlorine plasma. Calculations indicate that stable metal oxides can undergo chlorination and yield oxygen as a by-product, under the influence of such a plasma.

In spite of the name, "cold" plasmas are not necessarily cold. They can exist at molecular temperatures of up to 2,000°K. The name "cold" is meant to convey the idea that there is a significant difference between the molecular temperature and the kinetic temperatures of the electrons in the molecule, which may be thousands of degrees "hotter" than the overall kinetic temperature of the molecule. The reactive nature of a cold plasma is the direct result of this temperature difference between the electrons and the overall molecule.

Reaction rates may be substantially higher in a cold plasma than they are in a gas. The presence of broken bonds, partially-filled orbitals, and unbalanced charge makes the ions in a plasma highly reactive, and this may substantially enhance chemical processes.

The basic process is:

$$Cl_2 + \text{metal oxide} \Rightarrow \text{metal chloride} + \tfrac{1}{2}O_2$$

After the chlorine has released the oxygen from the metal oxide, the next step is to collect and dissolve the metal chloride. Then it can be electrolyzed to recover the reagent chlorine and obtain useful metals.

Construction of an experimental apparatus was begun (a schematic is shown in Figure E.4), but results have not been published. The design includes a microwave source, an applicator where the plasma is generated, a gas delivery system, a mass spectrometer for monitoring the extent of the reaction, an optical pyrometer for monitoring the temperature of the solid specimen in the plasma, a specimen holder that both rotates and allows for vertical translation of the solid specimen, and a vacuum system (Lynch, 1989).

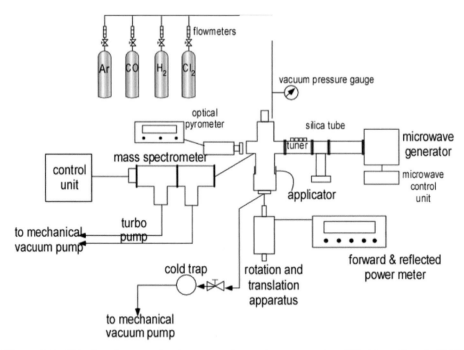

Figure E.4. Chlorine plasma reduction – experimental apparatus (modified from Lynch, 1989).

E.6 GAS/LIQUID PROCESSES

Products: oxygen, iron

Some processes that have been proposed involve a reaction between a gas and melted feedstock. Various kinds of lunar oxides can be used in this process, and it has been under study for several years (e.g., Rosenberg *et al.*, 1966). Lunar oxides are melted, then mixed with a carbon reductant (methane or native carbon, for example) and an anorthite fluxing[11] agent. In the case of ilmenite, this causes the following endo-thermic[12] reaction:

$$FeTiO_3 + C \Rightarrow Fe + CO + TiO_2$$

In the case of olivine (Mg_2SiO_4), the reaction is

$$Mg_2SiO_4 + 2CH_4 \Rightarrow 2CO + 4H_2 + Si + 2MgO \text{ (at } 1{,}625°C)$$

---

[11] A fluxing agent is any substance that reduces the melting point of a mixture, or that helps metals to melt.
[12] An endothermic reaction is one that requires heat in order to proceed.

In the case of pyroxene ($MgSiO_3$), the reaction is

$$MgSiO_3 + 2CH_4 \Rightarrow 2CO + 4H_2 + Si + MgO \text{ (at } 1,625°C)$$

The silicon produced during the reaction is a melt (the melting point of silicon metal is 1,410°C). The product carbon monoxide and hydrogen from this step are reacted at about 250°C as follows:

$$2CO + 6H_2 \Rightarrow 2CH_4 + 2H_2O$$

Finally, oxygen and hydrogen are produced by electrolysis, and the methane and hydrogen are recycled.

E.6.1 Carbothermal reduction of anorthite

Carbothermal reduction of anorthite (Bhogeswara et al., 1979) requires high temperatures (up to 1,800°C or so), which may present corrosion problems. The process chemistry is also complicated (Christiansen et al., 1988). The energy cost of the carbothermal process has been calculated by Lewis (1988) to be about 17,000 calories per gram of oxygen produced – over three times larger than the hydrogen-reduction-of-ilmenite process. Moreover, Hepp et al. (1994) also calculated higher thermal energy costs for carbothermal reduction than for most other processes.

Advantages of the carbothermal process include the fact that potentially many kinds of lunar oxides can be used, rather than just ilmenite. Less regolith would need to be mined per tonne of oxygen produced, and only size-sorting is needed. Terrestrial counterparts for the process exist and so there is some practical experience with it. In Rosenberg's (1966) scheme, beneficiation is not required. (Cutler's (1985) method requires ilmenite, and so this advantage disappears.) Possibly the biggest advantge is that this process requires less plant mass than most others. For this reason alone, NASA has chosen carbothermal reduction as one of the top three processes for further research.

Disadvantages include the fact that the feedstock must be molten. This disadvantage is shared by magma electrolysis (discussed below). Both involve corrosive liquids and high temperatures. Although steel making can be included as a part of the process (Cutler and Krag, 1985), it adds complicated steps that do not exist in the magma electrolysis process. Finally, recovery of the carbonaceous reductant is difficult. Some of the carbon alloys with the iron and much of the carbon is converted into CO, which in turn must be converted into CH_4, then finally back into pure carbon (Christiansen et al., 1988). Although the individual steps of the process are not difficult, the process of recovering the carbon adds several pieces of additional equipment to the process design.

E.6.2 Carbon-monoxide-silicate-reduction system

A variant of the carbothermal reduction process was described by Berggren et al. (2005). Soils are sequentially subjected to iron oxide reduction by carbon monoxide, in-situ deposition of carbon throughout the soil by carbon monoxide disproportiona-

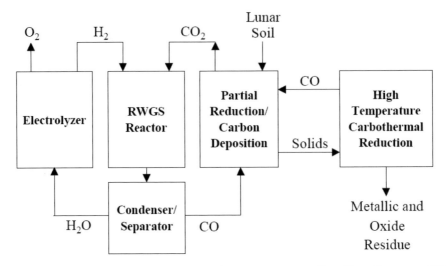

Figure E.5. Carbon monoxide silicate reduction system process flow (from Berggren, 2005).

tion catalyzed by metallic iron, and finally high-temperature reduction of silicates by the deposited carbon.[13]

Figure E.5 shows the process schematic. Carbon monoxide is used both to reduce iron and to deposit carbon evenly. This enables high-temperature carbothermal reduction of silicon oxide (described above) which yields five times as much oxygen from planetary regolith compared with hydrogen-reduction systems. Because it is not based on a specific rock or soil chemistry, it can also be used on Mars. The feedstock material is initially heated to temperatures where the iron-bound oxygen combines with carbon monoxide, a strong reducing agent (reductant). Simultaneously, the produced iron metal acts as a catalyst which causes the carbon monoxide to form carbon and carbon dioxide. The temperature is then raised for carbothermal reduction of the silicates, producing carbon monoxide (which is recycled back to the first-stage process) and silicon metal. The carbon dioxide created in the iron reduction/disproportionation step is processed with hydrogen to make carbon monoxide and water. After electrolysis, the oxygen is stored and the CO is recycled.

The process steps are:

Iron oxide reduction:

$$FeO + CO = Fe + CO_2 \qquad \Delta H = -15.7 \text{ kJ}$$

Carbon deposition (carbon monoxide disproportionation):

$$2CO = C + CO_2 \qquad \Delta H = -18.7 \text{ kJ}$$

---

[13] Disproportionation: a chemical reaction in which a single substance acts as both oxidizing and reducing agent, resulting in the production of dissimilar substances.

Carbothermal reduction:

$$FeO + C = Fe + CO \qquad \Delta H = 156.7 \, kJ$$

$$SiO_2 + 2C = Si + 2CO \qquad \Delta H = 689.8 \, kJ$$

Reverse-water-gas-shift reaction:

$$CO_2 + H_2 = CO + H_2O(l) \qquad \Delta H = -2.9 \, kJ$$

Electrolysis:

$$2H_2O(l) = 2H_2 + O_2 \qquad \Delta H = 571.7 \, kJ$$

Oxygen liquefaction:

$$O_2(g) = O_2(l) \qquad \Delta H = -96.8 \, kJ$$

The carbon-monoxide-silicate-reduction process requires temperatures up to 1,600°C. It might be possible to produce even more oxygen by increasing the mass of carbon deposited before carbothermal reduction.

E.7   BULK ELECTROLYSIS PROCESSES

Bulk electrolysis processes share a common problem of anode corrosion, hot gas collection, and molten metal containment. However, the equipment and processing steps are straightforward.

E.7.1   Magma electrolysis (feedstock-independent)

Products: oxygen, volatiles, metals, silicon, and molten slag for bricks

One promising method for producing oxygen from lunar soils is magma electrolysis. In this process, size-sorted lunar soil is fed into a containment structure that encloses an electrolysis cell. The material melts upon contact with a layer of molten rock in the cell, at a temperature of about 1,350°C. Electric current is passed between the electrodes, and oxygen gas is released as bubbles escaping from the melt. The bubbles move upward, and metal sinks to the bottom of the cell. Oxygen is piped from the enclosure to an oxygen liquefaction/storage facility. Slag and metal are removed and collected for further processing. The entire mass is converted to useful products, with no addition of reagents and no recycling of materials. The factory could be relatively small, and nothing would be wasted.

Lunar soils are silicates (they contain silicon and oxygen), and magma electrolysis is possible because these silicates, when melted, contain ions. A high proportion of the calcium, magnesium, and iron in these melts is present as the mobile cations Ca^{2+}, Mg^{2+}, and Fe^{2+}. Most of the silicon and aluminum atoms are combined with oxygen into complex, sluggish, polymerized[14] anions. The cations move easily through the molten lunar soil, so the melt is electrically conductive. Thus, if a sufficient electrical potential is established between two electrodes so that cations

---

[14] Polymer: a large molecule made by linking smaller molecules.

can be reduced to metal at one of them (the cathode), then the silicate anions can be oxidized at the other (the anode). The ions within the melt will carry electrical current to complete the circuit. The principal interfering reaction is the oxidation of Fe^{2+} to Fe^{3+} at the anode.

The best procedure for production-level magma electrolysis would be a continuous-feed system operating at about 1,450°C. A desirable steady-state composition for the melt arises naturally on electrolysis of lunar soils of most compositions sampled by the Apollo and Luna missions, with one exception being the feldspar-rich soils of Apollo 16. A low concentration of Fe^{2+} also occurs naturally in such a system, because Fe^{2+} is the most easily-reduced cation in the feedstock and it is quickly removed. This low concentration of Fe^{2+} improves the efficiency of oxygen production by reducing the probability for the interfering reaction which was discussed previously. The steady-state composition has a relatively high conductivity, which reduces energy loss from resistance heating. Oxygen production at a rate of 1 tonne per 24 hours would require an electrode potential of ~1.8 volts, and ~0.34 megawatts of electrical power.

The efficiency of the process can be measured in terms of electrical power required. This ignores the conversion of solar or nuclear power to electrical power and transmission losses between the generator and the electrolytic cell. The energy of dissociation of SiO_2 is the minimum electrical energy required for the process, because freeing oxygen from Si–O chemical bonds is the principal oxygen-producing reaction. Laboratory experiments have shown that three to five times as much electrical energy as this theoretical minimum is actually required. From those experiments, it was determined that under optimum conditions an energy of ~1.8 times the theoretical value, or ~8.2 megawatt hours of energy per tonne of oxygen, should be required. However, even at more than three times the theoretical value, its efficiency compares well with that of other proposed processes. As the molten metals and slag are themselves ready for casting or drawing into useful products, we may consider the 8.2 megawatt-hour as the shared energy cost for those products plus the oxygen.

In addition to the energy required to release the oxygen and metals, we must furnish heat to warm and melt the feedstock. Part of the electrical energy gets converted to heat, in the form of resistance heat produced by driving current through the silicate melt, which has a significant electrical resistance. The resistance heat warms the feedstock and helps keep it melted, so the energy is not wasted.

The high efficiency (47 percent) of conversion of feedstock into oxygen and metals means that the additional energy cost for feedstock delivery will be small compared with processes that use a specific mineral.

Silicate melts are hot and extremely corrosive. In one experiment, spinel (magnesium aluminate) was used as the container material for the melts. The steady-state melt composition is in equilibrium with spinel, so it is stable against corrosion. Platinum metal was used for the anodes, because platinum is nearly inert to reaction with the silicate melts. It was found, however, that if convection or stirring is too vigorous, particles of molten iron and silicon metal become dispersed in the melt, and they can alloy with the platinum electrode and destroy it. Thus, the flow pattern within a production cell must be carefully controlled if platinum or platinum-coated

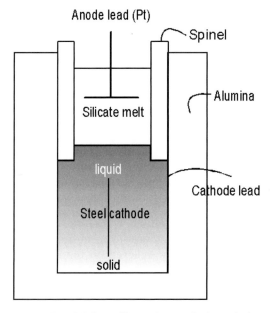

Figure E.6. Molten silicate (magma) electrolysis.

electrodes are to last. The cathode of the cell is the pool of molten ferro-silicon metal produced by electrolysis of the steady-state melt. The cathode grades from solid at the bottom to liquid at the top, as shown in the experimental configuration, Figure E.6.

It is an energy- and feedstock-efficient process for producing lunar oxygen, metals, and molten silicate, all valuable for use on the Moon and in space. There is an enormous gap, of course, between a small-scale demonstration of feasibility and showing that production by magma electrolysis is cost-effective on industrial scales. Nevertheless, the fundamental electrochemistry of the process has been validated. The next step toward evaluating magma electrolysis is at the kilogram level, where the problems are no longer the fundamental electrochemistry, but the dynamics and stability of a larger-scale system.

Magma electrolysis appears to be a relatively-simple operation with few steps involved. It is feedstock-independent, although there may be some limitations on the amount of Fe that can be tolerated. There is also the potential to recover 30 percent or more of the available oxygen.

Size sorting is required to remove pebble-size and larger particles, and fines may also need to be removed to prevent gritting up the valves. Experimental work has been done by Haskin (1990), and McCullough and Mariz (1990) have also studied this concept.

Magma electrolysis has been demostrated in laboratory-scale experiments. As mentioned earlier, the ferro-silicon alloy and depleted magma can immediately be formed into useful products. The alloy can be used to manufacture electrical conductors, transformer cores, photovoltaic sheets and mirrors, while the depleted magma can be cast into a number of useful shapes (e.g., Binder *et al.*, 1990; Capps and Wise, 1990). The Rockwell design (McCullough and Mariz, 1990) assumes that the furnace will be operated during daylight using solar power, and that during the night an auxiliary power system will be used to keep the furnace in hot standby. This may simply be a case of applying a small voltage to the electrodes, because resistive heating should generate significant heat.

E.7.1.1   *Technological challenges for magma electrolysis*

Several problems remain to be worked out with magma electrolysis. Early work by Boundy (1983) showed that, for this route to work, one must first melt the rock, using heat alone (without flux). This puts a more stringent design requirement on containment materials. This difficulty can be relieved on the full-scale plant by designing thermal gradients within the cell.

Another problem is that pipes must be very carefully thermally-controlled. Because the average operating temperature of the system is about 1,300°C, the pipes will have to be heated and insulated. If the molten product is not maintained at a high enough temperature, the liquid within the pipes would be likely to freeze first. This might pose a complicated plumbing problem. Thermal control in general may also be somewhat delicate. The Rockwell process envisions keeping the temperature very close to the liquid–solid phase boundary, so that solids collect around the walls of the cell to prevent thermal erosion. However, this solution creates another possible problem: an environment in which phase changes could easily go out of control. All of the liquid might solidify, stopping up the pipes and freezing the system, or all the material might melt, including the walls of the electrolytic cell. Furthermore, the amount of electrical resistance in the melt is a variable that strongly influences the heat balance of the process. One approach suggested by Haskin (per. commun., 1990) might be to add more feedstock than is actually required, so that its specific heat can be used to control the temperature.

In experiments conducted by Colson and Haskin (1991), using melts that were substantially more concentrated in $Al_2O_3 + SiO_2$, higher melt viscosity resulted in frothing that, in the worst case, caused high enough melt resistivity to raise the energy requirements to nearly ten times the theoretical value.

Experimental work by Boundy (1983), using basalt in the production of gram quantities of mixed metals, resulted in the identification of significant anode, cathode, and container problems. It was also found, however, that molten basalt is conductive enough to support electrolysis, even at temperatures just above the liquidus (du Fresne and Schroeder, 1983). Furthermore, the conductivity of the magma increases with increasing temperature. Another potential problem with metallic-ion conduction was also identified: if metal ions in different valence states exist in the melt, the effect of electrolysis might be to change that valence state rather than liberate free oxygen (du Fresne and Schroeder, 1983).

E.7.2   Fluxed electrolysis or molten salt electrolysis (feedstock-independent)

Products: oxygen, volatiles, iron, metal oxides, silicon, molten slag

Silicates can be melted or dissolved in a fused (melted) salt, then electrolyzed with oxygen evolved at the anode. As discussed in the previous section, direct melting and electrolysis is potentially a very simple process, but high temperatures of 1,400–1,500°C are required, which aggravates materials problems. By using a molten salt flux, operating temperatures can be lowered to about 1,000°C. In this case, however, losses of electrolyte components must be avoided.

A molten material such as salt or NaOH is used as a liquid solvent for an electrolysis process. LiF and CaF_2 are usually considered to be the most likely fluxing agents. When NaOH is considered, it has also been referred to as "caustic electrolysis".

In addition to oxygen, all the constituent metal oxides of the lunar regolith could also be produced (Binder, 1990). It is also predicted to be relatively simple and efficient compared with some other processes. The fine fraction of the regolith could be dissolved in hot NaOH and all the constituent oxides could be reduced by electrolysis.

Metal recovery occurs at two stages. As a result of the liberation of oxygen from the liquid, some free metals are produced which remain in the liquid state. Other free metals, such as Na, K, and P, vaporize as they form. These vaporized metals are pumped out of the electrolysis cell into a cooling column. There, fractional condensation is used to separate the metals. Because the free Mg which is formed is less dense than the liquid, this metal floats to the top, and can perhaps be mechanically removed. After this, another series of chemical and thermal separations are used to collect the metals in solution and to recover the original NaOH liquid for re-use in the oxygen-liberation step (Binder, 1990). Some experimentation with this process has been done by Keller (1991).

E.8   PYROLYSIS PROCESSES

Pyrolysis is defined as a chemical change, usually a break-up of molecules into fragments or elemental components, brought about by the action of heat.

E.8.1   Magma partial oxidation (mare only – depends on iron)

Products: oxygen, magnetite (FeO)

In contrast to direct magma electrolysis, the electrochemical step is in this case transferred to a low-temperature aqueous solution operation. Problems such as anode corrosion, high-temperature gas collection, electrical insulation, and molten metal cathodes should thus be reduced.

This process consists of five principal steps. First, iron-rich mare rock or regolith is melted in the presence of oxygen, thus it is allowed to oxidize. After controlled cooling to allow crystallization, the solidified mass is pulverized and spinels are extracted magnetically. The magnetic spinel phases are then dissolved in an electrochemically-stable aqueous mineral acid. The solution is then electrolyzed to recover iron and oxygen (Waldron, 1989). We are unaware of any experiments to test this process. Theoretically, it should be possible to separate a majority of the iron in the form of magnetite. Under the reducing conditions which are typical of the lunar surface, the iron is present in the form of Fe^{2+}, and it is almost always a constituent of the pyroxene, olivine, and ilmenite. However, if the feedstock is melted in the presence of a significant oxygen atmosphere, 90 percent or more of the iron should appear as magnetite, which may be magnetically separated from the other minerals which form alongside it.

In order for the desired magnetite to form, the cooling process must be carefully controlled. If the magma is cooled very slowly, equilibrium assemblages will form and change as the cooling process continues.[15] Annealing is seen as an alternative to this slow-cooling process. First, the magma is rapidly quenched to form a glass. Then the glass is heated to an intermediate temperature and allowed to re-crystallize at a moderate rate. The material is then ground to a grain size at which the magnetite crystals will be separate from the remaining material. The magnetite crystallites can then be separated by magnetic and/or electrostatic beneficiation.

Assuming 90 percent recovery of magnetite, Waldron (1989) calculates that 83 percent of the oxygen in the iron phases can be separated out in the form of magnetite.

If a throughput rate of 0.4403 tonne/hour of feedstock is assumed, Waldron (1989) calculates an output of 13.42 kg/hour of oxygen, which equates to about 118 tonnes of oxygen per year. Energy requirements (excluding mining and output gas liquefaction) are estimated at 350 kW at standard throughput, corresponding to 26.1 kWh/kg oxygen.

E.8.2   Vapor-phase reduction (feedstock-independent)

Product: oxygen

This term comprises a family of processes which employ high temperatures and the gaseous state. The feedstock is vaporized to transform oxygen-containing compounds into monoxides and oxygen, which can then be collected upon cooling. No consumables are needed. Among its advantages is the instant and complete consumption of the raw material. It is thought that the process can be turned on or off at any time, without any pre-heating or shut-down time (Steurer and Nerad, 1983). The major disadvantage is the high energy consumption. Figure E.7 shows a schematic of the process which has been proposed.

Steurer and Nerad (1983) classified both vapor-phase reduction and ion plasma separation in the same general category. The process they named "vapor separation" is now called vapor-phase reduction, and the process they called "selective ioniza-

---

[15] See Appendix D, Section D.2.2.1 and Figure D.8 for some explanation of phase relationships.

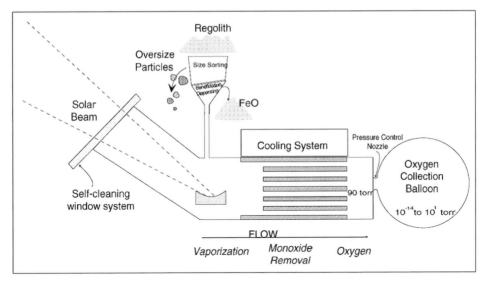

Figure E.7. Vapor phase electrolysis (modified from Steurer and Nerad, 1983).

tion" is now called ion separation. The vapor-phase reduction process is aimed specifically at the production of oxygen. The regolith is heated to such a high temperature that the molecular bonds are broken. Since FeO does not dissociate, some method of removing it (before or after processing) will be needed. After vaporization, the gas is rapidly cooled, so that everything besides the oxygen is condensed back into a liquid or solid state. Rapid cooling is essential, because the oxygen would otherwise recombine with the monoxides and metals that it was recently separated from. The free oxygen is then separated and removed. Although an oxygen collection balloon is shown in Figure E.7, it is likely that compression and liquefaction of the product oxygen would be more viable (Christiansen *et al.*, 1988). How one deals with the solidified by-products has not yet been addressed.

A temperature of 2,500–3,000 K (2,226–2,727°C) is required, which can be supplied by concentrated solar energy (Boundy, 1983). At 20 percent oxygen yield, 35.5 MW-hour of energy would be required per tonne of oxygen produced. Removal of the FeO before processing would result in a 24 percent oxygen yield, but would lower the energy requirements to 29.6 MW-hour per tonne oxygen. Neither of these estimates includes the energy requirements for beneficiation, nor do they include the energy required for mining or oxygen liquefaction. An advantage of vapor-phase reduction is that no reagents need be supplied from Earth. Bulk lunar soil can be used as a feedstock.

A disadvantage to the vapor-phase reduction process is the high energy requirement. Even if this can be supplied by solar energy, temperatures of 3,000 K will result in containment problems. Also, recombination of the dissociated constituents back into their original oxides may be an issue. Long-term fouling of condenser surfaces will lower the process efficiency (Christiansen *et al.*, 1988). Furthermore, low process pressures require larger equipment volumes (and masses). Finally, no terrestrial analogues for this process exist; thus, operational lessons learned from practical experience are unavailable (Christiansen *et al.*, 1988). Boundy (1983) reported some experimental results for vapor-phase reduction, with oxygen yield in weight percent of the starting material found to be in the order of 20 to 25 percent.

E.8.3   Ion (plasma) separation (feedstock-independent)

Products: oxygen, metals (?)

The "selective ionization process" first discussed by Steurer and Nerad (1983) is a variant of the vapor separation process. Figure E.8 illustrates the proposed process. Here, the oxygen is again separated from the metals, but at a much higher temperature, in the range of 7,000–10,000 K (6,727–9,727°C). In this case, the temperature is so high that the metals are largely stripped of their electrons – they are ionized – whereas the oxygen remains largely neutral up to a temperature of 9,000 K (8,727°C). There is an ionization gap between metals and oxygen when this temperature is reached. This means that the metals are more readily extracted by an electrostatic or electromagnetic field. Because the various metals will have differing ionization

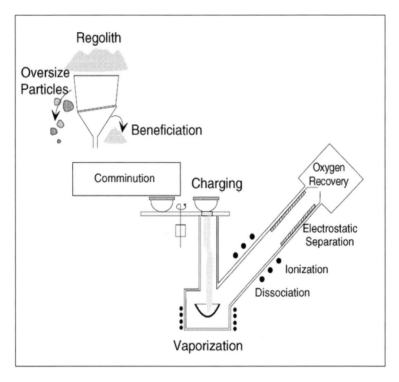

Figure E.8. Ion (plasma) separation (modified from Steurer and Nerad, 1983).

potentials at any given temperature, it may be possible to separate the metals from each other by taking advantage of this characteristic (Steurer and Nerad, 1983).

Hypothetical oxygen yields were calculated by Steurer and Nerad (1983). At 8,000 K (7,727°C), 28 wt% of the feedstock is converted into oxygen, and 37 wt% of the feedstock is converted into metals. At a temperature of 10,000 K, 38 wt% oxygen and 51 wt% metals are recoverable.

Advantages to the ion separation process include higher oxygen yield than the vapor-reduction process. As with the vapor-reduction process, it can be operated independent of re-supply of reagents from Earth.

On the minus side, the process is only a concept at present. The high temperatures will present extreme materials problems, and separation of condensed metals and oxides from electrostatic and condenser elements will be difficult. Finally, extra condensers will be required to remove the non-ionized metals and silicon which are not handled by the equipment discussed above. Only with this addition can pure oxygen actually be produced (Christiansen *et al.*, 1988).

Experimental work has shown that it is beneficial to remove iron-rich particles from the feedstock. Higher power densities, longer residence times, and smaller particles should also contribute to greater yields. It is expected that 34,500 kWh would be needed to produce a ton of oxygen, whereas 26,000 kWh would be needed to produce a ton of metal (Steurer and Nerad, 1983). An experimental plan was suggested by Steurer and Nerad (1983) to test these ideas. First, a pure argon plasma would be tested to improve techniques for starting the ionization process and to learn about the gas flow characteristics for plasma containment and sustainment. Next, metals would be mixed in with the argon, to test the cold-plate recovery of the metals. After that, oxide particles would be mixed in with the argon, to learn the degree of oxide ionization and to measure the oxygen and metal yields. Finally, oxide particles alone would be tested, to learn how to initiate the plasma condition, and to verify recovery of oxygen and metals.

E.9   SLURRY/SOLUTION PROCESSES

E.9.1   HCl dissolution and electrolysis (mare only – ilmenite only)

Product: oxygen, iron

Hydrochloric acid dissolution and electrolysis, Figure E.9, is a simple aqueous process which can be operated near ambient temperature to produce oxygen and iron from lunar ilmenite. Iron is leached from the ilmenite using hydrochloric acid, creating a ferrous chloride solution. This solution is then electrolyzed to produce oxygen and iron. The titania residue is then washed and dried in order to recover the reagents used.

Advantages to this process are that it operates at relatively-low temperature, has low thermal power demands, and is expected to be efficient. A disadvantage is that only ilmenite may be used, so that beneficiation of the feedstock will be required. This will add to the plant mass.

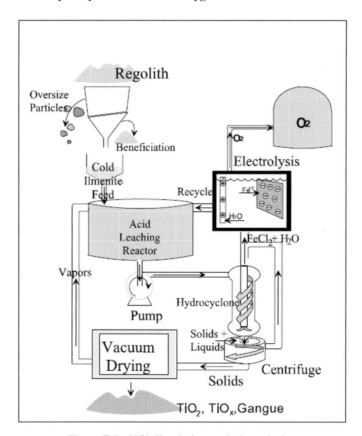

Figure E.9. HCl dissolution and electrolysis.

E.9.2   H₂SO₄ dissolution and electrolysis (mare only)

Products: oxygen, iron

This is a modification of an existing terrestrial process (Sullivan, 1990). Sulfuric acid is used to digest an ilmenite-rich feedstock, by the reaction

$$FeTiO_3 + 2H_2SO_4 \Rightarrow FeSO_4 + 2H_4O + TiOSO_4$$

Following this, the slurry is filtered to remove any unreacted solids, including silicates and other oxides. This step is called "clarification". The remaining "clean" solution is then sent through an electrolysis cell, whereby the oxygen and iron are separated in the reaction

$$FeSO_4 + H_2O \Rightarrow H_2SO_4 + Fe + O2$$

The oxygen and iron are both recovered, and the solids which were previously filtered out are calcined to recover entrained acid and volatile elements for re-use and production, respectively. Figure E.10 shows a schematic drawing of this process.

Although no experiments have yet been done to prove this concept, it is

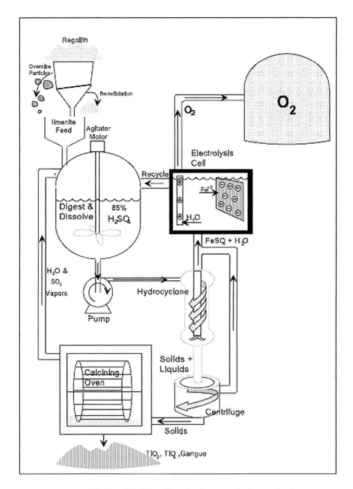

Figure E.10. H_2SO_4 dissolution and electrolysis.

nevertheless well-founded on commercial practices and procedures. In terrestrial applications, the processes are usually aimed at extracting iron or titanium dioxide as a primary product, whereas on the Moon, oxygen will be the most important product. The major unknowns occur because of this change in primary product, which necessitates bench-scale testing to establish process efficiency and reliability.

E.9.3   HF dissolution and electrolysis (feedstock-independent)

Products: oxygen, metals

The hydrofluoric acid leach process was described by Waldron and Criswell (1979) and by Waldron (1985). This involves dissolving bulk lunar regolith in hydrofluoric acid to create metal fluorides and water. The metal fluorides are separated from the slurry and from each other by a series of batch-mode reactors, which also produce steam and SiF_4 vapor in the process. The metal fluorides are reduced with sodium or potassium, creating free metals and NaF or KF. The NaF or KF is then mixed with water to regenerate the hydrofluoric acid. This is a complex procedure requiring multiple unit operations and high-temperature hydrolysis (1,000°C), ion exchange, distillation, centrifuges, and drying steps. The other product of the dissolution of lunar regolith in hydrofluoric acid is alkali hydroxide. This is melted and electrolyzed to liberate the oxygen.

Calculations suggest that 586 tonnes of plant mass will be needed to produce 1,000 tonnes of oxygen per year, assuming 40 percent recoverable oxygen and a 90 percent duty cycle. This is high in comparison to some other proposed processes. Furthermore, no specific feedstock is needed. Thus, less feedstock is required per unit oxygen product than in a reaction which utilizes only ilmenite, for example. Furthermore, because fluorine is recovered, it can be used to produce metals such as aluminum, silicon, and iron.

A disadvantage to the hydrofluoric acid dissolution and electrolysis process is that the acid leach reactors are operated in batch mode on a 30-minute leach cycle, making automation difficult. Although some experimentation has been done, additional research is required, especially regarding the separation and purification of the fluoro-compounds to recover fluorine. The large number of separate steps in the process require many small pieces of equipment, such as HF acid leach tanks, hydrolyzers,[16] strippers, distillation columns, ion-exchange beds, crystallizers, centrifuges, dryers, and molten sodium hydroxide electrolytic cells. This will require greater design, development, test, and evaluation costs than processes with fewer but larger unit operations. Furthermore, the reagent (as HF) is lost from the system at a rate of 1.5 kg per tonne of input feedstock.

Thorough (and energy-intensive) drying is required to reduce moisture and HF contents in residual solids. Finally, materials which are inert to HF, such as carbon brick, phenolic[17]/graphite, or CaF_2 liners, must be used in much of the process equipment.

Most of the process chemistry for HF dissolution and electrolysis has been investigated in the laboratory, and 75 percent of the process steps have been conducted in a comparable or equivalent pilot- or commercial-scale terrestrial process (Waldron and Criswell, 1982).

---

[16] Hydrolysis is a chemical reaction or process in which a molecule is split into two parts by reacting with a molecule of water. A hydrolyzer is thus a piece of machinery that subjects something to hydrolysis.

[17] Phenolic: of, relating to, or derived from a phenol, which is a caustic, poisonous, white crystalline compound, C_6H_5OH, derived from benzene and used in resins, plastics, and pharmaceuticals.

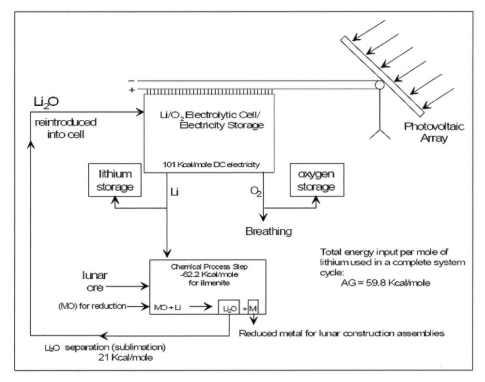

Figure E.11. Lithium or sodium reduction (from Christiansen, 1988).

E.9.4   Lithium, aluminum, or sodium reduction (feedstock-independent)

Products: oxygen, metals

Christiansen (1988) discussed this indirect electrochemical reduction of lunar oxides, which was first described by Semkow and Sammells (1987) and Sammells and Semkow (1988). Liquid lithium or sodium can be used to reduce lunar oxides to metal:

$$2Li + MO \Rightarrow Li_2O + M$$

(M stands for metal oxides such as SiO_2, FeO, TiO_2, etc.)

The resulting Li_2O can then be electrolytically separated into lithium and oxygen. The oxygen is used or stored, and the lithium is recycled. Figure E.11 shows a schematic diagram of the process.

Lithium will reduce FeO, TiO_2, and SiO_2, but it will not reduce Al_2O_3, CaO, or MgO. The expected reaction products are all solids at the reaction temperature, which makes their separation difficult. Li_2O may be sublimated at 700°C and near-vacuum pressure. After separation, the lithium oxide would be solidified and fed into a solid-state electrolytic cell which contains a melt consisting of LiF, LiCl, and Li_2O at about 900°C. The LiF and LiCl act as fluxing agents to reduce the melting temperature and viscosity, thus permitting higher ionic conductivity. Liquid lithium forms at the cathode, where it can be removed and recycled to the reduction reactor. Oxygen gas evolves at the anode.

Because silicates predominate over all other metal oxides on the lunar surface, it is advantageous that this process reduces silica as well as a few other metal oxides. Production would be independent of the site selected for the lunar outpost. Less feedstock is required per tonne of product oxygen than for processes which only operate on ilmenite or anorthite. Metals production should also be possible, although the exact strategy for doing this has not been worked out (Christiansen, 1988).

Among the disadvantages to this process is the fact that lithium oxide recovery will be difficult. Sublimation of Li_2O may require an extensive vacuum pump system, because process losses using lunar vacuum would be severe. A vapor system such as described would also be quite large. Re-solidification of the Li_2O will require energy for compression. Furthermore, flux is lost through degradation, and the stability of the electrolysis cell is uncertain. Corrosion of the anode and cathode may be severe, and the solid ceramic electrolyte material may be sensitive to mechanical damage.

E.9.5   Reduction by aluminum

Products: oxygen, aluminum, silicon

Anorthite, which is abundant in lunar highlands soils, can be reduced by a series of steps to produce oxygen, aluminum, and silicon. The basic flow chart is shown in Figure E.12. In the first step, the anorthite is dissolved in a molten mixture of cryolite (Na_3AlF_6), and another compound such as CaF_2, AlF_3, Al_2O_3 or BaF_2. The reaction takes place at about 1,000°C. Aluminum, added to the melt, reduces the silica to silicon:

$$3(CaO \cdot Al_2O_3 \cdot 2SiO_2) + 8Al \Rightarrow 6Si + 3CaO + 7Al_2O_3$$

Aluminum is added until all the available silica has been reduced to silicon. Then the solution is cooled to below 700°C which causes the formation of a solid silicon phase. The remaining melt contains an aluminum–silicon alloy of about 12.6 percent silicon (Keller, 1986; Anthony *et al.*, 1988). The solution will then require filtration by use of a centrifuge or other solid/liquid separator. This must be done in the same temperature range, 700–1,000°C, in order to separate the silicon cleanly. The alumina concentration must also be kept below 20 percent, in order to avoid precipitating any aluminum at the same time. The reaction takes about 1 hour to go to completion, so this must be a batch-mode process, unless a stirred-tank reactor can be used which is large enough to allow a residence time of one hour (Christiansen, 1988).

After removal of the silicon, the cryolite solution – which also contains CaO, alumina, and the unreacted aluminum – is pumped into an electrolysis cell where the alumina is reduced. Research is needed on the development of cermet[18] anodes which will be inert to the high-temperature oxidizing conditions.

[18] Cermet: a material consisting of processed ceramic particles bonded with metal and used in high-strength and high-temperature applications.

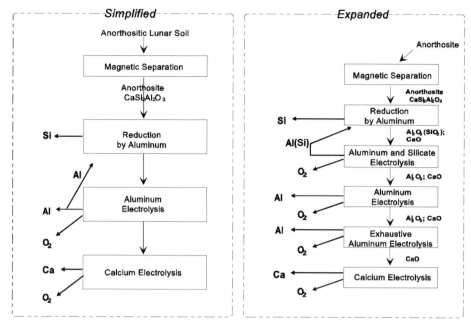

Figure E.12. Aluminum reduction of anorthite (modified from Keller, 1986 and Anthony *et al.*, 1988).

Electrolysis results in the formation of aluminum at the cathode, which sinks and must be collected from the bottom of the cell. This step may be difficult to automate. The production of 1,000 tonnes of oxygen per year would require an estimated 1.4 MW of electric power (Anthony *et al.*, 1988).

The reduction-by-aluminum process has the advantage that production of silicon and aluminum is possible; however, beneficiation will be required to separate pure anorthite from other regolith components, which is not easy. Another disadvantage is that high-temperature (700–1,000°C) solid/liquid filtering equipment will be required. Finally, recovery of the cryolite flux is difficult and expensive. This may result in significant penalties in electrical energy and equipment mass.

In experimental testing of the lithium-reduction process (Sammells and Semkow, 1988), the anode was made of strontium-doped lanthanum manganite ($La_{0.89}Sr_{0.10}MnO_3$). A solid electrolyte made of CaO or yttria-stabilized zirconia was used. Testing of the aluminum-reduction-of-anorthite process has been conducted by EMEC Consultants of Pittsburgh (Keller, 1986; Anthony *et al.*, 1988). They have shown the viability of the first step of the process (reduction of anorthite by aluminum), but additional research is required to quantify yields and optimum operating conditions. An efficient way to separate CaO from the electrolyte melt has not yet been developed.

E.9.6   Caustic dissolution and electrolysis

Products: oxygen, metals

As described by Christainsen *et al.* (1988), this process uses molten sodium hydroxide (NaOH) at 400°C to dissolve oxides from bulk lunar soil. The molten solution/slurry is electrolyzed, with oxygen forming at the anode and sodium forming at the cathode. The sodium then reduces the lunar oxides to produce metals, which deposit at the cathode, and Na_2O. Although originally proposed as a batch process (Dalton and Hohmann, 1972), continuous operation might be possible if separate units were operated for the dissolution and electrolysis steps. A mixed caustic and solids slurry could be continuously withdrawn from a pipe near the cathode. A third unit would be required to separate and recycle the caustic back to the solution tank.

Advantages of the caustic-dissolution-and-electrolysis process include the fact that all the constituent metal oxides of the lunar regolith are reduced by this technique (Binder, 1990). Metals production would then be possible, although additional steps would be required to accomplish this. Sodium exists in the lunar soil and this could possibly be used to replace the sodium lost in the course of processing.

It is expected that recovery of the caustic from residual solids will be difficult and may result in significant reagent make-up penalties (Christiansen *et al.*, 1988). Finally, dendritic growth of metals on the cathode could cause the electrolytic cell to short-circuit (Sammells and Semkow, 1988). Cutler (1985) proposed that KOH be used instead of NaOH, and another leach process, using NaOH and a non-electrolytic pyrochemical process, has also been suggested (Waldron and Criswell, 1982). However, that process appears to be particularly complex, with multiple steps at high temperatures, and creates a need for additional carbon reagents (Christiansen *et al.*, 1988).

Little experimental work has yet been done, beyond the preliminary results reported by Dalton and Hohmann (1972). They learned that nickel electrodes were consumed in the electrolysis process.

E.9.7   Ion sputtering (feedstock-independent)

Products: all elements

An ion-sputtering process was proposed by Curreri (1990). The process is expected to refine "sand-like" mineral ores (such as lunar anorthite, ilmenite, or olivine) and convert them directly into elements, or into precise technical materials such as semiconductor crystals and films.

The process consists of three basic steps: ore disintegration, element discrimination, and element collection. Ore disintegration is accomplished by a particle beam consisting of charged elemental ions (Figure E.13). Maximum target disintegration occurs at a beam energy of 5 MeV, regardless of the target ore composition or structure.

In the next step, element discrimination, the ions are driven by an electrical potential difference into an electrostatic/magnetic field that separates the elements. This is the principle behind the mass spectrometer. For example, if the elements are

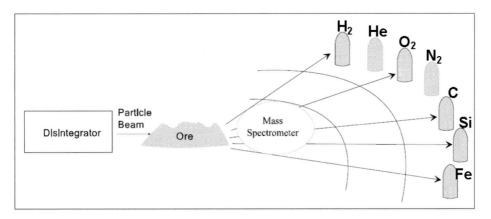

Figure E.13. Ion sputtering (modified from Curreri, 1990).

monocharged, a separation between H and He of about 70 cm would correspond to a separation between H and U of about 10 meters.

Advantages of the ion-sputtering process include the fact that it eliminates the need for beneficiation, including size-sorting. It is also designed to operate in hard vacuum, and thus is easier to use on the Moon than on Earth. Particle accelerators of the type described have been studied for use in the Space Defense Initiative. However, we are unaware of any published experiments which address the feasibility of this process.

E.10   PROCESS COMPARISONS

A major difficulty with making comparisons between the proposed processes is that each worker has based their yield and energy requirements calculations on arbitrary assumptions. For example, the amount of oxygen to be produced has a significant impact on the mass and volume of equipment that must be emplaced. Because there has been no standardization of this assumption, values from 4.4 tonnes/year to 19,500 tonnes/year have been used in the work done to date. Other processes have been calculated without any rate specified at all producing, for example, a batch of 10 tonnes (e.g. Astronautics Corp. of Am., 1987).

We propose that a standard value of 150 tonnes per year should be adopted, because there has already been a slight amount of standardization in this direction. A separate target of one tonnes/year would be useful to compare processes which could be emplaced early in the development of a lunar base. That level of production would be adequate for proof of concept.

Hepp *et al.* (1991, 1994) calculated the power requirements for eleven of the processes that have been proposed for oxygen-extraction. Their results are shown in Table E.1.

Table E.1. Power requirements for selected oxygen-extraction processes (from Hepp et al., 1994)

Process	Feedstock	Electrical power (kW per tonne of fuel per year)	Thermal power (kW per tonne of fuel per year)	Temperature (°C)	Products
H_2 reduction	Ilmenite ($FeTiO_3$)	0.72	0.18	900	O_2, Fe, FeO, TiO_2 Fe
Carbothermal	Enstatite ($MgSiO_3$)	0.82	3.28	1,625	O_2, Si, MgO, SiH_4 Si
Carbochlorination	Anorthite ($CaAl_2Si_2O_8$)	1.33 1.38	2.46 2.57	675–770	$CaCO$, Al, Si Si Al
HF acid leaching	Mare regolith	8.85	8.85	110	O_2, Al Al
Reduction by Li or Na	Mare regolith	2.15 3.30 14.55	2.15 3.30 14.55	900	O_2, Si, Fe, Ti Si Fe Ti
Reduction by Al	Anorthite ($CaAl_2Si_2O_8$)	2.56 2.64	0.64 2.64	1,000	O_2, Si, Al, Ca Si Al
Direct fluorination	Anorthite ($CaAl_2Si_2O_8$)	15.52 16.16	3.88 4.04	900	O_2, Al, Si, CaO Si Al
Magma electrolysis	Silicate rock	0.26	0.26	1000–1,500	O_2, Fe Fe
Fluxed electrolysis	Silicate rock	6.40 9.95 19.15	6.40 9.95 19.15	1,000–1,500	O_2, Al, Si, Fe Si Fe Al
Vaporization/Fractional distillation	Regolith	1.77 5.28	4.13 12.32	2,700	O_2, Al, Si Suboxides Si Al
Selective ionization	Regolith	7.90 12.20 23.60 52.00		7,700	O_2, Al, Si, Fe, Ti Mg Si Al Fe Ti

E.11   PROTOTYPE PLANT DESIGNS

E.11.1   Prototype robotic lander for ISRU

A prototype LUNOX processing plant was designed and built at the University of Arizona in the mid-1990s. It is a good example of the design considerations that would be appropriate to a robotic proof-of-concept mission. The design includes a rotating table, a solar furnace consisting of two mirrors and a small reactor, and two robotic arms. The electrical energy requirement is limited to powering the onboard computer and robotic motors, and a solar furnace is used for heating. The power needed can be generated by an array of solar cells and stored in a small bank of rechargeable batteries. Figure E.14 shows a side view of the prototype, and Figure E.15 shows a top view. The accurate and fully-operational prototype is a half-scale model, which includes computer-controlled robotics and tracking.

The primary arm gathers lunar soil which is sifted into crucibles. Then the secondary arm transfers each crucible to a furnace. The parabolic primary mirror tracks the Sun, while the secondary mirror focuses the solar thermal energy into the furnace, where the crucible has been placed. The gases released by the soil during heating in the furnace are collected and analyzed, then the spent material is formed into a small briquette that might be used as a building material. An onboard computer monitors the processing, while also providing command and control for the robotics.

As shown in Figure E.15, all components of the prototype are mounted on a round table, which is in turn mounted on top of the lander and can be rotated via an electric motor. A three-degree-of-freedom primary arm with a scoop attached collects the soil. The table rotates to align the primary arm with one of the crucibles that are positioned around the perimeter of the table. The primary arm dumps the soil through a sifter that separates out the rocks and allows small particles to enter the crucible. The table is rotated again so that the filled crucible is within reach of the secondary arm, which has six degrees of freedom. This secondary arm places the crucible in the solar furnace.

The primary mirror, one meter in diameter, rides on a quarter-circular track to allow vertical Sun tracking (Figure E.15). Mounted on top of the primary mirror is an array of four solar panels. The panels are arranged in a two-by-two formation, and separated from each other by plates that are perpendicular to the panels. These plates cast a shadow on some parts of the panels unless the arrangement faces the Sun directly. By comparing the voltage signals and moving the mirror, the Sun can be tracked.

At the focal point of the primary mirror is a smaller, secondary mirror that deflects the light vertically downward. Because the secondary mirror faces the center of the primary mirror's quarter-circle track, the image created by the primary mirror will always fall on the secondary mirror. If the rotation of the secondary mirror is coordinated with the vertical tracking motion of the primary mirror, the light will always be focused at the same spot – inside the top of a parabolic "light funnel" that sits on top of a vertical tube. The tube is mirror-coated on the inside, so the light rays will zig-zag down the tube to the bottom, where the furnace is located.

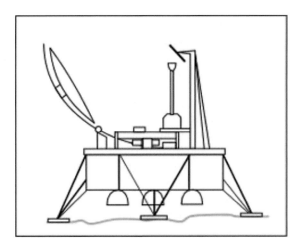

Figure E.14. Prototype LUNOX plant, side view (University of Arizona, Space Engineering Research Center).

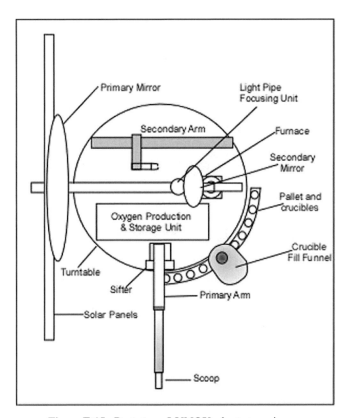

Figure E.15. Prototype LUNOX plant, top view.

The furnace consists of two pieces. The inner piece is 3.5 inches tall, with the lower part cylindrical in shape, and the upper part conical. The secondary arm inserts the crucible into the furnace from the bottom. The crucible is mounted on a plug which is used to seal the furnace. The outer piece of the furnace, which has a hole in the top to accommodate the light tube, is shaped such that there is a uniform clearance of 0.64 cm between the inner and outer pieces. The inside of the outer piece and the domed portion of the inner piece are mirror-coated, while the lower piece of the inner furnace has a rough surface finish. This will ensure that as the light enters the furnace through the top, it will travel down the clearance between the inner and outer pieces, and heat the crucible by heating the wall with the rough surface finish. The released gases are collected via a pick-off tube.

E.11.2   Suitcase-size hydrogen-reduction plant

An operational, "suitcase-size" lunar oxygen plant based on the hydrogen-reduction-of-ilmenite process was designed and built by Dr. Larry Taylor of the University of Tennessee in the early 1990s. Figure E.16 illustrates the LUNOX plant setup.

The idea was to show that a small, working LUNOX plant could actually be built, and that it is indeed possible to reduce ilmenite with hydrogen, thus creating water which can be separated into its component hydrogen and oxygen (the separation process was not included in this demonstration). Pure ilmenite was used for the demonstration, thus the actual yield of this plant, if operating on the Moon, would be less than what was shown in the demonstration.

One of the observations from the LUNOX plant was that a closed system is less efficient than an open system. An open system was used in the initial setup, which had a tank of hydrogen on one end. The hydrogen was pumped across the hot ilmenite, water was produced and collected, then the hydrogen was vented out the window. This would not be advisable on the Moon, but was done only to demonstrate the feasibility of the process. Since the open system had pure (dry) hydrogen flowing across the ilmenite at all times, the reaction rate was comparatively fast. When the system was closed so that the hydrogen was recycled, the reaction rate was slower. Because it is difficult to separate water from hydrogen with 100% efficiency, some water remained in the system and "diluted" the hydrogen gas. After a few cycles, a hydrogen/water mixture was passing over the ilmenite, which lowered the partial pressure of the hydrogen and thus lowered the efficiency. The yield of water per unit time was about 30 percent less in the closed system that it had been in the open system.

E.11.3   Roxygen

Work is underway at NASA to prepare for the first robotic missions to the surface of the Moon in over 30 years. Part of this planning involves the development of various pilot-scale oxygen-production plants. Carbon-monoxide-silicate reduction, fluxed magma electrolysis, and hydrogen reduction of lunar regolith are being studied.

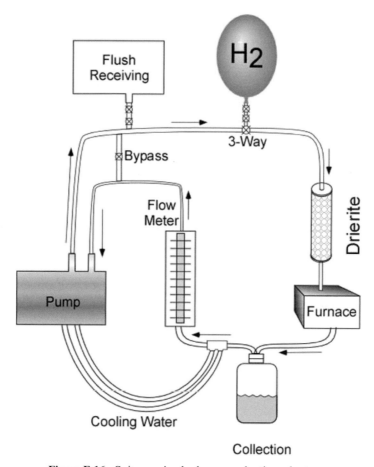

Figure E.16. Suitcase-size hydrogen-reduction plant.

Building on the work of the last 40 years, these designs are moving from the concept/ laboratory bench-top phase into the "engineering development unit" phase, in which working, full-scale versions are being built for testing. Testing and analysis over the next few years will determine which of these processes is selected for an oxygen production demonstration on the Moon.

E.12 SULFUR AND H₂S HAZARDS

In almost all of the processes that have been proposed for oxygen extraction from lunar materials, there is a potential for hydrogen sulfide (H_2S) contamination. Lunar regolith contains sulfur in various forms (Gibson *et al.*, 1973; Vaniman *et al.*, 1988). H_2S is released during the hydrogen-reduction process, by the reaction of hydrogen gas with troilite, FeS (e.g., Ness *et al.*, 1990), at temperatures above 800°C. Moreover, Gibson and Moore (1973) and Gibson *et al.* (1977) suggest that H_2S gas will be evolved by any process that heats the lunar soil.

H_2S is a highly-toxic and flammable gas. Because it is heavier than air it tends to accumulate at the bottom of poorly-ventilated spaces. Although very pungent at first, it quickly deadens the sense of smell, so potential victims may be unaware of its presence until it is too late. Hydrogen sulfide is considered a broad-spectrum poison, meaning that it can poison several different systems in the body, although the nervous system is most affected. The toxicity of H_2S is comparable with that of hydrogen cyanide.

H_2S has the potential to corrode and destroy plant equipment. It renders some steels brittle, leading to sulfide stress cracking. This would reduce the reliability and robustness of lunar oxygen-extraction systems.

In industry, ZnO is often used to scavenge H_2S, forming H_2O and ZnS – an approach which makes ZnO a consumable. With levels of FeS at approximately 0.1 percent in the lunar regolith, some of the processes discussed in this appendix could produce as much as 39 kg of H_2S per 100 tonnes of regolith feedstock. This would in turn require nearly 100 kg of ZnO. Supply of this material from Earth would add to the cost of producing oxygen on the Moon.

Some other approaches have been proposed to address the hazard of H_2S. Sullivan (1990) proposed to burn off the H_2S-containing gas, forming H_2SO_4, which could be used as re-supply for any losses in the sulfuric-acid-based process. E.K. Gibson (pers. commun., 2006) proposed using titanium metal as the scrubber. Titanium is more abundant than zinc on the lunar surface, and oxygen-extraction experiments could incorporate this concept in their prototype systems. This would serve two purposes: it would determine the amount and physical form of titanium to be used; and would provide protection to the researchers from the H_2S gas that is released by lunar regolith simulants, which also contain sulfur.

E.13 CONCLUSIONS

The general consensus of the trade studies is that oxygen production would be economically beneficial for lunar base development in almost any scenario. As a bonus, most of the processes that produce oxygen also provide other valuable materials.

We have reviewed several of the proposed processes for lunar oxygen production and discussed the benefits and possible drawbacks of each. Our conclusion is that the four most promising processes are: (a) hydrogen reduction of glass; (b) magma electrolysis; (c) H_2SO_4 dissolution/electrolysis; and (d) carbothermal reduction.

E.14   REFERENCES

Allen, C.C.; McKay, D.S.; and Morris, R.V. (1992a) Hydrogen reduction of lunar simulant glass [extended abstract]. *Lunar Planet. Sci.* **23**, Lunar and Planetary Institute, Houston, TX.

Allen, C.C.; McKay, D.S.; and Morris, R.V. (1992b) Lunar oxygen – the reduction of glass by hydrogen. In: *Engineering, Construction and Operations in Space III, Space '92, Proceedings of the Third International Conference*, American Society of Civil Engineers, New York, pp. 629–640

Allen, C.C.; Bond, G.G.; and McKay, D.S. (1994) Lunar oxygen production – a maturing technology. In: *Engineering, Construction and Operations in Space IV, Space '94, Proceedings of the Fourth International Conference*, American Society of Civil Engineers, New York, pp. 1157–1166.

Anthony, D.L, Cochran, N; Haupin, W.E,; and Keller, R. (1989) Products from Lunar Anorthite, in *Space manufacturing 7 – Space resources to improve life on earth*; Proceedings of the Ninth Princeton/AIAA/SSI Conference.

Astronautics Corporation of America (1987) *Lunar surface base propulsion system study*. Final report, Astronautics Corporation of America, NAS9-17468.

Berggren, M.; Zubrin, R.; Carrera, S.; Rose, H.; and Muscatello, S. (2005) Carbon Monoxide Silicate Reduction System [extended abstract]. In: *Space Resources VII*, Lunar and Planetary Institute, Houston, TX.

Bhogeswara, R.; Choudaray, U.; Erstfield, T.; Williams, R.; and Chang, Y. (1979) Extraction processes for the production of aluminum, titanium, iron, magnesium and oxygen from nonterrestrial sources. In: J. Billingham and W. Gilbreath, eds, *Space Resources and Space Settlements*, NASA SP-428, pp. 257–274.

Briggs, R.A. and Sacco, A. (1988) Oxidation and reduction of ilmenite: Application to oxygen production on the Moon (Preprint). In: W.W. Mendell, ed., *Second Symposium on Lunar Bases and Space Activities of the 21st Century*, paper no. LBS-88-170.

Burt, D.M. (1988a) Lunar mining of oxygen using fluorine (Preprint). *Proc. 2nd Symp. on Lunar Bases and Space Activities of the 21st Century, Houston.*

Burt, D.M. (1988b) Lunar Production of Oxygen and Metals Using Fluorine: Concepts Involving Fluorite, Lithium, and Acid-Base Theory [abstract]. *19th Lunar and Planetary Science Conference*, Lunar and Planetary Institute, Houston.

Calkins, D. (2006), unpublished personal communication.

Chang (1959) Process for preparing concentrated titanium mineral – Patent 3959465. U.S. Patent Office, Washington, D.C.

Christiansen, E.L.; Euker, H.; Maples, K.; Simonds, C.H.; Zimprich, S.; Dowman, M.W.; and Stovall, M. (1988) *Conceptual Design of a Lunar Oxygen Pilot Plant*, NAS9-17878, EEI Rpt. No. 88–182, Eagle Engineering Inc., Houston, TX.

Colson, Russell O. and Haskin, Larry A. (1991) Magma electrolysis: An update. In: *Resources of Near-Earth Space: Abstracts*, University of Arizona, Tucson, p. 7.

Cooper, B.; Sullivan, T.; Carter, J.; and McKay, D. (1989) *Proposed Processes for Lunar Oxygen Extraction: A Review*. Unpublished NASA Document.

Criswell, D.R. (1983) A transportation and supply system between low Earth orbit and the Moon which utilizes lunar derived propellants. *Lunar Planet. Sci.* **14**, Special Session Abstracts, Lunar and Planetary Institute, Houston, TX.

Curreri, P.A. (1990) *Ore nonspecific process for differentiation and the restructuring of minerals in vacuum*. Patent application, NASA in-house memorandum.

Cutler, A.H. (1985) An alkali hydroxide based scheme for lunar oxygen production (Abstract). *First Symposium on Lunar Bases and Space Activities of the 21st Century*.

Cutler, A.H. and Krag, P. (1985) A carbothermal scheme for lunar oxygen production. In: W.W. Mendell, ed., *Lunar Bases and Space Activities of the 21st Century*, Lunar and Planetary Institute, Houston, TX, pp. 559–569.

Cutler, A.H. and Wolff, A.R. (1990) Impact of silane fuel on lunar oxygen supply to low Earth orbit. *Proc. AIAA Space Prog. and Technologies Conf.* American Institute of Aeronautics and Astronautics, Washington, D.C. (in press).

Dalton, C. and Hohmann, E., eds. (1972) *Conceptual design of a lunar colony* (Systems engineering study for proposed lunar colony). NASA CR-129164.

Davis, H.P. (1983) Lunar oxygen impact upon STS effectiveness. *Lunar Planet. Sci.* **14**, Special Session Abstracts, Lunar and Planetary Institute, Houston, TX.

du Fresne, E. and Schroeder, J.E. (1983) Magma electrolysis. In: W.F. Carroll, ed., *Research on the Use of Space Resources*, NASA Jet Propulsion Laboratory, Pasadena, CA.

Gibson, M.A. and Knudsen, C.W. (1985) Lunar oxygen production from ilmenite. In: W.W. Mendell, ed., *Lunar Bases and Space Activities of the 21st Century*, Lunar and Planetary Institute, Houston, TX, pp. 543–550.

Gibson, M.A. and Knudsen, C.W. (1988) Lunar oxygen production from ilmenite (Preprint). In: W.W. Mendell, ed., *Second Symposium on Lunar Bases and Space Activities of the 21st Century*, paper no. LBS-88-056.

Gibson, E.K. Jr. and Moore, G.W. (1973) Carbon and sulfur distributions and abundances in lunar fines. In: *Proceedings of the Fourth Lunar Science Conference* (Supplement 4, *Geochim. et Cosmochim. Acta*, **2**), pp. 1577–1586.

Gibson, E.K. Jr.; Moore, G.W.; and Johnson, S.M. (1974) *Summary of analytical data from gas release investigations, volatilization experiments, elemental abundance measurements of lunar samples, meteorites, minerals, volcanic ashes and basalts*. NASA Johnson Space Center, Houston, TX.

Gibson, E.K. Jr.; Brett, R.; and Andrawes, F. (1977) Sulfur in lunar mare basalts as a function of bulk composition. *Proc. Lunar Sci. Conf. 8th*, Lunar and Planetary Institute, Houston, TX.

Gibson, M.A.; Knudsen, C.W.; and Roeger, A. III (1990) Development of the Carbotek process (TM) for lunar oxygen production. In: S.W. Johnson and J.P. Wetzel, eds., *Engineering, Construction and Operations in Space, Proceedings of Space '90*, American Society of Civil Engineers, New York.

Hepp, A.F.; Linne, D.L.; Landis, G.A.; Groth, M.F.; and Colvin, J.E. (1991) *Production and use of metals and oxygen for lunar propulsion*, AIAA paper 91-3481; E-6496; NAS 1.15105195; NASA-TM-105195.

Hepp, A.F.; Linne, D.L.; Landis, G.A.; Wade, M.F.; and Colvin, J.E. (1994) Production and Use of Metals and Oxygen for Lunar Propulsion. *Journal of Propulsion and Power*, 0748-4658 vol.10 no.6 (834-840).

Keller, R. (1991) Lunar production of oxygen by electrolysis. In: *Space Manufacturing 8 – Energy and Materials from Space; Proceedings of the 10th Princeton/AIAA/SSI Conference, Princeton, NJ*.

Lynch, D.C. (1989) Chlorination processing of local planetary ores for oxygen and metallurgically important metals. In: *Space Engineering Research Center for Utilization of Local Planetary Resources Annual Progress Report 1988–89*, University of Arizona, Tucson.

Maskalick, N.J. (1986) High temperature electrolysis cell performance characterization. *Int. J. Hydrogen Energy*; vol/issue: 11:9, Elsevier Publishing, Amsterdam.

McCullough, E. and Mariz, C. (1990) Lunar oxygen production via magma electrolysis. In: S.W. Johnson and J.P. Wetzel, eds., *Engineering, Construction and Operations in Space, Proceedings of Space '90*, American Society of Civil Engineers, New York.

McGannon, H., ed. (1964) The Making, Shaping, and Treating of Steel, 8th ed., United States Steel Corporation, Pittsburgh, PA.

Moulford, E.F.L. *et al.* (1996) A comparative evaluation of lunar oxygen production technologies. *26th International Conference on Environmental Systems, Monterey, CA*, SAE Technical Paper Series 961596.

NASA (1989) Proceedings of the 2nd Workshop on InSitu Resource Utilization. In *NASA Technical Memorandum 4170; Office of Exploration FY 1989 Annual Report; Exploration Studies Technical Report*; National Aeronautics and Space Administration, Washington, D.C.

Ness, R.O. Jr.; Runge, B.D.; and Sharp, L.L. (1990) Process design options for lunar oxygen production. In: S.W. Johnson and J.P. Wetzel, eds., *Engineering, Construction and Operations in Space, Proceedings of Space '90*. American Society of Civil Engineers, New York.

Rosenberg, S.D.; Guter, G.A.; and Miller, F.E. (1966) The on-site manufacture of propellant oxygen from lunar resources. *Aerospace Chemical Engineering, AIChE*, **62**, no. 61, 228–234.

Rosenberg, S.D.; Musbah, O.; and Rice, E.E. (1996) Carbothermal Reduction of Lunar Materials for Oxygen Production on the Moon: Reduction of Lunar Simulants with Carbon [extended abstract]. *Lunar and Planetary Science*, **27**, 1103.

Sammells, A.F. and K.W. Semkow (1988) Electrolytic cell for lunar ore refining and electric energy storage [abstract]. Proc. Second Symposium on Lunar Bases and Space Activities of the 21st Century, Houston.

Semkow, K.W., and A.F. Sammells (1987) The indirect electrochemical refining of lunar ores. *Jour. Electrochem. Soc.*, vol. 134, no. 8, pp. 2088–2089.

Simon, M.C. (1985) A parametric analysis of lunar oxygen production. In: W.W. Mendell, ed., *Lunar Bases and Space Activities of the 21st Century*, Lunar and Planetary Institute, Houston, TX, pp. 531–541.

Stump, B.; Babb, G.; Christiansen, E.; Sullivan, D.; and Rawlings, P. (1985) *Analysis of Lunar Propellant Production*. EEI Rpt. No. 85-103B, Eagle Engineering Inc., Houston, TX.

Stump, B.; Engblom, B.; D'Onofrio, M.; and Drake, K. (1989) *Support to Extraterrestrial Propellant Leveraging Trade Study*. EEI 89-251, Eagle Engineering Inc., Houston, TX.

Sullivan, T. (1990) Process Engineering Concerns in the Lunar Environment. *Proc. AIAA Space Prog. and Technologies Conf, 1990*, AIAA Paper 90-3753, American Institute of Aeronautics and Astronautics, New York.

Sullivan, T. (1991) *Method for producing oxygen from lunar materials*. U.S. Patent Number 5,227,032. U.S. Patent Office, Washington D.C.

Taylor, L.; Cooper, B.; McKay, D.; and Colson, R. (1991) Oxygen production on the Moon: Processes for different feedstocks. *Proceedings of the Annual Meeting of the Society for Mining, Metallurgy, and Exploration, Denver, CO*.

Vaniman, D.T. and G.H. Heiken (1990) Getting lunar ilmenite: from soils or rocks? Engineering, Construction and Operations in Space II, Space '90, Proceedings of the Second International Conference. S.W. Johnson and J.P. Wetzel, eds. American Society of Civil Engineers, New York.

Vaniman, D.T.; Pettit, D.R.; and Heiken, G. (1988) Uses of lunar sulfur. Second Symposium on Lunar Bases and Space Activities of the 21st Century, W. Mendell, ed. Lunar and Planetary Institute, Houston.

Volk, W. and Stotler, H.H. (1970) Hydrogen reduction of ilmenite ores in a fluid bed. *Journal of Metals*, vol. 22, no. 1, 50–53.

Waldron, R.D. (1985) Total separation and refinement of lunar soils by the HF acid leach process. *Proc. 7th Princeton Space Manufacturing Conf.*, American Institute of Aeronautics and Astronautics, New York.

Waldron, R.D. (1989) Magma partial oxidation: A new method for oxygen recovery from lunar soil. *Space Manufacturing 7: Space Resources to Improve Life on Earth*. American Institute of Aeronautics and Astronautics, Washington, D.C.

Waldron, R.D., and D.R. Criswell (1979) Overview of methods for extraterrestrial materials processing. In *Fourth Princeton/AIAA Conference on Space Manufacturing Facilities*, AIAA, New York.

Waldron, R.D., and D.R. Criswell (1982) Materials processing in space. Ch. 5, pp. 97–130, in *Space Industrialization, Volume 1*, B. O'Leary, ed., CRC Press.

Williams, R.J. (1985) Oxygen extraction from lunar materials: An experimental test of an ilmenite reduction process. In: W.W. Mendell, ed., *Lunar Bases and Space Activities of the 21st Century*, Lunar and Planetary Institute, Houston, TX.

Williams, R.J. and Mullins, O. (1983) Enhanced production of water from ilmenite: An experimental test of a concept for producing lunar oxygen. *Lunar Planet Sci.*, **14**, Special Session Abstracts, pp. 34–45.

Appendix F

Facilitating space commerce through a lunar economic development authority[1]

P.R. Harris   *United Societies in Space Inc. (USIS)*
D.J. O'Donnell   *World-Space Bar Association (W-SBA)*

The creation of space economic development authorities (SEDA) presents a near-term, viable strategy for the use of space resources. This is a legal and financial mechanism to encourage private investment in the formation of both a space infrastructure and off-world commercial opportunities, as well as to promote scientific research on the high frontier. If the twenty-first century is to realize the industrialization and settlement of space, humanity must establish innovative governance, management, and funding systems which move beyond sole support from public tax monies and operations only by national space agencies. An attitude change is necessary for space to become a place for the practice of synergy, so that planners promote global cooperation in human enterprises aloft, especially through public–private partnerships in space macroprojects.

Specifically, to return to the Moon permanently, utilize its resources, and facilitate both economic activity and settlement there the prototype SEDA program will be the founding of a lunar economic development authority (LEDA). The next step would be feasibility studies on applying this model in formation of a Mars economic development authority (MEDA) and then an orbital economic development authority (OEDA). Such legal entities would provide a means for international collaboration and support of new space markets, while fostering private-sector participation and investment. For LEDA to become a model in this process, legislation and agreements might enable it to:

- Issue bonds to underwrite lunar enterprises, such as a transportation system,

[1] This is an update of a paper prepared for the *International Symposium on New Space Markets at the International Space University, Illkirch/Strasbourg, France, in May 1997*. See *New Space Markets* (Dordrecht, the Netherlands, 1996), edited by G. Haskell & M. Rycroft, pp. 223–242.

industrial parks/bases, and projects ranging from beaming solar energy toward Earth to mining Helium-3.

- Coordinate and facilitate international endeavors on or near the Moon by space agencies, scientific organizations, universities, and private corporations or consortia.
- Create a leasing system and site permits for activity on or underneath the lunar surface, in order to build facilities, habitats, and other infrastructure.
- Provide an overall administration and security for lunar settlements and installations, as well as for commercial operations and tourism.
- Protect the lunar environment and ecology, possibly through regulations, zoning or inspection.
- The whole SEDA goal is to further private initiatives and entrepreneurs on the space frontier, so as to advance the creation of a spacefaring civilization for the new millennium.

F.1   INTRODUCTION

Expanding human enterprise on the space frontier during the twenty-first century will require attitudinal change on the part of planners and entrepreneurs, as well as new synergistic relationships and institutions. Dictionaries define enterprise as a plan, program, or project of some importance that is underway or to be undertaken, requiring readiness, boldness, and energy. Such an explanation admirably fits space enterprise, whether it is building spacecraft or robotic systems, off-world missions of exploration and discovery, or industrialization and settlement aloft. In his book on the subject, David Gump maintains that we are entering a new era of free enterprise aloft that will create an orbital economy. To open and develop new space markets also calls for people who are enterprising – that is, ready to engage in financial and technological ventures with imagination and initiative. To succeed in such endeavors demands strategic planning and management, characterized by thinking big and planning ahead by decades for the long term! One such astute insight relative to the Earth–Moon system came from Dr. David R. Criswell, co-inventor of the lunar solar power system, who said in a presentation on the two-planet economy:

> "By mid-century 2050, lunar power industries can be sufficiently experienced and profitable to diversify into a wide range of other products and locations, other than solar power beaming. Specialized industries on asteroids and other moons will arise. Mankind can begin the transition to living independently off Earth. People can afford to move to space and return the womb of the biosphere Earth to the evolution of other life." – Criswell (1994).

F.1.1   Attitudinal change

Attitudinal change implies that space market leaders explore funding prospects that go beyond the traditional public contracts and tax support. A cogent case for investment in space enterprises will attract new capital and participation by world corporations, venture capitalists, and individual financiers. In *Living and Working in Space* (Harris, 1996), there is considerable detail about why space is a place for synergy! Such a concept is one attitudinal change vital for space endeavors, which are very complex, costly, and involve high risk with people, technology, and finances. The practice should begin by attempting to integrate various visions and goals for space development. Furthermore, the private sector should be more involved in formulating and setting space policy, particularly with regard to establishing relative priorities as to joint undertakings on this frontier.

Global collaboration in space undertakings is demanded by the world market-place, as well as by current economic constraints on national space budgets. International space cooperation is necessary not only among nations and their space agencies, but also between the public and private sectors, as well as among universities, research institutions, corporations, and non-profit associations. One example of this synergy is the privatization of NASA's space shuttle system, resulting in a partnership between that agency and its contractor. That agreement was shaped by external demands for more flexibility and responsiveness, as well as for lowering the cost of access to space while improving performance.

The cultivation of emerging space markets calls for public–private partnerships. The National Construction Institute offers this helpful definition of the concept:

> Partnering is a long-term commitment between two or more organizations for the purpose of achieving specific business objectives by maximizing the effectiveness of each other's resources.

Sometimes this occurs by forming a *strategic* alliance or consortium. One such contemporary example in the Asia–Pacific launch vehicle business is Sea Launch. Established in 1995, this public–private partnership involves participation by the United States, Europe, and Russia (e.g., Boeing Commercial Space from the U.S.A., RSC Energiya of Russia, KB Yuzhnoye and PO Yuzhmash of the Ukraine, and Kvaerner of Norway) in building a ship and ocean launch platform for reliable Zenit rockets. On the high frontier, the need for synergistic partnering or alliance is essential for companies and industries so engaged, especially so for small businesses and start-up firms. It may help to explain why, in 1997, small-space-business leaders began a dialog about forming a trade group called the Space Entrepreneurs Association.

Potentially, billions of dollars may be earned in twenty-first century space markets. How this may come about, particularly with reference to the utilization of space resources, has been well documented over the last two decades. Certainly, the twentieth century prototype – the communication satellite industry – has proven the possibilities. There are many other challenges facing space enterprise, such as legal, political, financial, and macromanagement of large-scale orbital projects, which we have attempted to address in the past.

Throughout the initial fifty years of the Space Age, there have been seminal examples of creative solutions. One occurred in the first commercial space market, when in a rare demonstration of synergy 125 nations under the leadership of the

United States formed a commercial cooperative, INTELSAT, to own and operate a global satellite system. To be successful in space market expansion during the next century, humanity must replicate such enterprises by creating viable governance, financing, and management systems, especially through international consortia. NASA, too, has created strategic alliances with universities and for-profit entities that were mutually beneficial and may be worthy of imitation. These were intended to advance joint enterprises between federal and non-federal constituencies.

F.1.2   Commercial, legal, and political challenges in lunar enterprise

Among the multiple markets aloft, we focus next on the lunar venue as a case in point. To actualize the Moon's potential, there are hurdles to overcome and mechanisms to be put in place. Since 1967, the exploration and use of outer space has been the exclusive province of nations and their designated space agencies. Furthermore, international space treaties under the the United Nations have not been pro business development; individual citizens are not able to leave Earth and engage in off-world commerce unless licensed by their governments. Investors currently have little incentive to underwrite lunar enterprises: treaties preclude private ownership, transportation costs to and from the Moon are excessive, governments do not provide tax advantages for lunar expenditures, and ROI appears to be decades away.

Presently, there is no united vision of our Earth–Moon system and how it is to be developed. Despite past studies and missions to the lunar surface by the United States and Russia, neither country has immediate plans or official policy for obtaining lunar resources, or for using the Moon as a space station and launch pad into the universe. Even with contemporary lunar research and proposed initiatives by Europe and Asia, there is no international agreement in place for lunar development. The European Space Agency has encouraging proposals for lunar scientific projects yet to be funded, while rumors continue that China plans its own expeditions to the Moon in the near future. Japan alone has ambitious lunar plans involving both the private and public sector – the launch of Lunar-A for unmanned lunar orbit and penetration, while its Lunar and Planetary Society also recommended an evolutionary lunar program, including orbiters and landers, robotic rovers and tele-presence, astronomical projects, and habitat studies.

Typical of the synergistic planning that goes on in that island country, business leaders and scientists in 1994 requested that Japan's Space Activities Commission invest 3 trillion yen over 30 years to build a Moon station by 2024, entirely constructed by robots and maintained by solar power generation. With reference to the prospects for space tourism, large Japanese corporations have been planning orbital and lunar hotels, including on the Moon. Within the United States, 1996 saw the inauguration of a new journal, *Space Energy and Transportation*, devoting whole issues to space tourism, while in 1997, the Space Tourism Society was founded and held its first conference. Within Russia, a market is emerging through the Space Volunteer Project. Its founder, Dr. Sergey V. Krichevskiy, a cosmonaut, reminds us that only about 500 humans have actually been in orbit, so he seeks to create a critical mass of people who implement the space dream first on this planet and eventually mass-settlement of the universe.

The problem with all these schemes is that they are fragmented. Humanity lacks a common strategy, let alone a global institutional mechanism for developing the Moon's resources and markets. Previous international agreements of the United Nations have established a guiding principle that outer space belongs to all peoples, and that celestial bodies cannot be appropriated or owned by any one nation. As long as space was used for exploration and science, this approach worked reasonably well, but it is inadequate for industrialization and settlement. The second principle flowing from these UN treaties is that space, including the Moon, is the common heritage of humankind – that is, a province or venue reserved for the benefit of humanity.

Under these circumstances, there is limited incentive for private enterprise to invest in the development of space or lunar resources. However, the existing space law literature on these treaties does envision the establishment of an international regime to facilitate the latter. Scholars have proposed various means for accomplishing this, such as an authority to oversee and regulate space development, or some type of a space/lunar industrialization board. One proposal called for a UN International Space Center to better use extraterrestrial resources. But as we approach the end of the first decade of the new millennium, no agreement yet exists on an entity to coordinate and facilitate such efforts and markets off-world. Yet there is a growing consensus within the space community on the need for such a global institution.

The search for a solution to this problem is heightened not only because national space agencies have no unified strategy for lunar development, but because the leaders in space science and technology fields continue to think separately about the matter. Further, they seemingly ignore other necessary dimensions (e.g., legal, financial, and management). In the aerospace industry and within the astronautical field, representatives of various disciplines continue to press for their individual lunar enterprise.

For example, the University of Wisconsin's Fusion Technology Institute advocates He^3 as a potential billion dollar isotope market; Colorado State University's Center for Engineering Infrastructure and Sciences in Space promotes inflatable shelters covered with lunar regolith as the product of the future; other researchers push programs on the lunar surface for extracting oxygen, water, aluminum, iron, and even glass. What they do not seem to comprehend is that living and working safely and profitably on the Moon requires multi-disciplinary, multi-institutional, and multi-national mechanisms in place if their proposals are to be implemented.

F.2   SPACE ECONOMIC DEVELOPMENT AUTHORITIES

To resolve this impasse and further human enterprise in space, the United Societies in Space, Inc. (USIS) and its affiliated World-Space Bar Association (W-SBA) have devised a strategy for immediate implementation. Called the Space Economic and Development Authority (SEDA), the concept is to form a series of such authorities for both public and private international participation, aimed specifically toward providing the financing, governance, and management for free enterprise development of space resources and settlement. SEDA is being incorporated in the U.S. State of Colorado; this sponsoring organization will provide education, public information, and funding, as well as underwriting research and development, on future space authorities. Since W-SBA members include lawyers and paralegals interested in space, it is providing the legal research and lobbying capability to advance the strategy.

There is significant legal precedent and history on creation of quasi-public legal, as well as private, authorities and trusts to further infrastructure development on this planet. This is a respected, proven, and traditional approach to underwriting and managing both public and private undertakings, or combination thereof, across jurisdictions and borders. Such authorities have been used worldwide to construct major projects, from airports, bridges, and tunnels, to toll roads, power grids, and convention centers. In the United States, for instance, port authorities have been constituted from east to west, such as:

> The New York Port Authority, which crosses state lines to serve the metropolitan area's transportation and port needs, while issuing bonds to underwrite its activities. It has its own police force, which can arrest those who fail to comply with state, local, and authority regulations. The authority is justified because of the size, value and complexity of port needs and facilities relating to transportation, safety, docking, food supply and spoilage, as well as union contracts, especially for longshore personnel.

> The new Denver International Airport Authority, which has been financed by Municipal Revenue Bonds totaling $275 million. These are government-guaranteed to build one of the largest international air transportation centers with a full complex of retail services.

> The spaceport authorities, which have been established in various states from Florida to California, to provide launch facilities into orbit and related services which improve the economic viability of the region where they are located.

One example of a spaceport authority is the California Spaceport Authority, which manifests entrepreneurial innovation to take advantage of the launch facilities of the Vandenberg Air Force Base and other such nearby USAF bases. To protect the economy of Santa Barbara County, the Western Commercial Space Center was initially formulated as a non-profit corporation. It then succeeded in getting the State of California to pass legislation in 1993, which chartered the California Spaceport Authority.

CSA was authorized to further develop commercial launch, manufacturing, academic, and research operations, as well as to receive Federal grants for these purposes. WCSC has legal status as a non-profit business entity [501/c/6 under the U.S. Internal Revenue Code]. It retains an equity interest in spun-off, for-profit corporations related to CSA activities, such as Spaceport Tours, Inc. and the Space Access Corporation. Further, it is pursuing formation of a foreign free trade zone, a spaceport camp for youth, and a satellite control network. Both WCSC and CSA are powerful illustrations of how to open space markets by providing networking and integrating services to those wishing to use outer space for commercial, scientific, or educational purposes. Through these enterprises, the State of California exercises leadership in ensuring that the United States will remain a great spacefaring nation in the twenty-first century.

An interesting item of U.S. legal history regarding authorities is that Federal statutes inaugurating the American space program began with legislation on police authority (42 U.S.C 2456) relative to the power of its personnel to bear arms and arrest. That statute had a section creating an agency that was the forerunner to the present National Aeronautics and Space Administration. However, from the perspective of this proposal for a space economic development authority, the most significant model may be the following:

> The Tennessee Valley Authority was constituted to conserve assets for public benefit and to provide electric power in a region that crossed seven states and numerous local jurisdictions. TVA was authorized by the U.S. Congress in 1933 (16 U.S.C. 831 *et seq.*) with a board of governors appointed by the U.S. President and confirmed by the Senate.
>
> In its founding period, TVA was given donated land and facilities, as well as the power of eminent domain. This federal corporation, part business and part government, provides low-rate, competitive energy and other services related to flood control, navigation, and environmental quality. In 1994, it received Congressional authorization to designate some employees as law enforcement agents in the area of its jurisdiction (Public Law 103-322, Sec. 320929). For over seventy years, TVA has engaged successfully in power, resource, and economic development for the benefit of the citizenry, while improving the quality of life in the Tennessee Valley, particularly by attracting business and industry to the region.

From the viewpoint of SEDA, what is most interesting is that TVA is a profit-making enterprise, not funded from tax monies. To achieve its objectives, the TVA issued U.S.$50 million worth of non-governmental bonds in 1948 at 3.5 percent interest rate, secured by blanket debenture of its assets. Thirty years later, these debentures were retired with no defaults, rollovers, or commissions having been paid. Given the sorry state of NASA's underfunded budgets and annual budget fights in the U.S. Congress, this model seems a better alternative to consider to underwrite the financing, construction, and administration of a space infrastructure!

A three-stage development plan of SEDA calls for the formation of three separate and specific space authorities during the first half of the twenty-first century. The first is the prototype Lunar Economic Development Authority (LEDA). Next is the formation of a Mars economic development authority (MEDA) to encourage science, commerce, and settlement on the Red Planet. The third is an orbital economic development authority (OEDA), conceived for long-term construction of orbiting colonies or city-ships, as recommended by the late Gerard O'Neill and others.

Although OEDA was projected as a year 2050 solution, our Regent from USIS/Russia, Dr. Oleg Alifanov, proposed a more immediate use for this entity. Alifanov and his colleagues at the Moscow Aviation Institute ran a feasibility study based on an innovative possibility: rather than terminating the Mir space station as planned, why not upgrade and commercialize this orbiting facility under private auspices, such as OEDA? In that scenario, the International Space Station would be sponsored by the public sector through its fifteen-nation partnership, while a renewed Mir would have remained in orbit for private enterprise to develop new space markets, ranging from microgravity manufacturing to tourism! It was an intriguing idea for entrepreneurs to consider, although Mir was eventually de-orbited in 2001.

F.3   LUNAR ECONOMIC DEVELOPMENT AUTHORITY

Perhaps the possibilities of such space authorities can be appreciated best if we examine LEDA in some depth. As we have made clear in previous articles, we see it as the first demonstration model with others to follow. Space lawyer colleagues who have examined the plan have made these comments:

> "The Lunar Economic Development Authority, similarly, will be structured to create an international regime that would encourage, as well as regulate (rationalize), the habitation of and commerce on the Moon. The LEDA fills in blanks of an incomplete space law with details that can make space available for human development in a very short time." – Nathan C. Goldman, JD, Ph.D., author of *American Space Law* (Golden, 1996).

> "The USIS Lunar Economic Development Authority (LEDA) proposal is a serious step in the right direction. It is important that this proposal be viewed as being in accordance with international space law. The Outer Space Treaty, which has been ratified by over 90 nations, is particularly important. Few states, or their commercial entities would seriously consider participating in a venture which was not perceived as being in accordance with the Outer Space Treaty ... LEDA comports with the key principals of international law." – Milton L. Smith, JD, D.C.L., Sherman & Howard, P.C. (Smith, 1997).

We examine next two key questions related to this strategy:

F.3.1   How will LEDA facilitate new lunar markets?

Financing new ventures: If properly constituted, a lunar economic development authority would be in a position to issue bonds for public sale that would provide a new funding source for lunar enterprises, beyond the allocation of tax monies. It remains to be seen whether these bonds would or would not have government guarantees behind them. Of course, national legislative bodies, such as the U.S. Congress, might also create tax incentives for investors in these bonds, such as extra exemptions (e.g., double or even triple deductions).

These new funds might be used to contract in the global private sector for a reliable, low-cost lunar transportation system, not only to/from the Moon, but on/ under the lunar surface itself. LEDA would also underwrite the construction of a lunar infrastructure from a base to an industrial park, with all the suitable habitation and commercial facilities required. When such basic provisions are in place, LEDA might then be able to offer grants to universities, research institutes, corporations, and consortia that wish to undertake scientific, industrial, and educational programs on the Moon or beneath its surface. These projects might range from astronomical observatories to start-up ventures related to lunar mining, solar energy, and even tourism.

Attracting venture capitalists and investors: If a legal mechanism such as LEDA were in place, it could customize public information and education programs being pro- moted by SEDA – that is, it could finance media efforts to show the value of providing capital and investment in lunar enterprises. LEDA would act as a magnet to draw venture capitalists, financiers, and entrepreneurs to subsidize and engage in emerging lunar markets. Such activities for living and working on the Moon might extend from clothing, equipment and food systems necessary to survive and improve the quality of life in a low-gravity environment, through production of lunar oxygen and water, to building lunar ports for launching to Mars, other planets, and aster- oids. Parallel to its officially-sponsored macroprojects, LEDA should stimulate a wide range of scientific, technological, and commercial endeavors related to the Moon and its colonization.

Providing a legal regime for the lunar venue: Assuming LEDA came into being as a result of international agreement, it would offer a means for incentive to and co- ordination of lunar activity. For example, individuals and corporations might be willing to invest time, talent, energy, and money into technological venturing on or off the Moon, if a leasing system were in place. Thus, developers could theoretically obtain a 99-year lease to build a factory, high tech park, or settlement. The fees paid for such a service could be placed back into LEDA administration, so that it would be a self-supporting means of governance. In time, LEDA might offer zoning, regula- tory, and inspection services to further orderly development, while protecting the lunar environment and ecology.

Coordinating global efforts on the Moon and its vicinity: LEDA would have the capability of facilitating human enterprise in the use of lunar resources. Right now there exists no such entity to encourage international lunar cooperation, while avoiding overlap and chaos among various national space agencies, corporations, and organizations proposing lunar missions. LEDA would not only foster science and commerce on the Moon, it would also create a strategic, holistic lunar develop- ment plan!

Eventually, lunar dwellers will take over the administration of LEDA, and shape it to suit their own unique environment and needs. As attorney and space philosopher Dr. George Robinson reminds us, earthkind and spacekind in the human species will discover a growing interdependence, so that the governance principles of space societies will provide for home rule.

F.3.2 How will LEDA be legally constituted?

There are some who would leave lunar development to happenstance, much like the frontier days of the old American West. Their argument is to give enterprise aloft free reign; let law and order emerge gradually by the actions of spacefarers themselves. We see that attitude as constraining the lunar market by delay and disorder, by a mission-oriented approach where each nation or organization plots out its own course without regard to the common good. It is a piecemeal approach that is too costly to be advocated, especially if lunar resources are to accrue to the benefit of humanity. Better to plan change and to lay a foundation for how to move into space on a large scale. Ours is fundamentally a human space migration strategy that future lunar settlers may build upon. Our current research indicates four potential ways for formulating the prototype Lunar Economic Development Authority:

National legislation: One country could assume the leadership to establish LEDA and then invite other national partners to join in the venture. That is what happened with the present International Space Station when NASA eventually entered into agreements with other national participants. In this scenario, for example, the U.S.A. might pass an Act of Congress to formulate LEDA unilaterally, somewhat comparable with the way the TVA was established. It is conceivable that some assets of NASA, the Department of Defense, or the U.S. Air Force might be transferred to the new authority for the creation of a terrestrial lunar launch port (e.g., turning over a military or space base for this purpose). Those assets could then be used to back a bond issue, as was done in the TVA enactments. Another precedent exists in the Communications Satellite Act of 1962, which founded COMSAT, a private/public corporation, to develop a global satellite system in cooperation with other countries and to represent the U.S. in INTELSAT.

International consortium: A group of spacefaring nations and/or organizations that are interested in lunar development might enter into an international agreement that would allow or authorize LEDA to act on their collective behalf in financing and macro-managing resources on or about the Moon. INTELSAT is a precedent for such action, when some 28 nations signed an agreement through their designated public or private telecommunications entities to form this worldwide telecommunications satellite system. Although space is for the benefit of humankind, member states may own shares and share in the profits of this enterprise. In fact, the organizational structure of INTELSAT today, as it changes and becomes more commercially oriented, might be studied for lessons that might apply to the administration of LEDA.

Private transnational enterprise: A group of profit and/or non-profit entities might act synergistically to incorporate LEDA in one or more states. Under that "umbrella", they might combine their assets and plans for lunar development, and engage in joint ventures on or about the Moon. Many terrestrial macroprojects have been legally accomplished by free enterprise, with or without the participation of government.

Space metanation: The authors would prefer to see these space economic authorities develop under the auspices of a new "nation" in space, sponsored as a common-wealth of terrestrial nations. Ideally, this metanation should come into being for the twenty-first century under the United Nations charter, which has a provision for a trusteeship council of new territories. Under this scenario, the spacefaring nations might become the trustees for the next hundred years of a space metanation which has the power for self-governance, including the establishment of appropriate economic development authorities, starting with LEDA. Should any of these space authorities be inaugurated before metanation, it is our expectation that the latter body would assume administration of it.

Because of the international actors involved, we recognize that three of these four scenarios may take time to be implemented, and so for the moment the third strategy described above may be the most viable. In other words, private enterprise may be best positioned to move swiftly in the creation of the prototype Lunar Economic Development Authority. If one entrepreneur could create a profitable, worldwide broadcasting network called CNN, then certainly global space and business leaders can work together to found a mechanism like LEDA for the benefit of Earth's inhabitants.

F.4   CURRENT ENDEAVORS

Those seeking to establish LEDA held two organizing meetings in the United States during 1998. The first was in April at Albuquerque, New Mexico, while the second was in August at Denver, Colorado. A distinguished board of directors was confirmed, while other officials were chosen to lead the nascent enterprise. Dr. Michael Duke, a world-renowned planetary scientist, was appointed as the part-time LEDA president (e-mail: *duke@lpi.jsc.nasa.gov*). He is currently developing LEDA goals, policies, and plans, especially for revenue bonds. Brad Blair, a mining engineer at the Colorado School of Mines, is vice president and acting executive director (e-mail: *bblair@lynx.csn.net*). In a presentation on possible LEDA strategies to access the wealth of space, Blair offered insights for setting up an investment fund for outer space, especially the Moon. The proposed mission statement for LEDA includes these aims:

- To finance an international lunar colony through the worldwide sale of Lunar Development Bonds.
- To facilitate leading-edge space technology for the twenty-first century, such as in

transportation, agriculture, housing, medicine, industrial fabrication, construction, and mineral extraction.

- To create challenging and synergistic business opportunities on the Moon for international corporations, small high-tech firms, universities, and national laboratories.
- To lower the costs of space access and infrastructure through the creation of profitable mineral-processing facilities on the Moon.

F.5 CONCLUSIONS

To open and nurture vigorous space markets for the twenty-first century requires a new impetus, as well as a new appreciation of existing space law. We have suggested a general strategy to accomplish both called a space economic development authority, beginning with the prototype Lunar Economic Development Authority. This approach is consistent with existing space law and treaties that preclude appropriation of space resources by any one nation or entity. It is also compatible with the agreed principle that space be reserved for the benefit of humanity.

In our proposal, an international agreement entered into on a reciprocal, voluntary basis would permit LEDA to engage in rational lunar activities for the benefit of both Earth's inhabitants and our interplanetary commons, beginning with the Earth–Moon system. As Dr. Milton Smith has observed: "LEDA is an international regime for the orderly and safe development of natural resources on the Moon, and their effective management for the expansion of opportunities by their use." There is ample legal precedent for innovators to craft this mechanism for a stable legal environment that will promote both scientific and commercial use of lunar resources. We hope that a dialogue on the subject will result within the world space communities from this analysis. The authors welcome feedback and commentary, particularly for publication in the USIS journal, *Space Governance*.

Further information on Lunar Economic Development Authority developments may be obtained through its principal sponsor, United Societies in Space, Inc., a Colorado non-profit corporation with IRS 501(c)(3) tax-favored status. Its address is:

United Societies In Space
499 Larkspur Drive
Castle Rock
Colorado 80104
U.S.A.
Phone: 800-632-2828
Local phone: 305-688-1193
Fax: 303-663-8595
www.angelfire.com/space/usis/

F.6 ABOUT THE AUTHORS

Philip R. Harris, Ph.D., is a space psychologist and Associate Fellow of the American Institute of Aeronautics and Astronautics. Dr. Harris is Chairman Emeritus of the Board for United Societies in Space, Inc., and author/editor of forty books, including *Living and Working in Space*, Second Edition (Wiley/Praxis, 1996). (His address is 2792 Costebelle Dr., LaJolla, California, U.S.A.; tel: 1-619/453-2271; fax: 1-619/453-0788; e-mail: *philharris@aol.com*).

Declan J. O'Donnell, JD, is a tax attorney and president of both the World-Space Bar Association, as well as USIS, and publisher of its journal, *Space Governance*. A member of the International Institute of Space Law, he is on legal retainer to the State of Colorado's Office of Space Advocacy. (He may be reached at United Societies in Space, Inc., 3300 E. 14th Ave., Denver, CO 80206, U.S.A.; tel: 1-800/895-META or 1-303/ 780-0700; fax: 1-303/780-0732; e-mail: *USISpace@aol. com*).

F.7 ENDNOTES

1   For information on the Space Entrepreneurs Association, contact David Anderman at the Space Frontier Foundation (e-mail: *DavidA5625@aol.com*), or USIS Regent, Michael Simon at International Space Enterprises (4909 Murphy Canyon Rd., Ste. 220, San Diego, CA 92123, U.S.A; e-mail: *Michael@isecorp.com*). To get on list server for the electronic exchange on SEA contact Jim Benson (e-mail: *scia@SpaceDev.Com*).

2   Space Tourism Society, contact USIS Regent, John Spenser (326 S. Bundy Dr., Los Angeles, CA 90049, U.S.A. or e-mail: *jssdesign@aol.com*).

3   Agreement Governing the Activities of States on the Moon and other Celestial Bodies, 18 December, 1979, #1363 U.N.T.S., 3/18 I.L.M. 1434 (hereinafter referred to as the Moon Agreement entered into force July 11, 1984).

4   The non-profit organizations promoting this strategy may be contacted directly for further information: United Societies in Space, Inc. (USIS – 3300 E. 14th Ave., Denver, CO 80206, U.S.A.) and the World-Space Bar Association (W-SBA, 1390 Adams St., Denver, CO 80206, U.S.A.). Together they sponsor the journal, *Space Governance*, which is published twice a year, and conduct an annual conference of the same title during the first weekend in August. For further information, telephone 1-800/895-META or 1-303/780-0733; fax: 1-303/e-mail: *USISpace@aol.com*; Internet homepage: *http://www.USIS.org*. Currently, USIS has international branches in Russia and Mexico.

5   Tennessee Valley Authority Act, 1933–1983, 23 pp.; Information Statement – Tennessee Valley Authority, Dec. 23, 1996, 20 pp.; Tennessee Valley Authority 6.5% Global Power Bonds, September 11, 1996, 22 pp.; Violent Crimes and Control Act (PL 103-222, Section 320929 on TVA Law Enforcement), Sept. 13, 1994, 4 pp. These and other publications are available from TVA Communications (ET 7D-K, 400 West Summit Hill Drive, Knoxville, TN 37902, U.S.A.; tel: 423/632-6263).

6   For information about USIS/Russia and its Mir Feasibility Study, contact Dr. Oleg M. Alifanov, Director General, COSMOS International Center for Advanced Studies, Moscow Aviation Institute (Box 9, A-80, Moscow 125080, Russia; fax: 01-7-095-158-5865;

e-mail: *pm@glasnet.ru*), or USIS Regent, Derek Webber, Chair, Mir Privatization Task Force, Intonations Consultants, Inc. (4660 LaJolla Village Dr., Ste. 525, San Diego, CA 92122, U.S.A.; fax: 1-619/457-0938; e-mail: *dwspace@aol.com*).

F.8   REFERENCES

Benaroya, H. (1994) Lunar industrializtion. *Journal of Practical Applications in Space*, **VI**, no. I, 84–86.

Brown, A.S. (1997) Inflating hopes for a lunar base. *Aerospace America*, February, 21–23.

Center for Research and Education on Strategy and Technology (1993) *Partners in Space: International Cooperation in Space Strategies for the New Century*, May. CREST, Arlington, VA (1840 Wilson Blvd. #204, zip code: 22201; fax: 703/243-7175)

Criswell, D.R. (1994) New growth on the two planet economy! In: *Proceedings of the 45th International Astronautical Congress*. International Astronautical Federation, Paris.

Davis, N.W. (1997) New Asia-Pacific Players Launch Services Market. *Aerospace America*, February 1997, pp. 6–8.

ESA (1992) *Mission to the Moon*. ESA SP-1150, European Space Agency, Paris.

ESA (1994) *International Lunar Workshop*. ESA SP-1170, European Space Agency, Paris.

Finch, E.R. and Moore, A.E. (1984) *Astrobusiness: A Guide to the Commerce and Law of Outer Space* (NAPA report: Encouraging Business Ventures in Space Technologies). National Academy of Public Administration, Washington, D.C. (1120 G. St., NW, zip 20005).

Glazer, J.H. (1987) The expanded use of Space Act authority to accelerate space commercialization through advanced joint enterprises between Federal and non-Federal constituencies. *Rutgers Computer and Technology Law Journal*, **12**, no. 2, 338–405.

Golden, N.C. (1996) *American Space Law: International and Domestic*, 2nd edn. UNIVELT, San Diego, CA.

Goldman, N.C. (1985) *Space Commerce: Free Enterprise on the High Frontier*. Ballinger/Harper & Row, Cambridge, MA.

Goldman, N.C. (1996) A lawyer's perspective on the USIS strategies for metanation and the Lunar Economic Development Authority. *Space Governance*, July, **3**, no. 1, 16–17, 34.

Gump, D.P. (1990) *Space Enterprise Beyond NASA*. Praeger, New York.

Harris, P.R. (1987) Innovations in space management – macromanagement and the NASA heritage. *Journal of the British Interplanetary Society*, **40**, 109–116.

Harris, P.R. (1991) Challenges in space enterprise – Commercial, legal, political. *Space Commerce*, **1**, no. 2, 135–153.

Harris, P.R. (1994) A case for permanent lunar development and investment. *Space Policy*, **10**, no. 3, 187–188.

Harris, P.R. (1995) Why not use the Moon as a space station? *Earth Space Review*, **4**, no. 4, 7–10.

Harris, P.R., (1996) Human development and synergy in space. *Living and Working in Space, Second Edition: Human Beaver, Culture, and Organization*. Wiley/Praxis, Chichester, U.K., pp. 34–60.

Hayes, W.C., ed. (1980) *Space – New Opportunities for International Ventures*. UNIVELT, San Diego, CA (AAS #49).

Heppenheimer, T.A. (1979) *Colonies in Space*. Stackpole Books, Harrisburg, PA (PO Box 1831, zip 17105).

Howerton, B.A. (1995) *Free Space! Real Alternatives for Reaching Outer Space.* Loompanics Ltd., Port Towsend, WA (PO Box 1170, zip 98368).

Koelle, H.H. (1997) Steps toward a lunar settlement. *Space Governance*, January, **4**, no. 1, 20–25.

Kyodo News Service/News Net Press Release (1994) *Moon Station Blueprint Drawn up by Lunar Society*, 31 May.

Logsdon, T. (1988) *$pace Inc. – Your Guide to Investing in Space Expoloration.* Crown Publishers, New York.

McKay, M. F., McKay, D.S. and Duke, M.B., eds. (1992) *Space Resources.* U.S. Government Printing Office, Washington, D.C. (NASA SP-509, 6 vols.).

Mendell, W.W. (1985) *Lunar Bases and Activities of the 21st Century.* Lunar and Planetary Institute, Houston, TX.

O'Donnell, D.J. (1994) Founding a space nation utilizing living systems theory. *Behavioral Science*, April, **39**, no. 2, 93–116.

O'Donnell, D.J. (1994) Is it time to replace the Moon Treaty. *The Air and Space Lawyer*, Fall, 3–9 (available from the American Bar Association, 750 North Lake Shore Dr., Chicago, IL 60611, U.S.A.).

O'Donnell, D.J. (1994) Metaspace: A design for governance in outer space. *Space Governance*, **1**, no. 1, 8–27.

O'Donnell, D.J. (1994) Overcoming barriers to space travel. *Space Policy*, **10**, no. 4, 252–255.

O'Donnell, D.J. (1996) Commercialization by evolution of jurisdiction in outer space. *Proceedings of the 47th International Astronautical Congress, October 7–11, 1996*, International Astronautical Federation, Paris.

O'Donnell, D.J. (1996) Property law in outer space. *Space Governance*, July, **3**, no. 1, 14–15, 34.

O'Donnell, D.J. (1996) The model treaty of jurisdiction in outer space. In: P.R. Harris, ed., *Living and Working in Space*, 2nd edn. Wiley/Praxis, Chichester, U.K., Appendix D, pp. 360–368.

O'Donnell, D.J. and Harris, P.R. (1994) Space-based energy need: a consortium and a revision of the Moon Treaty. *Space Power*, **13**, no. 1/2, 121–134.

O'Donnell, D.J. and Harris, P.R. (1996a) Legal strategies for a lunar economic development authority. *Annals of Air and Space Law*, Centre of Air and Space Law, McGill University, Montreal, Vol. XXI, pp. 121–130.

O'Donnell, D.J. and Harris, P.R. (1996b) Strategies for a lunar economic development authority: Futures scenario for utilization of the Moon's resources. *Futures Research Quarterly*, Fall, pp. 1–11. World Future Society, Bethesda, MD (7910 Woodmont Ave. #450, zip 20814)

O'Leary, B. (2000) *Space Industrialization.* CRC Press, Boca Raton, FL (2000 NW 24th St., zip 33431).

O'Neill, G.K. (1989) *The High Frontier – Human Colonies in Space.* Space Studies Institute, Princeton, NJ.

Public–private Partnerships (1996) *San Diego Business Journal*, Special Supplement on "Wastewater Management," November.

Robinson, G. S. and White, H.M. (1986) *Envoys of Mankind: A Declaration of First Principles for the Governance of Space Societies.* Smithsonian Institution Press, Washington, D.C.

Robinson, G.S. (1995) Natural law and a declaration of humankind interdependence. *Space Governance*, Part I, **2**, no. 1, 14–17; Part II, **2**, no. 2, 32–35.

Simon, M.C., ed. (1994, 1995, 1996) *Proceedings of the International Lunar and Mars Exploration Conferences*, 1996 Vol. 3; 1995 Vol. 2; 1994 Vol. 1. International Space Enterprises, San Diego, CA (refer to endnote #1).

Smith, M.L. (1997) The compliance with international space law of the LEDA proposal. *Space Governance*, **4**, no. 1, January, 16–19.

Space Volunteers Project, contact Dmitry Pieson, Executive Director, USIS/Russia (Cosmos International Center, Moscow Aviation Institute, Box 9, A-80, Moscow 125080, Russia, or e-mail: *pm@glasnet.ru*).

Special Space Tourism Issue (1996) *Space Energy and Transportation*, **1**, no.1 (The High Frontier, 2800 Shirlington Rd., Ste. 204, Arlington, VA 22206, U.S.A.).

Wycoff, R.A. and Smith, D.D. (1996) Plans and strategies for a California Space Authority – A USIS case study. *Space Governance*, July, **3**, no. 1, 24–27.

Appendix G

Quality standards for lunar governance

David G. Schrunk

G.1 CREATION OF A LUNAR GOVERNMENT

A government must be created for the people who live and work on the Moon. The creation of the new government offers the opportunity to introduce quality standards for the design, evaluation, and disposition of its laws.

G.1.1 Jurisdiction

The jurisdiction of the lunar government will be the global structure of the Moon and lunar orbital space.

G.1.2 Purpose of government

The purpose of the lunar government will be to facilitate the responsible and peaceful exploration, development, and settlement of the Moon by effective, cost-efficient, and just means. It is assumed that the government will be a representative democracy[1] that serves the best interests,[2] and reflects the highest aspirations, of the citizens of the government.

---

[1] The alternatives to a democracy (where the citizens as a whole constitute the sovereign of the government) are a monarchy, an aristocracy, a theocracy, or an ideocracy. These alternatives place the interests of a royal family, a privileged class, a religion, or an ideology, respectively, ahead of the interests of the citizens as a whole, and are therefore considered to be unacceptable choices for the form of sovereign government of the Moon.

[2] The parameters that define the best interests of the people are human rights, living standards, and quality-of-life standards. (Quality of life standards refer to positive and negative cultural and environmental parameters such as the quality and availability of public use facilities and the incidence of crime and pollution.)

G.1.3   Constitutional rule of law

To establish the government, a constitution is created to define the structure, mechanisms, responsibilities, and limitations of the lunar government. The constitution is superior to all other directives, thus creating a rule of law[3] for the government and its citizens. The designated legislative branch of the government is empowered by the constitution to create a body of legislative laws, or statutes, by which the ends of government are attained.

The purpose of government is not realized if the government's laws are inadequate to their task. That is, if the laws of the government are cumbersome or ineffective, then the government itself will be cumbersome or ineffective. To reduce the possibility that the body of laws will be less than useful to the purpose of government, we recommend that quality standards for the design, follow-up evaluation, disposition, and improvement of laws[4] be incorporated in the constitution of the lunar government.[5] A code of ethics for the government, analogous to the code of ethics for business enterprises, is also recommended (see Table L.1, Appendix L).

G.2   THE NEED FOR QUALITY LAWMAKING

Science and engineering have produced tools and technologies that are characterized by high standards of quality and by step-wise improvements in performance. Products such as life-support systems, communication networks, and nuclear reactors are all marked by improvement from one generation to the next. Future space explorers will have access to tools that are even more powerful and sophisticated than those available today, and the ability of humankind to explore and develop space will continue to improve over time.

Historically, the laws of government have been an exception to the rule that the products of humankind improve from generation to generation. Laws are tools; they are the problem-solving means by which the ends of government are attained. However, the experience with the laws enacted by Earth governments has been that the bodies of laws tend to grow in size and complexity over time, but the societal problems they address (e.g., poverty, crime, illiteracy, health care, and energy issues, etc.) are rarely, if ever, solved in an efficient and just manner.

---

[3] The alternative to the rule of law is arbitrary rule, in which the desires of the "rulers" of the government are held to be superior to the written laws of the government.

[4] Quality standards apply to all of the laws of government, including constitutional law, administrative laws, judge-made laws, and every other enforceable written directive of government.

[5] At present, no national government observes standards of excellence for the design or improvement of their laws, including constitutional law. If the designers of the new lunar government observe standards of excellence in their design of the constitution then the constitution will in turn require standards of excellence for the design and improvement of all other laws.

The major reason for this relative lack of success of laws is that the present lawmaking process (the "traditional method of lawmaking") does not require the designers of laws to define and analyze societal problems, to conduct cost/benefit/risk analyses of law solutions, to base their law designs on reliable knowledge, or to cite references. Also, feedback evaluations of the performance of laws are virtually never performed – so no one can know the outcome of laws with certainty. Thus, outmoded, wasteful, and counter-productive laws remain in force indefinitely. In other words, lawmaking, as currently practiced by Earth governments, is devoid of the high standards of quality that characterize major productive industries.

The government of the Moon will be charged with the responsibility of providing a just and efficacious rule of law. The major concern is that the lunar lawmaking industry will copy the present method of lawmaking of Earth governments and thereby create a mediocre (or worse) body of laws that is unable to resolve societal problems. If that were to happen, the progress of space exploration will be hampered despite advances in every other field.

G.2.1 Quality standards for laws

The government of the Moon will be an entirely new government, thus a unique opportunity exists to redefine the lawmaking process. There will be no entrenched interests to prevent the lunar govenment from having quality systems to encompass laws and lawmaking. High standards of quality have proven to be effective and beneficial for every other productive industry and there is every indication that they will be equally effective and beneficial for lawmaking. Thus, an opportunity exists to upgrade the lawmaking process.

A quality assurance program for laws would maintain optimum government performance by periodically "weeding out" outmoded or ineffective laws that no longer provide a net benefit to the people of the Moon. Likewise, quality design standards would require new laws to be designed to solve societal problems, preventing the enactment of non-productive special interest legislation.

By these means the body of laws of the government would be kept to a minimum number of effective, cost-efficient, and user-friendly laws, and only those new laws that serve the best interests of the public would be enacted. A bonus of quality programs for laws is that their costs would be more than offset by the value they deliver. Suggested quality standards and programs for the design and disposition of laws are listed in Table G.1.

The application of quality standards to laws and lawmaking will cause the new lunar government to focus on the solution of societal problems for the benefit of the people as a whole. In other words, quality standards will convert the feed-forward lawmaking system of Earth governments, which only produces more laws, into a feedback control system that is self-correcting in the direction of optimum service to the people (see Figure G.1).

Table G.1. Quality standards for laws

Quality design (QD)/ quality improvement (QI) standards	*Quality assurance (QA) program*
Each law must contain validated statements of the following:	Following competent measurement and analyses of laws, repeal the following[6]
1. Definition of the societal problem that requires a law solution 2. Analysis of the problem 3. Priority assignment of the problem 4. Name and qualifications of law designer(s) 5. Measurable goal of proposed law solution 6. Cost/benefit/risk analyses 7. Prediction of results of law enforcement 8. Methods for measuring results 9. Citation of references	1. Laws that do not address problems 2. Laws that lack a stated goal 3. Laws that lack a citation of references 4. Laws that violate human rights 5. Laws that fail to achieve their goal 6. Laws whose results cannot be measured 7. Laws whose costs are greater than benefits 8. Laws whose costs and benefits are equal 9. Laws that are not enforced 10. Laws that have not been subjected to QA analysis within a ten-year period

G.3   SUMMARY

The technological advances that now make the human settlement of the Moon possible have all been the result of applying standards of excellence that are inherent in the methodologies of science and engineering. By applying those same quality standards to the design, evaluation, optimization, and disposition of laws, the performance of the laws of the new lunar government will parallel the patterns of success that characterize the products of other major enterprises.

---

[6] A regular quality assurance program for the identification and repeal of less than useful laws is not currently performed by any government on Earth.

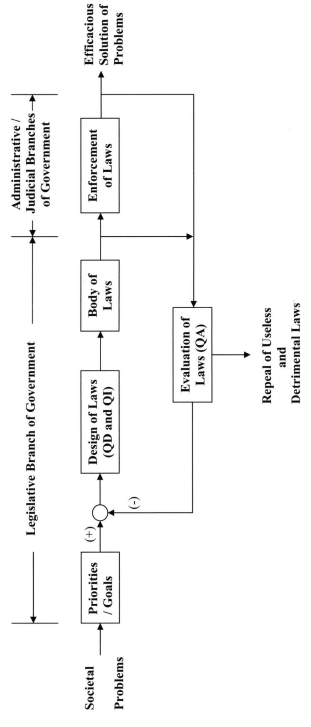

Figure G.1. Proposed quality control system of the lunar government. The application of quality programs (QD = quality design standards; QA = quality assurance standards; and QI = quality improvement standards) to laws and lawmaking for the lunar government will create a knowledge-based, problem-solving feedback control system of government that serves the best interests of the people (adopted from Schrunk, D.G., *The End of Chaos: Quality Laws and the Ascendancy of Democracy*. Quality of Laws Press, Poway, CA, 2005). (*Note*: the judicial branch of government will also perform a limited quality assurance role through the passive process of judicial review.)

G.4   REFERENCES

Beutel, Frederick K. (1975) *Experimental Jurisprudence and the Scienstate*. Fred B. Rothman & Co., Littleton, CO.

Clapp, Charles (1963) *The Congressman: His Work as He Sees It*. The Brookings Institution, Washington, D.C.

Crosby, Philip B. (1979) *Quality Is Free*. McGraw-Hill, New York.

Dickerson, Reed (1986) *The Fundamentals of Legal Drafting*. Little, Brown & Co., Boston.

Filson, Lawrence (1992) *The Legislative Drafter's Desk Reference*. Congressional Quarterly, Inc., Washington, D.C.

Froman, Lewis, Jr. (1967) *The Congressional Process*. Little, Brown & Co., Boston.

Hunt, Daniel V. (1992) *Quality in America: How to Implement a Competitive Program*. Technology Research Corporation, Homewood, IL.

Hutchins, Robert, editor-in-chief (1990) *The Great Ideas: A Syntopicon of Great Books of the Western World*, Vols. I and II. Encyclopaedia Britannica, Chicago.

Jewell, M. and Patterson, S. (1977) *The Legislative Process in the United States*. Random House, New York.

Juran, Joseph M. (1988) *Juran on Planning for Quality*. The Free Press, New York.

Kernochan, John (1981) *The Legislative Process*. The Foundation Press, Mineola, NY.

Mill, John S. (1990) Representative government. *Great Books of the Western World*, Vol. 43. Encyclopaedia Britannica, Chicago.

Offenau, Ludomir V. (1989) *Science and the Rule of Law*. P.C. Press, San Diego, CA.

Robertshaw, Joseph; Mecca, S.; and Rerick, M. (1978) *Problem Solving: A Systems Approach*. Petrocelli Books, New York.

Schrunk, David (1997) Lawmaking standards for space governance. *Space Governance*, January, 44 47.

Schrunk, David (2002a) The ideal law of government. *Proceedings of Space 2002*. American Society of Civil Engineers, Reston, VA.

Schrunk, David (2002b) The science and engineering of laws. *Proceedings of Space 2002*. American Society of Civil Engineers, Reston, VA.

Schrunk, David (2005) *The End of Chaos: Quality Laws and the Ascendancy of Democracy*. Quality of Laws Press, Poway, CA.

Weart, Spencer R. (1998) *Never at War: Why Democracies Will Not Fight One Another*. Yale University Press, New Haven, CT.

Willett, Edward, Jr. (1990) *How Our Laws Are Made*. U.S. Government Printing Office, Washington, D.C.

Appendix H

Helium-3

Bonnie Cooper, Ph.D.

H.1 INTRODUCTION

One of the most interesting possibilities for lunar resource utilization is related to the future development of nuclear fusion. Fission reactors face many problems, from public resistance to the storage of long-lived radioactive wastes to reactor safety questions. The *fusion* process involves combining small atoms (typically isotopes of hydrogen such as deuterium and tritium). This process can release enormous amounts of energy, as can be observed every day from the Sun. The fusion community appears to be within a few years of the first "breakeven" fusion milestone. If that goal is met, it is expected that fusion devices will be able to produce hundreds of megawatts of thermonuclear power in the coming decades.

Currently, the worldwide effort in fusion research is concentrating on the deuterium (D) and tritium (T) reaction,[1] because it is the easiest to initiate. However, 80 percent of the energy released in the reaction is in the form of neutrons. These particles not only cause severe damage to the surrounding reactor components, but also induce large amounts of radioactivity in the reactor structure. However, there is another fusion reaction, involving the isotopes of D and helium-3 (He^3)[2] that produces only 1 percent of its energy as neutrons. Such a low neutron production greatly simplifies the safety-related design features of the reactor, and reduces the levels of induced radioactivity such that extensive radioactive waste facilities are not

---

[1] Deuterium and tritium are *isotopes* of hydrogen. They have the same charge but different mass. Isotopes can be likened to the difference between one regular P.T. cruiser and another P.T. with a permanently installed extra weight inside: the characteristics of both cars are exactly the same except that one of the weighs more.

[2] Helium-3 is an isotope of helium in which a neutron is missing. In this case, the analogy would be that the "regular" P.T. Cruiser is the one with the extra weight, and the isotope has had the weight removed.

required. Furthermore, this energy can be converted directly to electricity with efficiencies of 70–80 percent.

However, there is no large terrestrial supply of helium-3. The amount of primordial He^3 left in the Earth is on the order of a few hundred kilograms. To provide a significant fraction of the world's energy needs would require hundreds of *tonnes* of He^3 each year.

Early studies of the lunar regolith showed that there is a relative abundance of helium-3 on the Moon, compared with Earth. A group of physicists from the University of Wisconsin's Fusion Energy Research Center has studied the possibility of using lunar helium-3, and they are convinced that it would be economically viable to do so (e.g., Kulcinski *et al.*, 1988). Over the 4-billion-year history of the Moon, several hundred million tonnes of He^3 have impacted the surface of the Moon from the solar wind. The analyses of Apollo and Luna samples showed that over 1 million tonnes of He^3 are loosely embedded in the grains at the surface of the Moon. Even a small fraction of this He^3 could provide the world's electricity for centuries to come.

H.2   HELIUM-3 FUSION

A D–He^3 fusion plant would be inherently safer than a D–T fusion plant. Calculations have shown that the consequences of a complete and instantaneous coolant loss are minimal, and that safety can be assured by passive means no matter what the accident sequence. A meltdown is virtually impossible in a D–He^3 reactor because they operate at lower temperatures and the maximum temperature increase over one month is only 350°C, even with no cooling and perfect insulation. Moreover, in the worst possible accident, exposure to the public would be only 0.1 rem, or roughly the equivalent of natural background radiation.

Because the D–He^3 reaction causes less damage to the walls of the energy plant, less plant maintenance would be required, again reducing the costs of the energy and increasing the availability. The total radioactivity associated with a D–He^3 plant is 20 to 80 times less than in a comparably sized D–T plant. Finally, the conversion efficiency for the D–He^3 reaction is about 60 percent, compared with 34–49 percent for D–T systems. Thus, the direct capital costs of D–He^3 reactors could be one-half that of D–T reactors. The added benefits of safety and reliability make the D–He^3 reaction far preferable to the D–T reaction. Because of the amount of safety-testing needed, it will be at least 50 years before the operation of the first commercial D–T plant; whereas with D–He^3, lessened risks would mean an overall time saving of 10 to 20 years.

H.3   REGOLITH RESOURCES OF HELIUM-3

It has been calculated that the Moon was bombarded with over 250 million metric tons of He^3 over the last 4 billion years. Because the energy of the solar wind is low, the He^3 ions did not penetrate very far into the surface of the regolith particles – only

0.1 m or so. The surface of the Moon is tilled as a result of meteorite impacts, and helium is trapped in soil particles to depths of several meters. Soil grains of the mineral ilmenite ($FeTiO_3$) are enriched in helium. Thus, the Sea of Tranquillity would be a prime target for initial investigations for a He^3 mining site. This area alone appears to contain more than 8,000 tonnes of He^3 to a depth of 2 meters.

Because the solar-wind gases are weakly bound in the lunar regolith, it should be relatively easy to extract them by heating the regolith to about 600°C. Because there seems to be a higher concentration of solar-wind gases in the smaller particles (presumably because of the high surface-to-volume ratio), it might to useful to size-sort the regolith, retaining only the smaller particles. The feedstock could then be pre-heated by heat pipes and fed into a solar-heated retort. In addition to the He^3, other solar-wind volatiles, such as H_2, He^4, C compounds, and N_2, would also be extracted. The spent feedstock would be discharged through the heat pipes, to recover 90 percent of its heat.

Once the volatiles are extracted, they can be separated from the helium by exposure to the temperatures of the lunar night. Everything except the helium will condense, and the He^3 can later be separated from the He^4. For every tonne of He^3 produced, some 3,300 tonnes of He^4, 500 tonnes of N, 400 tonnes of CO and CO_2, and 6,100 tonnes of H_2 gas are produced. The H_2 will be extremely beneficial on the Moon for making water and propellants. Moreover, the He^3 could be worth as much as ~$1 billion per tonne. Of the other volatiles, the N_2 could be used for plant growth, the C for the manufacture of plastics, and the He^4 as a working fluid for mechanical systems.

If the amount of available He^3 on the Moon is on the order of 1 million tonnes, that would amount to 10 times more energy than that contained in recoverable fossil fuels on Earth, and twice the amount of energy available from the most efficient fission process. To meet the entire U.S. energy consumption of 1986, 25 tonnes of He^3 would have been required, assuming that fusion technology were available. In that same year, the U.S. spent approximately $40 billion for fuel to generate electricity. If He^3 from the Moon were sold to Earth for $1 billion per tonne, then its use would have represented a saving in 1986 of $15 billion.

The concept of mining the Moon for He^3 ties together two of the most ambitious high-technology endeavors of the twenty-first century: the development of controlled thermonuclear fusion for civilian power applications, and the utilization of outer space for the benefit of humankind.

H.4   REFERENCES

Kulcinski, G.L.; Cameron, E.N.; Santarius, J.F.; Sviatoslavsky, I.N.; Wittenberg, L.J.; and Schmitt, H.H. (1988) Fusion energy from the Moon for the twenty-first century. In: W.W. Mendell, ed., *The Second Conference on Lunar Bases and Space Activities of the 21st Century*. NASA Conference Publication 3166, NASA Johnson Space Center, Houston, TX.

Schmitt, H. (2005) *Return to the Moon: Exploration, Enterprise, and Energy in the Human Settlement of Space*. Springer-Praxis, Chichester, UK.

Wittenberg, L.J.; Santarius, J.F.; and Kulcinski, G.L. (1986) Lunar source of 3He for commercial fusion power. *Fusion Technology*, **10**, 167–178.

Appendix I

NASA and self-replicating systems: Implications for nanotechnology[1]

Ralph C. Merkle

In the summer of 1980, NASA and the American Society for Engineering Education (ASEE) sponsored a summer study by fifteen NASA program engineers and eighteen educators from U.S. universities to investigate advanced automation for space missions. The resulting 400-pp. report included a 150-pp. chapter on "Replicating Systems Concepts: Self-Replicating Lunar Factory and Demonstration." This proposed a 20-year program to develop a self-replicating general-purpose lunar manufacturing facility (a self-replicating system, or SRS) that would be placed on the lunar surface. The design was based entirely on conventional technology.

The "seed" for the facility, to be landed on the lunar surface from Earth to start the process, was 100 tons (approximately four Apollo missions). Once this 100-ton seed was in place, all further raw materials would be mined from the lunar surface and processed into the parts required to extend the SRS. A significant advantage of this approach for space exploration would be to reduce or eliminate the need to transport mass from the Earth – which is relatively expensive.

The report remarks:

> The difficulty of surmounting the Earth's gravitational potential makes it more efficient to consider sending information in preference to matter into space whenever possible. Once a small number of self-replicating facilities has been established in space, each able to feed upon non-terrestrial materials, further exports of mass from Earth will dwindle and eventually cease. The replicative feature is unique in its ability to grow, in situ, a vastly-larger production facility than could reasonably be transported from Earth. Thus the time required to organize extraordinarily-large amounts of mass in space and to set up and perform various ambitious future missions can be greatly shortened by using a self-replicating factory that expands to the desired manufacturing capacity.

[1] From *Foresight Update No. 9*, p. 4, June 1990; a publication of the Foresight Institute.

The useful applications of replicating factories with facilities for manufacturing products other than their own components are virtually limitless.

Establishing the credibility of the concept occupied the early part of the chapter. The theoretical work of von Neumann was reviewed in some detail. Von Neumann designed a self-replicating device that existed in a two-dimensional "cellular automata" world. The device had an "arm" capable of creating arbitrary structures, and a computer capable of executing arbitrary programs. The computer, under program control, would issue detailed instructions to the arm.

The resulting universal constructor was self-replicating almost as a by-product of its ability to create any structure in the two-dimensional world in which it lived. If it could build any structure it could easily build a copy of itself, and hence was self-replicating.

One interesting aspect of von Neumann's work is the relative simplicity of the resulting device: a few hundred kilobits to a megabit. Self-replicating systems need not inherently be vastly complex. Simple, existing biological systems, such as bacteria, have a complexity of about 10 million bits. Of course, a significant part of this complexity is devoted to mechanisms for synthesizing all the chemicals needed to build bacteria from any one of several simple sugars and a few inorganic salts, and other mechanisms for detecting and moving to nutrients. Bacteria are more complex than strictly necessary simply to self-reproduce.

Despite the relative simplicity that could theoretically be achieved by the simplest self-reproducing systems, the proposed lunar facility would be highly complex: perhaps 100 billion to a trillion bits to describe. This would make it almost 10,000 to 100,000 times more complex than a bacterium, and a million times more complex than von Neumann's theoretical proposal. This level of complexity puts the project near the limits of current capabilities. (Recall that a major software project might involve a few tens of millions of lines of code, each line having a few tens of characters and each character being several bits. The total raw complexity is about 10 billion bits perhaps 10 to 100 times less complex than the proposed SRS.) Where did this "excess" complexity come from?

The SRS has to exist in a complex lunar environment without any human support. The complexity estimate for the orbital site map alone is 100 billion bits, and the facilities for mining and refining the lunar soil have to deal with the entire range of circumstances that arise in such operations. This includes moving around the lunar surface (the proposal included the manufacture and placement of flat-cast basalt slabs laid down by a team of five paving robots); mining operations such as strip mining, hauling, landfilling, grading, cellar digging and towing; chemical processing operations including electrophoretic separation and HF (hydrofluoric) acid-leach separation, the recovery of volatiles, refractories, metals, and nonmetallic elements and the disposal of residue and wastes; the production of wire stock, cast basalt, iron, or steel parts; casting, mold making, mixing and alloying in furnaces and laser machining and finishing; inspection and storage of finished parts, parts retrieval and assembly and subassembly testing; and computer control of the entire SRS.

When we contrast this with a bacterium, much of the additional complexity is relatively easy to explain. Bacteria use a relatively-small number of well-defined chemical components which are brought to them by diffusion. This eliminates the mining, hauling, leaching, casting, molding, finishing, and so forth. The molecular "parts" are readily available and identical, which greatly simplifies parts inspection and handling. The actual assembly of the parts uses a single relatively-simple programmable device, the ribosome, which performs only a simple, rigid sequence of assembly operations (there is no AI in a ribosome!). Parts assembly is done primarily with "self-assembly" methods which involve no further parts-handling.

Another basic issue is closure.

Imagine that the entire factory and all of its machines are broken down into their component parts. If the original factory cannot fabricate every one of these items, then parts closure does not exist and the system is not fully self-replicating.

In the case of the SRS, the list of all the component parts would be quite large. For a bacterium, there are only 2,000 to 4,000 different "parts" (proteins). This means that the descriptions of the parts are less complex. Because most of the parts fall into the same class (proteins), the manufacturing process is simplified (the ribosome is adequate to manufacture all proteins).

What does all this mean for humanity? The report says:

From the human standpoint, perhaps the most exciting consequence of self-replicating systems is that they provide a means for organizing potentially-infinite quantities of matter. This mass could be so organized as to produce an ever-widening habitat for man throughout the solar system. Self-replicating homes, O'Neill-style space colonies, or great domed cities on the surfaces of other worlds would allow a niche diversification of such grand proportions as never before experienced by the human species.

The report concludes:

The theoretical concept of machine duplication is well developed. There are several alternative strategies by which machine self-replication can be carried out in a practical engineering setting ... There is also available a body of theoretical automation concepts in the realm of machine construction by machine, in machine inspection of machines, and machine repair of machines, which can be drawn upon to engineer practical machine systems capable of replication ... An engineering demonstration project can be initiated immediately, to begin with simple replication of robot assembler by robot assembler from supplied parts, and proceeding in phased steps to full reproduction of a complete machine processing or factory system by another machine processing system, supplied, ultimately, only with raw materials.

What implications does the NASA study have for nanotechnology?

The broad implications of self-replicating systems, regardless of scale, are often similar. The economic impact of such systems is clear and dramatic. Things become cheap, and projects of sweeping scale can be considered and carried out in a reasonable time frame without undue expense.

The concepts involved in analyzing self-replicating systems – including closure, parts counts, parts manufacturing, parts assembly, system complexity, and the like – are also quite similar. The general approach of using a computer (whether nano or macro) to control a general-purpose assembly capability is also clearly supported. Whether the general-purpose manufacturing capability is a miniature cross-section of current manufacturing techniques (as proposed for the SRS), or simply a single assembler arm which controls individual molecules during the assembly process, the basic concepts involved are the same.

Finally, by considering the design of an artificial SRS in such detail, the NASA team showed clearly that such things are feasible. Their analysis also provides good support for the idea that a nanotechnological "assembler" can be substantially less complex than a trillion bits in design complexity. There are several methods of simplifying the design of the "Mark I Assembler," as compared with the NASA SRS. First, it could exist in a highly-controlled environment, rather than the uncontrolled lunar surface. Second, it could expect to find many of its molecular parts, including exotic parts that it might find difficult or impossible to manufacture itself, prefabricated and provided in a convenient and simple format (e.g., floating in solution). Third, it could use simple, "blind," fixed-sequence assembly operations.

Conceptually, the only major improvements provided by the Mark I Assembler over a simple bacterium would be the general-purpose positional control it will exert over the reactive compounds that it uses to manufacture "parts", and the wider range of chemical reactions it will use to assemble those "parts" into bigger "parts". Bacteria are able to synthesize any protein. The Mark I Assembler would be able to synthesize a very much wider range of structures. Because it would have to manufacture its own control computer as a simple prerequisite to its own self-replication, it would revolutionize the computer industry almost automatically. By providing precise atomic control, even the Mark I Assembler will revolutionize the manufacturing process.

I.1   FURTHER INFORMATION

Copies of *Advanced Automation for Space Missions* are available from NTIS. Mail order: NTIS, U.S. Department of Commerce, National Technical Information Service, Springfield, VA 22161. Telephone orders with payment via major credit cards are accepted; call 703-487-4650 and request "N83-15348. Advanced Automation for Space Missions; NASA Conference Publication (or CP) 2255." Publication date is 1982 (although the study was done in 1980). Purchase price is about $60.00, various shipping options are available.

Additional information about self-replicating systems is available at: *http://nano.xerox.com/nanotech/selfRep.html*

Appendix J

Human factors

J.1 HAZARDS IN THE LUNAR ENVIRONMENT

There are three hazards at the lunar surface that are related to the absence of an attenuating atmosphere: temperature extremes (of several hundred degrees celcius between lunar day and night), micrometeorites, and ionizing radiation. Spacesuits and capsules such as those used in the Apollo Program provide some protection against micrometeorites and temperature extremes, but offer virtually no protection against ionizing radiation. Fortunately, the solution to the problem of radiation also solves the problems of micrometeorites and temperature extremes. A fourth hazard is lunar dust, which, when inhaled, may cause damage to lungs (pneumoconiosis). The long-term biological effect of the Moon's reduced gravity ($\frac{1}{6}$ Earth's gravity) on humans is another unknown. Further studies of the physiological and possibly patho-logical effects of the Moon's gravity will begin when humans return for prolonged stays on the lunar surface.

J.1.1 Radiation

Hazardous radiation reaches the surface of the Moon from the Sun and from deep-space sources. The Sun produces the "solar wind", which is an ionized, plasma that is spewed from the Sun's outer atmosphere into space. The solar wind is mostly comprised of electrons and protons (hydrogen nuclei); ions of low-atomic-number elements such as helium and carbon. A small amount of high-energy photons (X-rays and gamma rays), are also present. On occasion, a transient increase in the volume and energy of the solar wind, known as a solar flare, produces a surge in radiation levels in space and on the Moon. Solar flares typically reach a peak of activity in several hours and return to the baseline levels over a period of several days. The second source of radiation that reaches the Moon originates in deep space (galactic and extra-galactic), and is known as cosmic radiation. Cosmic rays are comprised of highly-energetic nuclei of atoms (90 percent hydrogen, 9 percent helium, and 1 percent of all of the other elements) and a small number of electrons.

The deleterious effects of ionizing radiation on living tissue are a function of both the dosage and the duration of exposure. Radiation damage shortens biological life spans and induces neurological and genetic defects, as well as cancer. Radiation is cumulative, but biological systems have some ability to repair radiation damage over time. Individuals who work in a radiation environment can reduce their radiation dose by the use of shielding and by limiting the time of exposure.

Ionizing radiation[1] is measured in terms of the energy that it deposits on a given target; the unit of radiation energy transfer is the rad, which represents the deposition of 100 erg of energy in one gram of mass. Some forms of radiation, such as protons, produce greater damaging effects on living tissue at a given energy level than others, such as gamma rays. To compensate for the relative biological effectiveness (RBE) of different forms of radiation, dose rate measurements for living tissue are calculated in terms of the equivalent damaging effects of radiation, or rem (radiation equivalent man), rather than the energy deposited in rads.

A newer unit of radiation dosage is termed the Sievert (after Rolf Sievert, a medical physicist), abbreviated as Sv. One $Sv = 100$ rem and one milliSievert $(mSv) = 0.1$ rem. For the general public, the maximum permissible whole-body radiation dose rate is 5 mSv/year. For individuals who work with radiation, the maximum permissible dose is 50 mSv/year. The energy of the solar wind is low enough that the particles do not produce an unacceptable radiation risk to people who are protected by a spacesuit while performing EVAs of relatively short duration. However, during solar flares, proton energies in the solar wind may exceed 1 million electron volts (1 MeV), and can deliver a fatal dose of many Sieverts to individuals (wearing spacesuits) on the lunar surface in a matter of hours.

The radiation dose from cosmic rays is a constant figure of approximately 300 mSv/year. Because cosmic and solar-flare radiation are highly penetrating, spacesuits and space capsules provide virtually no protection to astronauts against those sources of radiation. If a permanent human settlement is to be established on the Moon, it must have sufficient shielding to reduce radiation levels to the Earth standard for radiation workers of less than 50 mSv/year.[2]

One solution for protecting the occupants of the lunar base is to use the bulk material of the lunar regolith to attenuate solar-flare and cosmic radiation to permissible levels. By placing sufficient regolith on top of the habitable structures, radiation can be attenuated to required levels.[3] It must be recognized that cosmic rays

---

[1] The term "ionizing radiation" means that the radiation, when striking an atom, has sufficient energy to displace an electron from the atom, thus causing it to be ionized.

[2] It should be pointed out that a condition of zero radiation can never be reached. The materials that are used for shielding – and the humans they are intended to protect – are all radioactive. The objective is to minimize radioactivity, not eliminate it.

[3] Attenuation of radiation is a function of the mass of material that is placed in the path of the radiation. The units of attenuation are expressed in mass per unit area, or grams/square centimeter.

produce secondary radiation when they collide with matter, and the regolith that is used for shielding thus becomes a *source* of secondary radiation when it is used to attenuate cosmic rays. It has been estimated (Silberberg *et al.*, 1985) that shielding of approximately 400 gm/square cm, or two meters of densely-packed regolith, would reduce radiation levels to less than 50 mSv/year, a level that is permissible for radiation workers. As the lunar population grows, shielding will be increased to reduce radiation levels in working and living spaces to less than 5 mSv/year – the equivalent of sea-level conditions on Earth.

The first human-habitable structures that are erected on the Moon can thus be covered with sufficient regolith to reduce radiation doses to permissible levels. These measures will also eliminate the threat of micrometeorites, and will provide the thermal insulation that is needed to protect the occupants against temperature extremes.

J.1.2   Lunar dust

The Moon has been bombarded by micrometeorites for eons, with the result that the surface has been pulverized into a fine talc-like powder. This "lunar dust" is abrasive and tends to cling to the objects that it comes in contact with, such as spacesuits. The problem for humans (and all other air-breathing organisms) is that airborne particles of the lunar dust can enter the lungs and pass to the smallest units of the respiratory tree (alveoli) where they are deposited and produce a disease called pneumoconiosis (a category of disease that includes asbestosis). The disease is characterized primarily by scarring of the lungs, which interferes with respiratory function. In addition, the inhaled lunar dust may be carcinogenic.

To reduce the hazard of airborne lunar dust, preventive measures must be taken to keep it from entering habitable areas. Monitoring and air-filtration systems that are specifically designed for lunar dust will be required.

J.1.3   Lunar gravity

A third challenge to human presence on the Moon is its gravity ($\frac{1}{6}$ of the Earth's gravity), which may produce deleterious biological effects. A great deal of knowledge has been gained on the adverse effects of long-term exposure to zero gravity, including loss of bone and muscle mass, increased incidence of kidney stones, and degradation of cardiac performance. However, no long-term human exposure to $\frac{1}{6}$ G has been conducted, and the effects of lunar gravity can only be learned by making appropriate physiological measurements of humans after they have arrived on the Moon.

Despite the reduced gravity, moving about on the Moon may provide sufficient neuro-muscular stimulation so that a satisfactory level of muscle tone and bone mass will be maintained, and the deleterious effects of microgravity environments will thus be avoided. For the present, it can only be speculated that the Moon's lower-than-Earth gravity will have some adverse effects on human physiology that are similar to, but very probably less severe than, the effects of zero (micro) gravity that have been observed in Earth orbit. On the positive side, the low lunar gravity may cause the Moon to become an important rehabilitation site for individuals who have decreased mobility as the result of trauma or debilitating diseases. The next several teams of astronauts who go to the Moon will be, perforce, experimental subjects who will generate data on the physiological and possibly pathological effects of low gravity.

J.2   PHYSIOLOGICAL NEEDS OF HUMAN HABITATION

The second challenge for human habitation is to provide the oxygen, water, and food that are needed for survival.

J.2.1   Oxygen

An oxygen-bearing atmosphere is a fundamental requirement for human survival. Fortunately, oxygen is the most abundant element on the surface of the Moon, comprising 50 percent or more of the lunar regolith. The oxygen is chemically bound to other elements, and is not conveniently available in a gaseous form. However, it can be extracted by various methods (see Appendix E) to supply the atmosphere for the first manned operations.

J.2.2   Water

Water is also essential for human survival; adult humans require approximately 3 kg of water (including water in liquid form and water contained in food) per day. Water that is used for agriculture, personal hygiene, and other purposes related to human presence may result in water usage of 2 to 4 orders of magnitude greater than the amount that is needed for survival.

Evidence of hydrogen (which may imply water ice) has been discovered by the Clementine and Lunar Prospector satellites, in the polar regions of the Moon.[4] Because water is critical for human habitation, the amount of water that is putatatively accessible at the poles will influence the size of the first permanent settlements. Therefore, a high priority will be placed on finding out whether water ice is present at the south pole, and, if so, finding the means to recover it.

[4] Lunar Prospector instruments detected hydrogen ions in the polar regions of the Moon. If this hydrogen is in the form of water ice, as was hypothesized by Watson (1961), it would be a valuable resource.

In addition to whatever may be present at the poles, water can be formed by combining lunar-derived oxygen with the hydrogen that is adsorbed to the grains of regolith. Hydrogen atoms are deposited in the lunar regolith by the solar wind (see Chapter 3). There are sufficient quantities of trapped hydrogen atoms in the regolith to produce approximately 1.5 million liters of water per square kilometer of lunar surface that is mined to a depth of 1.8 meters (Haskin and Colson, 1990). The water that is thus created can supply the needs of the first human settlements. Small amounts of water can also be collected from the exhaust of chemical rocket engines at rocket launch/landing pads. Eventually, space mining of asteroids, comets, and other moons in the solar system will supply the Moon with its water needs.

J.2.3 Food

While an oxygen atmosphere and water can be made available for human habitation from existing lunar resources, food is not available in any form on the Moon and will need to be transported from the Earth for initial manned operations. In order to achieve autonomy of lunar settlements, a means for producing food on the Moon will need to be devised. Food production experiments will have a high priority for the first human settlements on the Moon.

Agricultural experiments on Earth that use lunar regolith simulants have grown a variety of food crops such as wheat, soybeans, potatoes, and lettuce. Experiments with the growth of food crops from "seed to seed" have also been attempted in Earth orbit. When sufficient power and water become available at the first lunar base, seeds of wheat and other food crops will be transported to the Moon and lunar agricultural experiments can begin. Artificial light will be required, because food crops such as wheat cannot adapt to the lunar day/night cycle of 14 Earth days of sunshine and 14 days of darkness.[5] The food crops will also need nitrogen, carbon[6] (e.g., carbon dioxide), and nutrients such as phosphorus and sulfur – all of which are present in the lunar regolith and can be extracted during the same mining and processing operations that are used to recover oxygen and other elements from lunar ores.

All biological sources of food are subject to diseases (viruses, bacteria, fungi). If food crops on the Moon were to be infested with diseases, the result would be potentially catastrophic for the human crews that depend on them. To minimize the threat of crop diseases, a careful disease-screening procedure will need to be developed for all materials that are imported to the Moon. Redundant, biologically-isolated crop-growing regions should also be developed. The lack of an atmosphere on the Moon will make it possible to quarantine a food-growing area that has contracted an infectious disease.

---

[5] If the first lunar base is at Mons Malapert, sunlight will be available for agricultural experiments and projects for as much as 90% of the lunar year.

[6] It is possible that the "cold-trap" mechanism that may have resulted in the concentration of water-ice in the polar regions of the Moon may have caused other volatile materials, including carbon-based compounds, to be deposited in those same areas.

Based upon the data that has been gained from experiments that use lunar soil simulants and artificial lighting, there appear to be no significant technological hurdles to the production of wheat[7] and a variety of other foods in sufficient volume to support human settlements. After it has been demonstrated that one species of food crop can thrive on the Moon, another species will be introduced and tested for its suitability as a source of food. Over time, many species of crops will be grown on the Moon, to minimize the threat of disease and to provide a wide variety of foods for the human population. After the successful demonstration of plant life cultivation, marine life, including lobsters, and livestock such as chickens and other animals, may also be imported.

Another possibility for the production of food on the Moon would be to "grow" basic food groups and nutrients (carbohydrates, fats, proteins (amino acids), and vitamins) by inorganic processes. For example, methane (CH_4), ammonia (NH_3), nitrogen (N_2), hydrogen (H), and water (H_2O), all of which are available or can be produced on the Moon, can be combined to make amino acids, vitamins, and fatty acids (Miller, 1974). Variations of the Fischer–Tropsch chemical synthesis process can be used to produce a wide variety of organic materials such as glucose, fructose, and galactose sugars from carbon, oxygen, and hydrogen (Mizuno and Weiss, 1974). Of some interest to astronauts, perhaps, ethyl alcohol (a mood elevator and source of carbohydrates) can also be manufactured from water and carbon dioxide by a modification of the Fischer–Tropsch process (Rosenberg, 1996). From these inorganic processes, a wide variety of food may be manufactured without the need for biological intermediaries.

The major objection to the use of inorganic processes for the production of food has been the high energy costs and low yields of desirable products. However, the proposed solar-powered lunar electric grid will produce sufficient energy. If an inorganic process provides low yields of edible fats, sugars, or amino acids, that low yield will be of little concern because the process can be repeated as often as required until the desired quantity has been produced. The excess non-edible organic products that are produced can potentially be used as feedstock for the manufacture of plastics and polymers, and for industrial bacterial processes. The production of food on the Moon is a problem that awaits solutions from many promising technologies.

J.3 CONTROLLED ECOLOGICAL LIFE-SUPPORT SYSTEM

The physiological and psychological needs of human habitation will be provided by a controlled ecological life-support system (CELSS). A CELSS is, in theory, a

---

[7] Experiments on Earth have demonstrated that the caloric requirements of one person can be satisfied by the wheat output that is grown on an area of approximately 20 square meters (Salisbury and Bugbee, 1985). For a lunar base of 50 people, the area of lunar soil under cultivation would need to be 1,000 square meters; a wheat field of one square kilometer could theoretically support as many as 50,000 people.

completely-closed system wherein continuous waste-recycling and regenerative systems provide 100 percent of the food, water, and breathable atmosphere in a psychologically-acceptable human environment. In practice, a completely-closed system is not possible, because some loss of resources (such as leakage of oxygen and water out of the habitat) is inevitable. The goal, then, of CELSS is to approach, as closely as possible, a condition of self-reliance for humans on the Moon, where a minimum re-supply from Earth is necessary.

The goal of self-reliance will be much more achievable for a lunar base than for an Earth-orbiting station because the lunar base will have access to local resources, such as water and oxygen, that are not available in Earth orbit. Furthermore, the Moon has a gravity that simplifies waste management, food preparation, and water purification procedures, and the lunar electric power grid will provide the lunar base with power.

The CELSS concept may be divided into four components: biological (plant growth) and inorganic food production; food processing; crew space; and waste processing. In addition to maintaining the proper oxygen-bearing atmosphere, the major challenges to CELSS on the Moon are the provision of food for the crew and the management of wastes.

J.3.1   Food crops

In a functioning CELSS operation, plant biomass such as wheat, lettuce, soybeans, and potatoes is grown in a separate compartment where the lighting, temperature, humidity, and carbon dioxide levels are optimized for plant growth. Oxygen that is produced by plant growth is recovered for use in the waste-processing system and for crew breathing air. Other basic foods (sugars, fats, amino acids, and vitamins) are separately manufactured by both organic and inorganic processes.

Mature plant products are passed to the food-processing system, where edible materials are harvested and, in combination with inorganically-produced foodstuffs, converted into prepared foods. The crew then consumes the food and metabolizes it into carbon dioxide, urine, and feces. These are transferred to the waste-processing system, which converts the waste into nutrients for crops, carbon dioxide, water, and mineral residues. The carbon dioxide and waste are recycled, and food, water, and oxygen are regenerated.

J.3.2   Waste management

The treatment and recycling of human, plant, and industrial wastes is of critical importance to lunar settlements. Waste material, if not properly treated, would threaten the base with disease and toxic materials. Perhaps just as important, wastes are a resource of low-atomic-number elements that are scarce on the Moon, and they must be recovered and reused.

Each adult human produces approximately 1,500 grams of urine and 200 grams of feces per day.[8] The agriculture and industrial base that provides food and other organic materials for human settlements will also yield waste organic materials (trash) of 10–100 kilograms or more per person per day, which must be recycled. Municipal waste treatment systems on Earth rely upon bacteria to break down solid organic wastes into simple compounds such as water and carbon dioxide, and an adaptation of existing waste treatment technology may provide an adequate means of waste treatment for lunar bases. Bacterial-based waste treatment systems have the disadvantage of being living systems that are susceptible to failure as the result of disease or toxins.

An alternative waste management technology is termed the supercritical water oxidation system (Sedej, 1985). This waste treatment system is based upon the physical and chemical properties of water at conditions above its supercritical pressure (250 atmospheres) and temperature (670°F). Under these conditions, complete combustion of organic materials occurs in the presence of oxygen, yielding sterile water, carbon dioxide, and nitrogen. Dissolved inorganic salts also precipitate out of the solution, and can thus be removed by solid separation techniques. The advantages of this system are multiple:

- It can accept gaseous, solid, and liquid wastes.
- It destroys all microorganisms (e.g., bacteria, viruses, fungi).
- Combustion products such as sterile water are resources that can be separated and recycled.
- It is a mechanical waste treatment system that is not susceptible to disease or toxins.

The power requirements[9] of supercritical water oxidation waste treatment systems have been considered too high for lunar bases, especially those that relied on battery power during the lunar night. However, if power requirements are satisfied by the planned solar-powered lunar electric grid, this obstacle is removed. The only possible disadvantage is that the supercritical water oxidation waste treatment systems are still in the development phase. However, the same is true for many of the systems that will eventually be used on the Moon.

---

[8] Humans also produce approximately 1 kilogram of carbon dioxide per day. The carbon dioxide "waste" output of humans is a resource for both agricultural and industrial processes, and the "waste management" of carbon dioxide is therefore a matter of transferring it from the atmosphere that humans breathe to the atmospheres of food crops which thrive on high levels of carbon dioxide.

[9] Power consumption for an eight-person crew has been estimated to be 300–400 watts, continuous (Sedej, 1985).

The integration of all of the elements of a CELSS into a stable operating system is a complex task. For example, a reliable CELSS must monitor thousands of parameters that define the physical environment of multiple interconnected systems and compartments, and provide reliable controls that maintain critical components within normal limits, including atmospheres and waste treatment systems, within normal operating limits. Because of its complexity, a regenerative, fully-operational controlled ecological life-support system has not yet been developed.

J.3.3   Experiments

The problems associated with the design of controlled ecological life-support systems are illustrated by the experience of the Biosphere 2 experiment, which was an attempt to provide food, water, and air for an eight-person staff in a closed environment from 1991 to 1993 (Cohen and Tilman, 1996). During the two-year experiment, many of the plant and animal (both vertebrate and invertebrate) species that had been brought into the enclosure became extinct. Oxygen levels fell from 21 percent to 14 percent, and additional oxygen had to be added from the outside before the end of the experiment in early 1993. Other problems, such as high and fluctuating carbon-dioxide and nitrous-oxide levels, and high levels of nutrients in water systems, were also encountered. One of the conclusions of the experiment was that, "No one yet knows how to engineer systems that provide humans with the life-supporting services that natural eco-systems produce for free."

However, the lessons that were learned from the Biosphere 2 experiment are being applied to life-support system designs, and experiments with various stages of life-support systems are being conducted at sites such as the Advanced Life Support Test Chamber at the Johnson Space Center (JSC) in Houston, Texas. A 60-day study at JSC that evaluated several critical life-support components such as oxygen, carbon dioxide, and water recycling systems was successfully completed in March 1997 with an eight-person crew.

Plans are underway to expand the tests to include food production and waste treatment systems with human crews for as long as a year. This series of tests in the "Bioregenerative Planetary Life-Support System Test Complex", or "Bioplex", will simulate full-scale controlled ecological life-support systems for a lunar (or Mars) base.

J.4   PSYCHOLOGICAL NEEDS OF HUMAN HABITATION

On long-duration scientific missions such as wintertime Antarctica expeditions, stress is a hazard that can degrade the performance of the crew and jeopardize the mission objectives. Stress produces adverse effects on human behavior, cognition, physiology, and organization (Harrison et al., 1991; Harris, 1996). Its symptoms include insomnia, depression, irritability, decreased intellectual and physical performance, and decreased attention span. Stress is produced by many factors, including the following:

- Forced confinement.
- Isolation.
- Lack of privacy.
- Noise and vibration.
- Forced socialization with fellow crew members.
- Increased or decreased sensory input.
- Fear of equipment failure.
- Monotony (of food, recreation, tasks, etc.).
- Complexity of mission tasks.

The living and working quarters at the first lunar base will subject the crew to multiple stress factors. The base will be spartan, with little more than the barest essentials that are needed for survival and experimentation. Because the weight and volume of materials that can be delivered from the Earth to the Moon is limited and because transportation expenses are large, the living spaces for the first crews will be small, prefabricated modules that have been built on the Earth and transported to the Moon[10] (see MALEO, Appendix S). After they have been delivered to the lunar surface, the modules will be connected and covered with regolith to provide shelter from temperature extremes, meteorites, and ionizing radiation. The first lunar base will thus provide the crew with little more than the minimum requirements for life-support and working conditions, analogous to the living and working conditions of a scientific base in Antarctica[11] or a space station in Earth orbit.

The first lunar inhabitants will need to conduct many important tasks, such as the maintenance of food, water, atmosphere, and waste treatment systems that are critical for their survival. They will begin experiments with lunar agriculture, perform daily medical tests, conduct geoscience investigations on the lunar surface, and perform many other tasks such as minor repairs of mining and manufacturing equipment. The first human missions will typically last for 90 days, during which time the crew members will be in constant contact with one another, but will be physically isolated from all other humanity.

To summarize, the first humans on the Moon will be thrust into an environment that has virtually every significant stress factor. They will be required to work in confined, barren, and noisy quarters (most experiments and environmental control

---

[10] The first manned missions will be performed before lunar manufacturing autonomy has been achieved. For this reason, the first human habitats must be constructed on the Earth and transported to the Moon.

[11] Interestingly, the Moon base will not be as isolated as previous Antarctica missions, which operated for long periods of time without direct contact with the outside world. The Moon base, in contrast, will have constant contact with Earth; in fact, the next generation of astronauts on the Moon will undoubtedly be inundated with e-mail messages from people all over the Earth.

systems will operate continuously), yet perform highly-important tasks that relate directly to their survival and to the success of their mission and future missions. The first manned lunar base will thus present habitat designers with the challenge of minimizing stress, and will, inadvertently, provide psychologists and other investigators with an excellent opportunity to gain further data on the human responses to stress.

To minimize the possibility of mission-threatening reactions to stress, screening procedures will be used to select crew members that have a high tolerance for stress. The design of the first lunar base must provide as much privacy as possible, as well as pleasant surroundings for the crew.[12] Time must be set aside for recreation, and work assignments from mission control on the Earth must be adjusted so that they conform to the day-to-day performance capabilities and needs of the crew.

Psychological monitors will be employed to measure and record behavior, sleep patterns, cognitive ability, and other stress markers. These data will also be used to indicate any deterioration in the performance of the crew so that corrective measures can be taken. Because a significant amount of the stress is related to the limited size of the first manned base, efforts will be made to increase the living and working spaces. The first option for adding living space to the base is to "fly in" additional habitable modules from the Earth and connect them to the base.

The second option, which makes the Moon a logical place for the expansion of humankind into space, is to use the existing on-site mining and manufacturing capabilities to create new living spaces. In the same time frame as the first humans establish a base on the Moon, the lunar industrial base will have the capability of producing glass bricks, metal support beams and other construction materials from the lunar regolith. Tele-operated robots will use these materials to create underground structures and connecting tunnels that will have a breathable atmosphere.[13] These expanded habitable spaces will become extensions of the original lunar base, and will provide the crew with space for work, scientific experiments, and recreation, away from the confines of the original modular base. By this process, large living spaces, and eventually cities, will be created, and the threat to the psychological health of lunar inhabitants from isolation and confinement will be reduced.

---

[12] Windows to the outside world help to reduce stress. The habitable areas of the lunar base will be covered with several meters of regolith; however, a combination of windows and mirrors can be designed that will permit the crew to have panoramic views of the lunar surface and the sky (see Appendix T for an example).

[13] The creation of large living spaces that are filled with a breatheable atmosphere will be possible because oxygen and nitrogen can be extracted from the regolith.

J.5   REFERENCES

An advanced life-support bibliography may be found on the Internet at *http://pet.jsc. nasa.gov/ alssee/demo_dir/nscort/generalbiblio.html*

Advanced Life Support Requirements Document (2003) NASA document CTSD-ADV-245C, Lyndon B. Johnson Space Center, Houston, TX, February.

Ahif, P.; Fogleman, G.;and Tomko, D. (2000) Human Exploration and Development of Space (HEDS) activites in preparation for future exploration missions. *Proceedings of Space 2000, Albuquerque, NM*, pp. 24–30.

Allen, J. and Nelson, M. (1987) *Space biospheres.* Orbit Books/Krieger Publications, Melbourne, FL.

Averner, M. and Blackwell, C. (1996) The development and operation of life-support systems for long-term space exploration missions. *26th International Conference on Environmental Systems, Monterey, CA, July 8–11*, SAE Technical Paper Series 961494.

Barta, D.J. and Dominick, J.S. (1995) Early human testing of advanced life-support systems, phase I. *25th International Conference on Environmental Systems, San Diego, CA, July*, SAE Technical Paper Series 951490.

Batsura, I.D.; Kruglikov, G.G.; and Arutiunov, V.D. (1981) Morphology of experimental pneumoconiosis developing after exposure to lunar soil. *Biulleten Eksperimentalnoi Biologii i Meditsiny*, **92**, no. 9, 376–379, September.

Belkin, V.I. *et al.* (1983) Biological activity of the lunar soil from the Sea of Abundance after intratracheal administration. *Izvestiia Akademii Nauk SSSR, Seriia Biologicheskaia*, **3**, 461–465, May–June.

Carrasquillo, R.L.; Wieland, P.O.; and Reuter, J.L. (1996) International space station environmental control and life-support system technology evolution. *26th International Conference on Environmental Systems, Monterey, CA, July 8–11*, SAE Technical Paper Series 961475.

Chamberland, D. and Carpenter, S. (1996) Ocean habitats as analogs for space habitats. *26th International Conference on Environmental Systems, Monterey, CA, July 8–11*, SAE Technical Paper Series 961397.

Cohen, J.E. and Tilman, D. (1996) Biosphere 2 and biodiversity: The lessons so far. *Science*, **274**, 1150–1151, 15 November.

Couch, H.T.; Birbara, P.J.; and Grin, W. (1996) Steam gasification and reformation of spacecraft wastes. *26th International Conference on Environmental Systems, Monterey, CA, July 8–11*, SAE Technical Paper Series 972273.

Dornheim, M.A. (1997) Station life-support advances technology. *Aviation Week*, 56–58, December 8.

Drysdale, A. and Sager, J. (1996) A re-evaluation of plant lighting for a bioregenerative life-support system on the Moon. *26th International Conference on Environmental Systems, Monterey, CA, July 8–11*, SAE Technical Paper Series 961557.

Drysdale, A.E. *et al.* (1994) A more completely defined CELSS. *24th International Conference on Environmental Systems*, SAE Technical Paper Series 941292

Eckart, P. (1996) *Spaceflight Life Support and Biospherics.* Microcosm Press, Torrance, CA.

Edeen, M. and Barta, D. (1996) *Early Human Testing Initiative, Phase I.* Final report, JSC 33636, Doc. No. ADV-208, Crew and Thermal Systems Division, Johnson Space Center, Houston, TX, March 25.

Fielder, J. and Leggett, N. (1988) A second generation lunar agricultural system. *Journal of the British Interplanetary Society*, **41**, 263–268.

Fielder, J. and Leggett, N. (1990) Composting for lunar agriculture. *Proceedings of Space '90, Albuquerque, NM, April*, Aerospace/American Society of Civil Engineers, New York, pp. 1233–1241.

Fielder, J. and Leggett, N. (1998) Lunar base plant propagation and supply. *Proceedings of the Sixth International Conference and Exposition on Engineering, Construction and Operations in Space, Space '98, Albuquerque, NM, April*, pp. 417–421.

Fielder, J. and Leggett, N. (2005) Requirements and opportunities for the successful development of lunar and Martian agricultural systems. *Proceedings of Space 2002, Albuquerque, NM, March*, pp. 170–175.

Gitelson, J.I. and Shepelev, Y.Y. (1996) Creation of life-support systems for the lunar outpost and planetary bases: History, present state of research in Russia (former Soviet Union) and prospects. *26th International Conference on Environmental Systems, Monterey, CA, July 8–11*, SAE Technical Paper Series 961553.

Harris, P.R. (1989) The influence of culture on space development. *Behavioral Science*, **31**, no. 1, 12–28, January.

Harris, P.R. (1993) Personal challenges in constructing a lunar power system and base. *Futures Research Quarterly*, Spring, **9**, no 1., pp 19–61 (World Future Society, Bethesda, MD).

Harris, P.R. (1996) Challenges in the space environment – personnel deployment systems. *Proceedings, International Conference on Environmental Systems, Monterey, CA, July 8–11*.

Harris, P.R. (1996) *Living and Working in Space: Human Behavior, Culture and Organisation*, 2nd edn., Wiley-Praxis, Chichester, UK.

Harris, P.R. and Moran, R.T. (1996) *Managing Cultural Differences*, 4th edn. Gulf Publishing, Houston, TX.

Harrison, A.A. (2001) *Spacefaring*. University of California Press, Berkeley, CA.

Harrison, A.A.; Clearwater, Y.A.; and McKay, C.P., eds., (1991) *From Antarctica to Outer Space: Life in Isolation and Confinement*. Springer-Verlag, New York.

Helmreich, R.L.; Wilheim, J.A.; and Runge, T.E. (1983) *Human Factors in Outer Space Productions*. Westview Press, Boulder, CO.

Henninger, D.L. (1989) Life-support systems research at the Johnson Space Center. In: *Lunar Base Agriculture*, ASACSSA-SSSA, Madison, WI.

Hunter, J. and Drysdale, A. (1996) Concepts for food processing for lunar and planetary stations. *26th International Conference on Environmental Systems, Monterey, CA, July 8–11*, SAE Technical Paper Series 961415

Hurlbert, K. *et al.* (1997) *Lunar–Mars Life-Support Test Project, Phase II*. Final Report JSC-38800, Doc No. CTSD-ADV-307, August. NASA Johnson Space Center, Houston, TX.

Hypes, W.D. and Hall, J.B. (1992) The environmental control and life-support system for a lunar base – What drives its design? *2nd Conference on Lunar Bases and Space Activities for the 21st Century*, pp. 503–511.

Kirby, G.M.; Tri, T.O.; and Smith, F.D. (1996) Bioregenerative planetary life-support systems test complex: Facility description and testing objectives. *26th International Conference on Environmental Systems, Monterey, CA, July 8–11*, SAE Technical Paper Series 972342

Laws, B.A. and Foerg, S.L. (1995) Early human testing of advanced life-support systems, phases II and III. *25th International Conference on Environmental Systems, San Diego, CA, July*, SAE Technical Paper Series 951491.

Miller, S. and Orgel, L. (1974) *The Origins of Life on the Earth*. Prentice Hall, Englewood Cliffs, NJ.

Ming, D.W. *et al.* (1997) Lunar–Mars life-support test project, phase II: Human factors and crew interactions. *27th International Conference on Environmental Systems, Lake Tahoe, NV, July 14–17*, SAE Technical Paper Series 972415.

Ming, D.W. (1989) Manufactured soils for plant growth at a lunar base. In: *Lunar Base Agriculture*, ASA-CSSA-SSSA, Madison, WI.

Mizuno, T. and Weiss, A.H. (1974) Synthesis and utilization of formose sugars. In: Tipson, R.S. and Horton, D., eds., *Advances in Carbohydrate Chemistry and Biochemistry*, Academic Press, New York, pp. 173–227.

Nicogossian, A. *et al.* (1993, 1994, 1995) *Space Biology and Medicine*, Vols. 1–5. American Institute of Aeronautics and Astronautics, Washington, D.C.

NRC (1987) *A Strategy for Space Biology and Medical Science*. National Research Council, National Academy Press, Washington, D.C.

Possehl, S. R. (1997) Keeping life sciences alive in Russia. *Aerospace America*, 42–47, March.

Reuter, J.L. and Reysa, R.P. (1996) International space station environmental control and life-support system design overview update. *26th International Conference on Environmental Systems, Monterey, CA, July 8–11*, SAE Technical Paper Series 972333.

Rosenberg, S.D. (1997) *The happy astronaut*. Private communication on a proposal to use a modification of the Fischer–Tropsch process to produce ethyl alcohol from carbon dioxide and water.

Sager, J.C. (1996) KSC advanced life-support breadboard: Facility description and testing objectives. *26th International Conference on Environmental Systems, Monterey, CA, July 8–11*, SAE Technical Paper Series 972341.

Salisbury, F.B. and Bugbee, B.G. (1985) Wheat farming in a lunar base. *Lunar Bases and Space Activities of the 21st Century*, Lunar and Planetary Institute, Houston, TX, pp. 635–651.

Schwartzkopf, S.H.; Kane, D.G.; and Stempson, R.L.(1988) Greenhouses and green cheese: Use of lunar resources in CELSS development. *18th Intersociety Conference on Environmental Systems*, SAE Technical Paper Series 881057.

Sedej, M.M. (1985) Implementing supercritical water oxidation technology in a lunar base environmental control/life support system. *Lunar Bases and Space Activities in the 21st Century*, Lunar and Planetary Institute, Houston, TX, pp. 653–662.

Silberberg, R. *et al.* (1985) Radiation transport of cosmic ray nuclei in lunar material and radiation doses. *Lunar Bases and Space Activities of the 21st Century*, Lunar and Planetary Institute, Houston, TX, pp. 663–670.

Taylor, L. and James, J. (2006) Potential toxity of lunar dust. *Space Resources Roundtable VIII, Golden, CO, November.*

Vander, S.T. *et al.* (1996) Operational psychological issues for Mars and other exploration missions. *26th International Conference on Environmental Systems, Monterey, CA, July 8–11*, SAE Technical Paper Series 972290.

Volk, T. and Cullingford, H. (1992) Crop growth and associated life support for a lunar farm. *2nd Conference on Lunar Bases and Space Activities for the 21st Century*, Lunar and Planetary Institute, Houston, TX, pp. 525–530.

Appendix K

Maglev trains and mass drivers

K.1 ELECTROMAGNETIC TRANSPORTATION

Magnetic levitation ("maglev") trains and mass drivers are electromagnetic transportation systems that operate at high speeds. They are comprised of tracks or coils
that use magnetic fields to suspend and accelerate (or decelerate) a cargo along the
track. Because the cargo is suspended by magnetic lines of force, it generates no
friction with the rail as it travels. The technology of mass drivers has been under
development for close to a century, and it is the basis of magnetic levitation trains
that are now being tested in Germany and Japan, and are operational in China.

The Moon will be an ideal place to adapt electromagnetic transportation technology for high-speed surface maglev railroad transportation and for the launch and
recovery of lunar satellites. Electromagnetic propulsion systems require large power
sources and sophisticated energy management systems; however, abundant and
continuous electric power will be available with the development of the lunar power
system. There is no atmosphere on the Moon, so the primary limitation of the
velocities that can be reached with these systems is the acceleration tolerance of
the cargo, not its air resistance.

K.2 MAGLEV TRAINS

The first railroads on the Moon will be very simple two-track systems, as described in
Chapter 6. As an increasingly-sophisticated lunar manufacturing capability develops,
it will become possible to create a magnetic levitation rail system on the Moon.
The Moon is well suited for maglev rail operations, particularly in the flat mare
regions. Because there is no atmosphere on the Moon, maglev "express" trains will be
able to travel at almost orbital velocities on long-distance routes. The maglev transportation system may become the primary mode of global transportation on the

Moon (see Figure 9.24). The technologies developed for lunar use will have potential applications on Earth as well. Maglev roadbeds that collect solar energy and serve high-speed surface vehicles may eventually provide alternatives to highway, rail, and air travel on Earth.

K.3 MASS DRIVERS

One of the barriers to lunar development is the risk, cost, and complexity of the chemically-fueled rockets that are used to transfer cargoes between the surface of the Moon and lunar orbit. In particular, a major concern of space planners is the "energy cost" of transporting a payload from lunar orbit to the surface of the Moon, and returning a similar payload back to lunar orbit from the "gravity well" of the Moon with conventional rocket systems.

When a payload is transferred between lunar orbit and the surface of the Moon (decelerating on landing or accelerating at launch), it must undergo a change in velocity (delta-V) of 1.7 kilometers per second (km/sec).[1] For a chemically-fueled rocket, a significant proportion of the total mass of the vehicle is the mass of the propellants that are consumed in this process. As a result, the mass of the payload that is transferred between the surface of the Moon and lunar orbit is considerably less than the total mass of the rocket – an inefficient and expensive means of transportation. Re-usable chemical rockets are also complex machines that must be inspected and refurbished after each use. They use fuels that are both hazardous and valuable (fuels such as liquid oxygen and hydrogen have a high value for other purposes on the Moon), and their exhausts create a transient atmosphere that can interfere with Moon-based telescopic observations.

Electromagnetic mass drivers bypass these problems because the "fuel" of a mass driver is the electric energy that is present in the lunar electric grid. The concept of using mass drivers to launch material from the lunar surface was discussed in some detail by O'Neill (1989). In a typical launch sequence, a sled or capsule that carries the payload is accelerated by electromagnetic forces to the desired orbital or escape velocity, then the sled releases the payload on its trajectory. The sled is then decelerated along the remaining segment of the track and is returned to the base of the mass driver, where it is loaded with another payload[2] and the launch sequence is repeated. The deceleration of the sled by electromagnetic forces converts the kinetic energy of the sled into electric energy that is fed back into the lunar electric grid. The mass driver operates entirely on electrical power without the need to consume valuable chemical propellants.

---

[1] The minimum delta-V for escape from the Moon's gravity is 2.37 km/sec.
[2] Alternatively, the sled may be an integral part of the payload structure, or it may be caused to enter lunar orbit before it separates from the payload, so that it can pick up an "incoming" payload for return to the lunar surface.

Figure K.1. Mass driver on the Moon (Gerard K. O'Neill, Space Studies Institute)

Mass drivers have few moving parts and they use no hazardous materials for their operations. The only limitation on the launch rate of cargo is the time that it takes to place the next payload in position at the base of the mass driver. If desired, payloads could be launched by a mass driver from the Moon to lunar (or other) orbit at the rate of one every few minutes.[3] Figure K.1 is an illustration of a mass driver on the Moon.

K.3.1 "Capture" operation of mass drivers

In addition to their ability to launch payloads from the surface of the Moon, mass drivers can also be made to operate *in reverse* and thus capture and return payloads from lunar orbit to the surface of the Moon. That is, a satellite in orbit around the Moon can be maneuvered so that the perilune[4] segment of its orbital path precisely coincides with the long axis of the mass driver (Figure K.2). The satellite then enters

---

[3] See discussion of the use of a mass driver to launch one-ton containers of liquid oxygen every two hours from the Moon to low Earth orbit in *A Lunar Superconducting Quenchgun Design* by William R. Snow and Henry H. Kolm, pp. 128–133, Space Resources, NASA SP-509, Vol. 2, Government Printing Office, 1992.

[4] Perilune is the term for the point of closest approach of a satellite to the lunar surface as it orbits around the Moon.

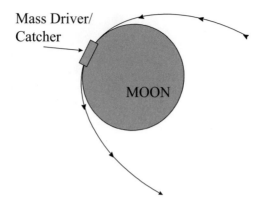

Figure K.2. Schematic of mass driver launch and capture operation (Sharpe/Schrunk).

the high-speed end of the mass driver, and the electromagnetic propulsion system works in reverse to decelerate the satellite until it comes to a stop at the base of the mass driver on the lunar surface. When a mass driver is used in reverse to capture satellites, the orbital kinetic energy of the satellite is converted into electrical energy which is then fed back into the electric power grid[5] (see Figure K.2). With their ability to launch and recover satellites, mass drivers could replace rockets as the principal mode of space transportation on the Moon.

K.3.2   Human-rated mass drivers

In addition to launching and recovering hardware and raw materials, mass drivers can operate over long distances at low levels of acceleration, thus enabling them to deliver spacecraft with human cargoes between the lunar surface and cislunar space.[6] For example, a single passenger unit of a maglev train could slowly accelerate to orbital velocity along the maglev track from point A to point B on the Moon. However, instead of continuing on along the lunar surface, the passenger unit would be shunted onto a straight "launch" track where it would undergo additional acceleration, be released from the track, and coast into low lunar orbit. Depending upon the acceleration forces and length of track employed, the maglev launch system could be used to propel passenger capsules into lunar orbit or on escape trajectories to the Earth, Mars, or other rendezvous points. In its reverse mode of operation, the maglev rail system would capture passenger-carrying capsules that entered lunar orbital space.

---

[5] If the total tonnage of materials that is delivered to the lunar surface by the capture operation of mass drivers is greater than the tonnage that is launched into space, then a net surplus of energy will be generated in the lunar electric grid.

[6] A mass driver that is approximately 50 kilometers in length could reach lunar orbital velocity, and thus deliver human cargoes from the lunar surface to lunar orbit, with an acceleration force no greater than Earth's gravity (1 G).

K.4   SUMMARY

Magnetic levitation trains and mass drivers may eventually be an important transportation mode on the lunar surface, and between the surface and cislunar space. Lunar cities could be connected by a global magnetic levitation rail system, and mass driver transportation centers would be the spaceports where freight and human cargoes are transported between the Moon and cislunar space.

K.5   REFERENCES

Arnold, W.H. *et al.* (1985) *Mass Drivers I: Electrical Design.* Space Resources and Space Settlements, NASA SP-428, NASA Scientific and Technical Information Branch, Washington, D.C., pp. 87–100.

Bilby, C. *et al.* (1987) *A Lunar Electromagnetic Launcher.* Final Report for the Large Scale Programs Institute, Center for Space Research & Center for Electromechanics, University of Texas at Austin, Many.

Clarke, A.C. (1950) Electromagnetic launching as a major contributor to space flight. *Journal of the British Interplanetary Society*, **9**, 261–267.

Coffey, H.T.; Chilton, F.; and Hoppie, L.O. (1972) *The Feasibility of Magnetically Levitating High Speed Ground Vehicles.* Stanford Research Institute, Palo Alto, CA, DOT-FRA-10001.

Eastham, T.R. (1995) High-speed ground transportation development outside the United States. *Journal of Transportation Engineering*, **121**, no. 5, 411–416, September/October.

Eastham, T.R. (1995) High-speed rail: Another golden age? *Scientific American*, 100–101, September.

Graneau, P. (1980) Self-energized superconducting mass driver for space launching. In: *Proceedings of the Conference on Electromagnetic Guns and Launchers, November 4–6*, Part II, U.S. Army, Picatinny Arsenal, NJ, 1980.

Henderson, B. (1990) Sandia researchers test "coil gun" for use in orbiting small payloads. *Aviation Week and Space Technology*, May 7.

Kolm, H.H. and Thornton, R.D. (1973) Electromagnetic flight. *Scientific American*, **229**, no. 4, 17–25, October.

Kolm, H.; Fine, K.; Mongeau, P.; and Williams, F. (1979) Electromagnetic propulsion alternatives. In: *Space Manufacturing Facilities III, Proceedings of the 4th Princeton/AIAA Conference*, pp. 299–306, American Institute of Aeronautics and Astronautics, New York.

Morgan, J.A. and Robinson, E.Y. (1998) A long shot for satellite launch. *Aerospace America*, 32–39, April.

O'Neill, G.R. (1989) *The High Frontier: Human Colonies in Space.* Space Studies Institute Press, Princeton, NJ, pp. 139–150.

Snow, W.R. and Kolm, H.H. (1992) Electromagnetic launch of lunar material. *Space Resources: Energy, Power, and Transport*, Vol. II, NASA SP-509, U.S. Government Printing Office, Washington, D.C., pp. 117–135.

Snow, W.R.; Kubby, J.A.; and Dunbar, R.S. (1981) A small scale lunar launcher for early lunar material utilization. In: *Space Manufacturing 4, Proceedings of the 5th Princeton/AIAA Conference*, American Institute of Aeronautics and Astronautics, New York, pp. 157–171.

Thom, K. and Norwood, J. (1961) *Theory of an Electromagnetic Mass Accelerator for Achieving Hypervelocities*. NASA Langley Research Center, Hampton, VA, June.

Appendix L

Development of the lunar economy

L.1 INTRODUCTION

Space exploration in the twentieth century was treated as the exclusive province of governments, in which commercial enterprises were relegated to the role of government contractors. While this approach met with some notable successes, it is self-limiting because the percentage of gross domestic product that governments have available for space exploration is expected to continue falling.

To overcome the limitations of the government-only approach to space exploration, the "Planet Moon Project" will be significantly different from past efforts, in that commercial (for-profit) and other non-government institutions will make significant contributions to lunar exploration and development. The combined technological/management expertise and financial resources of governments *and* commercial enterprises are far greater than those of governments alone.

L.2 COMMERCIAL SPACE OPERATIONS

Private companies now make significant contributions to the exploration and development of space. Space Adventures Inc. has sucessfully flown paying passengers to the International Space Station. The X-Prize Foundation awarded a prize for the first privately crewed vehicle to reach space, return safely and repeat the mission in less than two weeks. This feat was accomplished by Spaceship One, a reuseable suborbital vehicle designed by Scaled Composites of Mojave, California. Virgin Galactic, a subsidiary of the British airline Virgin Atlantic, has contracted the builders of Spaceship One to produce a fleet of such vehicles and it hopes to fly passengers on suborbital missions as early as 2008.

Bigelow Aerospace plans to place an inflatable habitat-hotel into Earth orbit as a follow up to its successful subscale habitat orbital mission. It also plans to place habitats on the lunar surface. Orbital Sciences Corporation has been providing space launch services using their Pegasus and Minataur vehicles. Celestis, Inc. has a successful cremation and memorial services industry, and Ecliptic, Inc. manufactures a variety of cameras and sensors that can be mounted on launchers and spacecraft. Space X, a private rocket builder, has successfully flown its Falcon 1 rocket and it hopes to achieve the goal of low-cost launches to Earth orbit. SpaceDev, a private space development company, has had several successful space missions and is a provider of hybrid rocket engines for government missions. It too is looking at space tourism operations in collaboration with Bigelow Aerospace (see the color plate section). These and other private space ventures bode well for a vigorous and growing commercial space industry.

L.3   FUNDS FOR LUNAR DEVELOPMENT

At the first lunar base, the lunar government (or some combination of national space agencies) will be the initial source of funds (the pooled tax-derived funds from participating Earth governments and the funds that are raised from the sale of Lunar Development Bonds – see Appendix F) for scientific investigations. The lunar government will facilitate the creation of markets on the Moon for commercial goods and services, thus initiating the lunar economy. For example, geologic expeditions in the south polar region and the operation of astronomical observatories will require transportation, power, and communication services, all of which can be provided by commercial vendors.[1]

If the framework of laws of the lunar government enables the lunar economic system to function efficiently, corporations that offer infrastructure goods and services, such as electric power and telephone services, will be profitable, and their financing will come from investors rather than from government. Ideally, there will be a significant benefit to the public from lunar science projects (sponsored by national space agencies), and corporations will prosper, provide high-quality jobs, pay dividends to shareholders, and pay taxes to governments.

L.4   BUSINESS OPERATIONS EXAMPLE: THE "LUNAR ELECTRIC POWER COMPANY"

The steps leading to the manufacture, construction, and operation of the first solar electric grid on the Moon will inaugurate the "Lunar Electric Power Company". The

---

[1] The economic model by which commercial vendors supply goods and services at a profit to the space agencies and other government institutions on the Moon is the same system that is now in place on the Earth. No space agency owns and operates its own electric power system, railroad, or telephone system – these services are provided by commercial institutions.

company will establish manufacturing facilities on the Moon for the production of solar cells, transmission lines, and other components of the electric power system, and will distribute power to end users.

At first, the electric power that is generated will be sold to governments and other consumers on the Moon. As power-generating capacity increases, it will become possible to sell excess electric power (via microwave beaming) to consumers on the Earth, as discussed in Chapters 7 and 10. The sale of electric power from the lunar electric grid to consumers on the Moon, and the export of energy to the Earth, will be the basis for individuals and commercial institutions to continue investing in shares of stock of the Lunar Electric Power Company.[2] The funds that are raised from investors and the income that is derived from the sale of electric power will be used to expand the grid and increase power output.

Return on investments in the Lunar Electric Power Company will be high. The demand for electric energy on the Earth is expected to rise from approximately 4 terawatts (one terawatt equals one million megawatts) in the year 2010 AD to more than 20 terawatts by the middle of the twenty-first century. This demand is expected to cause a corresponding increase in the price of electric power on the Earth from approximately U.S.\$0.1/kW-hr to as high as U.S.\$0.25/kW-hr. The continued production of electrical power from traditional sources on Earth also threatens to cause significant disruptions to the Earth's environment from the mining, processing, transportation, and consumption of fossil and fission[3] fuels.

If the energy needs of the Earth in the following decades were to be generated on the Moon by photovoltaic (solar) cells that are constructed from lunar materials, and beamed to Earth by microwave energy, the pollution "cost" of electricity production would be negligible. The dollar cost of producing and exporting solar electric power on the Moon at the 20-terawatt level over an 80-year period has been estimated to be as low as U.S.\$0.002/kW-hr (Criswell, 1991), which offers a high profit potential. The combined advantages of virtually-zero pollution, low production costs, and steadily-increasing energy markets on the Earth (as well as on the Moon and other locations in space) will assure the success of lunar electric power companies. Lunar electric power will be popular with consumers, environmentalists, and investors alike.

To meet the electric power needs of lunar development and the growing energy markets on Earth, the Lunar Electric Power Company will need to purchase a wide range of goods and services from other companies that do business on and with the Moon. For example, the Lunar Electric Power Company might purchase:

[2] The use of investor funds to finance the construction of infrastructure projects will be advantageous for the lunar government: the infrastructure will be built without the use of taxpayer funds; the government will derive tax revenues from investor-financed commercial operations; and it will use those funds for new scientific projects such as geoscience expeditions to the poles.

[3] The use of nuclear reactors to generate electricity may also be used to create plutonium, which can be used to make nuclear weapons – an undesirable side-effect of fission fuels.

- Solar cells that are made on the Moon by solar cell manufacturers.
- Construction materials that are made on the Moon by glass-, brick-, and metal-working companies.
- Robotic devices from the lunar robotics industries.
- Transportation services from the lunar railroad company.
- Telecommunications services from the lunar telephone company.[4]
- Robotic tele-operations services from Earth-based companies.[5]
- Information systems from the lunar computer industries.
- Oxygen and other gases from lunar gas and pipeline companies.
- Insurance, legal, and financial services from both Earth- and Moon-based companies.

L.5   A SELF-DEVELOPING LUNAR ECONOMY

The Lunar Electric Power Company example could apply equally well to other basic industries, such as oxygen production (see Appendices D and E, ISRU/LUNOX), telecommunications, railroad operations, metal fabrication, and helium-3 mining and processing. As government institutions purchase goods and services from commercial vendors, those vendors will in turn purchase goods and services from one another. The demand for additional goods and services and the potential for making profits will give rise to new lunar markets, and an Earth/Moon economy will evolve (see Chapter 5, Figure 5.5).

L.6   ETHICAL STANDARDS

The major benefit of the lunar government is that it will create and oversee the rule of law for the people and institutions of the Moon (Chapter 9). By providing clarity about what is and is not legal, by protecting human rights (which include property rights), and by upholding the validity of contracts, the government will greatly facilitate the involvement of commercial entities in the exploration, development, and settlement of the Moon. While the rule of law is clearly useful, legality does not encompass all ethical issues for commercial enterprises. Businesses may be able to

---

[4] For the beginning stages of lunar development, it is likely that the lunar government will grant limited-term monopolies for electric power, telephone, and rail transportation services. Monopoly status and government-created markets will assure the initial success (and investor support) of start-up companies. Once a self-sustaining lunar economy has been established, competition (e.g., by creating multiple small companies from a parent company) will be introduced to foster innovation and to keep consumer costs at a minimum. The goal of lunar government will not be to provide favors to commercial enterprises, but to assure the expedient and responsible development of the Moon

[5] Eventually, hundreds, then thousands of tele-operated robots will be working on the Moon, thus creating a demand for Earth-based companies that specialize in tele-operation services.

Table L.1: Code of ethics for off-Earth commerce[6]

1. No principle in this code of ethics will be construed in such a way as to be a cost or regulatory burden upon commercial space activities.

2. Businesses will give consideration to the effects on future generations of all off-Earth development.

3. Business dealings in space and on Earth will be of the highest level of integrity, honesty, fairness, and ethics.

4. Businesses will commit themselves to ensuring a free-market economy off-Earth.

5. Space will be treated with respect, concern, and thoughtful deliberation, regardless of the presence or absence of life forms.

6. Businesses will be responsible stewards of outer space and all its economic resources.

7. Businesses will support the environmental protection of designated areas on the Moon and other celestial bodies, just as there are designated protected areas on Earth.

8. All employees of a business, as well as other people working with the business, agree to be responsible and accountable for the ethical conduct of business activities. Executives, in particular, agree to demonstrate ethical leadership and compliance with this code of ethics.

9. Businesses will be managed with consumer and product safety in mind.

10. Businesses will establish a corporate ethics committee to address issues of an ethical nature and to approve all off-Earth business ventures.

11. Conflicts of interest will be fully disclosed to the ethics committee.

12. Each business will make full and immediate public disclosure of any contributions made to political candidates or organizations. The ethics committee must approve all such contributions.

13. Businesses will work within their industry groups to create legitimate supervisory organizations, either public or private, designed to monitor and support the ethical development of space.

---

operate within the government's framework of laws but still undertake practices that are unethical. Even though it is in the best interests of commercial businesses to manage resources in a manner that benefits present and future generations, and to minimize pollution, some businesses may nevertheless make unethical decisions.

To facilitate the responsible and beneficial development of space resources, a code of ethics for commercial enterprises has been proposed by Dr. David Livingston (see Table L.1). The code of ethics is intended as a guide for businesses that develop

[6] Adapted from Livingstone, David, *A Code of Ethics for Off-Earth Commerce*.

space resources for the benefit of shareholders, customers, and the citizens of the Moon. In addition to commercial entities, the code of ethics should apply to the legislative, executive, and judicial branches of the lunar government. The code of ethics is analogous to quality standards for lawmaking (Appendix G) in the sense that the evolution of a new world for human habitation offers the opportunity to experiment with and establish standards of excellence for the conduct of human affairs.

L.7   SUMMARY

The establishment of a viable lunar economy is a worthwhile goal. The abundance of lunar material and energy resources, combined with scientific opportunities, investors, technological advancements, a reliable and just framework of laws, and a code of ethics, constitute the foundation of a productive and beneficial lunar economy. Lunar businesses and investors will prosper; the lunar government will be self-supporting; and the people of the Earth and the Moon will benefit from the expanding lunar economy that has access to the limitless energy and material resources of space.

L.8   REFERENCES

Benaroya, H. (2000) Commerce at a lunar base. *Proceedings, Space 2000, Albuquerque, NM,* pp. 234–251.

Carrier, W.D. III (1993) Resources exploration: Industry perspective. The franchise model. *Proceedings, Fourth Annual SERC Symposium,* University of Arizona, Tucson, pp. 102–113

Criswell, D.R. (1991) Results of analyses of a lunar-based power system to supply Earth with 20,000 GW of electric power. *Proceedings of SPS '91, Power from Space: The Second International Symposium, Paris,* 11 pp.

Criswell, D.R. (1993) Lunar solar power system and world economic development. *World Solar Summit, UNESCO and UNO,* Paris Solar Energy and Space Report (Chapter 2.5.2), July.

Criswell, D.R. (1993) World energy requirements in the 21st century. *Proceedings of 1st Annual Wireless Power Transmission Conference,San Antonio, TX,* p. 285.

Criswell, D.R. (1994) New growth in the two-planet economy. *Proceedings of the 45th International Astronautical Congress, Paris.*

Criswell, D.R. (1995) Lunar solar power system: Scale and cost versus technology level, bootstrapping, and cost of Earth-to-orbit transport. *46th International Astronautical Federation Congress, Oslo, Norway, October 2–6.*

Criswell, D.R. (1996) World and lunar solar power systems costs. *Proceedings of the Fifth International Conference on Space '96, Albuquerque, NM,* Vol. 1, pp. 293–301.

Gibson, R. (1990) Commercial space activities: An overview. *Space Commerce,* **1,** no. 1, 3–6.

Good, M.L. and Calhoun-Senghor, K. (1997) New space era is here – and it's commercial. *Aviation Week,* p. 90, June 9.

Greenberg, J.S. and Hertzfeld, R., eds. (1992) *Space Economics,* American Institute of Aeronautics and Astronautics, Washington, D.C.

Harris, P.R. (1994) A case for permanent lunar development and investment. *Space Policy*, **10**, no. 3, 187–188.

Harris, P. and O'Donnell, D. (1997) Facilitating a new space market through a lunar economic development authority. *Space Governance*, **4**, no. 2, 122–130, July.

Hayes, W.C., ed. (1980) *Space – New Opportunities for International Ventures*. Univelt, San Diego, CA.

Jones, E.M. (1992) A basis of settlement: Economic foundations of permanent pioneer communities. *2nd Conference on Lunar Bases and Space Activities of the 21st Century*, NASA Conference Publication 3166, Vol. 2, pp. 697–702.

Koelle, H.H. (1993) A frame of reference for extraterrestrial enterprises. *Acta Astronautica*, **29**, nos. 10/11, 735–741.

Livingston, David (2003) *A Code of Ethics for Off-Earth Commerce*, ISDC, Houston, TX.

O'Leary, B., ed. (1994) *Space Industrialization*, Vols. I/II. CRC Press, Boca Raton, FL.

Schmitt, H.H. (1994) Lunar industrialization: How to begin? *Journal of the British Interplanetary Society*, **47**, 527–030.

U.S. Department of Commerce (1990) *Commercial space ventures – Financial perspective*. U.S. Department of Commerce, Washington, D.C., April.

Woodcock, G.R. (1986) Economic potentials for extraterrestrial resources utilization. *37th Congress of the International Astronautical Federation, Innsbruck, Austria*, IAA-86-451.

Appendix M

Lunar mysteries
Bonnie Cooper, Ph.D.

Studies of the Moon have produced a wealth of information to date, but they have also uncovered unusual phenomena that remain unexplained. When the *in-situ* exploration of the Moon resumes, one of the goals will be to solve the mysteries of the Moon.

M.1 LUNAR TRANSIENT PHENOMENA

Lunar transient phenomena (LTPs) are temporary brightenings and darkenings seen on the lunar surface. The first documented sighting of an LTP occurred in 1540, when observers in Worms, Germany, reported that a red star was visible (with the naked eye) on the dark face of the new Moon. It was depicted as a star played between the eyebrows of a bearded face (the "Man in the Moon"), the eyes of which probably represent the large maria, Imbrium and Serenitatis.

The first piece of recorded instrumental evidence was a spectrum taken from the central peak of the crater Alphonsus. This spectrum demonstrated the presence of hydrogen during a lunar transient event there. At least twenty other instrumental records exist, mostly obtained accidentally during observation programs with other aims.

In recent years, very little discussion of LTPs has been seen in the scientific literature. The main reason is that most of these sightings, reported by amateurs, are extremely difficult to verify. The tendency of the science community is sometimes to dismiss these sightings as wishful thinking on the part of overzealous amateur astronomers.

Before the subject became virtually taboo in lunar science circles, two important summaries were published in *Physics of the Earth and Planetary Interiors*, **14**, 1977. The summary papers, by B.M. Middlehurst and W.S. Cameron, gave recapitulations of the reported lunar transient phenomena sightings as of that time.

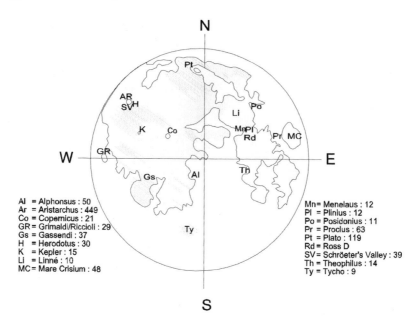

N

W — E

S

Al = Alphonsus : 50	Mn= Menelaus : 12
Ar = Aristarchus : 449	Pl = Plinius : 12
Co = Copernicus : 21	Po = Posidonius : 11
GR= Grimaldi/Riccioli : 29	Pr = Proclus : 63
Gs = Gassendi : 37	Pt = Plato : 119
H = Herodotus : 30	Rd = Ross D
K = Kepler : 15	SV= Schröeter's Valley : 39
Li = Linné : 10	Th = Theophilus : 14
MC= Mare Crisium : 48	Ty = Tycho : 9

Figure M.1. Locations and number of sightings per location for lunar transient phenomena.

They attempted to categorize these findings by locality on the Moon, phase of the Moon, and position of the lunar terminator with respect to the LTP location, among other statistical tests. Figure M.1 shows the near side of the Moon with the locations and number of sightings per location for LTPs, modified from these two papers.

LTPs seem to have a periodic cycle, which suggests that tidal forces can trigger disturbances on the Moon. Periodicities are also seen in deep and shallow moonquakes. Tidal stress changes, which are about 80 times stronger on the Moon than on the Earth, are still probably too small to be the single cause of the events, but they may add their cumulative effect to pre-existing stresses within the Moon. This in turn may cause rock slides, and gas may be released.

Nearly all of the reported LTP sites show features such as hummocky floors, central peaks, and irregular collapse depressions that are likely to be of endogenic origin – in other words, caused by forces within the Moon rather than by meteorite impacts.

Cameron analyzed 1,400 reported LTP sightings. There were 201 sites reported at least once; about half had two or more reports. A dozen sites contain 70 percent of all observations, and one site, Aristarchus, provides about 30 percent of the total (449 reported sightings). Of the dozen most-reported sites, half are rayed craters and half are dark-floored craters. The distribution of sites strongly favors the mare/highlands border areas, and the few reported locations in the highlands correspond to dark, flat areas.

Cameron identified five general categories of LTP: brightenings, darkenings, gaseous, reddish, and bluish events; and four proposed causes: tidal effects, low-illumination thermoluminescence, magnetic-tail effects, and solar-flare effects. Her analyses suggested that different kinds of LTP have different causes. Gaseous and darkening phenomena occur most frequently at apogee, and thus may be tidally related. Reddish LTP have the strongest correlation with sunrise. Sightings at Aristarchus, Plato, Ross D, and all of the bluish phenomena have the strongest correlations with solar-flare activity that produces magnetic storms on Earth. There was also a correlation with the phase of the Moon. The fewest LTP were reported between the New Moon and Full Moon, with a larger number reported from the Full Moon to the following New Moon. Neighboring sites were seen to behave differently; for example, observations in Aristarchus do not correspond with observations in Herodotus, an older crater which is adjacent to Aristarchus.

As a result of the above observations, Cameron concluded that tidal effects could be the cause of only a very few LTPs, with gaseous and darkening LTPs having the strongest tidal correlation. Almost all the locations and kinds of LTP had a correlation with lunar sunrise/sunset. Most categories and features had strong peaks when the Moon was in the Earth's magnetopause, and there was a surprisingly high percentage of observations that correlated with terrestrial magnetic storms.

M.2 LUNAR HORIZON GLOW

The Surveyor 1 spacecraft that landed on the Moon in 1966 photographed the western horizon shortly after sunset. The images showed an unexpected phenomenon: a glowing light was seen along the horizon, almost as if an atmosphere were scattering the vanished sunlight across the horizon toward the observer. Because it was well known that the Moon had no atmosphere, the images were assumed to be the result of an aberration in the camera system (Norton *et al.*, 1967). However, the same horizon glow was seen on Surveyor 5 (Batson *et al.*, 1974).

Perhaps as a result of this second observation, Surveyors 6 and 7 were specifically programmed to observe the post-sunset horizon. Shoemaker *et al.* (1968) were the first to actually identify lunar horizon glow as a real phenomenon. Figure M.2 shows a Surveyor 7 image, taken about 1 hour after local sunset. It clearly shows the bright, linear nature of the horizon glow. Something was occurring that had not been predicted.

The Surveyor images provided the impetus for further study of the lunar horizon glow, and attempts during the Apollo missions to image the effect met with varying degrees of success. Photography using 35-mm and 70-mm cameras from lunar orbit failed to capture any trace of the horizon glow (MacQueen *et al.*, 1973; Mercer *et al.*, 1973). However, the Apollo 17 astronauts all observed and made sketches of the lunar horizon glow, as they orbited the Moon. Figure M.3 shows the sketches made by astronaut Gene Cernan over the course of about six minutes, while watching "orbital sunrise" from the Command Module.

In 1972, an explanation for this phenomenon was offered by David Criswell, then at the Lunar Science Institute. According to Criswell, the horizon glow could be explained by electrostatic transportation of dust grains being levitated 3 to 30 cm

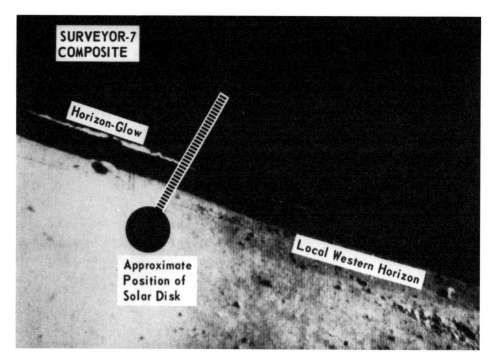

Figure M.2. Two Surveyor images of the lunar horizon, superposed and offset vertically to show the glow that followed the horizon after lunar sunset.

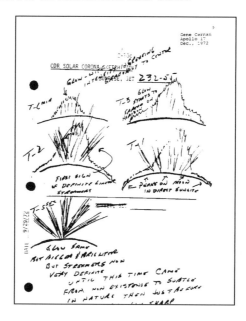

Figure M.3. Lunar horizon glow as sketched by Apollo 17 astronaut Gene Cernan.

above rocks or surface irregularities in the lunar terminator[1] zone. The extreme ultraviolet and soft X-rays from the Sun might be responsible for the high-voltage electrostatic charging that causes the dust to levitate. The charging would occur on the daylight portion, causing the dust to levitate, and the absence of charging on the night portion of the surface would create large electrical potentials that would cause the dust to stay suspended for some time (Criswell, 1972, 1973). There was supporting evidence, from the Apollo 17 surface experiments package, that the amount of dust in the near-surface environment did increase during the passage of the morning and evening terminators (Berg *et al.*, 1976).

Zook and McCoy (1991) published a new analysis of the lunar horizon glow, based on astronaut sketches from the Apollo 17 mission. They showed that the glow was likely to be caused by sunlight reflected by dust grains, and that the effect was probably not caused by gases in the lunar exosphere, as some had supposed. Furthermore, they calculated the scale height of the lunar dust exosphere at about 10 km, and asserted that the dust grains that are charged and electrically ejected from the lunar surface must be small (less than 0.1 m in radius), as advocated by Criswell.

Further attempts to observe the lunar horizon glow from Earth-based telescopes met with only limited success. For now, lunar horizon glow remains somewhat of a mystery. Does it occur all over the planet or only in certain areas? Does it occur throughout the year or only at certain times of year? Perhaps when a permanent observation facility is located on the Moon, these questions can be answered.

M.3   MYSTERY OF THE RUSTY ROCKS

One of the unsolved mysteries of the Apollo lunar sample collection is the origin of the Apollo 16 "rusty rocks". These rocks were uncovered from a depth of about 2 meters, and returned to Earth as part of the "deep drill core" section. These were obtained by the use of a manual coring device. Upon first opening the core, scientists noticed a rusty red "halo" around two particles in particular. Particles such as these have been dubbed the "rusty rocks", and there is still some controversy about their origins.

Because water is so scarce on the Moon, it seemed unlikely that the rust would have formed while the samples were in the lunar subsurface. Instead, most people thought that the samples were contaminated during the return trip or during subsequent handling.

The Apollo 16 deep drill core penetrated approximately 2.2 meters into the lunar subsurface at station 10, which was about 105 m southwest of the Lunar Module. The core, diagrammed in Figure M.5, has four major lithologic[2] units: Unit A, at the base of the core (190–224 cm), has abundant rocks and yellow glass; Unit B (55–190 cm) has fewer rocks and an abundance of green glass; Unit C (6–55 cm) contains

---

[1] The terminator is the line separating light from dark on the disk of the Moon or other planetary body.

[2] Lithologic: literally translated, "rock-study". Adjective of lithology, which means the description of rocks, especially on the basis of color, mineralogic composition, and grain size.

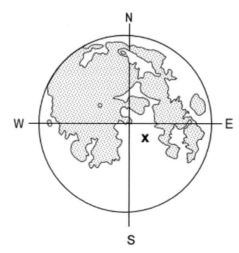

Figure M.4. Location of the Apollo 16 landing site, where the "rusty rock" was found.

many fragments of the mineral plagioclase, and lacks orange glass; Unit D (0–6 cm) has fewer rocks and fewer glass particles than any of the rest of the core. Each of these sections may represent different depositional events on the lunar surface – different cratering events that tossed rocks and soils from other areas and deposited them here.

Most of the core consists of highlands-derived materials, which is not surprising because the Apollo 16 landing site is in a highlands region (see Figure M.4 for location). The components that are not highlands material are restricted to glass droplets. All of the Apollo drill cores were retrieved in sections, and each section of the core has a unique sample number. The bit section has the smallest number, then the numbers increase going upward in the drill stem. In this case the bit section is sample #60001, and the core sections are numbered 60002 through 60007. The only part of the Apollo 16 deep drill core that contains the rusty rocks is section 60002. It is the lowermost section of the drill core, excluding the bit section.

There were two samples that had rust on them at first examination (see Figure M.5). These are particles which were retrieved from the base of the section as a solid clod (60002,178), and at the center of the section as a discrete metal sphere (60002,108). These particles came from depths below the lunar surface of approximately 216 cm and 200 cm, respectively, and they are part of Unit A. The interface between Unit A and Unit B is probably an ancient regolith[3] surface. The oxidation results from the interaction of water with the mineral lawrencite ($FeCl_2$), which is found in meteorites (Taylor, 1974).

Lawrencite is very unstable under terrestrial conditions and it oxidizes rapidly

---

[3] Regolith: fragmental and unconsolidated rock material that overlies bedrock. The particles range in size from microscopic to blocks more than a meter in diameter. It is formed by repeated meteoritic impact over a long period of time.

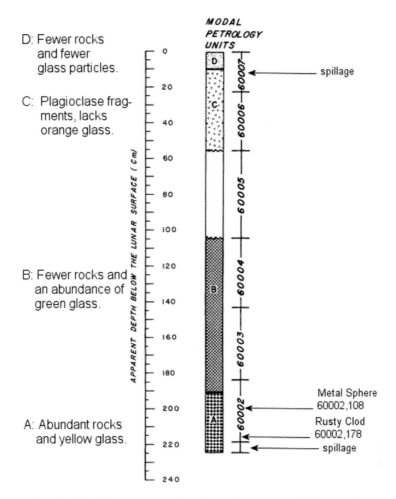

Figure M.5. Core Section Diagram showing where the "rusty rocks" were found. Spillage occurred in sections 60001 and 60007, but no rust was seen there. The locations of the two samples that had oxidation, 60002,178 and 60002, are shown.

into the mineral akaganéite (FeO·OH), which is unstable under lunar conditions (Taylor, 1974). It would seem natural that lawrencite would be transformed into akaganéite only when the sample was transferred from the Moon to the Earth.

A study published by Haggerty (1978) showed that there is more than just akaganéite present in the rust. There is a stratified layering, from the metal substrate to the outermost surface of the particle, which may include several types of minerals. Carter (1972) also noted that the crystalline akaganéite is not in direct contact with the metallic Fe–Ni particle. This suggests that the formation of akaganéite is at least a two-stage or even perhaps a multi-stage process.

Several factors argue in favor of reopening the question of small quantities of lunar water, based on the oxidized particles in the Apollo 16 deep drill:

1. The metal spherule shows little evidence of surface exposure in the form of micro-craters. If it was rapidly buried, then any akaganéite that might have formed would be buried in the subsurface where it might be somewhat more stable than it would have been if left exposed on the lunar surface.

2. The variety of distinctive and stratified minerals that were identified by scanning electron microscope suggests that the oxyhydrate (rusty material) on particle 60002,108 is not entirely akaganéite, but a complex mixture of microcrystalline minerals that appear to be well ordered with respect to the nickel–iron substrate. This again argues that the instability of akaganéite may not apply to the entire assemblage. An oxygen isotope study suggested that at least some akaganéite samples from the Apollo 16 site are contaminated by terrestrial water. However, it is not clear that this contamination can account for all of the rust.

3. Although the drill string was not returned to Earth in a vacuum container, the rusty particles appear at an apparent depth of 2 meters or more from the surface, and they occur in a densely compacted horizon. Dissection of the cores took place in dry-nitrogen cabinets at the curatorial facility; nevertheless, the rusty particles were observed to be oxidized, with a surrounding red halo, at the instant that these areas were uncovered. Some other parts of the core (60001 and 60007) suffered from a spillage accident. It is noteworthy that the spilled sections, which were more exposed to contamination, have not been characterized as containing "rusty" particles.

4. There are a large number of metal particles elsewhere in the drill core that are not oxidized. Lawrencite or a chlorine-rich component may be necessary to trigger the onset of rust. However, it is also possible that oxyhydration took place only in specific soil horizons, because these horizons were relatively enriched in small concentrations of water vapor. The oxidized metal particles are restricted to the lowermost horizon, which is radically different from the remainder of the core. This basal unit is coarse-grained and immature, whereas the units above it (B through D) are more mature and of finer grain size.

5. Undoubtedly the strongest point is that lawrencite was identified in the sample by Carter (1975) three years after the core was returned from the Moon. Lawrencite is highly reactive in the presence of water vapor. Considering the large concentrations of water vapor that are potentially available during the return from the Moon and storage on Earth, all of the lawrencite should have reacted to completion if contamination was serious enough to create any of the observed rust in the first place.

6. Splashes of glass covered some of this oxidized material, which could only happen while the material was exposed on the lunar surface, such that it could experience micrometeorite bombardment. Thus, some oxidation must have occurred while the particles were exposed to the surface, 1 billion years ago.

Carter and McKay (1972) proposed that small amounts of water can be produced on the Moon by reduction of metal oxides, based on their observations of metallic mounds on lunar glass beads. They found that metallic mounds could be produced in the laboratory by passing either heated hydrogen or carbon gas over the glass samples. The reduction reaction removes oxygen from iron oxide compounds, and if hydrogen gas is used as the reducing agent, steam, H_2O, is produced:

$$FeO(glass) + H_2(gas) \Rightarrow Fe(metal) + H_2O(gas).$$

Carter and McKay (1972) concluded that hydrogen and carbon could have acted as reducing agents to create metallic mounds on glass beads. Hydrogen and carbon are present in the solar wind, and would therefore be present in small amounts in the lunar regolith. Therefore, solar-wind hydrogen reduction could have been the source of the water in the lunar rusty rocks. It is also possible that a comet impacted the lunar surface and that its icy debris, spreading out from the impact site, was in contact with some lunar particles. That would also account for the sporadic sightings of rusty material.

Keller and McKay (1993) demonstrated that the lunar soil is much richer in volatiles than was previously supposed. It contains micrometer-sized mineral grains surrounded by thin glassy rims. For many years, it was thought that these rims were the result of radiation damage. Electron microscope studies have now shown that the glassy rims are compositionally distinct from the host mineral grains, and that the rims consist largely of vapor-deposited material. This material was probably generated by micrometeorite impacts into the lunar regolith.

The presence of these vapor-deposited rims suggests that there is a much greater abundance of some volatiles in the lunar soil than was previously supposed. Although these volatiles are mostly gaseous sodium, potassium, and sulfur, it now seems more likely that oxygen or even minute amounts of free water could have existed on the Moon, and that oxyhydration could have occurred there. Still, no one knows for sure how the rusty particles were formed. The mystery of the rusty rocks remains a mystery after all these years.

M.4 MYSTERY OF THE REINER GAMMA MAGNETIC ANOMALY

The Moon's total magnetic field is extremely weak – about 0.1 percent that of the Earth. However, there are small areas on the Moon with much stronger magnetic fields. These include the Reiner Gamma region, on the near side (see Figure M.6), and a region near Van de Graaff Crater on the far side. At Reiner Gamma, the anomaly is associated with a peculiar pattern of light- and dark-colored swirls.

One model has been proposed to explain the higher magnetism: as the magma ocean cooled after the Moon was formed,[4] and while the crust was still only a few kilometers thick, in-falling material could have penetrated it, exposing the magma beneath and forming many lava-filled basins. Two lava basins, with different surface temperatures, might have been connected beneath the surface by magma. This

[4] It is hypothesized that early in the Moon's history the heat generated by the formation of the Moon was sufficient to melt the entire surface of the Moon. As the surface liquid cooled, some types of minerals solidified more quickly than others, and these either rose to the surface or sank to the bottom, depending on their relative density.

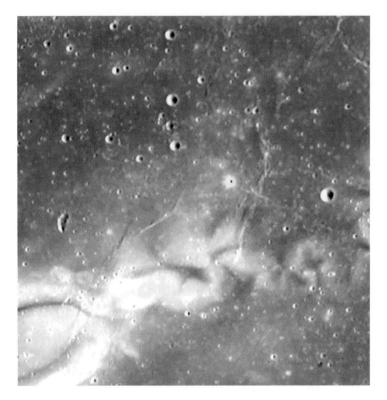

Figure M.6. The Reiner Gamma magnetic anomaly with a light-colored swirl patter. Image courtesy of European Space Agency SAMRT-1 mission.

configuration has the basic elements of a thermoelectric circuit: two dissimilar conductors joined at two junctions that are at different temperatures. Fields generated by this model are large enough to account for the magnetism in most of the returned lunar samples. The thermoelectric mechanism is compatible with the high degree of inhomogeneity[5] found in measured remnant magnetic fields and with the absence of a measurable net global magnetic moment.

Hood and Williams (1991) have proposed a different mechanism for the Reiner Gamma anomaly, as well as for the other anomalies that have the two-tone swirl pattern. These areas tend to be located antipodal to relatively young, large impact basins such as Imbrium, Orientale, Serenitatis, and Crisium. Perhaps both the magnetic anomalies and the swirl patterns originated because of compression of the local magnetic field by an ionized vapor cloud at the antipode of these impacts.

Other models have also been proposed; thus, at the present time the final answer is still unknown. A good radar map of the subsurface in this area might help to determine whether the anomaly is associated with any buried structures.

---

[5] Inhomogeneity: a state of being poorly mixed; lack of homogeneity.

M.5   SUMMARY

There is still much to learn about the Moon. When humans return there on a permanent basis, areas where Lunar Transient Phenomena have been reported can be observed constantly with cameras and spectrometers to learn what may be causing them. We can learn whether lunar horizon glow is stronger at the equator than at the poles, or if geography affects how they are seen. By examining many more lunar samples from widely dispersed areas, we may find other rusty rocks, and thus determine that small pockets of water have indeed existed on the Moon in the past. We can explore Reiner Gamma, and obtain detailed maps of the magnetic fields, subsurface structures, and topography to get clues about its history. We may even find new mysteries to solve that push our science beyond its present bounds. We have much to look forward to.

M.6   REFERENCES

Batson, R.M.; Jordan, R.; and Larson, K.B. (1974) *Atlas of Surveyor 5 Television Data*. NASA SP-341, National Aeronautics and Space Administration, Washington, D.C.

Berg, O.E.; Wolf, H.; and Rhee, J. (1976) Lunar soil movement registered by the Apollo 17 Cosmic Dust Experiment. In: H. Elsasser and H. Fechtig, eds., *Interplanetary Dust and Zodiacal Light*, Springer-Verlag, New York, pp. 233–237.

Cameron, W.S. (1972) Comparative Analyses of Observations of Lunar Transient Phenomena. *Icarus*, vol. 16, p. 339.

Carter, J.L. (1975) Surface morphology and chemistry of rusty particle 60002,108. *Proc. Lunar Sci. Conf. 6th*, Lunar Science Institute, Houston, TX, pp. 711–718.

Carter, J.L. and McKay, D.S. (1972) Metallic mounds produced by reduction of material of simulated lunar composition and implications on the origin of metallic mounds on lunar glasses. *Proc. 3rd Lunar Sci Conf.* (Supplement 3, *Geochim Cosmochim. Acta*, **1**), MIT Press, Cambridge, MA, pp. 953–970.

Criswell, D.R. (1972) Lunar dust motion. In: *Proc. 3rd Lunar Sci. Conf.*, MIT Press, Cambridge, MA, pp. 2671–2680.

Criswell, D.R. (1973) Horizon-glow and the motion of lunar dust. In: *Photon and Particle Interactions with Surfaces in Space*, D. Reidel Publishing Company, Dordrecht, Holland, pp. 545–556.

Haggerty, S.E. (1978) Apollo 16 deep drill core: A review of the morphological characteristics of oxyhydrates on rusty particle 60002,108, determined by SEM. *Proc. Lunar Planet Sci. Conf. 9th*, Lunar and Planetary Institute, Houston, TX, pp. 1861–1874.

Hood, L.L. and Williams, C.R. (1991) Lunar swirls: Distribution and possible origins. *Proc. 19th Lunar and Planet. Sci. Conf.*, Cambridge University Press and Lunar and Planetary Institute, Houston, TX.

Keller, L. and McKay, D. (1993) Discovery of vapor deposits in the lunar regolith. *Science*, vol. 261, no. 5126, pp. 1305–1307.

MacQueen, R.M.; Ross, C.L.; and Mattingly, T.K. (1973) Observations from space of the solar corona/inner zodiacal light. *Planet. Space Sci.*, **21**, 2173–2179.

Mercer, R.D.; Dunkelman, L.; and Evans, R.E. (1973) Zodiacal light photography. In: *Apollo 17 Preliminary Science Report*, NASA SP-330.

Middlehurst, B.M. and Moore, P.A. (1967) Lunar Transient Phenomena: Topographical Distribution. *Science*, vol. 155, issue 3761, pp. 449–451.

Norton, R.H.; Gunn, J.E.; Livingston, W.C.; Newkirk, G.A.; and Zirin, H (1967) Surveyor 1 observations of the solar corona. *Journal of Geophysical Research*, **72**, no. 2, 815–817.

Shoemaker, E.M.; Batson, R.M.; Holt, H.E.; Morris, E.C.; Rennilson, J.J.; and Whitaker E.A. (1968) III. *Television Observations from Surveyor VII*. JPL Technical Report 32-1264, NASA Jet Propulsion Laboratory, Pasadena, CA, pp. 9–76.

Taylor, L.A.; Mao, H.K.; and Bell, P.M. (1974) FeOOH, akaganéite, in lunar rocks. *Proc. 5th Lunar Conference* (Supplement 5, *Geochim et Cosmochim. Acta*, **1**), pp. 743–748.

Vaniman, D.T. and Heiken, G.H. (1990) Getting lunar ilmenite: From soils or rocks? In: S.W. Johnson and J.P. Wetzel, eds., *Engineering, Construction and Operations in Space, Proceedings of Space '90*, American Society of Civil Engineers, New York.

Vaniman, D.; Pettit, D.; and Heiken, G. (1988) Uses of lunar sulfur (Preprint Abstract). *Proc. 2nd symposium on Lunar Bases and Space Activities of the 21st Century*, Lunar and Planetary Institute, Houston, TX..

Zook, H.A. and J.E. McCoy (1991) Large scale lunar horizon glow and a high altitude lunar dust exosphere. *Geophys. Res. Lett.*, **18**, no. 11, 2117–2120.

Appendix N

Milestones of lunar development

1   Establishment of the first permanent unmanned lunar outpost.

2   Production of the first solar cell on the Moon.

3   Production of oxygen and metals from the lunar regolith.

4   Production of first lenses and mirrors.

5   First person to stand at the south/north pole.

6   Operation of first telescope on the Moon.

7   Establishment of lunar repository of human knowledge.

8   Formation of a lunar government.

9   Power level in the lunar electric grid reaches one megawatt.

10   Production of the first rail segment for the lunar railroad.

11   Production of first lunar food crop, seed to seed.

12   Permanent human habitation of the Moon.

13   First generation of animals born on the Moon.

14   Completion of first circumferential utilities infrastructure (i.e., driving of the "golden spike" to close the first power, communications, and transportation loop around the Moon).

15   Operation of first interferometer telescope on the Moon.

16   First person to circumnavigate the lunar surface at the equator.

17   First mass driver operation; launch of the first spacecraft from the Moon.

18   First baby born on the Moon.

19   Lunar population reaches 100 … 1,000 … 1,000,000.

Appendix O

International Lunar Observatory/Association[1]

O.1 INTRODUCTION

The International Lunar Observatory (ILO) is a multi-national, multi-wavelength astrophysical observatory, power station, and communications center that is planned to be operational near the south pole of the lunar surface as early as 2009 (see color plate section).

The International Lunar Observatory/Association (ILO/A) is the organization that supports the ILO and its follow-on missions through the timely, efficient, and responsible utilization of the human, material, and financial resources of spacefaring nations, enterprises, and individuals who are stakeholders in this enterprise.

O.2 HISTORY

The ILO mission was conceived during the *International Lunar Conference 2003* – the fifth meeting of the International Lunar Exploration Working Group (ILEWG). Lunar scientists, entrepreneurs, policy makers, advocates and others gathered to discuss the next step in human exploration of the Moon. The Hawaii Moon Declaration, a one-page *Ad Astra per Luna* manifesto, written and signed by conference participants. Soon after, the ILO mission began to take shape.

[1] This appendix is reproduced by permission of International Lunar Observatory Association/Space Age Publishing Company/Lunar Enterprise Corporation (Hawaii Office, 65-1230 Mamalahoa Highway, D-20480 Kamuela, Hawaii 96743, U.S.A., phone 808-885-3473, fax 808-885-3475, *www.iloa.org*; California Office, California Ave. #303, Palo Alto, CA 94306, phone 650-324-3705, fax 650-324-3716, *www.spaceagepub.com*).

An ILO Advisory Committee was established in 2005, consisting of about fifty supporters from the international science, commerce, and space agency communities. The Committee convened for the ILO Workshop in November 2005 on Hawaii Island to discuss the organizational, scientific/technical, and financial/legal direction of the ILO.

O.3 INTERNATIONAL, COMMERCIAL, AND INDIVIDUAL SUPPORT

Pivotal to the success of the ILO is the cooperation and support of the world's major spacefaring powers, most notably the U.S.A., Canada, China, India, Italy, Japan, and Russia. Fundamental to ILO advancement are the private and commercial enterprises – which catalyze, organize, and help attract the support of nations. The impetus behind a unified international mission is to engage the human and material resources of these nations, enterprises, and individuals into a pioneering and peaceful mission that benefits all of humanity.

O.4 CURRENT PROGRESS WITH NATIONS

O.4.1 Hawaii

The formation of the ILO/A in 2006 has initiated multiple outreaches, strategies, and communications with supporters of the ILO. Relationships with Mauna Kea Support Services, Canada France Hawaii Telescope (CFHT), West Hawaii Astronomy Club, the Onizuka Space Center and the Hawaii Island Space Exploration Society have been established.

O.4.2 Canada

The ILO was presented to the Canada Astronomical Society (CASCA) in Calgary in June 2006, as well as the *Canada Aeronautics Space Institute and the International Institute for Space Law Conference* at McGill University in Montreal in June 2006. The ILO/A has neighborly relations with CFHT, and allies in the Canada Space Agency.

O.4.3 China

For several years, the ILO/A has shared a strong rapport and mutual support with chief Chinese lunar scientist Ouyang Ziyuan, who has explicitly advocated a telescope on China's first lunar-lander Chang'e-2 mission. An ILO astronomy MOU was established with National Astronomical Observatories R&D Planning and Funding Director Suijian Xue in April. Additional rapport and ILO interest exists with the Shanghai Astronomical Observatories, the Chinese Academy of Sciences, the China National Space Administration and the Chinese Society of Astronautics.

O.4.4   India

The ILO has direct support within the Indian Space Research Organization (ISRO) and the Indian Institute of Astrophysics, most notably with ISRO PRL Council Chairman and cosmologist U.R. Rao, lunar scientist Narendra Bhandari, and current ISRO Chair Madhavan Nair.

O.4.5   Italy

Italy was the site of the *2005 Moonbase Symposium*, which received widespread Italian industry and academia support. Italy is also the site of the *2007 ILEWG Conference*. The Italian Space Agency (ASI) has established agreements with China and Germany on lunar rover mission planning.

O.4.6   Japan

Support for ILO efforts exists within the Japan Aerospace Exploration Association, notably with lunar scientists Kohtaro Matsumoto, Susumu Sasaki, Yoshisada Takizawa, and Hitoshi Mizutani.

O.4.7   Russia

The ILO/A has a rapport with prominent Russian lunar scientists Vlacheslav Shevchenko, Vladislav Ivashkin, and Erik Galimov.

O.5   FINANCING

A study performed by ILO prime contractor SpaceDev states that the first ILO mission can be launched, landed and operated within a modest budget of under U.S.$50 million. The strategy to fund this mission includes securing $10 million in investments (possibly at $2 million a year for five years) from various supportive government entities. These entities include the science and space agencies/institutions of Hawaii (U.S.A.), Canada, China, India, Europe, Japan, and Russia. Commercial sponsorship from high-tech companies such as Google, Cisco, Yahoo, and others is also sought, as well as endowments from philanthropic groups and individuals.

O.6   PRESUMED FACTS

Mons Malapert is located about 122 kilometers from the lunar south pole, and the adjacent mare plain just north of the 4.6-km-high mountain is the intended landing region for the ILO. Near-constant sunlight (thought to be 89 percent full, 4 percent partial) provides an energy-rich environment, and the lunar peak enjoys a continuous line of sight to Earth and direct Earth–Moon communications. The mountain dominates its surrounding area giving an excellent vantage point and is near enough to putative water ice deposits – which (if they exist) could be utilized for oxygen, drinking water, and rocket fuel – around the lunar south pole (Shackleton Crater, Aitken Basin). Given these factors, Mons Malapert is considered to be the most suitable location for the ILO to conduct astronomy and to catalyze commercial lunar development and human lunar base build-out.

In 2003 the Lunar Enterprise Corporation (LEC) hired SpaceDev of Poway, CA to serve as the prime contractor for the ILO. A Phase A feasibility study conducted by SpaceDev concluded that "it is possible to design and carry out a private commercial lunar landing mission within the next several years." The Phase B study, conducted the following year, recommended researching a safe and accurate lunar-landing navigation system that can deliver the ILO to a "Peak of Eternal Light". Also concluded was the possibility of landing the ILO at a specific target within an accuracy of about 100 meters using "currently available commercial technology". The ILO will utilize a mix of leading-edge propulsion, inertial navigation, and celestial navigation, together with established Earth-based deep-space tracking, to achieve the required accuracy.

O.7 RICHARDS' MASTER PLAN

An International Lunar Observatory/Association (ILO/A) master plan was completed in February 2006 by Optech Space Division Director and ISU co-founder Bob Richards. The plan outlines how to build the ILO/A as a science, organizational, and commercial entity that operates within the scope of investor markets, equity players, management, industry, and customers. Key players include the Space Age Publishing Company, Lunar Enterprise Corporation, Space-X (of Europe), the Canadian Space Agency, the University of Hawaii, and countries such as China, India, and Russia. International MOUs would be established between the ILO/A and the key players, exchanging science data and commercial communications for principal funding.

The ILO will serve as a multi-wavelength astrophysical observatory that will utilize VLF, millimeter, sub-millimeter, and optical wavelengths. The scientific objectives of the ILO include imaging the galactic center; analyzing interstellar molecules to determine the origin of the solar system; searching for NEOs and Earth-like planets; Earth observations; planetary and solar observations; and searching for extraterrestrial intelligence.

The ILO has the potential for interferometrical build-out, much like the Harvard–Smithsonian Sub-millimeter Array on Mauna Kea that operates the first such observatory (eight mobile antennas at six meters each operating interferometrically). Another example of cutting-edge interferometry is the 64-dish Atacama Large Millimeter Array being constructed at a 5,000-m altitude in Chile's ultra-dry high desert. Sub-millimeter and millimeter astronomy is forefront, cutting-edge science. Interferometry, operating in the 0.25 to 1.7 mm wavelengths, offers a multitude of beneficial applications.

O.8   ILO FEATURES AND BENEFITS

The ILO/A strives to pioneer and innovate several niches of human science and technological endeavor. For the benefit of science, commerce, and humanity, the ILO will serve as:

- An astrophysics facility: unique and pioneering for the next frontier of astronomy, with special emphasis on advancing Hawaii's twenty-first century astronomical leadership.
- A power station: a solar device will be required to supply power to the ILO and its instruments, and may offer additional capacities for other energy-related functions.
- A communications center: transmission of astrophysical data; Space Age Publishing Company's (SPC) *Lunar Enterprise Daily* flagship publication; commercial communications, broadcasting, advertising, and imaging. A series of lunar commercial communications workshops are being conducted in Silicon Valley through the support of the SPC California office.
- A lunar property rights precedent: automatically raises the question of "who owns the Moon?"
- A site characterizer: to gather and report data of the surrounding area, including solar-wind and radiation measurements, temperature, altitude, and seismic and meteoric activity.
- A toe-hold for lunar base build-out: to serve as a precursor to future missions.
- A virtual nexus dynamic website: developing a website that delivers real-time astrophysical data, lunar video, earthrise imagery, broadcast communications and other viable information to institutions, popular media, schools and the general public.

Appendix P

Cislunar orbital environment maintenance

Abdullah Alangari, Adrienne Catone, Divya Chander, Andrey Glebov, Matthew Marshall, Michael Nolan, Brian Phail, Allen Ruilova, and Madhu Thangavelu   Conductor, Space Exploration Architectures Concepts Synthesis Studio[1]

This project was conducted in the AE599 Space Exploration Architectures Concepts Synthesis Studio, Department of Aerospace Engineering, University of Southern California, in the Fall semester of 1995. These conceptual ideas were created employing a unique architectural approach to complex engineering concepts generation. Eight participants from the industry as well as the school were exposed to poignant scientific and engineering concepts in space exploration, and lectures by visiting experts. They were instructed in rapid concept generation and visualization methods. The resulting ideas were presented to a panel of experts, who reviewed their work and provided feedback. This paper is the combined result of that activity over a period of fifteen weeks which included forty-five hours of meetings that were equally split among instruction in complex systems concepts generation, expert lectures in space exploration concepts and space systems, and debate and discussion. Thanks are due the members of the expert panel and visiting lecturers who included Dr. R.F. Brodsky, Dr. M. Gruntman, Dr. G. Friedman, Dr. L. Friedman, Dr. M. Cohen, Bill Haynes, G. Noreen, R. Ridenoure, and D. McCall.

Special thanks are due to Prof. E.P. Muntz, Chairman of Aerospace Engineering for nurturing this course, Prof. Elliot Axelband, Associate Dean for Engineering Research for assisting in the curriculum development, Prof. Eberhardt Rechtin for his global support and feedback in creating the conceptual and operational framework for this course. Thanks are also due to Prof. G.G. Schierle and M. Schiler from the School of Architecture for their pedagogic guidance in creating the curriculum and to

---

[1] Team Project, Fall 1995, Space Exploration Architectures Concepts Synthesis Studio, Department of Aerospace Engineering, University of Southern California, Los Angeles, CA 90089, 1–8.

the enthusiastic team of participants who were a delight to instruct, work, and debate with.

P.1   ABSTRACT

This project depicts an evolving vision of an Earth orbital facility that would gradually build up our capability to repair and service satellites in distress. Starting with orbital debris mitigation operations and missions such as placing spacecraft in their intended orbits, station relocation, and plane change operations, or safely decommissioning aging and useless stations in orbit, the proposed orbital infrastructure will grow to be able to handle tasks such as the specific replacement of malfunctioning or degraded subsystems aboard sub-optimally-operating satellites in order to enhance their performance and life expectancy as well as evolve their capability, using a modular strategy. Preliminary studies are presented as to the nature of the projected Earth orbital environment by the turn of the century and expected mission opportunities for this architecture are explored. Alternative concepts for satellite repair facility configurations are depicted and related issues regarding the maintenance and evolution of such an infrastructure are discussed. Potential projects stemming from this open-ended architecture are also portrayed.

P.2   INTRODUCTION

The Earth orbital regime, stretching from the low Earth orbit (LEO) to the geo-synchronous orbit (GEO), is already a rather crowded environment, with over 8,000 operational, aging, and decommissioned spacecraft and related deployment hardware objects. Less than 10 percent of this number are functional stations. Artificial debris arising from the degradation of discarded hardware are threatening the safety of operations in this environment. Projecting into the next decade and into the new millennium, several planned constellations of commercial communication satellites in low and medium Earth orbits (MEO) will most definitely increase the traffic in the LEO to MEO Earth orbital environment as well.

Extrapolating from past experience, we can expect to see a proportional increase in spacecraft failures and anomalies during all phases of deployment, including launch to orbit, orbital transfer maneuvers, orbit changes, station-keeping operations, and decommissioning procedures.

In the context of increased activity in this regime, it may be prudent and economically feasible now to develop and establish a permanent space-based infrastructure in order to repair, service, and evolve spacecraft in a systematic effort to enhance performance, as well as to maintain the order and stability of these stations in the increasingly-crowded Earth orbital environment.

P.3   EARTH ORBITAL ENVIRONMENT 2001

Looking at the present trend, as well as the unprecedented applications blitz at the International Telemetering Union (ITU) for satellite licenses, we speculate ever-increasing traffic in the Earth orbital environment for the next decade. Commercial telecommunications and defense surveillance satellites designed to deliver a plethora of ingenious new services will dominate the arena. Unless entirely-radical and unforeseen methods of Earth-to-orbit (ETO) deployment are employed, we will continue to see a rise in the number of failures due to normal accidents during deployment and other phases of satellite operations. In addition, a large number of stations in the so-called "Clarke belt" along the geostationary orbit will reach their end of life by the end of the coming decade, thereby becoming potential targets for additional debris production. Consequently, unless prudent measures are taken now, we will most definitely see an even more polluted and dangerous orbital debris environment all the way from low Earth to medium and geosynchronous orbits and beyond.

Though ETO insurance premiums might become more and more affordable to users due to the increased activity in this business, the monetary aspects of deployment and operations do not suggest any long-term beneficial effect at all on this orbital debris scenario. In fact, all indications are to the contrary. Affordable premiums and the like are no substitute for an on-schedule, healthy, operating satellite, because time lost in mounting another campaign (two to three years) can leave a service provider without a viable business in this fiercely-competitive industry.

Unless innovative new systems are put in place, even as we plan to exploit the benefits of high-bandwidth global personal communications and a host of other applications, we can expect to see substantial impairment of station performance on a global scale, and, consequently, an eventual deterioration in the quality of ubiquitous services provided by stations located in the Earth orbital regime. Architectures are needed that will continually monitor, assist, correct, repair, remove, and dispense of stations and debris in a manner that will enhance traffic in this regime. Several recent studies by NASA and contractors suggest the need for this capability. The Orbital Maneuvering Vehicle (OMV) proposed some time ago by NASA falls into this category. More recently, the European Space Agency (ESA) has also conducted a study for a satellite for on-orbit repair and maintenance of stations. The proposed Satellite Service Facility (SSF) is envisaged as such a system architecture, which draws on the latest developments in technology to achieve its goals.

P.4   ARCHITECTURAL ELEMENTS OF THE SATELLITE SERVICE FACILITY (SSF)

The elements that make up the SSF architecture are as follows:

1. The Space Tug Mk-1 (ST-1). The ST-1 is a basic Phase 1 inter-orbital maneuvering vehicle with a protected payload bay. Otherwise costly plane changes are done using a unique aerodynamic assist maneuver, coupled with a high specific-impulse

solar-electric propulsion system. Its primary mission is to boost incorrectly-injected or stranded spacecraft to prescribed station orbits. Its secondary mission is to rendezvous payload satellites with the STS (Space Transportation System) for manned intervention or return to Earth, as well as to return repaired payloads to desired orbits. Deorbiting and safely disposing of large decommissioned objects (like spent stages and other spacecraft deployment articles) through controlled atmospheric ablation and disintegration is also part of ST-1 mission operations.

2. The Space Tug Mk-2 (ST-2). The ST-2 is a more sophisticated system that is evolved as Phase 2 of the SSF architecture. This larger vehicle is capable of all ST-1 missions, as well as on-site repair and maintenance of stations in all orbits. The ST-2 toolkit offers on-the-spot diagnostics, orbital replaceable units, and a high bit rate communications link that allows tele-operation of a highly-dexterous set of robotic manipulator systems (RMS).

Both ST-1 and ST-2 are capable of large plane changes, employing a unique, detachable aerodynamic plane change assist shell that protects all subsystems and the payload from the extreme thermal loading during these maneuvers. All exposed systems including solar arrays, communication system antennas, and observation cameras, are designed to fit within or retract into the shell during the plane change operations. The results from the scheduled Aeroassist Flight Experiment (AFE) scheduled to fly soon aboard the NASA space shuttle, would be used to optimize the shell design. Data from the recently-retrieved Long-Duration Exposure Facility (LDEF), as well as the low-cost Advanced Photovoltaic and Electronics Experiment (APEX), will be used in the selection of materials and system components for the synthesis of this vehicle.

Both ST-1 and ST-2 have similar spacecraft buses in that they share the same kind of photovoltaic array power system with back-up batteries, and use a modular hybrid electric/cold-gas propulsion system and a high-bandwidth communications system. A set of cameras and remote manipulators are also common to both spacecraft. They are both designed to be refueled and serviced on-orbit. The aerodynamic plane change assist shell has a small avionics package that supports its maneuverability during the parking orbit period when it is separated from the rest of the spacecraft and payload.

P.4.1   Advanced technologies identification

Advances in electric propulsion, as evident in the Hughes XIPS station-keeping system as well as the NSTAR system for the New Millennium Program (NMP) Deep Space 1 (DS1), are pushing the envelope in high-specific-impulse electric thruster technology. Advanced cascade solar cell arrays will deliver the power needed to drive these engines. Modular clusters of these engines, coupled with the unique aerodynamic maneuver using an advanced aerobraking shell, should provide sufficient energy for plane change operations. Optical communication systems are employed for compact, high-efficiency, interference-free, inter-tug mission control and related high-bandwidth links. Autonomous guidance and navigation and high-fidelity tele-operation (made possible by the advent of supercomputers, biologically-accurate, self-educating neural algorithms, highly-dexterous mechanical manipulators and end effectors, as well as high-strength, light-weight alloys and composite materials and the design of smart/hot structures) make possible an intelligent spacecraft structure that is able to maneuver, articulate, capture, and redeploy target satellites and objects efficiently in space and at the Earth's atmospheric boundaries, enabling it to conserve fuel and power for repeated cyclic operations.

P.5   A SYNERGETIC SUPPORTING ARCHITECTURE

Seven major systems, both existing and envisioned for the coming decade and beyond, are expected to be of synergetic relevance to the SSF architecture evolution. They are:

P.5.1   The existing manned national Space Transportation System (STS)

Though SSF will provide its service mainly in a robotic mode, we expect manned intervention for certain tasks. In the primary stages of evolution, Phase 1, these manned tasks will be performed in the vicinity of the STS. The SSF/payload configuration will rendezvous with the STS, and the crew can access the payload on EVA. The crew may also retrieve payloads for a return trip to Earth. All of these mission scenarios, less the SSF, have been accomplished in past STS missions. The Intelsat rescue mission, the LEASAT mission and the Long-Duration Exposure Facility (LDEF) retrieval mission all provide precedents to ST-1 mission profiles.

P.5.2   Advanced Tracking and Data Relay Satellite System (ATDRSS)

The command, control, communication, and intelligence (C3I) system is crucial to the successful operation of any complex architecture. By allocating a schedule that does not conflict with other NASA activities, the ATDRSS constellation and associated ground segment would have the capability to provide mission control for ST-1 and ST-2 missions. It is envisaged that as the demand for bandwidth increases with the need for multiple missions and in order to accomplish more complex tasks, the SSF architecture will transition to a dedicated telecommunications system. It is also foreseen that radio interference can be expected while conducting operations in the vicinity of target stations. Optical links are suggested as a way to avoid this problem.

Recent experiments using the NASA Advanced Communications Test Satellite (ACTS) promise to open up the Ka band for high-bit-rate communications. Optical communications offers the promise of even more bandwidth. Along with advances in the field of real-time tele-robotics, we expect to be able to employ very sensitive remote manipulator systems on board the ST-2 to do highly-dexterous tasks like handling orbital replaceable units (ORU) on stations without manned intervention.

P.5.3   The orbital debris removal system

A system capable of continually keeping track of and eliminating debris is needed. As the orbital environment becomes more and more crowded with stations of all kinds (scientific, defense-related, and commercial), debris hazard to these stations will also become worse. The ST-1 and the ST-2, between missions, may be used in coordination with an orbital debris removal system architecture to speed up the debris-detection-and-mitigation process.

P.5.4   An artificial gravity facility in Earth orbit

The International Space Station (ISS) is being designed as a facility devoted to the study of human biology in prolonged weightlessness and other microgravity research in the physical sciences. In the decade after the commissioning of ISS, a facility to study the effects of induced gravity on the human system may become operational. The research conducted on board would be invaluable in designing truly-long-duration spaceflight vehicles. Crew aboard this orbiting facility may be synergetically employed for enhancing SSF missions. In the advanced phase of operations, we speculate that more complex and time-consuming manned satellite service missions could be accomplished on board this platform.

P.5.5   Lunar infrastructure development architecture

Growth, transformation, and the budding of new architectures is a dynamic part of the normal process of evolution in any complex system architecture. Using the advanced SSF ST-2 as a stepping stone, it is possible to leverage that infrastructure to begin the exploitation of the Moon. Starting with the removal and disposal of orbital debris and useless, decommissioned stations in high-energy orbits (including GEO) to safe, allocated sites on the lunar surface, we envisage the development of a polar-orbiting lunar station that would eventually become an orbital logistics gateway to the lunar surface. This unique orbit provides total accessibility to the lunar surface during the lunar period of rotation. Since the Moon orbits around the Earth in a 28-day cycle, this lunar polar orbit gateway station also can be used to inject payloads into a variety of useful Earth orbits without the costly penalty of plane change maneuvers. The ST-2 and other orbital transfer vehicles would ply cislunar space between Earth and this lunar polar orbit, while lunar lander shuttles would rendezvous periodically with this gateway station.

P.5.6   Spacecraft salvage operations architecture

The deployment of the SSF architecture has immediate ramifications for global satellite operations. A spacecraft that is able to deliberately alter the position of another has far-reaching legal implications. New regulations and the creation of an international body to oversee the activities of such a system will be needed. Issues dealing with ownership of salvaged spacecraft, their reuse or the manner of their disposal, the effect of such activity on other stations, and the guarantee of operational safety in orbital regions allotted or licensed to other parties will all have to be examined and modified to accommodate the smooth operation of the SSF architecture.

P.5.7   Lunar environment maintenance

Just as Earth orbit will continue to see ever-increasing activity through satellite deployment, shuttle operations, and space station assembly, the same will be true when lunar-base-building activity is in progress. It will be necessary to regulate payload orbital insertion, docking, and de-orbiting technologies in order to minimize the accumulation of debris surrounding the Moon. Without an atmosphere to dis-integrate and vaporize decaying orbital debris, and with the non-uniform lunar gravitational field, matter ejected by maneuvering orbital spacecraft in low lunar orbit will end up rather quickly on the lunar surface. Possessing very high kinetic energies, paint flakes and propellant ejecta can cause catastrophic events to occur on lunar base elements.

Therefore, the space tug architecture and orbital debris removal system will find application in the lunar regime as well. Identical to electric systems employed in Earth orbit, but lacking the aerobraking shell, these tugs will rendezvous with cargo and move them about between desired orbits, remove spent tanks and stages from orbit, routinely track and monitor the condition and surrounding environment of every object in lunar orbit and the ascending and de-orbiting landers, and assist vehicles in orbit collision avoidance.

P.6   MISSION DESIGN AND OPERATIONS

P.6.1   ST-1 mission operations

In Phase 1, ST-1 is launched into orbit on a Titan IV class vehicle. After check-out, it is parked in orbit. It could be called upon to assist in a global orbital debris cleanup campaign while not being mission-tasked for satellite rescue.

Immediately after an accident where a satellite is injected into a wrong orbit due to a launch or upper-stage failure, ST-1 mission operations are initiated. First of all, the ST-1 engages in an aero-assisted plane change maneuver to approximate the orbital plane of the target satellite. Next, it disengages from the aero-assist shell and leaves it in a parking orbit. Communications antennas/telescopes and solar arrays are unfurled and cameras are exposed. A high-bandwidth communication link is secured with mission control and the vehicle proceeds to capture the satellite.

Using well-known rendezvous procedures that are routinely employed today, the SSF will use phasing orbits, homing, and terminal approach maneuvers to gradually rendezvous with the target satellite using onboard electric thrusters. All operations in the vicinity of the target satellite will be conducted carefully, using selected systems, without polluting the immediate environment of activity. Using a small, clean, inert, non-polluting cold-gas attitude control system, remote manipulators and grapples will firmly attach the satellite to the payload bay of ST-1.

After integrity check-out, the total system gradually returns to dock with the aero-assist shell. (Using this approach, the additional mass of the aero-assist shell does not compromise the payload capacity of the ST-1.) After docking with the shell and performing an integrity check-out, the entire assembly does another aero-assisted plane change maneuver to bring the payload into the desired orbital plane. Again, the shell is disengaged and parked while the rest of the ST-1 and payload is slowly thrust into desired orbit. Maneuvers are repeated at the end of the operation to unite ST-1 and the aero-assist shell at the end of the mission.

An application that needs to be studied further is the possibility that a fleet of such vehicles parked in LEO might be used regularly and frequently, as tugs are in the maritime industry, to slowly boost all spacecraft to their desired orbits in a very clean and reusable operation. All pollution and debris associated with orbit transfer maneuvers, their motors, casings, ullage gases, and unspent tankage could be eliminated. Such an application of the SSF architecture could have wide-ranging implications for the scope and economics of this infrastructure.

This staging procedure in LEO would also allow us to thoroughly check out a payload that has just survived the severe ETO launch environment, before sending it gently away to its destination using the low but constant thrust of the ST-1 electric engines. This concept of using low-thrust engines also opens up a new area of spacecraft design, in which very large yet fragile craft like optical interferometric telescopes and extensive solar-sail-powered vehicles, or even wide fields of arrays for solar-power satellites, may be assembled in low Earth orbit and slowly placed in their proper orbits.

However, if we use these low-power electric thrusters, we would have to deal with the aspect of the payload being severely exposed to the Van Allen radiation belts through the course of this procedure. Since stations headed to GEO, for instance, are sufficiently radiation-hardened, this should not pose a big problem.

P.6.2   ST-2 mission operations

In Phase 2 the ST-2 vehicle may be carried up on a Titan-class vehicle, deployed fully and checked out with assistance from the crew of a co-orbiting space shuttle in a dedicated STS mission.

The ST-2 is envisioned as a more capable vehicle. It is a larger version of the ST-1, with a more dexterous remote manipulator system and a kit of tools allowing it the ability to do maintenance and delicate repair and replacement operations on spacecraft located in a range of orbits from LEO to GEO. Lessons learned during ST-1 operations will be used in the design of ST-2. It may be initially commissioned to be parked in GEO, where it might slowly move along the GEO belt to assist the constellation with such activities as temporary station keeping, visual examination of stations for various purposes, stabilizing anomalous conditions aboard stations, providing auxiliary power, removing and replacing ORUs, cleaning and maintaining thermal louvers and blankets, and removing decommissioned satellites from orbit.

The prevailing technique of moving useless stations to graveyard orbits where they eventually deteriorate and add to debris through accidental exposure to micro-meteors or other debris will be totally eliminated. It is the consensus of the USC group that it is a prudent philosophy to de-orbit spent stages and other large deployment-related hardware, before they break up and scatter more debris over large regions.

Both ST-1 and ST-2 are envisioned as fully-reusable spacecraft. Depending on the energy needed for each mission, these vehicles are designed to carry out between five and eight missions before refueling and service operations. These operations are carried out in LEO, possibly in the vicinity of an SSF-dedicated STS mission or on board the advanced artificial gravity facility. Modular cold-gas and electric thruster propellant tankage units are replaced, all docking and other mechanical hardware, including solar array and camera drives, are checked out, the aerodynamic shell is tested for integrity, and the SSF is made ready for its next mission.

P.6.3   Merits and limitations

A reusable SSF like the ST-1 and ST-2 offers a new way to manage our resources in Earth orbit. Though the demand for satellite rescue missions is quite unpredictable at this time, we can expect to see a sizable number of events in the coming decade, given the projected activity. Such rescue missions could save an otherwise perfectly-healthy satellite and circumvent the need for mounting another long launch campaign that could cost the owner their business.

The ability to replace subsystems on board sub-optimally-operating stations is a highly-desirable activity for which an infrastructure is lacking at the present time. The recent spate of subsystem anomalies aboard stations like the Anik, the Palapa C1, AsiaSat 3, JCSat 3, and the AMSC-1 clearly suggest the growing need for such a system.

Besides filing insurance claims for loss or damage, the current strategy, at least in the deployment of multi-satellite constellations like the Global Positioning System (GPS) and the proposed big LEOs like Iridium and Teledesic, is to provide several fully-functional spares on orbit that are turned on by tele-command to take over from a faulty station. However, until that need arises, as mentioned before, these spares have the potential to cause more debris. The SSF offers a different "fix it or safely dispose of it" philosophy and could be a timely architecture for the coming decade and beyond, given the number of spacecraft planned for launch as well as those expected to reach end of life during this period.

The propellant penalty of plane change operations has been a prohibitive factor against the development of this sort of architecture. However, by combining the emerging technologies of high-specific-impulse, clustered, low-thrust electric engines powered by high-efficiency solar arrays employing cascade cell technology with the concept of aerodynamic steering using light-weight, high-strength, composite, hot-structures technology, it should be possible to overcome this difficulty. The term reusability is still not fully applicable to the aerospace world, but it is the key to more economically-viable architectures. The STS employs a partially-reusable architecture.

We have to strive for more reusability in all systems, and the SSF is designed for reusability. The ST-1 and ST-2 will need to be serviced and refueled on orbit. This technology has yet to be perfected and is on the critical path for any major achievement in space.

To assure the success of the SSF architecture, it is imperative that new standards in the design and engineering of spacecraft for serviceability be drawn up and followed closely. Once the design and engineering parameters for the SSF elements at the components level are established, satellite production lines will have to begin the transition to incorporating serviceability features into the assembly sequence. These include the provision of hooks and scars to accommodate orbital replaceable units (ORUs), multiple hard points for grabbing, anchoring, and latching on to the SSF, and mechanisms that enable deployed satellites to retract solar arrays and antennas back to a stow-away configuration, as well as compatible designs for a variety of remote manipulator systems (RMS) and other end effectors on board the SSF elements, just to name a few features. The need for ambient lighting control in the proximity of the target payload during multi-orbit operations is also a factor of importance.

P.7   RECOMMENDATIONS

1   New and tighter international regulations are needed to curtail the production of orbital debris from current satellite deployment procedures. Though providing spare stations in orbit makes sense economically, it is a dangerous practice from the environmental point of view, however slight the mathematical probability of explosive disintegration due to a variety of reasons.

2   Begin deployment of orbital debris mitigation systems as soon as possible. It is the consensus of the USC team that all potential debris-producing bodies be de-orbited before they the disintegrate and scatter debris over large regions. This includes de-orbiting spent stages and other deployment-related hardware. The proposed architecture can assist in this mission.

3   New debris-free deployment techniques are needed for satellites to supra-LEO orbits. Furthermore, clean and non-polluting systems should be employed during target satellite proximity operations. The SSF staging and operations architecture suggests one strategy. Insurance premiums should reflect the cleanliness of deployment operations.

4   Start designing spacecraft that provide standardized hooks and scars to facilitate serviceability in orbit. This includes easily-accessible hard points for grabbing and anchoring, protected fixtures that will not fail during service/decommissioning procedures and the incorporation of mechanisms on payload satellites to retract deployed elements like booms and arrays during service and redeployment operations.

5   On-orbit refueling technology demonstration and aerodynamic braking/steering are high-priority items on the critical path for this architecture to

succeed. Build an SSF fleet of vehicles, in a phased manner akin to ST-1 and ST-2 evolution, which can be serviced and refueled on orbit.

6   Design versatility into fleet architecture for rescue operations and ORU change-outs, as well as for orbital debris mitigation.

7   Design the SSF fleet with interfaces to accommodate interaction with STS, as well as with an advanced manned station in orbit.

8   Establish an international body to oversee the activities of the SSF architecture. This same body would gather a pool of resources to subsidize the orbital debris mitigation program for companies willing to invest in this mission.

9   Multi-tasking may be the key to satisfying the economic requirements of such an architecture. The creative coupling of orbital debris removal and station decommissioning activities, satellite maintenance, inspection, modular upgrades, and satellite rescue, may be the answer to an economically-viable, versatile architecture.

10 Satellite deployment/operations insurance company policies should be reviewed and revised to provide new standards and more incentives for cleaner, safer campaigns all the way from "cradle to grave".

P.8   ECONOMICS OF THE SATELLITE SERVICE FACILITY

Communication satellites cost U.S.$100–$300 million or more per unit to design, build, and test. A typical launch campaign from satellite integration procedures through launch preparation, lift-off, transfer orbit injection, circularization, and final check-out and hand-over to operators can easily amount to another $100–$200 million. A typical conservative underwriter will insure the payload for between 20–30 percent of the value of this total campaign. At a minimum, a full campaign from start to safe operation will cost the owner between $200–$250 million.

Satellite deployment failures are hard to predict. However, in the last five years alone, several satellites have failed to reach prescribed station orbits. Launch failures and upper-stage malfunction were responsible for many of these events. Though we cannot yet portray any definitive statistical model on how often a satellite might end up in this situation, a communication satellite that does not reach or cannot maintain its prescribed orbit, geostationary or any other, is considered a complete loss because it cannot operate effectively in any other regime, technically or legally. Such a stranded station, besides being useless and a liability, can threaten the orbital environment by explosively scattering debris upon accidental impact with other debris or micrometeorites.

Under these circumstances, the SSF becomes an attractive option in order to safely and efficiently place the spacecraft in its intended orbital slot. The typical rescue mission costs would be set to provide both a profit for the SSF service provider and a bargain for the satellite owner, both parties understanding that lost time in mounting another campaign from the beginning could cost the satellite owner their business in this very lucrative but fiercely-competitive industry.

We estimate that a phased SSF architecture design, development, test, and engineering (DDT&E) program comprising three ST-1s and two ST-2s, along with mission control infrastructure, would cost about $3.5 billion. This capital investment is well within the reach of a large private aerospace company or a consortium to field. Building on high-power satellite buses nearing production, the ST-1 could be built and flight-tested in two years. The ST-2 might evolve in another three. A five-spacecraft full-up SSF fleet could be operational shortly after the turn of the century.

Furthermore, between these unpredictable but highly-profitable rescue missions, the SSF architecture would pay its way by assisting in debris mitigation activities, salvage operations, visual examination of spacecraft for verification of system or subsystem status for various technical and legal purposes, and for maintaining stations by regular operations, including cleaning solar panels and thermal control louvers, fixing thermal blankets, assisting in temporary station keeping, and decommissioning useless stations. These secondary, less dramatic, but continually-available missions should at least pay for the upkeep of the SSF architecture.

P.9   CONCLUSION

The Earth orbital regime is getting crowded with satellites providing a host of services. Judging by the applications being filed for new licenses, we can expect to see a great number of stations being deployed in the next decade alone. Barring unforeseen new developments in deployment technology, we can also expect to see a sizable number of deployment failures, resulting in fully-functional, yet useless, stranded satellites that will threaten the orbital environment by producing more debris. The orbital satellite service facility architecture will provide the infrastructure and elements to rescue these satellites and deliver them to their intended orbits. Though such rescue events are considered quite unpredictable, every such mission can be very profitable to both the SSF service provider and the satellite owner. Between such missions, the SSF would pay its way by providing a host of other valuable functions, including assisting in orbital debris mitigation and other satellite maintenance and salvage activities. Multi-tasking is seen as the key to economic feasibility.

We anticipate that the nature and scope of the SSF architecture and related activities will infringe on the technical and legal domain of satellite operators worldwide. We suggest the creation of an international body to gather the resources and to oversee the smooth operation of such a complex system. The development of such an infrastructure can then be leveraged to include activities on the Moon by first establishing a lunar orbital station for various purposes and then continuing to expand to our activity to the lunar surface and beyond.

P.10   REFERENCES

Alexander, C. (1964) *Notes on the Synthesis of Form*. Harvard University Press, Cambridge, MA.

Armstrong, M. (1996) *Hughes Electronics Herald*, Hughes Electronics Corp., Los Angeles, CA.

Beech, M.; Brown, P.; and Jones, J. (1995) The potential danger to space platforms from meteor storm activity. *Q. J. Royal Astron. Society*, **36**.

Bekey, I. (1985) Concepts for space stations and platforms. *Space*. American Institute of Aeronautics and Astronautics, New York.

Berry, Robert (1996) *Plenary Address AIAA International Satcom '96, Washington, D.C.*

Brackey, Thomas (1996) *Plenary Address AIAA International Satcom '96, Washington, D.C.*

Brodsky, R.F. (1995) *Spacecraft Systems Architecture Lecture to Participants of the Concepts Studio, October.*

Brown, D.L.; Ashford, E.; De Peuter, W. *et al.* (1995) *Satellite Servicing in GEO by Robotic Service Vehicle*. ESA Bulletin
78, European Space Agency, Technical and Telecommunications Directorate.

Burch, G.T. (1967) *Multi-plate Damage Study*. AFATL-TR-67-116, Air Force Armament Library, Eglin Air Force Base, Florida.

Chabildas L.C. (1996) Techniques to obtain orbital debris encounter velocities in the laboratory. *Space '96*, American Society of Civil Engineers, New York.

Chobotov, V. (1991) Orbital dynamics. *Progress in Astronautics*, AIAA Education Series, ISBN 1563470071, Washington DC.

Curtis, Anthony R. (1994) *Space Satellite Handbook*. Gulf Publication, Houston, TX.

Davis, B. (1989) *Drawing on the Right Side of the Brain*. ISBN 0874770882 Trade Paperback (Davis, CA).

Dorfman, Steven D. (1996) *Address accepting the "Satellite Executive Of the Year" Award at Satellite '96: Emerging Global Markets, February, Washington, D.C.*

Fuller, B.F. (1983) *The Inventions of Buckminster Fuller*. Michael Denneny (editor), St. Martin's Press, New York.

Griffin, M.D. and French, J.R. (1991) Spacecraft vehicle design. *Progress in Astronautics*, AIAA Education Series, ISBN 0930403908, Washington DC.

Hawkes, N. (1993) *Structures: The Way Things Are Built*. Collier Books, Maximillian Publishing Co, New York.

Haynes, William F. (1996) *A Preliminary Concept for Orbital Debris Removal*, oral communication with editor, February.

ISU (1995) *Vision 2020*, Report of the International Space University, Strasbourg, France.

Johnson, H.T. (1995) *Hyper-velocity Impact Testing of Habitation Module Wall Materials*. Report No. 98-28980, NASA White Sands Facility.

Kessler D.J. *et al. (1988) Orbital Debris Environment for Spacecraft Designed to Operate in Low Earth Orbit*, NASA TM 100471, NASA, Washington, D.C.

Leaven, G.; Hauck, F. *et al.* (1993) The NASA/INTEC Satellite Salvage Repair Study. *ESA Conference, June.*

Macaulay, D. (1998) *The Way Things Work*. Walter Lorrained Books, Houghton Mifflin Publishers, ISBN 0395938473.

Mendell, W. (1985) *Lunar Bases and Activities of the 21st Century*. Lunar and Planetary Institute, Houston, TX.

Miller, R. (1993) *The Dream Machines*. Orbit Books, Malabar, FL.

NASA (1988) *Space Exploration Reports TM4075*, Vols. 1 & 2, NASA, Washington, D.C.

NRC (1995) *Orbital Debris: A Technical Assessment*. National Research Council, National Academy Press, Washington, D.C.

NSTAR (1995) Project plan. *Ion Propulsion Begins a New Era in Solar System Exploration*. NASA SEP Technology Application Readiness, Spectrum Astro/JPL, NASA Jet Propulsion Laboratory, Pasadena, CA.

Orbital Sciences Corp. (1995) Small satellites come of age at OSC. *Space News*, Corporate Profile, September.

OSTP (1995) *Interagency Report on Orbital Debris*. Office of Science and Technology Policy, NSTC Publications, Washington DC, November.

Paine, T.O. (1990) *The Next Forty Years in Space*, IAF, Malaga, Spain.

Phail, Brian A. (1996) A preliminary concept for orbital debris removal. *Space '96, Albuquerque, NM*.

Rechtin, Eberhardt (1991) *Systems Architecting: Creating and Building Complex Systems*, Prentice Hall, Englewood Cliffs, NJ.

Rechtin, Eberhardt (1994) *Collection of Student Heuristics in Systems Architecting, 1988–1993*, University of Southern California, Los Angeles, CA.

Ridenoure, R. (1995) The New Millennium Program. *A System Architect's Overview: Space Systems Architecting Lecture to Participants of Concepts Studio*, NASA Jet Propulsion Laboratory, Pasadena, CA, November.

Ruoff, Carl F. (1989) Overview of space telerobotics. *Progress in Astronautics and Aeronautics*, **21** (AIAA, Washington, D.C.).

Schonberg, W.P. and Williamsen J.E. (1996) Space station module wall hole size and crack length following orbital debris penetration at 6.5 km/sec. *Space '96 Proceedings*, American Society of Civil Engineers, New York.

SSFPO (1988) *Common Module Shell Unzipping Due to Meteoroid/Orbital Debris Strikes*, Station Program Office, Festoon, VA, GS-04.05-PT-6-001.

Thangavelu, M. (1993) Modular assembly in low Earth orbit: An alternative strategy of lunar base establishment. *Journal of the British Interplanetary Society*, January.

Thangavelu, M. (1993) *Concepts: Generation Class Notes*, University of Southern California, Los Angeles, CA.

Viasat (1995) *Survey of Satellites*, Viasat, Carlsbad, CA

Wertz, James R. (1992) *Orbit and Constellation Design Seminar: AIAA/Utah State University Sixth Annual Small Satellite Conference, September*.

Wertz, James R. and Larsen, W. (1991) *Space Mission Analysis and Design*, Kluwer, Dordrecht, The Netherlands.

Whipple, F.L. (1947) Meteorites and space travel. *Astronomy Journal*, **52**.

Williamsen J.E. (1994) *Vulnerability of Manned Spacecraft to Crew Loss from Orbital Debris Penetration*. NASA TM 108452, NASA, Washington, D.C.

Williamsen, J.E. and Serrano, J. (1994) Atmospheric effects in spacecraft interiors following orbital debris penetration. *SPIE*, 2483-06.

Woodcock, G. (1993) *Space Stations and Platforms*. Orbit Books Company, Malabar, FL, ISBN 0894640011.

Wortman, Mary F. (1992) Gotcha! Awakening a slumbering giant: From the brink to ... the Intelsat VI F-3 rescue mission. *Hughes SCG Journal*, August.

Appendix Q

The Millennial Time Capsule and L-1 Artifacts Museum

Madhu Thangavelu   *Conductor, Space Exploration Architectures Concepts Synthesis Studio*

The basic blocks behind the concept architecture[1] for the Millennial Time Capsule and L-1 Artifacts Museum first occurred in the final course work of a participant, John C. Fujita, in this class, in the Spring of 1995. A version was presented at *Space '98, Albuquerque, NM, 1998.*

Q.1   MILLENNIAL TIME CAPSULE

Scientists and scholars have been preoccupied with concepts for space stations, lunar and planetary bases for much of the space age. Their primary objective in this effort has been to advance science and technology and our knowledge of the universe. Architects and engineers have assisted this endeavor, focusing on exploration missions, lunar astronomical observatories, and science platforms.

They have often suggested the establishment of mineral enrichment facilities, laboratories and factories for activities ranging from processing helium-3 and conducting hazardous biological experiments to manufacturing propellant, solar arrays, and building components. Recently, a new wave of entrepreneurs have been contemplating using the Moon for profitable ventures, including tourism, recreation, and other cultural activities, in an effort to embrace the cislunar domain as part of the landscape for routine human activities in the new millennium. Radio stations and even lunar cemeteries have been proposed of late. A new rationale for lunar bases is being sought.

[1] Space Exploration Architectures Concepts Synthesis Studio, Department of Aerospace Engineering, University of Southern California, Los Angeles, CA 90089-1191.

Q.2   SPACE ACTIVITIES: A BROAD GLOBAL
HUMANITARIAN PERSPECTIVE

It must be noted at the outset that, in spite of all the high-profile publicity associated with space flight, when one realistically portrays the affairs of humanity on a global, all-inclusive arena of human endeavor (economic and otherwise), science and technology accounts only for a small percentage of our activities, and space-related activities are but a fraction of that percentage. It may be argued that, at this point in time, the events and processes that shape our history are based on other activities and value systems as much as they are on science and technology. Even the historic Apollo program derived its impetus from policy decisions that had little to do with the advancement of science. However, great strides were made in science, technology, and our understanding of the Moon, and spin-off technologies have enhanced our lives on Earth because of it.

Art, culture, commerce, politics, government, military posture, and international affairs are all important driving forces in societal affairs, and more attention should be paid to their immediate needs when high-technology mega-projects like space stations and lunar bases are considered. Vast changes in the way humanity perceives and allocates resources for development continue to erode support for the space program. Can we continue to support the civil space program as an ivory tower of high technology that was closely associated with a bygone era which used it expressly as a tool for military muscle flexing? It seems that our scientific and technical education in aerospace engineering and allied disciplines, and the projects we pursue thereafter, somehow removes us gradually from the mainstream arena of human activities to a niche profession that is still closely tied to the defense industry. It has constant difficulty in justifying its immediate value to humanity, let alone maintaining its budget without being threatened with drastic reduction year after year in Congress.

Q.2.1   A humanitarian concept based on space activities

Is it possible that we need to take a fresh look for imaginative new connections and create large aerospace projects that the taxpayer can truly appreciate and proudly support on a routine basis? Are there deep-rooted symbolisms, metaphors, and allegories in our history and the humanities that we could exploit to weave together meaningful, innovative projects in space? Maybe there is yet another powerful, non-scientific constituency out there that is waiting to be tapped to accelerate the peaceful development of outer space? Is it possible for those potential driving forces to be incorporated into the design of space stations and lunar bases in a direct way?

Is there a very specific space project that we can create to celebrate the new millennium? The International Space Station is still under construction. A manned mission to Mars may be difficult to orchestrate, given the time constraint. The technology required for the direct imaging of a habitable extra-solar planet may not be ready by then, and the possibility of detecting an intelligent extraterrestrial signal is still left to chance. Several robotic space exploration missions are planned, but they may not provide the broad and immediate humanitarian appeal that is necessary of such a project, aspects that are often lacking in high-technology projects today.

The obvious constraints include the following. The project should have significant symbolic and metaphorical content for all of humanity. It must broadly appeal to all segments of society. The project should be economically viable without jeopardizing other programs and executable in two years with minimum investment in research and technology. Finally, it should bring the peoples of the world together to celebrate the true spirit of humanity.

The rationale for the project is derived from the following background. Humanity's preoccupation with immortality is very evident from time immemorial. The yearning to transcend our rather short life cycle has pronounced effects on the way we live and operate. For this reason among others, the history of civilizations, their life styles, cultural activities, their civil architectures and processes leading to their rise and fall, have always played a dominant role in the aspirations of great societies. While it is customary among cultures and nations to preserve their history through books, documents, oral tradition, and other media including time capsules, so cultural upheavals and weathering by the elements and other environmental circumstances continue to erase hard-earned collective knowledge and experience in short order.

The selective, continuing decimation of minorities all over the world and the burning of books at Alexandria, in China or in Nazi Germany are examples of how the collective cultural heritage and knowledge of peoples can be wiped out by short-sighted individuals, juntas, nations at war, international terrorists, and opportunistic, oppressive, and tyrannical governments operating on short term agendas. Natural calamities like those brought on by earthquakes, volcanic eruptions, deluges, pestilence, disease, and climatic changes, both abrupt or gradual, have all taken their toll on the collective cultural heritage of humanity as well. The present difficulty in deciphering the activities of lost civilizations of the Maya, the Egyptians, the Mesopotamians or the Mohanjo Daro peoples offer examples of how, through the continuous, prolonged effects of sheer elemental weathering, the sands of time can totally and completely erase all signs of human activity even without any violent human intervention. Scientists now believe that powerful forces external to the Earth may also play a major role in this activity, such as, for example, cometary impacts that have caused widespread global destruction in the past.

In light of these unavoidable processes, one way to preserve collective human knowledge may be through the prudent use of space exploration activity for peaceful purposes. Places exist in the vicinity of our planet where much of the above-mentioned environmental weathering activities and man-made calamities are all but absent. The Moon is such an example. The lunar surface presents an environment that is not subject to such dynamic and scouring weathering processes as are experienced on Earth. While it is constantly bombarded by meteorites and experiences tremendous diurnal temperature swings between the lunar day and night cycles, the lack of dynamic aeolic and aqueous processes has left lunar surface features much the same for millions of years, a short period in geological time, but very significant in the time scale associated with the rise and fall of civilizations.

This observation leads to the concept behind the Lunar Humanity Repository. The idea is to encode the collective knowledge of the cultures of the world into an extensive electronic database that is then safely and securely emplaced on the Moon. This interactive multimedia database system can then be accessed by the peoples of the Earth, and periodically updated, like an evolutionary, celestial time capsule.

Q.2.2   Technologies at the threshold of maturity

The confluence of several key technologies in these past decades make it possible for humanity to create and operate a compact repository of all human knowledge on the Moon. Information storage, retrieval, and affordable mass media delivery systems (including the emergence of the World Wide Web, interactive multimedia technologies, efficient interference-free high-bandwidth communications), light-weight high-density power generation, conversion and storage devices, highly-dexterous robotic and tele-operated systems (employing artificial intelligence with autonomous capabilities), and related technologies are all mature enough to be able to support such a project. High-power satellites and their communications systems already offer the capability to service such a lunar facility.

Despite the data rate limitations of operating a commercial off-the-shelf communications system at ten times the distance from the Earth than to geo-synchronous orbit, it is still possible to reap the benefits of such an architecture with minimum investment in research and development. By increasing the power and heat rejection capability of existing geosynchronous station designs, higher transmission power levels and wider bandwidth could easily be accommodated on the proposed lunar station.

A highly-capable, fully-robotic lunar lander, about half the size of the Apollo lander of yore, would carry all of this information and make a soft landing at a suitable predetermined spot on the Moon. Large, oversized modular solar arrays delivering 30 kilowatts of power, and a variety of communications antennas including an optical system would be deployed in the vicinity. Over the course of the next few months, operating during the lunar day and hibernating at night, a specially-designed auger mechanism on the nadir of the lander would slowly burrow and emplace a part of the lander containing the electronic database deep into the lunar regolith, thus protecting the core of the facility from micrometeorites and large solar events that might otherwise destroy it over the course of time.

Q.3   THE LUNAR HUMAN REPOSITORY ARCHITECTURE

For the first few years we may opt to operate the station at full capability only during lunar daylight hours, allowing a compact rechargeable battery system to take care of housekeeping functions during the long lunar night. By adding larger batteries during modular system upgrade procedures, we might then operate the system around the clock. On the other hand, if the station were located close to the poles where there is near continuous access to sunlight, then it may be possible to operate continuously with solar arrays alone.

The data is stored on highly-stable, extremely-long-life solid-state memory chips and using state-of-the-art, digital CD-ROM technology. About 300 disks and several hundred chips would suffice to contain the first installment of all essential human knowledge in duplicate. An interactive multimedia architecture could be adopted for the format. Another 100 recordable optical disks will allow Earth operators to load current data of historical and cultural significance on a periodic basis for some years to come. The system is designed to be critically redundant so that it can suffer mechanical system malfunction and still perform for years without totally collapsing over a single point failure.

Once emplacement procedures are complete and the system checked out and certified for operation, the Lunar Humanity Repository would be accessible to users on Earth through a variety of channels of differing bandwidth depending on user capability. Pointing the L, Ku, Ka, or V-band receiver antenna would be as simple as sighting the Moon through a boresight on the receiver dish and switching on a lunar clock drive, similar to any astronomical telescope. For those with high-bandwidth laser communication capability, more precise alignment and tracking systems would be made available. When the lunar disk is invisible to the eye, as during new Moon or daylight moonrise, we could acquire the station using conventional lunar ephemeris data. Optical communications would be favored for both up- and downlinks in order to eliminate the problem of interfering with an ever-increasing Earth-orbiting station traffic.

An alternative architecture for data handling and distribution would be possible by hooking up the lunar facility to a few World Wide Web nodes so that users might access this site as they would any other, using ordinary browsers and search engines. The line-of-sight constraint is eliminated for round-the-clock access. The penalty to be paid in this configuration are the data rate limitations imposed by terrestrial links. However, it is possible to imagine a hybrid architecture allowing users to access the lunar repository in many ways.

The typical user might have random access to information regarding nations, their history, culture, and current news, both domestic and international. They might be able to scan through museum artifacts, music, and fine art, and compare technologies and commerce much like the Internet and the World Wide Web but with one salient difference. Information stored in the Lunar Humanity Repository on the Moon is preserved for posterity in the most advanced and stable manner. The lunar station could also broadcast programs following a long-term schedule, for those viewers and listeners who may be interested. Programs are picked and scheduled by user demand and interest. By employing a "bent pipe" architecture, it is possible to broadcast or communicate between points on the Earth, using the lunar station as a passive or active relay.

Q.3.1   Merits of the architecture

A lunar library would be far more permanent and easier to maintain than one in Earth orbit, for instance. All activities associated with station keeping would be reduced or eliminated. There would be no need to worry about de-orbiting at end of life, which is fast becoming a problem with decommissioned stations in Earth orbit. Since the periodic librations of the Moon are very well understood and highly predictable, the programmed antenna drivers on the station would smoothly and precisely compensate for the lunar wobble while pointing toward the Earth. There would be no need for corrosive station-keeping propellants on board the station and thermal cycling would not be an issue. By staying buried under a few meters of regolith, the station would be protected from the constant bombardment of meteorites and solar-particle events, and would maintain a constant temperature, which would be key to the long-term survivability of sensitive electronic and mechanical components.

The orbit of the Moon is inclined to the Earth's geostationary orbit. Therefore, any interference with stations operating in the Clarke belt is minimized. However, if we plan to operate the lunar station in already-crowded frequencies, a method to mitigate interference with proposed LEO and MEO orbiting stations in inclined planes will be needed.

Q.3.2   A natural evolution scenario for the facility

It is possible to imagine a natural evolutionary progression of this facility as time goes by. Starting as a collection of electronic databases, more and more modules would be attached to this facility as more information is documented for posterity. The entire architecture is designed to operate remotely and reliably for a few decades at least. Certain maintenance activities will require human crews to visit the facility periodically. They would carry modular systems to the site for replacement and to service the facility. Modular replaceable units would keep the solar power at optimum levels, tracking devices might be replaced periodically, laser optics and transmitter horns might require modification, and recorders might need periodic maintenance. As they require more and more complex servicing assignments, the crew will need a place to stay for longer tours of duty.

As the repository evolves to contain more and more specialized knowledge, we might consider emplacing satellite facilities around the main repository. By dispersing the facility over various regions of the Moon, we might also avoid being instantly knocked out of service by a direct meteoritic hit. At some point in time, this temple of human knowledge would begin to attract other humanitarian events. Maybe, as the human-tended facilities grow large enough to support several crew over a tour of duty of many months, in addition to building observatories and factories they could also be tasked to build habitats and larger dwellings where the leaders of the spacefaring nations of the world might one day decide to hold summits.

Q.3.3 The case for international subsidies

What might such a project cost? Less than a billion (in 1998 U.S. dollars). How might we pay for such an architecture? Taxpayer dollars might fund the initial setup. When the concept of preservation for posterity is demonstrated over a period of time, several candidates may become interested. Museums of the world would like to preserve electronic files regarding their artifacts. The great libraries would be clients wanting to keep their records safely. Private foundations and trusts would most likely want to use this facility. And the list goes on. If international subsidies are orchestrated to take advantage of it, we can hope to have a facility on the Moon for a fraction of the budget indicated above. For instance, Russia might provide an Energiya launch vehicle, the United States might build the translunar injection stage and lander, while Europe might coordinate the activities to put together the facility that is built component-wise all over the world. Japan would work on the system to emplace the facility safely under several meters of regolith, while China, India, Africa, Australia, South America, and Russia would accelerate the ground station/global information infrastructure (GII) development to hook into this facility.

Q.4 ARCHIVES OF HUMANKIND

Long before establishing the facilities of the Lunar Government, a lunar archives facility may be emplaced at a landmark region on the Earth-facing side of the Moon. A spacecraft containing data about all the achievements of humankind to date could be landed at a site such as Aristarchus Plateau. An auger mechanism would burrow into the lunar regolith and lower the facility to a depth of several meters, thereby providing a stable thermal environment and protection against the hazards of micrometeorites.

Powered by long-life RTGs, the system would turn its whole range of advanced communication antennas (including wideband optical systems) toward the Earth and be accessible to the people of the world, directly or through the Internet. Every country would have partitioned space for storing information that would be updated with events of historical and cultural interest. An early-warning tracking system would allow the entire facility, antennas, optics, and all to move completely underground during a meteor shower threat and then to "bloom" again when the threat has passed. More modules would be added as this library grows and, eventually, the adjacent area might become the seat of the Lunar Government. The lunar archives facility would thus provide a secure reservoir of knowledge of humankind on another world – the Moon – for safekeeping in the event of a natural or human-made catastrophic event on Earth (see Figure Q.1).

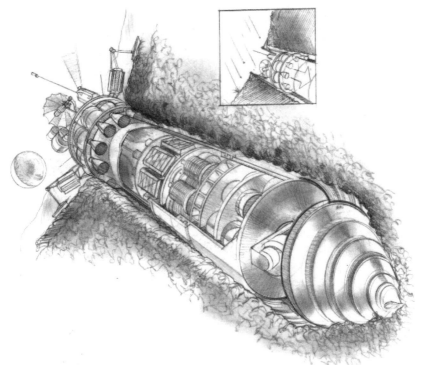

Figure Q.1. Archives of humankind storage (Thangavelu/DiMare).

Q.5   CONCLUSION, MILLENIAL TIME CAPSULE

A highly-specific, goal-oriented mission to celebrate the new millennium using peaceful space activities and international collaboration as the theme is proposed. The lunar library and archiving facility will employ state-of-the-art spacecraft and communication technologies to create and maintain a long endurance facility for all of humanity. Though we may not discover a black monolith on the Moon that sends cryptic messages off to Jupiter, as Arthur Clarke suggests in his famous novel, we could still have a unique, resourceful, updatable, evolutionary interactive multimedia facility buried under the lunar soil that beams the achievements of humanity back to Earth, for those who wish to access, interpret, and cherish the evolution of our species, the history of our civilizations, our rich and diverse cultural heritage, and life on Earth, unperturbed by the natural or artificial calamities that might befall us here on Earth, for generations to come.

Q.6   L-1 ARTIFACTS MUSEUM

Great civilizations of the past are studied and appreciated by the artifacts they leave behind. Technologies and processes are often lost over millennia but their buildings

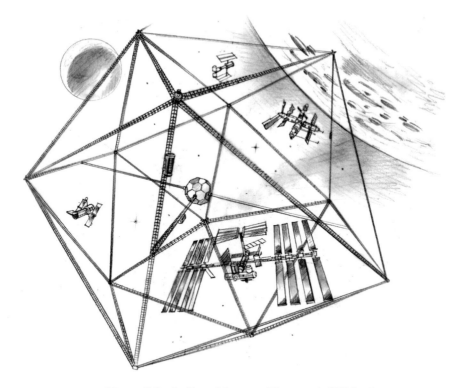

Figure Q.2. Artifacts Museum (Thangavelu/DiMare).

and monuments offer a glimpse into their life and times. Space activities are at the pinnacle of modern civilization and yet many of the artifacts that served us well are decommissioned and meet their end of lives in a fiery reentry into the Earth's atmosphere. We have lost a significant number of historically-valuable spacecraft, such as the Mir space station, and may continue to do so unless a new philosophy of saving and preserving them is adopted.

One such concept is the International Space Artifacts Museum. A large skeletal space structure co-located with the transit hub at the L-1 Earth–Moon Lagrange point would serve as a superstructure in which decommissioned and cocooned space stations like the International Space Station and the Hubble Space telescope are preserved (Figure Q.2). Over time, a variety of spacecraft may be exhibited there for future generations to research and appreciate.

Appendix R

MALEO: Modular assembly in low Earth orbit
Madhu Thangavelu

R.1 ABSTRACT

Modular assembly in low Earth orbit (MALEO) is a new strategy for building an initial operational capability lunar habitation base, the main purpose of which is to initiate safety and to sustain early lunar base build-up. The architectural drivers for the strategy are the dust-free and less harsh radiation environment of low Earth orbit. These permit safer manned extravehicular activity (EVA), the maximum inheritability from International Space Station (ISS) technology, infrastructure and associated assembly experience, and minimize the risky primary manned EVAs and associated vehicles and assembly equipment suggested in conventional strategies. Modular lunar base components are brought up to low Earth orbit by the Space Transportation System/Heavy Lift Launch Vehicle fleet, and are assembled to form the complete lunar base. Modular propulsion systems are used to transport the MALEO lunar base intact to the lunar surface. On touchdown, a substantial, safe, and completely-self-sustained MALEO lunar habitation base is operational.

Modular, erectable, deployable, inflatable and indigenous architectures for lunar base evolution could be developed, site-tested and certified for long-term human occupancy while the astronaut assembly crew live in the MALEO base. Advantages of the MALEO over conventional strategies are pointed out. It is concluded that, using space-station-derived experience and technology, the MALEO strategy holds promise for rapid lunar base deployment about the turn of the century.

R.2 INTRODUCTION

On the twentieth anniversary of the Apollo 11 mission, President Bush charged NASA with developing plans for a permanently-manned lunar base and then extending the human domain on to Mars. NASA had been working on lunar return concepts for some time. Recent reports from the Office of Exploration suggest several case studies for both manned lunar and planetary missions under consideration. Assembly in low Earth orbit has been suggested for the mission to Phobos and Mars.

Several competing technologies and strategies have been proposed for building primary lunar base habitats. Some include erectable and deployable modular concepts, inflatable structures, indigenous *in-situ* construction of magma and ceramic structures, and the modification of natural formations like lava tubes for habitation. All these need substantial manned EVA for erection and eventual habitation. Permanent lunar base development entails long-term tour duties (six months or more) for the assembly crew and the establishment of complex CELSS systems which will have to perform flawlessly over comparable periods of time. These systems would have to be tested, optimized or "tweaked", and certified on site before the habitats they support could be safely occupied by the astronaut crew. Until that time, the assembly crew will need a safe haven from which to operate.

The MALEO LHB-1 is meant to be that safe haven/site office workshop. A Phase 1 lunar habitation base (LHB-1) is conceived as the first permanently-manned facility on the lunar surface. The facility will provide a test-bed for extended habitation, exploration, and scientific investigation. It is considered analogous to a forward base camp in the Antarctic or on Mount Everest, or a conventional terrestrial site office from which major construction is supervised. LHB-1 will also be the nucleus, if needed, for further expansion and experimentation which would be necessary during the evolution of what might be the first fully-self-sustained, permanently-manned lunar colony. This lunar colony will support spacecraft operations in cislunar as well as interplanetary space by providing propellants and other material manufactured on the lunar surface.

The first priority in a Phase 1 extended duration mission of this nature is the provision of an assured safe haven for the astronaut crew (which will alleviate astronaut anxiety associated with build-up operations), followed by a comfortable environment within the facility to enhance crew productivity.

R.3 DEVELOPMENT OF THE MALEO STRATEGY

Space station-like modules are the major components which need to be assembled to form a Phase 1 lunar habitation base (LHB-1). Conventional strategies suggest launching these modules separately from Earth, landing them one at a time on the lunar surface and assembling them there, using robots and astronauts in extra-vehicular activity (EVA). Precursor missions are employed to land crew and assembly equipment. If Earth-based robotic teleportation is suggested for assembly operations, a complex cislunar telecommunications network is required before automated assembly operations could commence (Figure R.1).

The idea of assembling the components for the lunar base in low Earth orbit was proposed by the author in 1988. Modular propulsion systems would be employed to transport the entire base, complete and intact, directly to the lunar surface for immediate occupation by the astronaut crew (Figure R.2). This strategy would avoid

Table R.1. Acronyms

ACRC	Assured Crew Return Capability
ACSP	Attitude Control System Pallet
CELSS	Controlled Ecological Life-Support System
ETO	Earth To Orbit
IOC	Initial Operational Capability
LEO	Low Earth Orbit
LHB-1	Lunar Habitation Base-1
LOI	Lunar Orbit Insertion
LOR	Lunar Orbit Rendezvous
LPO	Lunar Parking Orbit
LLS	Lunar Landing System
MALEO	Modular Assembly in Low Earth Orbit
MOTV	Modular Orbital Transfer Vehicle
TLI	TransLunar Injection

Several ETO Launches
Several component wise TLIs and LOIs
Several lunar landings
Assembly on lunar surface
Substantial precursor missions

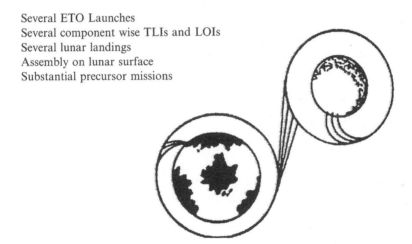

Figure R.1. Lunar surface assembly.

the cost and risk associated with manned extra-vehicular activity on the lunar surface, which had been proposed in several earlier concepts.

The MALEO philosophy may be summarized as follows:

1. Exploratory mission first ... establish validity ... build up infrastructure. The Apollo mission helped us get to know our neighbor. The exploratory aspects of the MALEO extended-duration mission are substantially different and more difficult. Though unmanned precursor missions will provide invaluable information on lunar base location, detailed manned exploration and analyses of terrain and resources will be necessary. More than one region may have to be explored in

Several ETO Launches
Assembly in LEO
One or two TLIs and LOIs
One lunar landing
Minimum precursors

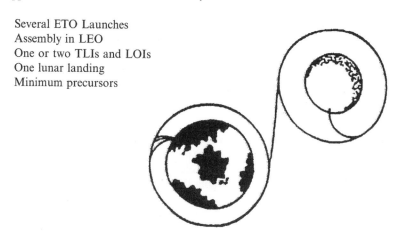

Figure R.2. The MALEO strategy.

detail (e.g., equatorial versus polar base). Several on-site demonstrations of technologies will be required before the validity of the base can be established. It is probably better to postpone extensive infrastructure build-up like orbital transfer vehicle systems and complex satellite communications networks until then.

2. Lunar/Mars mission must evolve directly from the LEO infrastructure. Evolution will require using what is available for the reason of engineering economy, reliability, and for quick development of the base. Commonality and inheritability should be maximized.

3. Substantial improvement in ETO launch (as well as lunar-landing payload capability) is required before safe and permanent manned lunar basing can commence. The HLLV will be an indispensable component for lunar base build-up.

4. The primary extended-duration mission must be completely self-sufficient. Redundancy is low, and all the extra-base build-up must be considered strictly experimental. Extra-base habitats and expanded indigenous facilities all require substantial manned assembly operations, system interaction, test "tweaking", and optimizing in order to support reliable manned habitation and operation. Until then, the primary base must provide the crew with a completely-self-sustained, safe haven/site office/workshop.

5. The primary mission must be able to support alternate architectures during lunar base evolution (erectable, deployable, indigenous). Competing technologies and schemes are offered for lunar base build-up. Many suggest an "incremental architecture" (straightforward organic expansion of the primary habitat). Erectable, deployable, modular, inflatable, and indigenous habits have all been suggested. All require substantial "tweaking" before reliable operation can begin. The primary base, then, must support the assembly/test activity safely while these technologies are being site-tested and certified for human occupation

and operation. As lunar build-up proceeds, unforeseen events may dictate further evolution along certain preferred technologies or hybrids most economically suited for the purpose. Therefore, the primary mission should have the versatility to evolve in any way without expending more than the necessary investment in time and money. The MALEO strategy suggested is an "adaptive incremental architecture".

6. The primary extended-duration mission must minimize manned EVA for assembly operations. Much more EVA would have to be carried out for science and technology experiments during the primary mission. Exploration will also require substantial EVA time. This activity is risky in an extended-duration mission due to the paucity of redundant systems at the phase of lunar base development. Present strategies suggest substantial manned EVA for assembly operations, which are even more difficult than exploratory operations, which are in turn even more difficult than exploratory or science-related manned EVA. The primary base should therefore minimize EVA for assembly operations.

7. A highly-accessible, serviceable emergency crew escape system must be emplaced at commencement, or preferably prior to astronaut crew activity on the lunar surface. Assembly crew safety is of the first priority. In the event of an assembly accident, injury or loss of life is possible for a variety of reasons but is highly unacceptable for reasons witnessed in the aftermath of the Challenger disaster or the Apollo launch pad accident. A highly-accessible emergency rescue system must be in place to deal with such a crisis. It seems reasonable to assume from the Skylab/Mir experience, as well as terrestrial analogue studies, that astronaut crew anxiety and stress related to long-duration missions will be a critical factor in determining lunar base design, assembly techniques and evolution, and for assembly crew productivity in particular. The provision of a quickly-deployable rescue vehicle would ease astronaut stress associated with build-up operations.

8. A solar storm shelter must be operational at commencement of, or preferably prior to, astronaut activity on the lunar surface. For the same reason as that given in item (6), the solar storm shelter is essential before long-term or permanent manned presence on the lunar surface.

9. Long-duration missions are a "learn as we go" experience, so the environment will have to be tailored as the mission proceeds. Limited data are available on long-duration spaceflight, and much less on extended astronaut performance under partial gravity conditions. From the Skylab/Mir experience, it is clear that the major design driver will be the maintenance of astronaut health and psyche. The mission architecture will be highly influenced by this factor, and it follows from item (5) that MALEO architecture will help towards this goal.

R.4 CONFIGURATION OF THE LUNAR HABITATION BASE-1 (LHB-1)

Several configurations are possible for the design of LHB-1. Two or more modules might be employed, as required by the mission objective (Figure R.3). In the past, horizontal and vertical configurations have been proposed for lunar habitation bases.

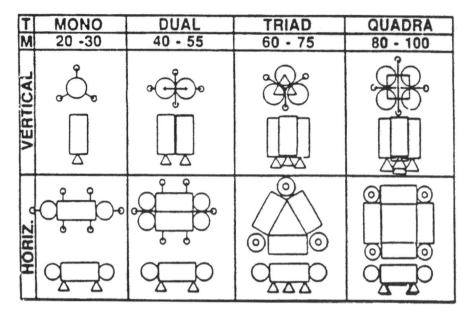

T	MONO	DUAL	TRIAD	QUADRA
M	20 -30	40 - 55	60 - 75	80 - 100

Figure R.3. Schematic lunar base module configuration study.

In the MALEO strategy, the horizontal configuration was preferred over the vertical one for the following reasons:

1. Commonality with ISS enhances design and engineering economy.
2. Commonality enhances crew adaptation and productivity.
3. Better and larger work spaces for extended-duration missions, which minimizes circulation spaces.
4. Wider footprint for better landing stability.
5. Ease of expansion during evolution by attaching additional horizontal modules.

Using space-station-derived modules as a base line, it would be possible to contain all the necessary systems required for the Phase 1 LHB (for a crew of four astronauts/mission specialists) in four modules and four nodes. They would be interconnected in a rectangular closed-loop configuration, which will facilitate dual egress for the astronaut crew. This configuration is also best suited for the central, symmetric location and clearance required for attaching the Lunar Landing System to the MALEO spacecraft.

R.5   COMPONENTS OF THE LUNAR HABITATION BASE-1 (LHB-1)

The major accompaniments of the LHB-1 are the modules and nodes which constitute its manned core (Figure R.4):

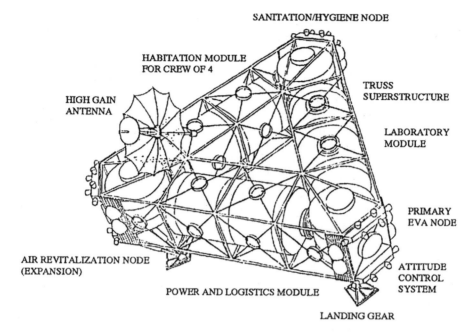

SANITATION/HYGIENE NODE

HABITATION MODULE
FOR CREW OF 4

HIGH GAIN
ANTENNA

TRUSS
SUPERSTRUCTURE

LABORATORY
MODULE

PRIMARY
EVA NODE

AIR REVITALIZATION NODE
(EXPANSION)

ATTITUDE
CONTROL
SYSTEM

POWER AND LOGISTICS MODULE

LANDING GEAR

Figure R.4. Components of a three-module lunar habitation base (LHB-1).

1. The habitation module houses the four astronauts/mission specialists and contains crew sleeping quarters, an exercise/recreation facility and a galley.
2. The sanitation/hygiene node is conveniently located adjacent to the habitation module and houses water-recycling and solid-waste management systems, as well as sanitation.
3. The physical laboratory module is also adjacent to the sanitation/hygiene node. This is equipped with interchangeable racks which are partially outfitted so that modifications and new setups are possible as the base evolves.
4. The life sciences laboratory module will study the effect of partial gravity on the human system as well as provide an array of agricultural and life-support experiments needed for further lunar base evolution.
5. The power and logistics module contains the power generation, storage, and regulation equipment. The module contains solar-power panels which are deployed externally. It can accommodate a nuclear power source that could be deployed externally.
6. The Controlled Ecological Life-Support System (CELSS) node is also designed so that the base may be extended by attaching additional modules. The concept is similar to the logistics module proposed for ISS, except that upon depletion, the modules are used to expand the base rather than being decommissioned or sent back to Earth.
7. The primary EVA node is the airlock used for all extra-base activity. It contains the EVA suits and equipment to prevent regolith back-tracking and contamination.

Table R.2. MALEO LHB-1 mass summary

	Weight (Mt)
Habitation module	15–17.5
Laboratory module	15–17.5
Power/logistics module	15–17.5
Primary EVA node	5–7
Air revitalization node	5–7
Sanitation/hygiene node	5–7
Truss superstructure	6
Landing shocks/airbags	4
Solar arrays/comm. ant	3
Lunar Rover X2	2
Miscellaneous	10
Total	*100*

8. The command centre node monitors and regulates the activity on the MALEO LHB-1. It will serve as the cockpit for the operation if a manned landing approach is adopted. Components which integrate the modules as a complete spacecraft are the following.

- The truss superstructure, which has three functions:
  - (a) it supports the thrust structure of the Modular Orbital Transfer Vehicle (MOTV);
  - (b) it distributes the forces transmitted from the MOTV uniformly through the entire LHB-1 during TLI, LOI and lunar landing;
  - (c) it offers the primary attachment points for the attitude control system pallet (ACSP), the landing gear/airbag system, and the storage pockets.
- The landing gear/airbag deployment system is designed to absorb the shock on impact, and may be conventional lunar-excursion-module-type absorbers, controlled gas escape airbags, or a hybrid system.
- The three attitude control propulsion pallets are assembled and fueled on Earth, brought up to LEO by the STS, and attached to the three corners of the LHB-1 after the truss superstructure has been built around the modules. They stabilize the attitude of the spacecraft during transit and landing operations.
- The storage racks, which may be configured as required, are placed symmetrically about the MALEO truss superstructure to maintain thrust structure symmetry and balance. They contain the rovers, the solar panels, and other EVA equipment, which are essential for the effective initial operational capability (IOC) of LHB-1.

R.5.1   The Modular Orbital Transfer Vehicle (MOTV)

The MOTV is used to transport the complete MALEO lunar base to the prescribed lunar parking orbit. It provides the required ΔV for TLI, LOI, and mid-course impulse and correction maneuvers. Two current or near-term options available for the design of the MOTV propulsion system are chemical cryogenic propulsion and nuclear electric propulsion. Advanced space shuttle main engine (SSME) technology would be applicable for MOTV development.

R.5.2   The Lunar Landing System (LLS)

The LLS is used to de-orbit and soft-land the MALEO LHB-1 on the lunar surface. Chemical cryogenic propulsion is favored for the descent and landing operation. Advanced RL-10 rocket engine technology would be applicable for LSS development.

R.6   THE LUNAR MALEO ASSEMBLY AND DEPLOYMENT OF LHB-1

At least four options exist for the MALEO assembly of LHB-1:

1. Free-space assembly using the STS as the work platform.
2. Free-space assembly using the STS as the primary assembly platform with assistance from ISS (Figure R.5).
3. MALEO LHB-1 assembly attached to ISS.
4. MALEO assembly connected to the manned core of ISS (Figure R.6).

The fourth option utilizes the ISS infrastructure most effectively by providing the assembly facilities and crew required for MALEO LHB-1 assembly and operation. The two options possible for the deployment of MALEO LHB-1 are:

1. Single-phase direct-lunar-landing option.
2. Two-phase lunar-orbit-rendezvous (LOR) option.

In the simplest MALEO single-phase direct-lunar-landing option, the Lunar Habitation Base (LHB-1) and the Lunar Landing System (LLS) are integrated in low Earth orbit (LEO), and the entire assembly is delivered directly to the lunar surface. The operation is expected to take a few days from TLI to touchdown. This option favors a fully-cryogenic MOTV propulsion system.

In the two-phase option, the LHB-1 is first transported to a lunar parking orbit (LPO) by the MOTV. In the second phase, the Lunar Landing System is transported by a similar MOTV to dock with the LHB-1 in LPO. After the LHB-1 and the LLS are securely interconnected in lunar orbit, the LHB-1 and LLS descend and touch down on the lunar surface as in the first option. The two-phase LOR option effectively reduces the TLI mass of the MALEO operation by one-half, and might be

Figure R.5. ISS-assisted MALEO LHB-1 assembly using the STS as the primary platform (four-module configuration) (Pat Rawlings).

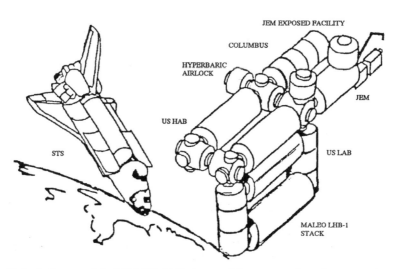

Figure R-6. A four-module MALEO LHB-1 assembly connected to the manned core of the ISS.

advantageous for cryogenic propellant management during the MALEO operation. This option is particularly suited for the slower, electric MOTV orbital transfer, where cryogenic propellant boil-off might be a prime concern. Manned EVA is required in lunar orbit.

Figures R.7 to R.15 illustrate the MALEO strategy two-phase option for the deployment of a three-module lunar habitation base.

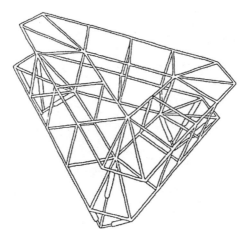

Figure R.7. MALEO LHB-1 truss superstructure.

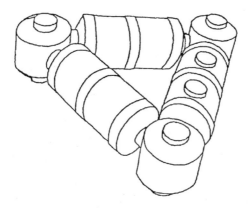

Figure R.8. MALEO LHB-1 module assembly.

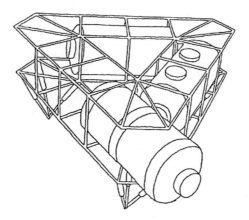

Figure R.9. MALEO LHB-1 module insertion.

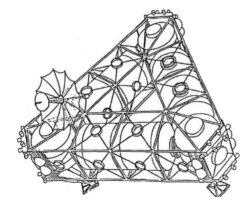

Figure R.10. MALEO LHB-1 assembly complete.

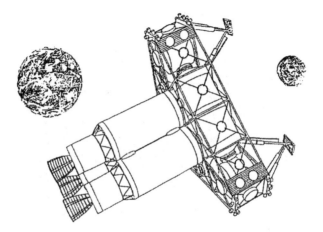

Figure R.11. Phase 1 LHB-1/MOTV translunar injection.

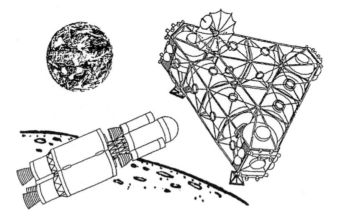

Figure R.12. LLS/LHB-1 lunar orbit rendezvous (LOR).

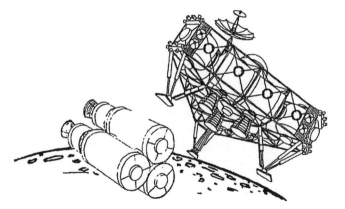

Figure R.13. LLS/LHB-1 de-orbit and descent (expended MOTV in background).

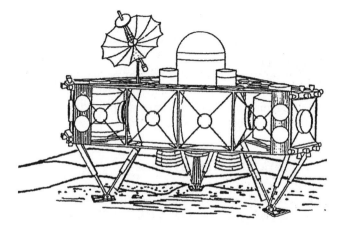

Figure R.14. MALEO LHB-1/LLS lunar surface touchdown.

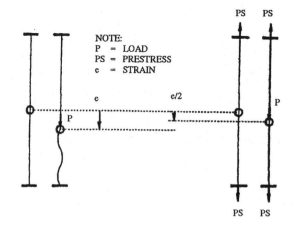

NOTE:
P = LOAD
PS = PRESTRESS
e = STRAIN

Figure R.15. The effect of pre-stress on strain.

R.7   PRINCIPLES OF PRE-STRESSED TRUSSES

It is evident from the operations listed above that the truss superstructure of the LHB-1 and the complementing structures of the Lunar Landing System (LLS) and the Modular Orbital Transfer Vehicle need to be highly efficient, light-weight, and reliable. Historically, spacecraft have used few, if any, members in tension.

Inherently-stiff tension members are suggested in the MALEO truss superstructure pre-tensioning system to conserve mass, minimize structural deflection and distribute thrusting forces optimally among the truss members. Compared to conventional trusses, pre-stressed trusses have some or all of the rigid bars substituted by flexible tension members.

The effect of pre-stress is as follows:

1. It reduces structural strain deformation by as much as 40–50 percent.
2. Compressive forces applied on the tension members are absorbed through a reduction of pre-stress.
3. Slack members are prevented if design pre-stress is at least 50 percent of the design load.
4. Pre-stressing the truss structure over 50 percent of the design load will compensate for creep and temperature changes, the latter being significant for structures in space where the temperature fluctuations of up to 200 gray may be expected due to thermal-cycling effects in low Earth orbit.

R.8   MALEO LHB-1 STRUCTURAL SYSTEM

The MALEO configuration is structurally strengthened by suspending the LHB-1 modules with a truss superstructure, so that LHB-1 will be able to absorb the stresses induced on it uniformly during injection (TLI), lunar orbit insertion (LOI), and lunar surface touchdown. The thrust structure, which includes the truss superstructure, the Lunar Landing System, and Modular Orbital Transfer Vehicle interconnections, is selected so that the forces are applied symmetrically about the truss superstructure. The MOTV and the LLS share the same thrusting point on the truss superstructure. The truss superstructure is designed to facilitate quick and easy real-time tele-robotic/ EVA-assisted assembly in LEO. The structure is designed and built with the limits and capabilities of the astronaut assembly crew in mind. An erectable/deployable philosophy is employed so that the structure is partially assembled even before the launch to LEO. This philosophy is applied again upon lunar surface touchdown, after which much of the truss superstructure is disassembled and used for other building purposes on the lunar surface.

Though a stressed skin module configuration could be designed to take the stresses arising from TLI and LOI, which are estimated to be about 2 G maximum, module safety and other factors favor the use of a separate truss superstructure to be employed for the thrusting load distribution.

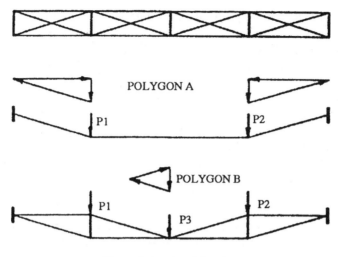

POLYGON A

P1

P2

POLYGON B

P1

P2

P3

Figure R.16. Load bearing.

The primary forces applied to the structure during transit and landing on the lunar surface determine the structure and its configuration. These forces are computed to the first order from data available on the state-of-the-art propulsion systems and from cislunar orbital dynamics, which are well known. Figure R.16 illustrates the approximate velocity changes (ΔV) required for the various cislunar maneuvers in order to accomplish the MALEO mission. From these ΔVs it is possible to calculate the propellant required for each of those maneuvers, and thereby the mass of the MALEO spacecraft/propellant tankage/engines can be estimated.

R.9 MALEO: TRANSPORTATION LOADS AND FORCES

To first order, using state-of-the-art technology, for every unit of payload to be landed on the lunar surface, an equal amount of propellant is required in lunar orbit. Seven units of propellant are required to ship one unit of payload all the way from low Earth orbit to the lunar surface, and approximately 20 units of propellant are required to lift one unit of payload from the Earth to low Earth orbit. Table R.3 illustrates the mass of the vehicle/payload during cislunar transit and lunar landing.

From Table R.3, the mass of the MALEO assembly during cislunar transportation is known. Using Figure R.16, the velocity changes during cislunar transportation of the MALEO spacecraft are also known. Using space shuttle main engine (SSME) technology, the accelerations during the orbital maneuvers may be computed. Given the same set of engines, these accelerations are proportional to the velocity change required to conduct the maneuver. From Figure R.16, it is evident that translunar injection (TLI) forces are expected to be the most severe forces during transit. A force equivalent to twice that of the Earth's gravitational attraction (2 G) is expected to be imparted to the MALEO truss superstructure from the Modular Orbital Transfer

Table R.3. Vehicle payload mass during cislunar transportation

Earth (Mt)	Low Earth orbit (Mt)	Low lunar orbit (Mt)	Lunar surface (Mt)
160	8	2	1
16,000*	800	200	100 (MALEO)

* Eight STS-equivalent missions to LEO.

Vehicle (MOTV) during TLI. The TLI maneuver would last for about five minutes. New technologies such as electric propulsion may substantially decrease these forces by virtue of the slow but continuous thrusting of these engines. Lunar orbit insertion, de-orbit, and lunar touchdown forces are expected to be typically less than the TLI forces. The gravitational force on the Moon is about $\frac{1}{6}$ that of the Earth. The MALEO spacecraft would weigh about 20 Mt on the lunar surface. Assuming an impact loading factor of 50 percent in the event of an unsymmetrical or heavy impact landing, the forces encountered by the MALEO structure are in the order of 30 Mt. These loads and forces are distributed among the three sets of landing gears, or are uniformly distributed by the multicellular airbags which would cushion the touchdown.

R.10 ADVANTAGES OF THE MALEO STRATEGY

If space-station-derived modules are to be employed in the construction of a Phase 1 lunar base, then the MALEO strategy offers the following advantages:

1. Safer LEO radiation environment for EVA, which is also less risky and more economical than EVA on the lunar surface.
2. LEO offers the possibility of Earth-based, real-time, tele-robotic assembly operations.
3. Spares and replacements are easily and more economically flown to LEO, or borrowed from the LEO ISS infrastructure.
4. LEO offers ISS-assisted MALEO assembly operations. ISS offers a stable platform/scaffolding required for assembly.
5. The clean LEO assembly environment avoids dealing with the lunar soil, which has undesirable and cohesive properties. Regolith has interfered with manned EVA systems in the past.
6. In the event of an assembly accident, a crew rescue is more feasible from LEO using the assured crew return capability (ACRC) of ISS than from the lunar surface.
7. The inheritability and commonality with ISS hardware enhances cost/unit economy, as well as spares and replacement unit, for both the MALEO and the ISS program.

8. It avoids the risk of the repeated Earth-to-orbit (ETO) launchings, TLIs, LOIs, and lunar landings associated with the typical component-wise sequential build-up prescribed by the assembly-on-lunar-surface strategy.
9. The LHB-1 is substantial and safely configured for habitation upon touchdown. Conventional strategies involving lunar surface assembly cannot offer a comparably-safe environment during base build-up without additional investment.
10. In this primary extended-duration mission, the MALEO LHB-1 offers an assured and substantial safe haven at IOC which will help to diminish anxiety and enhance productivity among the astronaut crew.
11. The MALEO strategy will provide the necessary experience to assemble and operate the manned Mars mission vehicle.
12. It offers the possibility for substantial international cooperation for the peaceful development of space.

R.11 THE CHALLENGES

A disadvantage of the MALEO strategy is the risk of losing the entire LHB-1 in the event of an accident during transportation and landing of the base. However, this should be evaluated carefully against the repeated component-wide ETO launches, TLIs, LOIs, and lunar surface landings associated conventional strategies in which operations are arranged sequentially, and are therefore susceptible to delay and operational collapse in the event of a failed component launch or landing.

The MALEO strategy suggests a prefabricated Phase 1 lunar base, which deprives the assembly crew of the initial experience of learning to work on the lunar surface. A substantial safe haven is the first priority for the astronaut crew who are on an extended mission to establish a permanent base on the Moon. Lunar surface EVA experience will be gained as exploration and experimentation continue, and the assured safety of the MALEO base would also help to relieve astronaut crew anxiety while the base is operating in IOC mode.

As MALEO is a large manned spacecraft, all systems – the LHB-1, the LLS and the MOTV – need to be studied, their dynamic characteristics examined, and their limitations confirmed. Though the TLI and LOI can be effectively controlled using an electric propulsion MOTV for the orbital transfer maneuver, the vibration and resonance characteristics of such a large structure during descent and heavy impact of an unsymmetrical landing require detailed analysis.

R.12 CONCLUSIONS

The systems described here need to be designed and built for assembly in space. In an extended-duration primary mission of this nature, crew safety and provision of a safe haven and comfortable spaces for work and rest require the highest priority. The MALEO strategy minimizes the initial, risky, manned EVA associated with lunar surface assembly operations and provides a safely-configured working environment for the astronaut crew on touchdown. The LHB-1 will alleviate crew anxiety and enhance productivity. The MALEO strategy maximizes the commonality and inheritability from the LEO infrastructure that will be in operation by the year 2000. The experience gained while assembling the International Space Station will be invaluable for the MALEO assembly and deployment of LHB-1. More study is required to confirm the feasibility of clustering existing propulsion systems to build Modular Orbital Transfer Vehicles, the modifications which might be necessary, and if these vehicles might be launched from the Earth fully loaded with propellant, so that a cryogenic depot might not be necessary in LEO for the MALEO operation. High-power electric propulsion like the SRPS ammonia arcjet offers promise for an MOTV for use in the MALEO deployment of LHB-1.

MALEO LHB-1 is essentially a large manned spacecraft that could be used as an orbiting station for the Moon and Mars. With its lack of aerodynamic contours and its control and propulsion requirements, the concept is particularly suited to landing on low-gravity bodies without an atmosphere, with the Moon an ideal first choice. MALEO bases could land on Phobos, or could be used for prospecting the near-Earth objects in the future.

The importance of highly-reliable structural systems for the truss superstructure of the LHB-1, the Lunar Landing System and the Modular Orbital Transfer Vehicle, are clearly evident for the success of such a mission. Advances in structural materials, on-orbit assembly techniques acquired while assembling and operating ISS, advances in electric propulsion for orbital transfer applications, and the year 2000 time frame for execution all seem to coincide favorably for the successful application of the strategy for a lunar base build-up.

Improving U.S.–Russian superpower relations, the return to flight of the STS, the imminent United Europe and the maturing space programs of China and Japan all show a trend toward accomplishing major near-term achievements in the space frontier. A project like the MALEO deployment of LHB-1 could become a catalyst for the peaceful international development of space.

R.13   ACKNOWLEDGMENTS

This appendix is dedicated to my mother, Nanoo Saraswathy Thangavelu. I would like to thank P. Diamandis and T. Hawley at the International Space University. Thanks are due to G. Dorrington of Cambridge University, and to R. Shaefer and P. Rawlings for concept development. It is the patience and encouragement of the faculty at the University of Southern California which caused the concept to come to fruition. Thanks also to R. Harris, Dean of Architecture, L. Silverman, Dean of Engineering, R.K. Miller, Associate Dean of Engineering, G.G. Schierle, Director of Building Science, E.P. Muntsa, co-chairman of Aerospace Engineering, E. Rechtin, R.F. Brodsky and D. Vergun.

R.14   REFERENCES

Alred, J.W. and Bufkin, A.L. (1987) *Lunar Surface Construction and Assembly Equipment*. NASA, Washington, D.C., Eagle Engineering Report NAS9-17878.

Alred, J.W. *et al.* (1989) *Lunar Outpost*, Systems Definition Branch, Advanced Programs Office, NASA Johnson Space Center, Houston, TX.

Alred, P. *et al.* (1989) Development of a lunar outpost: Year 2000–2005. *AIAA 2nd Symposium on Lunar Bases*.

Bekey, I. (1989) Personal correspondence with Dr. E. Rechtin, Professor of Engineering, University of Southern California, Los Angeles, CA.

Bufkin, A.L. *et al.* (1989) EVA concerns for a future lunar base. *Lunar Base Symposium*, LBS-88-241.

Bush, G. (1989) *Presidential Address on 20th Anniversary of the Apollo 11 Moon Landing*, The White House, 20 July.

Cohen, M.N. (1987) *Light Weight Structures in Space Station Configurations*, NASA Ames Research Center, Unisearch Ltd., Moffet Field, CA.

Deringer, W.D. *et al.* (1989) *Arc Jet Propulsion System for an SP-100 Flight Experiment*, NASA Jet Propulsion Laboratory, Pasadena, CA, Recon. 89A24247.

Duke, M.; Wendell, M.; and Roberts, B. (1985) Strategies for a permanent lunar base. *Lunar Bases and Activities for the 21st Century*, Lunar and Planetary Institute, Houston, LBS 88-7232.

Horz, F. (1985) Lava tubes: Potential Shelters for habitats. In: Mendell, W., ed., *Lunar Bases and Activities for the 21st Century*, Lunar and Planetary Institute, Houston, TX.

Hypes, D.W. and Hall, J.B. (1988) *ECLS Systems for a Lunar Base: A Baseline and Some Alternate Concepts*. NASA Technical Report, NASA HQ, SAE 881058.

Iwata, T. (1988) Unmanned surface developments for manned surface activities. *Lunar Base Symposium*, LBS-88-7232.

Johnson, S.W. and Leonard, R.S. (1985) Evolution of concepts for lunar bases. *Lunar Bases Conference*, Lunar and Planetary Institute, Houston, TX.

Kaplicky, J. and Nixon, D. (1985) A surface assembled structure for an IOC base. *Lunar Bases and Activities for the 21st Century*. Lunar and Planetary Institute, Houston, TX.

Khalili, N. (1985) Magma Ceramis and fudes Adobe structures generated in-Situ. In: Mendell, W., ed., *Lunar Bases and Activities for the 21st Century*, Lunar and Planetary Institute, Houston, TX.

Kline, R.; McCaffres, R.; and Stein, D. (1985) Potential designs of space stations and space platforms. In: Bekey, I and Herman, H., eds., *Progress in Astronautics and Aeronautics*, American Institute of Aeronautics and Astronautics, New York, AIAA-99-985.

Loftus, J.P. and Patton, R.M. (1988) *Astronaut Activity*, NASA Johnson Space Center, Houston, TX.

Lowman, P.D., Jr. (1985) Lunar bases: A post Apollo evaluation. In: Mendell, W., ed., *Lunar Bases and Activities of the 21st Century*, Lunar and Planetary Institute, Houston, TX.

MacElroy, R.D. *et al.* (1985) The evolution of CELSS for lunar bases. In: Mendell, W., ed., *Lunar Bases and Activities for the 21st Century*, Lunar and Planetary Institute, Houston, TX.

Mikulas, M. (1988) In-space assembly of large space structures. *International Space University Lectures, July*, MIT Press, Boston.

NASA (1988) *Exploration Studies Technical Report FY 1988 Status*, Vols. 1 and 2. Office of Exploration, NASA, Washington, D.C., TM-4075.

National Commission on Space (1986) *Pioneering the Space Frontier,* Bantam Books Inc., New York, 1986.

NRC (1990) *Human Exploration of Space: A Review of NASA's 90-Day Study and Alternatives,* National Research Council, National Academy Press, WA.Schierle, G.G. (1990) Prestressed trusses: Analysis, behavior and design. *Proceedings of the Forth Rail Bridge Centenary Conference,* Chapman & Hall, London.

Thangavelu, M. (1989a) *MALEO: Modular Assembly in Low Earth Orbit.* Masters thesis in Building Science, University of Southern California, Los Angeles, CA.

Thangavelu, M. (1989b) *MALEO: A Space Station Freedom Based Architecture for an IOC Lunar Base* (Aerospace Systems Architecting AE599 for Prof. Rechtin). University of Southern California, Los Angeles, CA.

Thangavelu, M. (1990a) MALEO: Modular Assembly in Low Earth Orbit – Strategy for an IOC lunar habitation base. *Vision 21 Symposium: Space Travel for the Next Millennium,* NASA Lewis Research Centre, Cleveland, OH.

Thangavelu, M. (1990b) MALEO: Modular Assembly in Low Earth Orbit – An alternate strategy for lunar base development. *Proceedings of Space '90: 2nd International Conference on Engineering, Construction and Operations in Space,* ACSE, Albuquerque, NM.

Thangavelu, M. and Dorrington, G.E. (1988) *MALEO: Strategy for Lunar Base Build-Up,* IAFC, Bangalore, India, ST-88-15.

Thangavelu, M. and Schierle, G.C. (1990) Structural aspects of a lunar base assembled in low Earth orbit, recent advance. In*: Structural Engineering: Proceedings of the Forth Rail Bridge Centenary Conference, Herriot Watt University, Edinburgh.*

Weaver, L.B. and Laursen, E.F. (1988) Techniques for the utilization of extraterrestrial resources. *IAFC Proceedings.*

Wood, L. (1990) *The Great Exploration Plan for the Human Exploration Initiative.* Lawrence Livermore National Laboratory, Livermore, CA, LLNL Phys. Brief 90-402.

Woodcock, G. (1986) Logistic support for lunar bases. *Proceedings of IAFC 1986.*

Woodcock, G. (1983) *Mission and Operations Modes for Lunar Basing,* Boeing Aerospace Corporation, Huntsville, AL.

Appendix S

Logistics for the Nomad Explorer assembly assist vehicle

Madhu Thangavelu   *Conductor, Space Exploration*
Architectures Concepts Synthesis Studio

S.1 ABSTRACT

The traditional concepts of lunar bases describe scenarios where components of the bases are landed on the lunar surface, one at a time, and are then assembled to form a complete stationary lunar habitat. Recently, some concepts have described the advantages of operating a mobile or "roving" lunar base. Such a base vastly improves the exploration range from a primary lunar base. The Nomad Explorer is such a mobile lunar base. This paper describes the architectural program of the Nomad Explorer, its advantages over a stationary lunar base, and some of the embedded system concepts which help the roving base to speedily establish a global extra-terrestrial infrastructure. A number of modular autonomous logistics landers will carry deployable or erectable payloads, and service and logistically re-supply the Nomad Explorer at regular intercepts along the traverse. Starting with the deployment of science experiments and telecommunications networks, and the manned emplacement of a variety of remote outposts using a unique EVA Bell system that enhances manned EVA, the Nomad Explorer architecture suggests the capability for a rapid global development of the extraterrestrial body. The Moon and Mars are candidates for this strategy.

S.2 INTRODUCTION

Permanent lunar base establishment will entail detailed terrain exploration, sampling, and analyses. Current studies expect all the site analyses to be performed robotically. Though unmanned precursors would provide valuable information on possible sites for lunar base location, in order to "live off the land", as President Bush directed, detailed manned exploration and formal site analyses of selected candidate sites will be required. Since site selection is a critical task, initial manned missions will have to

Table S.1. Acronyms

AMCL	Autonomous Modular Common Lander
DIPS	Dynamic Isotope Power System
EVA	Extra-Vehicular Activity
ECLSS	Environmental Control and Life-Support Systems
GNC	Guidance, Navigation, and Control
LOI	Lunar Orbit Insertion
MOSAP	MObile Surface APplications Traverse Vehicle
POS	Point Of Start
RFC	Regenerative Fuel Cell
RMS	Remote Manipulator System
TLI	TransLunar Injection
VLTV	Very-Long-range Traverse Vehicle

"site-hop" before settling on the prime candidate site or sites. Global mobility would then become imperative in the initial manned mission. Examining the program requirements which have already been developed by NASA, it might be possible to design a single manned mission assisted by an autonomous logistics lander system that would conduct all the required tasks, as well as establish a global infrastructure of remote science outposts, telecommunications networks, and pilot projects. All of these activities could be simultaneously manifested in a combined manned/unmanned mission architecture.

S.3 DEVELOPMENT OF THE NOMAD EXPLORER STRATEGY

NASA exploration studies indicate the need for manned rovers to assist in the development of extraterrestrial bases. Three classes of rover have been identified and studied. Point design concepts have also been proposed in all the categories. Studies include the short-range rover for around-the-base activity with a range of 50–100 km; the long-range vehicle with an operating range of 1,000 km; and finally the very-long-range traverse vehicle (VLTV) capable of covering 3,000–10,000 km on a single traverse. The mobile surface application traverse vehicle (MOSAP) is a vehicle that NASA has proposed for this purpose. Capable of all the exploratory functions and normal EVA that is carried out from a conventional stationary base, the VLTV, by virtue of its long range and enhanced manned crew systems capability, is in essence a "roving lunar base". Unlike a stationary base, from which detailed manned terrain exploration is limited by the range of the rovers, this mobile concept for a primary exploration-oriented base offers an unlimited range for exploratory traverses (Figure S.1). It is an extension of the program requirements of this very long-range traverse vehicle that leads to the possibility of the Nomad Explorer strategy.

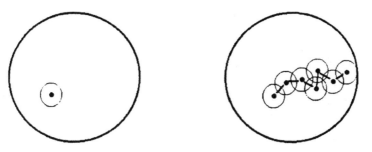

Figure S.1. Stationary extraterrestrial base and mobile extraterrestrial base.

The Nomad Explorer strategy is a synthesis of two major system architectures:

1. *The Nomad Explorer vehicle.* The Nomad Explorer vehicle is derived from the MOSAP vehicle developed by NASA. Several other studies carried out by NASA contractors on long-range rovers are also available. Very-long-range traverses are possible using this vehicle. The architecture of this vehicle is adaptable to both manned or autonomous unmanned operations. The configuration studied in this paper portrays a manned Nomad Explorer vehicle (Figure S.2).

2. *The Autonomous Modular Common Lander system.* The Autonomous Modular Common Lander (AMCL) is a common vehicle that can deliver both crew or cargo, depending on the modular configuration employed. Logistics, consumables, and erectable/deployable science outpost components that are launched from Earth are delivered to the lunar surface using the AMCL system. The concept is derived from the lunar lander and the Common Lander Study which NASA has developed. The AMCL is outfitted with a modular payload on the Earth, landed autonomously on the lunar/Mars surface, intercepted by the Nomad Explorer, and the payload unloaded and deployed or transferred to the roving vehicle. The modular capability allows the lander to be sized for any mission, ranging from 5to 25 Mt, depending on the requirements during the traverse of the Nomad Explorer vehicle (Figure S.2).

The Nomad Explorer strategy is aimed at establishing a global scientific and telecommunications network, even during the initial "trail-blazing" permanent base selection run. The Nomad Explorer, besides being a VLTV, is also conceived as a manned EVA and assembly assist vehicle.

Exploratory activity like soil sampling, soil mechanics, and locating natural formations that enable habitation (such as lava tubes) is followed by the deployment of science outposts and habitation facilities during the course of the same mission. During the course of its very long traverse, the vehicle will intercept autonomously-landed payloads ("modular common lander" payloads) along the traverse route, assemble the payloads or deploy them, check them out at the site, and certify them for operations, carrying out all these functions using the crew of the vehicle. During the later stages of extraterrestrial base evolution, the strategy could be used for

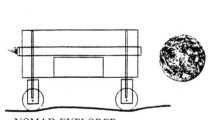 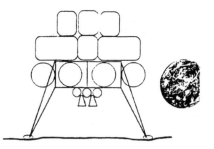

NOMAD EXPLORER

Salient Features
1. Very long range traverse vehicle.
2. Crew of three, six months, 10,000 km.
3. Manned assembly assist vehicle
4. Bell attachment for enhancing manned EVA.
5. Nuclear power.
6. Logistically resupplied by modular replaceable units delivered by the autonomous Lander.

Mission
1. 'Site-hop' to locate best site for permanently manned lunar base.
2. During traverse, establish science outposts along the way, deploy experiments, and establish extraterrestrial global telecommunication infrastructure using fibre optics and laser technology.

AUTONOMOUS MODULAR COMMON LANDER

Salient Features
1. Autonomous landing capability.
2. Modular structural/propulsion design.
3. Common carrier for Nomad Explorer crew transport, logistics, eretable or deployable science payloads.
4. Payload capability – 5.10, 15.20.25 mt depending on modular configuration.

Mission
1. Provide crew transport. Logistics support for the Nomad Explorer.
2. Carry science/other payloads to be deployed/erected by crew or Nomad Explorer.

Figure S-2. Schematic of Nomad Explorer and the Autonomous Modular Common Lander (AMCL).

maintenance and repair activity of these remotely-based highly-sensitive scientific experiments, such as optical interferometry observations, pilot plants for manufacturing lunar indigenous materials and components, telecommunications relay platforms, and so on.

The Nomad Explorer global extraterrestrial strategy is as follows:

1. Precursory high-resolution mapping of the entire lunar/Mars surface using polar-orbiting satellites.
2. Analysis of terrain information, followed by determination of alternate likely candidate sites for lunar/Mars base location. Alternate rover traverse routes established with the aim of maximizing scientific returns along the route while conducting detailed terrain surveys for locating alternate sites for a permanent base. A lunar polar traverse might be considered to explore trapped volatiles.
3. One or more Nomad Explorer VLTVs launched from ETO, coupled with lunar lander tankage in LEO, and landed on the lunar surface at the predetermined point of start of traverse.

4. Lunar lander with crew accompany Nomad Explorer to the point of start (POS) of traverse.
5. Crew transfer to Nomad Explorer and begin traverse. First telecommunication Earth-link established at POS.
6. Modular autonomous landers deliver modular payload along traverse route. Alternate sites surveyed. Crew intercept payload, retrieve consumables/logistics modules, prepare and carry out assembly and deployment of science experiments and telecommunication relay stations, using manned EVA enabling systems which are part of the Nomad Explorer vehicle architecture.
7. Nomad Explorer completes traverse. Establish global fiber-optic telecommunications network. Install scientific outposts. Conduct detailed geological surveys and prospecting. Determine location for a permanent lunar settlement.
8. If more than one Nomad Explorer is landed, parallel activity would further speed up base site selection and global infrastructure development. If required, the teams could assist each other by congregating at a particular site of interest, for exploration, for additional manpower, or for establishing a permanent manned lunar base at the most suitable location.
9. At end of traverse, moth-ball Nomad Explorer into energy-conserving "hibernation mode", or set up vehicle traverse operations. Permanent base site established.
10. Crew transfer to Earth-return vehicle that has been autonomously landed at end of traverse. Crew depart for Earth. Mission complete. Permanent base establishment activities commence.

The Nomad Explorer strategy for lunar basing is depicted in Figure S.3.

S.4 THE NOMAD EXPLORER VEHICLE SYSTEMS ARCHITECTURE

The Nomad Explorer is a VLTV with an essentially-unlimited operating range, which depends entirely on the number of logistic re-supply missions that are flown to it during the course of the mission. Assisted by the AMCL system, the Nomad Explorer carries only the consumables, logistics, and spares it would require between regular intercepts of the AMCL; much like the optimum pit stops and refueling operations carried out in automobile racing. A 10,000-km traverse could be used as an example to demonstrate the proposed capability of the Nomad Explorer. The duration of traverse could be six months to a year, with the possibility of a complete crew change-out in the middle of the mission. Two regenerative fuel cell (RFC) power plants with a total peak output in the range of 50 kW are required for powering the drive train and all of the manned and unmanned systems. Mission architecture dictates a crew of three for optimum performance, and the manned systems and life support are provided for three crew members. A simple exploded schematic in Figure S.4 shows the basic systems of the manned Nomad Explorer.

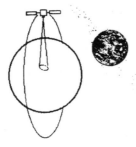 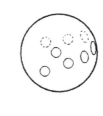

1. High-resolution imaging of extraterrestrial surface. Detailed global maps prepared

2. Several candidate base sites identified both on lunar near side and far side.

 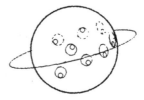

3. Alternate Nomad Explorer traverse routes examined. Final traverse route identified.

4. Autonomous lander (AMCL) delivers logistices to candidate sites.

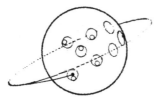 

5. Nomad Explorer landed at point of start of traverse. All systems checked out from Earth. Crew lands next to vehicle in AMCL configured for crew transport. Transfers to Nomad Explorer

6. Detailed manned exploration of natural terrain formation (lava tubes, rilles etc.). Soil mechanics and other 'hands-on' experiments conducted. Telecommunication link deployed as mission proceeds.

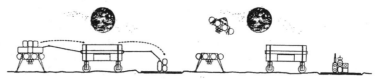

7. As mission proceeds, Nomad Explorer regularly rendezvous with AMCLs carrying site specific payloads of science experiments and related hardware. Nomad Explorer intercepts AMCLs, assemblies and deploys science packages. Logistics replenshied. If mission duration exceeds safe stay times, second full crew replacement arrives during middle of mission and the first full crew returns to Earth in AMCL.

8. Nomad Explorer completes traverse. Vehicle switched to 'hibernation mode' until next mission. Crew return to earth in AMCL that is ready and waiting at end of traverse.

Figure S.3. The Nomad Explorer mission plan.

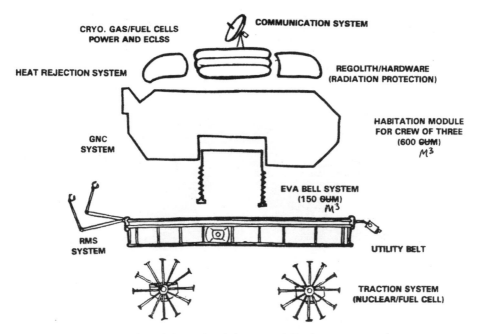

Figure S.4. Schematic of the Nomad Explorer systems.

The main features of the Nomad Explorer are as follows:

1. *Habitation.* The pressurized volume of the vehicle is about $600 \, \text{m}^3$. This volume contains long-term accommodation facilities for three crew, work and conferencing areas, a command and control center, and ample storage space. Besides a galley, hygiene and waste management facilities, the long-term accommodation includes a health maintenance facility and a recreation space.
2. *Controlled Ecological Life-Support System (CELSS).* The CELSS will handle the needs of the three crew members. Cryogenic nitrogen, oxygen, hydrogen, and water are available from RFC operations. Though complete closure of the CELSS is not envisaged, it may be possible to operate with a 90 percent efficiency. The International Space Station (ISS) will provide the basis for the Nomad Explorer CELSS.
3. *The EVA Bell.* The EVA Bell for enhancing manned EVA is a prominent feature of the vehicle, and is later described in detail. The invention allows the astronaut crew to perform EVA in a more comfortable manner by providing a shirt-sleeve environment around the payload to be assembled and deployed (Figure S.6)
4. *The utility belt.* A utility belt is strapped around the perimeter of the vehicle. This belt has modular racks carrying "plug-on" modules for logistics, consumables, and waste management. These are replenished by replacement modules arriving on autonomous landers which the vehicle intercepts from time to time along the traverse. The belt also carries tools and accessories for EVA. Two small

Mass	45MT
Pressurized volume	276m³
Traverse capability	3,000 km
Logistic resupply	AMCL
Speed of traverse	20 kmph
Power (nuclear/fuel cell)	50 kw

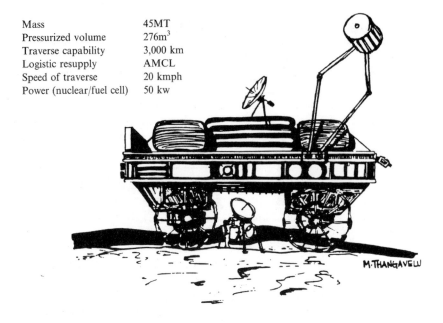

A Longitudinal Section

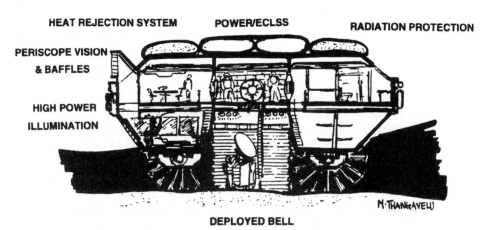

DEPLOYED BELL

Figure S.5. The Nomad Explorer: a possible configuration.

unpressurized rovers for "around-the-base" traverses are also part of the EVA accessories. These rovers have a range of about 100 km, and are powered by fuel cells and photovoltaics.

5. *The remote manipulator system.* Two remote manipulator systems, capable of assisting manned EVA, loading and unloading cargo, and providing anchoring or scaffolding support during assembly/deployment operations, run along two

tracks on the top and bottom of the entire length of the utility belt. High-resolution cameras mounted on the track provide video support during the traverse, as well as assembly operations.

6. *Traction system.* Traction could be provided using several options. The Nomad Explorer configuration in Figure S.5 depicts independently-powered and steer-able, large, variable-diameter wheels which are unfurled and deployed after landing. The Figure S.7 schematic shows a telescopic traction system that is capable of adjusting the height of the Nomad Explorer chassis for enhancing traverse, as well as assembly operations.

7. *Radiation protection.* Radiation protection is provided by skilful placement of system hardware on the vehicle, so that they provide sufficient mass for protec-tion. Tankage might be employed to enhance radiation protection. Regolith bags could be packed and laid in areas that require additional protection during solar-particle events which might occur during the traverse.

8. *Guidance, navigation, and control system.* Guidance, navigation, and control (GNC) of the vehicle is achieved through appropriate systems. Visual feedback could be direct or augmented by video support. Real-time telecommunication is possible through a 6-m Earth-pointing antenna and a chain of surface-based relays or low-orbiting satellites that could be employed for contact when line-of-sight communications is not possible. Fiber-optics/free-space laser communica-tion and related technologies could enhance Nomad Explorer operations.

9. *Power system.* Advanced regenerative fuel cell (RFC) technology, as well as nuclear technology, are suggested for the Nomad Explorer power system. The nuclear power option using the dynamic isotope power system (DIPS) needs further study. Shielding requirements need to be considered for nuclear power systems. A set of RFC batteries could power the Nomad Explorer between AMCL intercepts. At each intercept of the AMCL, these RFCs could be recharged. In addition, advanced photovoltaic arrays could be employed to provide support to the RFCs. In generating about 50 kW of power (25 kW for drive train and 25 kW for the manned systems including the ECLSS), the heat rejection system would have to handle about 16 kW. High-efficiency heat rejec-tion systems are required. Dust contamination of radiator surfaces will require study and appropriate design. Recent developments in "power beaming" – a technique whereby a microwave/laser beam from an external source is used to transmit power to the vehicle during traverse operations – could substantially improve the performance of the Nomad Explorer by reducing the payload associated with power generation and storage equipment.

10. *Mass and payload configuration.* All these systems are designed to fit within an HLLV (Energiya or revived Saturn V-B technology) payload shroud that is 30 m tall and 10 m in diameter. The payload mass at launch is about 35 Mt + propul-sion, tankage, and structure for TLI, LOI, and lunar landing. This mass is above the NASA lunar landing capability of 25 Mt. Though this Nomad Explorer study suggests that larger landers are required for the mission, it would be possible to scale down the vehicle and still preserve the mission architectural strategy for a smaller vehicle. The unmanned version of the Nomad Explorer is such a small

Table S.2. The Nomad Explorer extraterrestrial assembly assist vehicle for rapid lunar and martian global infrastructure development.

Summary specifications
Vehicle dimensions:	$16 \times 4.5 \times 10$ m
Touchdown mass:	45.0 Mt
Cruising speed:	20 km/h, maximum 30 km/h
Dynamic suspension system	
Traction:	120 4-hp electric motors
Traverse logistics/service interrupt:	3,000 km
Power system:	primary power cart, 50 kW maximum
Auxiliary system in crew vehicle:	10-kW solar/fuel cell

EVA Bell specifications
Pressurized volume: $(4 \times 4 \times 3m)$, 48 m^3
Nominal operational pressure/O$_2$ atmosphere: 4 psi
Alternative concepts for floor-plate/mat to compensate for uplift:
    (a) Internal stabilization used if stiff floor-plate is adopted for mat system
    (b) External multiple anchoring suggested for flexible mat system
    (c) Pressure-seal technology:
        double neoprene O-ring pressure lock;
        leakage rates <10–12 litres/m

Power
50 kW from solar array/fuel cell cart supplied through drag-line cable
10-kW solar array nickel–hydrogen fuel cell battery in the Explorer vehicle
    Also proposed is a more compact nuclear-power cart trailer, employing either cable or intermittent power-beaming to Explorer vehicle. Autonomous trailer guidance and navigation is proposed. If nuclear, there will be a regolith water shield around the cart – 1,000-m nominal trailing distance. Deep-space radiator heat rejection system, 20 kW rated.

Pressurized interior
Volume: 276 m^3
Command and control cockpit (periscope for radiation mitigation + through visor + stereo camera navigation); vehicle capable of high-fidelity tele-operation/GNC from Earth (between crew duty?)
Meeting facility and galley
Airlock/EVA toolkit and optional suit don/doff
Six crew quarters and logistics attachment module access
Hygiene facility

Communications
Earth links: optical, using a 12-inch Schmidt Cassegrain; bandwidth for high-fidelity tele-operations on multiple channels
Local communications: several options

Radiation protection
Water tanks/regolith bags to form shelter area

Traction
Four 120-hp dual (redundant) high-torque independent electric units; integral veriable/meshed gear transmission employing actively variable-diameter wheels to compensate for high-terrain bank angles; active damping

Lighting
High-pressure sodium/halogen up to 1 kW for activity ranging from specimen collection to long-range visibility

vehicle. Substantially smaller (5–7 Mt), the robotic Nomad Explorer would employ the same mission strategy. However, reliability and the capability to handle contingencies require demonstration. A possible manned Nomad Explorer configuration is depicted in Figure S.5.

The next section discusses certain special features of the Nomad Explorer vehicle which were developed to enhance mission capabilities.

S.5   THE PROBLEM WITH CONVENTIONAL EXTRA-VEHICULAR ACTIVITY

Conventional EVA is a time-consuming, hard, and inefficient, yet essential part of human activities in space or on the extraterrestrial surface. Complex and lengthy preparatory procedures are part of EVA. Present studies aimed at establishing extraterrestrial bases will require substantial EVA during build-up operations. Long hours of continuous EVA are expected during the early development phase of these projects. It is well known that present-day designs for spacesuits are simply inadequate for these operations. When inflated to the optimum operating pressure of about 8 psi, the suit becomes very stiff and quite difficult to flex at the required joints. The astronaut then has to work against this suit pressure stiffness as well as the forces which are required of the task to be performed. Loss of dexterity is the result, and almost every component in a payload package to be assembled or deployed has to be designed to adapt to the limitations imposed by this loss of dexterity (Figure S.6). In past EVA missions, astronauts and cosmonauts have complained about the difficulty of working in the EVA suit. Lunar dust is notorious for degrading astronaut as well as vehicle performance. Compounded by the fact that future missions are expected to be more complex and substantially longer in duration, it is imperative that alternative methods for conducting EVA be studies. Hardsuits – where the fabric is substituted for a metallic shell with articulation mechanisms at the essential joints – have been studied, but they have their limitations too. Compact modules with appropriate life-support and remote manipulator systems (the so-called "man-in-a-can" concept) have also been suggested as an alternative means for enhancing long-duration EVA. Fully-robotic systems have yet to prove their ability to handle contingencies, and until then manned EVA will continue to play the leading role in assembly operations in space and on the extraterrestrial surface. It is also possible that hybrid concepts employing both robotic and manned systems may prove to be more effective, using a mutual support strategy, during extra-vehicular build-up activity.

S.6   RATIONALE FOR AN ALTERNATIVE MANNED EVA SYSTEM

The EVA Bell concept proposed in this appendix is a concept for enhancing manned EVA operations on the extraterrestrial surface. The rationale for the concept are as follows:

CONVENTIONAL EVA PROBLEMS

- COMPLEX, LENGTHY PREPARATION
- ASTRONAUT FATIGUE
- CLUMSY&POOR DEXTERITY
- DUST CONTAMINATION
- SHORT DURATION LIMITS

NOMAD EXPLORER BELL SYSTEM ADVANTAGES

- MINIMAL PREPARATION
- SHIRT SLEEVE ENVIRONMENT
- DUST FREE ENVIRONMENT
- LONG DURATION EVA

TO NOMAD EXPLORER
ECLSS

LCVG liner (tricot)

LCVG outer layer
(nylon/spandex)

ARM

Pressure garment
(urethane coated

Pressure garment co
(dacron)

TMG liner
(neoprene coated nylon ric

LCVG water transport tubing

TMG insulation layers
(aluminized mylar)

TMG cover (ortho-fabric)

BELL PROTECTS CREW

SEAL

DEPLOYED SURFACE SEAL FABRIC

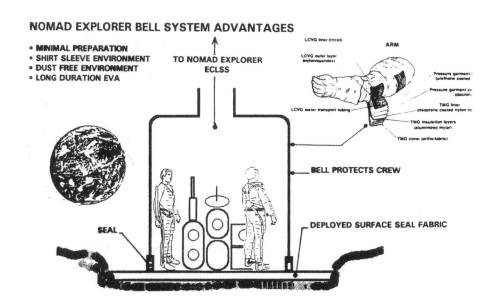

Figure S-6. A comparison of conventional EVA and EVA Bell systems.

1. Conventional EVA is an extremely-inefficient way of utilizing astronaut capabilities. EVA time is expensive and must be used more efficiently.
2. Conventional EVA requires that components to be assembled/deployed be designed to respond to the limitations of EVA-suited astronauts. Such a strategy limits efficient design and operation and therefore should not be a design driver.
3. Extraterrestrial base development will surely involve much more complex and arduous EVA tasks that will heavily tax the physical and mental capabilities of the best astronauts. Though conventional EVA suits may be ample for some of the envisaged activity, alternate concepts for EVA are required to handle different and diverse EVA scenarios, which are a natural implication of the plethora of necessary EVA tasks required that will eventually lead to final establishment and operation of the base.
4. Dust contamination from the extraterrestrial surface has and will continue to pose a serious threat to successful assembly and maintenance operations. EVA concepts are needed that will effectively combat this problem during assembly/repair operations.
5. It may not be possible to tackle all EVA scenarios using the same strategy (i.e., use of only the conventional EVA suit). Therefore, at the planning stage, it is prudent to have as many alternative concepts for manned EVA as possible, so that tasks may be designed for efficient execution.
6. Many payloads to be deployed and maintained remotely on an early lunar base are quite small in their physical dimensions (e.g., remote data relay stations, small scientific experimental platforms, photovoltaic arrays, optical interferometry array components).
7. It should be possible to provide an EVA environment for the assembly/repair/maintenance crew which is less taxing and more comfortable. Concepts are required which would enable the astronaut crew to perform more precise and delicate tasks on the site without the strain imposed by EVA suit constraints.
8. Establishment of a rapid and highly-flexible global extraterrestrial telecommunication/scientific experimental station network infrastructure may be possible if a manned and robotic hybrid architecture is adopted during the primary phase of build-up activities.

It is on the basis of these premises that the Nomad Explorer architecture for rapid lunar/Mars global infrastructure development and the EVA Bell concept for enhancing manned EVA are developed.

S.7 THE EVA BELL ARCHITECTURE

In its simplest manifest, the idea is to separate the astronaut from the suit and try to provide as close to a shirt-sleeve environment as possible during EVA. In order to accomplish this, we will provide a pressurized shack, referred to as the "Bell" (programmatically similar in many ways to the diving bell used underwater), at the place where the EVA is to be performed (Figure S.6) Obviously then, this Bell becomes an integral part of the Nomad Explorer! Though only one configuration of the EVA Bell is addressed in this paper, several other ways exist in which to provide this protective enclosure around the payload and the astronauts during assembly activities.

The Bell works in the following manner (Figure S.7):

1. The Nomad Explorer drives to the EVA/assembly site. The payload to be deployed could have been landed at the site separately (common lander concept?) or carried in the vehicle.
2. The surface is roughly leveled by the RMS on the vehicle. A surface seal fabric is unrolled on the smoothed-out terrain (this fabric could be part of each common lander payload).
3. The payload/experiment to be assembled/deployed is then unloaded on top of the prepared surface seal fabric.
4. The Nomad Explorer then aligns itself with the prepared surface seal and gently lowers the Bell so the complete payload is covered by it with space around the payload to spare. The volume inside the Bell is about $150\,\mathrm{m}^3$.
5. The Bell is lowered till it uniformly contacts the surface seal fabric all around the payload. The Bell is then secured to the surface seal fabric in such a way as to produce a nominal pressure seal between the inside of the Bell and the extraterrestrial surface. Studies are underway that examine several ways of establishing this pressure seal.
6. The Bell is then pressurized (8 psi nominal).
7. After assuring that the seal is operational and that the nominal leakage rates are not exceeded, an airlock into the Bell allows the astronaut crew to access it from inside the Nomad Explorer.
8. The assembly/deployment activity is performed by the crew wearing minimal EVA garments (an emergency pressure suit?). After a test and check-out of the experiment/setup or system (e.g., optical IF, VLBI components, relay stations, remote monitoring equipment), the crew get back into the vehicle.
9. The Bell is depressurized and retracted, and the Nomad Explorer is on its way to the next assembly assist/experiment setup/maintenance/repair site.

S.8   CHALLENGES POSED BY THE EVA BELL

1. How do we mitigate counter-pressure? At 8 psi, a 4×6-m Bell footprint would produce a total uplift of nearly 3,000,000 lb! However, we have a substantial surface contact perimeter of 20 m to devise an anchoring mechanism. External as well as internal anchoring mechanisms for the EVA Bell are being explored.
2. How do we make sure of the seal? We will require 100 percent reliability on the seal mechanism if a shirt-sleeve environment is the goal.
3. How large a Bell can we practically build and operate? It has to be compatible with payloads that we intend to assemble and deploy of course!
4. The surface seal fabric will have to be left in place after the assembly activity.
5. The payload cannot contact the surface during the assembly operation.

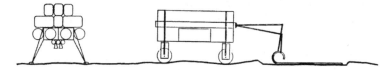

1.   Nomad Explorer intercepts and parks close to autonomous modular common lander (AMCL) which carries logistics and site specific science payloads. Modular logistics retrieved. Supplies replenished. Depleted logistics modules/waste discarded.
2.   Nomad Explorer prepares terrain for payload assembly/deployment

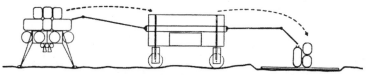

3.   Nomad Explorer lays surface seal fabric over prepared terrain. This fabric will help to contain the pressure inside the deplyed EVA Bell

4.   Nomad Explorer unloads payload from AMCL over the surface seal fabric at the payload assembly site.

5.   Nomad Explorer is elevated, slowly moves to prepared site and aligns EVA Bell over the payload and the surface seal fabric.

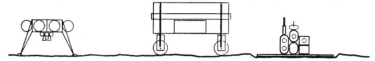

6.   Nomad Explorer deploys EVA Bell which provides crew a shirtsleeve environment for manned EVA. Crew assemble, check out, certify payload inside dust, thermal, micrometeoritic and radiation protected and pressurized EVA Bell.

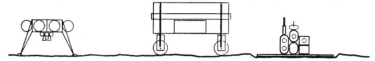

7.   Assembly complete. EVA Bell retracted. Nomad Explorer moves away from payload assembly. Payload is operational. Proceeds to next site.

Figure S.7. Nomad Explorer payload assembly assist sequence.

S.9   ADVANTAGES OF THE EVA BELL SYSTEM

1. Shirt-sleeve environment/deployment activity. No pre-breathing or associated EVA preparations are required.
2. The fully-enclosed EVA Bell system provides a completely dust-free, contamination-free environment during assembly and check-out of sensitive experimental equipment.
3. Lacking post-landing serviceability, conventional missions carrying sensitive scientific equipment need to be designed to accommodate the shock of lunar lander impact. This translates as additional mass or design for shock absorption. Alignment and recalibration of equipment could pose a problem. The EVA Bell strategy provides a way to avoid this problem.
4. If the 8 psi pressure is too much to handle, then the crew could still work inside the Bell wearing a pressure suit that is just enough to combat the differential between the Bell and the suit. For example, if the Bell can withstand up to 4 psi, then the crew needs roughly another 4 psi inside the suit. Such a decrease in suit pressure will make the suit less stiff during operation and enable more comfortable EVA. This method is also very safe in the event of Bell pressure seal failure. New, more maneuverable and comfortable pressure suits could be designed for use with the Bell.
5. All of the multilayer insulation used in the conventional EVA garment for radiation and micrometeoritic protection are eliminated from the suit, which then retains only the pressure garment.

S.10   ADVANTAGES OF THE NOMAD EXPLORER STRATEGY

1. The Nomad Explorer strategy eliminates the need for conventional build-up equipment and infrastructure. All the heavy machinery and associated equipment, roads, launch and landing pad facilities associated with previous studies are not required.
2. Nomad Explorer technology is mature and does not require heavy investment to realize. NASA has studied long-range vehicles like MOSAP in enough detail to be able to build and test Nomad Explorer prototypes.
3. Prototypes based on existing vehicles used for special terrestrial purposes (like the MX missile transporter and other advanced recreational vehicles) could also be modified and studied in order to design and build the Nomad Explorer economically.
4. The Autonomous Modular Common Lander (AMCL) is not a new concept. NASA has been working on several lunar landers including the "common lander" and Artemis using existing RL-10B or equivalent engine technology. Modular clustering of engines and tankage need more study in order to realize the AMCL concept. Innovative ways of landing bulk payload that are not so sensitive to higher-than-average terminal impact velocities (95–10 m/sec) need further study.

5. If the AMCL system is employed to land modular payloads containing experiments, logistics, and consumables at various locations along a predetermined traverse route, then the Nomad Explorer could assemble and deploy the payload at regular intercepts along the path. Such a strategy will provide unlimited range for the Nomad Explorer.

6. The Autonomous Common Lander approach would bring to the Nomad Explorer architecture a powerful design flexibility. During the course of the traverse, if unforeseen events require different logistics, science, or consumable payloads, the landers could be outfitted and flown to intercept the Nomad Explorer with the required mission-specific hardware at short notice.

7. During the course of the Nomad Explorer global traverse, if a globally-accessible fiber-optic cable could be laid by the vehicle along the traverse, then it may be possible to eliminate the need for the deployment and maintenance in orbit of a constellation of telecommunication satellites.

8. This flexible open-ended mission architecture can be tailored as the mission proceeds. The "if–then" philosophy is best suited for exploration-oriented missions and can be used to alter the traverse to best suit the exploration and extraterrestrial base site selection as well as the science mission goals.

9. The EVA Bell on the Nomad Explorer will provide a less stressful environment for astronaut crew on long-duration EVA missions. The crew will have more freedom to alter traverse routes on the basis of their own exploration results, making the mission more exciting and eventful for both astronaut crew as well as mission control.

10. This ultra-dynamic strategy will also hold the fascination of the public because of the continuously-changing terrain visits, the regular rendezvous with the AMCL system, the number of mission goals which are rapidly met, and the spontaneous nature of the tasks that may have to be performed by the crew during this mission.

11. In this way a speedy global infrastructure may be established on the lunar/Mars surface.

S.11 TECHNOLOGY FOR THE NOMAD EXPLORER STRATEGY

The Nomad Explorer vehicle and the Autonomous Modular Common Lander system employ state-of-the-art and mature technologies. NASA has been working on similar concepts and sufficient data exists within the US which could be used to build and test prototypes.

The Nomad Explorer vehicle requires a heavy-lift launch vehicle (HLLV). HLLVs are required in order to minimize otherwise costly and risky on-orbit assembly and rendezvous procedures for which the infrastructure is not in place yet. Reviving the Saturn V-B and incorporating new modifications to the 20-year-old technology could result in the work-horse launcher that NASA needs and that is essential for any permanent manned presence on the Moon or Mars. The Russian

Energiya is a typical example of an HLLV that could provide support for Nomad Explorer operations.

Communications technology has come a long way since Apollo. Fiber-optics and free-space laser communication systems can provide dependable and very-high-bit-rate communications for the Nomad Explorer strategy. Furthermore, if the Nomad Explorer is used to lay a fiber-optic network as the traverse proceeds, eventually it might prove possible to have a global extraterrestrial communication system that would eliminate the need for a constellation of satellites in unstable orbits (in the lunar case) and poor life times. Optical line-of-sight free-space laser communications without atmospheric disturbances and associated attenuation is possible on the lunar surface. The technology for laying such extraterrestrial fiber-optic cable/free-space links requires serious study.

S.12   THE NOMAD EXPLORER BUDGET

Project Apollo cost $100 billion in 1990 dollars. Much of the technology base had to be built up from scratch. Though much of the hardware and infrastructure associated with the project is no longer with us, much wealth in hardware and experience from the project remains dormant within NASA and the space industry. In addition, NASA and the space industry have already carried out much study on long-range rovers and landers. MOLAB and MOSAP are some of the long-range rovers which have been studied in depth. Several designs have also been developed by NASA for lunar landers. Artemis, lunar lander, and the common lander are some of the NASA studies under way. The Nomad Explorer strategy will rejuvenate NASA and the manned spaceflight hardware builders of the world by tapping into research that is already underway. Using Apollo as the gauge for the Nomad Explorer, it should be possible to return to the Moon using the Nomad Explorer and the Autonomous Modular Common Lander for about the same price tag.

S.13   CONCLUSION

The Nomad Explorer strategy for extraterrestrial base evolution is an alternative strategy for global extraterrestrial infrastructure establishment. Capable of all the functions normally conducted from a stationary base, a mobile base like the Nomad Explorer, when coupled with an autonomous payload/logistics lander system like the AMCL system, has unlimited exploration range as well as a highly-tailorable mission plan. Maximum flexibility is the chief attribute of the "open architecture". The "if–then" capability allows the mission to be tailored as it proceeds while maintaining close contact with mission control. Autonomous Modular Common Landers carry mission-specific hardware as dictated by the crew. The EVA Bell, a new system concept for enhancing manned EVA in the Nomad Explorer vehicle, will help to assist assembly and deployment of science experiments and pilot projects while the vehicle traverses from site to site, examining them in detail for establishing a permanent manned base. An extraterrestrial fiber-optic/free-space global telecommunications infrastructure may be laid during the course of the exploratory traverse, which will eventually eliminate the need for expensive operations and maintenance of the constellation of telecommunications satellites which would otherwise be required as complex extraterrestrial projects evolve. Rapid extraterrestrial development that will hold the interest and excitement of the public as the mission proceeds from site to site is the consequence of using such a mobile strategy. The various systems constituting this architecture require further study and analysis.

S.14   ACKNOWLEDGMENTS

This study was carried out as a task for the extraterrestrial surface transportation group at Rockwell International, Space Systems Division, Downey, CA. Thanks are due to Ed Repic, Project Manager for the Space Exploration Initiative, Steve Kent, who supervised my work, John Pulley, manager of the Surface Transportation Group, and all the others, including Todd Wise and Ron Jones, who helped to critique the concept. Thanks are also due to the participants of the Institute of Aerospace Systems Architecture and Technology at the University of Southern California who are embarked on a brave new voyage, exploring and charting new territory in the field of Aerospace System Architecture.

S.15   REFERENCES

Alred, John (1989) *Lunar Outpost*. Systems Definition Branch, Advanced Programs Office, NASA Johnson Space Center, Houston, TX.

Augustine, N. (1991) *Report of the Commission for the Future of the Space Program*. NASA HQ, Washington, D.C.

Brown, W.C. (1969) Experiments involving a microwave beam to power and position a helicopter. *IEEE Transactions, Aerospace Electronics Systems.*, **AES-5**.

Burke, F.B. (1985) Astronomical interferometry on the Moon. In: Mendell, W.W., ed., *Lunar Bases and Space Activities in the 21st Century*. Lunar and Planetary Institute, Houston, TX.

Burke, J.D. (1985) Merits of a lunar polar base location. In: Mendell, W.W., ed., *Lunar Bases and Activities in the 21st Century*. Lunar and Planetary Institute, Houston, TX.

Cintala, M.J.; Spudis, P.D.; and Hawke, B.R. (1985) Advance geologic exploration supported by a lunar base: A traverse across the Imbrium–Procellarum region of the Moon. In: Mendell, W.W., ed., *Lunar Bases and Activities in the 21st Century*. Lunar and Planetary Institute, Houston, TX.

Cohen, M. *et al.* (1989) *Report of the 90-Day Study on Human Exploration of the Moon and Mars*. NASA Johnson Space Center, Houston, TX.

Coomes, E.P.; Dagle, J.E.; and Wise, J.A. (1992) *Design Study of a Laser Beam Powered Lunar Exploration Vehicle*. Pacific Northwest Laboratory, Richland, WA, IAF-92-0742.

Crawley, Ed. (1988). Aerospace systems analysis and costing. *Lecture to the International Space University at MIT, Boston*.

Glaser, P.E. (1974) *Feasibility Study of a Satellite Power station*. NASA, Washington, D.C., NASA CR-2357.

Glaser, Peter (1991) Dust contamination and control. *IDEEA Conference, Houston, Texas*.

Gobanov, B.I. (1989) *Energiya: A New Versatile Rocket Space Transportation System*, IAF Congress, Malaga, Spain, IAF-89-202.

Griffin, B. and Appleby, M. (1991) *The Pressurized Lunar Rover: Radiation Analysis Using the Boeing Radiation Exposure Model (BREM)*. Advanced Civil Space Divisions, Boeing Co., Huntsville, AL.

Haskin, L.A. (1985) Toward a spartan scenario for use of lunar materials. In:Mendell, W.W. Mendell, W.W., ed., *Lunar Bases and Space Activities in the 21st Century*. Lunar and Planetary Institute, Houston, TX.

Horz, Fredrich (1985) Lava tubes: Potential shelters for habitats. In: Mendell, W.W., ed., *Lunar Bases and Space Activities in the 21st Century*. Lunar and Planetary Institute, Houston, TX.

Humphries, R. (1991) Life-support and internal thermal control system design for space station freedom. *4th European Symposium on Space Environmental Control Systems*, ESA SP-324 (NASA MSFC, Huntsville, AL).

Hypes, W.D. and Hall, J.B. (1988) ECLS systems for a lunar base: A baseline and some alternative concepts. *18th Intersociety Conf. on Environmental Systems, San Francisco*, SAE 881058.

Kuznetz, L.H. (1990) *Space Suits and Life Support Systems for the Explorations of Mars*. NASA Ames R.C., Moffett Field, CA.

Landis, G.A. (1989). Solar power for the lunar night. *9th Biennial SSI Conference on Space*, NASA TM 102127.

Landis, G.A. (1991). A proposal for a Sun-following Moon base. *Journal of the British Interplanetary Society*, vol. 44, 125–126.

NASA (1988) *Lunar Reports*, Vols. 1 and 2. Exploration Studies Technical Report TM 4075, NASA Office of Exploration/Eagle Engineering Inc., Webster, TX, NASA contract NAS9-17878.

NASA (1989) *Planetary Surface systems*, Vol. III. Exploration Studies Technical Report TM 4170, NASA Office of Exploration, NASA, Washington, D.C.

NASA (1991) *Common Lander Study*, NASA Office of Exploration, NASA Johnson Space Center, Houston, TX.

NRC (1990) *Human Exploration of Space: A Review of NASA's 90-Day Study and Alternatives*. National Research Council, National Academy Press, Washington, D.C.

Rummel, J.D. and Arvrner, M. (1991) *Regenerative Life Support: The Initial CELSS Reference Configuration*. Life Sciences Division, NASA Hq. Regenerative Life Support Systems and Processes, Society of Automotive Engineers, NASA Technical Publication, NASA HQ, Washington DC, SP-873, paper ~911420.

Schwartzkopf, S.H. and Brown, M.F. (1991) *Evolutionary Development of a Lunar CELSS*. Life Sciences Division, NASA Hq. Regenerative Life Support Systems and Processes, Society of Automotive Engineers, SAE HQ, Washington DC, SP-8743, paper ~911422

Sellers, W.O. and Keaton, P.W. (1985). The budgetary feasibility of a lunar base. In: Mendell W.W., ed., *Lunar and Space Activities in the 21st Century*. Lunar and Planetary Institute, Houston, TX.

Stafford, T.P. (1991). *America at the Threshold* (report of the Synthesis Group on America's Space Exploration Initiative). HASA HQ, Washington, D.C.

Stone, C. (1991) *ALPS: Advanced Lunar Power System*. Space Systems Division, Rockwell International, Downey, CA.

Thangavelu, M. (1991) *The Multicellular Airbag Landing System: A Concept for Land Extraterrestrial Payloads*. Space Systems Division, Rockwell International, Downey, CA.

Thangavelu, M. (1991) *The Very Long Range Traverse Vehicle: Option for Primary Lunar Base Establishment*. Space Systems Division, Rockwell International, Downey, CA.

Wise, Todd (1991) Dust contamination control. *IDEEA Conference, Houston, Texas*.

Appendix T

Beyond our first Moonbase: The future of human presence on the Moon

Peter Kokh

T.1 BEGINNINGS

Humankind's first outpost on the Moon will start to become a reality around 2020, an historic event that, were it not for politics, might have happened decades earlier.

The vision outlined in *The Moon: Resources, Future Development and Settlement* by David Schrunk, Burton Sharpe, Bonnie Cooper, and Madhu Thangavelu (i.e., the current book) is a bold one, showing how we could set up our first outpost so that, far from being a goal in itself, it would become the nucleus from which human presence would spread across the face of the Moon.

NASA itself has such a vision, but the agency can only do what it is authorized to do, and there is the catch. If the history of the International Space Station offers clues, NASA's official goal, which only includes setting up a first limited outpost as a training ground for manned Mars exploration and nothing more, will be under increasing budgetary pressures to slim down into something with no potential for growth at all. The intended crew size, the planned physical plant, and the capabilities that such a plant will support will all be tempting places to cut budget "fat" by those who cannot see, or appreciate, the possibilities beyond. But that is the risk of publicly-supported endeavors in space. It is difficult to get our political leaders, and the public itself, to look beyond the next election or the next payday. The chances that our first outpost will be born sterile cannot be dismissed.

However, if private enterprise is involved to the extent that it is ready to take over when and where NASA's hands are tied, there is a bright future for humanity on the Moon. Much of that promise may involve finding practical ways to leverage lunar resources to alleviate Earth's two most stubborn problems, both inextricably intertwined: generating abundant clean power, and reversing the pressures of human civilization on Earth's irreplaceable environmental heritage.

T.2   CRADLEBREAK

The Moon has enormous resources upon which we can build a technological civilization. But first we must overcome the limitations inherent in depending on government to do all the work. A humble start can be made by demonstrating the easier, simpler ways to start lessening the outpost's dependence on Earth. Resources such as solar power and rocket propellant must come first. Oxygen production will be an early goal, followed by hydrogen production. Hydrogen may come from hypothesized lunar polar ice deposits, or it may come from the implanted solar wind particles found in the upper layers of the regolith everywhere on the Moon.

T.2.1   Cast and sintered basalt

If we have access to basalt soils, found mainly in the lunar maria (described in Chapter 1), we can cast or sinter this material into many useful products. These include pipes, sluices, and other components of regolith-handling systems, as cast basalt is abrasion-resistant. If we expand the outpost using inflatable modules shipped from Earth at significant weight savings over hard-hull modules, we can use cast-basalt products, including floor tiles and tabletops, to help outfit these spaces. There is a thriving cast-basalt industry on Earth to learn from.

T.2.2   Lunar concrete, glass–glass composites (GGC), and silicon

Experiments on Earth with lunar simulant, of similar chemical and physical composition to lunar regolith, then repeated with Apollo lunar dust samples, give us confidence that concrete and glass composites can be used in any future construction and manufacturing activity on the Moon. We could make additional pressurizable modules from fiberglass-reinforced concrete or GGC. We can make spars for space frames and many other products out of these composites as well. The Moon's abundant silicon will allow us to make solar panels for generating power. Production of usable metal alloys will come later. The Moon is rich in four of the common engineering metals: iron (steel), aluminum, titanium, and magnesium.

T.3   A STRATEGY FOR INDUSTRIAL DIVERSIFICATION

We refer to our stategy as industrial "MUS/CLE". If we concentrate on producing things on the Moon that are Massive, yet Simple, or which are small but needed in great numbers (Ubiquitous) so as to provide the major tonnage of our settlement's need on the Moon, we will be making significant progress towards lessening the total tonnage of items needed from Earth to support the expansion effort.

Until we can learn to make Complex, Lightweight, and Electronic items on the Moon, we will continue to import them from the Earth. We could design everything needed on the Moon as a pair of subassemblies, the MUS assembly to be manufactured locally, and the CLE assembly to be manufactured and shipped from Earth, both being mated on the Moon.

Examples of this are a TV set, with the electronics manufactured on Earth and the cabinet built on the Moon; a steel lathe would be built on Earth and the heavy table on which it is mounted would be manufactured on the Moon; steel pipe and conduit could be made on the Moon, all the fittings and connectors imported from Earth.

As the population of pioneers and settlers grows, and our lunar industrial capacity becomes more sophisticated and diversified, manufacturing of many of the "CLE" items will begin on the Moon as well. Making clear and steady progress in assuming an ever-greater share of manufacturing is essential if we are going to encourage both continued governmental support and attract ever-greater participation by private enterprise.

T.3.1   Paying for the things we must import

Many writers have identified "zero-mass products" such as energy, to provide the lunar settlements with export earnings. The need for exports is indeed vital. As long as the settlement effort must still be subsidized from Earth, there will always be the risk of unrelated budgetary pressures on Earth fueling support for those who would pull the plug on lunar operations.

It is thus urgent that the settlers develop products for export that will help them pay for what they must still import. Only when we reach that import–export parity will the lunar settlement have earned the right to "permanence". Permanence cannot simply be declared; it must be earned. Tagging NASA's first Moonbase as "a permanent presence on the Moon" is unrealistic without planning for import–export parity. If we do not begin developing and using lunar resources seriously and aggressively, the effort will fail of its own costly weight.

Plans have been proposed to use lunar resources to build giant solar-power satellites in geosynchronous orbit about the Earth; or to build giant solar power farms on both the east and west limbs of the Moon to beam power directly to Earth; or to harvest precious helium-3 from the lunar topsoil or regolith blanket (see Appendix H). But none of these schemes will materialize right away.

However, we propose that Earth itself is not the market. Developing alongside an upstart settlement on the Moon will be tourist facilities in Earth orbit. And that is something the lunar settlement effort can support. Anything future Lunan pioneers can make for themselves, no matter how unsophisticated in comparison with the vast variety of terrestrially-produced alternatives, can be shipped to low Earth orbit at a fraction of the cost that functionally-similar products made on Earth can be shipped up to orbit. It is not the distance that matters, but the depth of the gravity well that must be climbed. It will take one-twentieth of the fuel cost to ship a table and chairs, a bed frame, interior wall components, floor tiles, even water and food, from the Moon, 240,000 miles away, than from Earth's surface, 150 miles below.

Thus, in the near term, the future of lunar settlement will be closely tied to the development of tourist facilities, hotels, casinos, gyms, and so on, in Earth orbit. This sort of development will start to bloom about the same time as a lunar settlement effort starts to break out of an initial limited Moonbase egg. But the linkage will become visible much earlier: it is possible that the first space tourist will loop the Moon (without landing) before the first astronaut arrives on the surface. The Russians say that they can provide such a tourist experience (skimming low over the Moon's mysterious far side) in just two years after receiving a pre-payment of $100 million. That event would create a benchmark that others would want to follow, inevitably bringing the price down for a ride to an orbiting resort.

T.4   THE MOON FROM A SETTLER'S POINT OF VIEW

Magnificent desolation? Yes. Harsh and unforgiving? That too. Alien and hostile? Of course it has always been so from the time our ancestors on the plains of East Africa started pushing ever further into unfamiliar territory: the lush, dense jungles, the hot dry deserts, waters too wide to swim, high mountain ranges, and eventually, the Arctic tundra. Judged by the pool of past experience, each new frontier was hostile, unforgiving, and fraught with mortal dangers ... until we settled it anyway.

Once we learned how to use unfamiliar resources in place of those left behind, once we learned how to cope with those new dangers, as if by "second nature", then the new frontier became as much home as the places we left behind. A person raised in a tropical rain forest, suddenly transported to the north slopes of Alaska, might soon perish, at a loss for how to cope. But the Eskimo never gives it a second thought. How to cope with ice, cold, the Arctic wildlife, the absence of lush plant life have all become second nature.

And future Lunans will reach that point as well. Yes, there is sure suffocation outside the airlock. Yes, the Sun shines hot and relentlessly with no relief from clouds for two weeks on end. Yes, the Sun stays "set" for two weeks at a time while surface temperatures plunge. Yes, the moondust insinuates itself everywhere. The litany goes on and on. So? Lunans will learn to take it all in their stride. How to take the right precautions for each of these potential fatal conditions will have become culturally-ingrained second nature. The Moon, to Lunans, will become a land of promise.

T.4.1   Making themselves at home

Even in the first initial human outpost, crew members could bring rock inside the habitat as adornment in itself, or perhaps carve it into an artifact. An early cast-basalt industry, early metal alloys industries, early lunar farming will all supply materials out of which to create things to adorn private and common spaces alike. Learning to do arts and crafts on the Moon may seem useless and irrelevant to some, but it will be the first humble start of learning to make the Moon "home". And so it has been on every frontier humans have settled.

We will also learn to schedule our activities and our recreation in tune with the Moon's own rhythms. We'll do the more energy-intensive things during dayspan, the more energy-light, manpower-intensive things saved for nightspan. With no three-month-long seasons on the Moon, the monthly dayspan–nightspan rhythm will dominate. The pioneers may bring some holidays with them, but will originate other festivities and monthly and annual celebrations.

Getting used to lunar gravity will also help pioneers settle in. They will abandon trying to adapt familiar terrestrial sports, which can only be caricatures of the games of Earth. Instead, they will invent new sports that play to the $\frac{1}{6}$ gravity and traction, while momentum and impact remain universally standard. Alongside the development of lunar sports will be forms of dance. Can you imagine how ethereal a performance of Swan Lake would be on the Moon? How many loops could an ice-skater do before finally landing on the ice?

T.4.2   But they have to live underground, for heaven's sake!

On Earth, our atmosphere serves as a blanket which protects us from the vagaries of cosmic weather: cosmic rays, solar flares, micrometeorite storms. On the Moon, eons of micrometeorite bombardment have pulverized the surface and continue to "garden" it into a blanket of dust and rock bits 3 to 15 meters thick. Tucking our pressurized outpost complex under such a blanket will provide the same protection, along with insulation from the temperature extremes of dayspan and nightspan.

Will our outposts look like somewhat orderly mazes of molehills? To some extent, perhaps; but the important thing is that we do not have to live as moles. We have ways to bring the sunshine and the views down under the blanket with us. In the spring of 1985, I had the opportunity to tour a very unique Earth-sheltered home some 32 km northwest of Milwaukee. Unlike other earth-sheltered homes, TerraLux (EarthLight) did not have a glass wall southern exposure. Instead, large mirror-faceted cowls followed the Sun across the sky and poured sunlight inside via mirror-tiled meter-wide tubes that traversed a 2.5-meter-thick soil overburden (Figure T.1).

Meanwhile, picture-window-scale periscopic windows provided beautiful views of the Kettle Moraine countryside all around. I had never been in a house so open to the outdoors, so filled with sunlight, as that "underground" one. I at once thought of the lunar pioneers, and how they could make themselves quite cozy amidst their forbidding, unforgiving magnificent desolation. The point is that, yes, the Moon is a place very alien to our everyday experience, but, nonetheless, human ingenuity will find a way to make it "home" (Figure T.2).

T.4.3   What about the outdoorsmen amongst us?

While Lunans will find plenty to do within their pressurized homes, workplaces, and commons areas, many will miss the pleasures of outdoors life on Earth: fishing, swimming, hunting, boating, flying, hiking, mountain climbing, and caving. The list goes on and on.

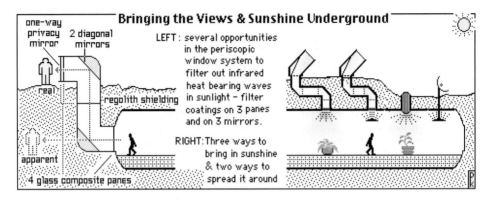

Bringing the Views & Sunshine Underground

one-way privacy mirror | 2 diagonal mirrors

real

regolith shielding

apparent

4 glass composite panes

LEFT: several opportunities in the periscopic window system to filter out infrared heat bearing waves in sunlight – filter coatings on 3 panes and on 3 mirrors.

RIGHT: Three ways to bring in sunshine & two ways to spread it around

Figure T.1. Channeling of light in underground habitats.

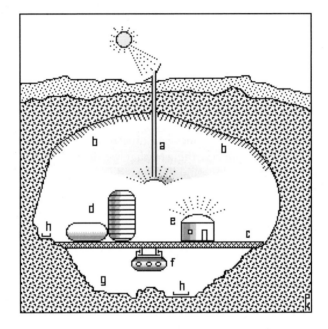

Figure T.2. Concept for early lavatube outpost. KEY: (a) sunshine access and defuser system; (b) white-washed "firmament" for best sunlight reflection; (c) "town deck" on tube-spanning beams; (d) assorted structures; (e) "yurt/hogan" type home with translucent dome to flood interior with firmament-reflected sunshine; (f) monorail transit system; (g) lavatube floor left natural; (h) nature walks.

Yet some of these pleasures we may be able to recreate indoors – fishing in trout streams, for example. We will want an abundant supply of water, and waste water in the process of being purified can provide small waterfalls and fountains, even trout streams for fishing and boating. In large, high-ceiling enclosures, humans may finally be able to fly with artificial wings, as Icarus tried to do.

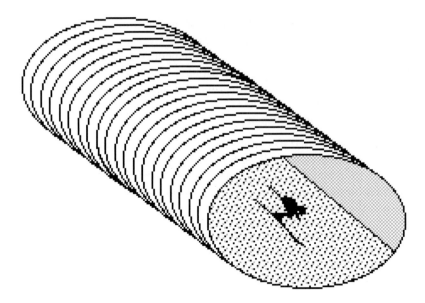

Figure T.3. Luna-tube skiing on the Moon.

Out on the surface of the Moon, sporting activities will be more of a challenge. Present spacesuits are very cumbersome and clumsy. We must develop suits that give us more freedom of motion, that tire us less easily. Once we do, hiking, motor biking, mountain climbing, and caving (in lava tubes) will become possible. Sporting events, rallies, races, and games will follow. Over the years, we will learn to take the Moon's conditions for granted, and to "play to them", just as northerners have invented skiing, ice skating, snow boarding, and other winter sports, that people in the tropics would never have imagined (Figure T.3).

T.4.4   Agriculture and mini-biospheres

We must approach the creation of living space on the Moon as one of mating modular architecture with "modular biospherics". Every pressurized module should have a biosphere component, so the two – living space, and life in that space – grow apace, hand in hand. Lunar settlements will be "green" to the core, and that above all will make us feel at home.

T.4.5   One settlement, a world "doth not make"

The Moon's resources are not conveniently and homogeneously located. A settlement located to take advantage of possible polar ice reserves may not be able to take advantage of mare basalts, iron- and titanium-rich ilmenite, or lava tubes. As the lunar economy expands, it will be necessary to establish a number of settlements in various areas, and that will make the Moon a new human world. Lunans will be able to travel elsewhere, get away from it all, experience cultural, artistic, and enterprise variations. Again, permanence has to be earned. Even as an outpost cannot be declared permanent, neither can a first true settlement. When thinking of the Moon, no matter where we choose to begin, we must have a global vision. The authors have this vision, and their brilliant concept of a lunar railroad network illustrates that well.

T.4.6   Getting through the nightspan

To many people, overly used to energy "on demand", the need to store up enough energy during the two-week-long dayspan to allow the outpost not only to survive the nightspan, but to remain productive, may seem challenging. But all of human progress is built on utilizing various forms of power storage, starting with firewood. Even in nature, the spread and survival of species has turned on this point, from bear fat to squirreling away nuts. The problem is one of attitude. Those who have the right attitude will find a way – many ways in fact. The same goes for managing the thermal differences between lunar high noon and predawn. Since we first began to move out of our African homeworld to settle the continents of Eurasia and the Americas, we have tackled harder problems. Those who are not intimidated by the challenge will lead the way.

T.4.7   The pattern emerges

Lunan pioneers will make progress in all these areas together: learning to provide for the bulk of their material needs by mastering lunar resources; becoming ever more at home through lunar-appropriate arts, crafts, sports, and hobbies; developing a truly Lunan culture. Those things shipped from Earth should be designed and manufactured in MUS/CLE fashion, so that the components at first made on Earth can be replaced with components made on the Moon. If things shipped from Earth are made of elements hard to produce on the Moon (e.g., copper or thermoplastics), that will help spur infant lunar industry at a quicker pace.

T.5   THE NECESSARY GAMBLE

People will be concerned about pregnancy and bearing children on the Moon. Yet, this is something that cannot be postponed forever. We will learn from the animals that we bring with us whether reproduction results in viable offspring. Eventually, we may gain confidence that it is also possible to have human children on the Moon.

   The development of lunar civilization will depend on the decisions of individuals to remain on the Moon and raise families. Nations may build outposts, but only people pursuing their own personal and economic goals can give us settlement.

T.5.1   Token presence or real settlement

The paradigm obvious to many, of human activity in Antarctica, is a dead-end example of what our presence on the Moon could become. On the seventh continent, there is no real use of local resources, no economic activity, and no real society. For the Moon, we see instead a real human frontier in which a first outpost of a few persons will in time become a self-supporting frontier of hundreds of thousands of pioneers in several settlements. Many of these new Lunans will be native-born, their ranks swollen with recruits from Earth seeking the promise of starting over, starting fresh, getting in on the bottom floor of things. Throughout human history, those who were doing well stayed put. Frontiers have always been pioneered by the talented but "second-best", for whom the frontier promised a more open future. The Moon will become a truly human place.

Peter Kokh, President
The Moon Society
www.moonsociety.org
Editor, *Moon Miners' Manifesto*
www.moonminersmanifesto.com
kokhmmm@aol.com

Appendix U

Lunar rock structures

Madhu Thangavelu *University of Southern California*

Rock structures have been used on Earth for habitats and related infrastructure for millennia. Moon rocks, both raw and machined, can also be used to build elements of a permanent lunar colony.

U.1 INTRODUCTION

Efficient and durable habitats and related structures built of rock have been erected on Earth for thousands of years. While specimens abound all over the world, fine and lasting examples include the pyramids of Egypt, the pre-Celtic and cave structures of Europe, the ziggurats of central and south America and the temples of Greece, south east Asia and India. They were all built centuries ago, and some of them are serviceable even today.

A few meters below the lunar surface, the regolith becomes increasingly indurated, and large rocks are found at depths of 3 to 15 meters below the regolith. It should be possible, with the proper tools and methods, to utilize this resource for building substantial portions of the permanent lunar colony infrastructure.

U.2 COMBATING THE LUNAR ENVIRONMENT

Hard vacuum, Galactic Cosmic Radiation (GCR), high-energy solar particles, large thermal variations (up to 400°C) and micrometeorite showers make the lunar surface a harsh environment for both humans and machines. It is possible to provide a pressurized environment using either inflatable or rigid pressure vessels. However, three to five meters of regolith would be needed to effectively shield against temperature swings, radiation and the constant bombardment of micrometeorites. Carrying the required mass of shielding materials along from Earth is prohibitively expensive.

A lunar colony might locate its pressurized habitats inside lava tubes, within a rille or trough, or perhaps under an overhang within a crater (if such a formation were found). Alternatively, they could pile up the required thickness of loose material around and over a pressurized habitat. However, routine inspection and maintenance of the exterior of the pressurized vessel is hampered by this oft depicted concept of shielding.

U.3   ROCK STRUCTURES

Rock structures consist of those which are naturally occurring as well as those that are manufactured. Naturally occurring rock structures include randomly distributed, hollows or cavities on the terrain such as lava tubes, rilles and graben. These features are formed by volcanic and seismic activity on the Moon, as well as by meteorite impacts. Artificially built rock structures for the Moon would employ raw or unimproved rock in various forms as well as tooled or machined rock for building permanent habitat structures and surroundings.

U.4   ROCKS-TOOLS – USES

Data about lunar rocks comes primarily from specimens returned by Apollo and Luna missions. Moon rocks fall into five groups. They are basaltic volcanic rocks, pristine rocks, breccias, impact melts and lunar soil.

Using appropriate quarrying, transportation, tooling and laying/setting systems, some of them adapted from terrestrial heavy machinery and rock processing equipment, that include rock splitters, transport vehicles, crushers, shapers, polishers, connectors and robotic integrators, a very useful building capability may be evolved on the Moon and other planets, that might find extensive and highly effective use in the expedient and permanent build-up of extraterrestrial colony elements. Components to be made from raw rock include stone columns, slabs, blocks, tiles, and a variety of grades of aggregate offering different textures and albedos. Using the age old technology called "dry packing" (where stones are cut and fit against each other without mortar or other glueing agent) shade walls, aprons, towers, dome exteriors, roads and tunnels, landing pads, zone markers and other exposed platforms are some of the elements that might make use of this technology. More than half of all the materials needed for building an extraterrestrial settlement, such as the one depicted in the Figure U.1, could be derived from extraterrestrial rock and soil. As lunar settlement activities progress, building projects including systems for advanced Mars bases, such as wind energy farms, may be erected in order to empirically test and verify assembly and erection procedures. See Figure U.2.

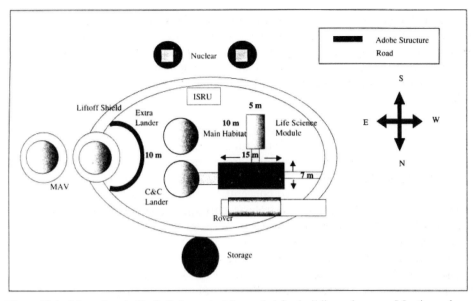

Figure U.1. More than half of all the materials needed for building a lunar or Martian colony could be derived from unprocessed or machined extraterrestrial rocks and soil.

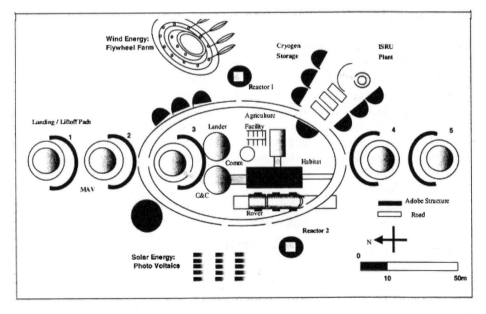

Figure U.2. In phase ll, evolved settlement continues to stress the use of unprocessed rocks for building platforms, landing pads, ejecta shields and zone markers. A wind farm is erected to simulate harvesting energy for future Mars bases.

U.5   TECHNOLOGY CONTINUUM DILEMMA

Humankind has been building habitable architectures for millennia. Most of the structures that have survived are made of rock. The resilience of such structures is partly due to the materials used, but their habitability and serviceability are mainly attributable to empirically tested, evolutionary methods that are employed in their design and creation.

Space architects, designers, engineers and builders may all do well to expand their world view about man-made lunar structures. The highly constrained, narrow, often trendy, state-of-the-art high-technology driven mindset will have to accept and be inspired by a more broadminded "technology continuum" philosophy, seeking solutions from age old and time tested concepts for building spaces for habitation and adapting them appropriately for extraterrestrial applications.

U.6   MERITS

1.   Rocks of varying compositions and sizes are abundant on the Moon.
2.   The low gravity allows for easier handling of massive blocks from factory to placement.
3.   Raw, unmachined rocks may be sorted/graded and neatly arranged/packed to create a variety of useful structures including shadow walls, radiation shielding, paving, glare and backscatter mitigation, dust traps and covers, and elevated foundations and rough hewn platforms using the age old "dry packing" technology.
4.   Machined blocks may be precision-cut into slabs, blocks, bricks, columns and beams that can be used as exterior and interior load bearing and finishing elements in lunar colony buildings and related infrastructures.
5.   Having been created and subjected to the space environment for geological time, they are very durable in the extraterrestrial environment.
6.   Existing quarrying and tooling machines including rock splitters, bridge saws and wire and chain saws may provide insight into the design and operations of robotic equipment that would manufacture these rock products and fabricated architectural elements from them in advance of human arrival to permanently settle the Moon.
7.   The combination of low gravity, the absence of wind loads and precipitation, (forces that constrain the size and form of Earth bound structures) may allow lunar structures much more morphologic freedom, with the possibility of constructing tall and slender structures, long, uninterrupted canopy spans and so on.

U.7   CHALLENGES

1.   Quarrying, tooling and finishing create substantial high-energy debris that must be curtailed at the source. Particularly on the Moon, these particles can easily

approach or exceed orbital velocities and thus create environmental havoc. So, it is important to process rock in a carefully contained environment.

2.   Having been created and exposed to vacuum for geological time, lunar rock and derived products are stable in their natural, anhydrous condition. Their use in relatively humid habitat interiors may require some study and possibly special processing to maintain strength and stability.

3.   Extensive testing alone can provide the data needed before commissioning these rock products for lunar building applications. So, more testing of lunar material is essential to identify merits and limitations, and appropriate use of rock for building Moon colonies and related infrastructure.

4.   Space architects, designers, engineers and builders will all have to expand their world view about man-made lunar structures. The highly constrained technology driven mindset must accept and be inspired by the alternatives that exist within the lunar environment, seeking solutions from age old and time tested concepts for building spaces for habitation and adapting them appropriately for extra-terrestrial applications where the unique characteristics of lunar materials can be used to great advantage.

5.   The use of lunar rocks, as suggested in this appendix, is specifically for structures subject to axial compression in lunar gravity. While it is possible to imagine carved hollow structures that might act as pressure vessels, it is perhaps much more practical to employ pressurized membranes for that purpose separately and then use rock structures around them to combat the lunar thermal, radiation and micrometeoritic environment.

U.8   CONCLUSION

Humanity has successfully built rock structures and habitats on Earth since the earliest cultures known from prehistory. We are continually discovering prehistoric civilizations through hewn and processed rock structures still standing all over the world, the only surviving remnants of their culture. This is indeed a testament to their durability.

Lunar rocks and soils with magnetic and microwave susceptibility are ubiquitous on the Moon. They may be used for lunar colony buildings. Sinterd soil has several exterior uses while cast basalt may find extensive application both for load-bearing structures and also for finishing exteriors and interiors.

Existing tools and machinery used on Earth may be adapted for robotic operation on the Moon that could build up a lunar rock products warehouse inventory of colony-building materials well in advance of human arrival.

Space architects, engineers, designers and builders may do well to adopt a "technology continuum" philosophy and broaden their palette of high-technology materials and artifacts to include not only timeless ways of building, but new opportunities brought about by the unique lunar environment. We must utilize the best materials and practices handed down over the millennia in concert with these advanced concepts as we proceed to build meaningful, efficient and aesthetically pleasing permanent structures on the Moon.

U.9   REFERENCES

Eckart, P., *The Lunar Base Handbook*. McGraw Hill, New York, 2000.

Khalili, E.N., Lunar structures generated and shielded with on-site material, *Aerospace Journal*, American Society of Civil Engineers, July 1989.

Khalili, E.N., *Ceramic and Fused Adobe Structures Generated In-Situ*, Lunar Bases and Space Activities of the 21st Century Ed., Wendell Mendell, Lunar and Planetary Institute, 1985.

Lin T.D., Proposed remote control solar powered concrete production experiment on the Moon, *J. Aerosp. Engrg.*, Volume 10, Issue 2, 104–109. ASCE Publications Reston VA. (April 1997).

Schrunk, D., Sharpe, B., Cooper, B., and Thangavelu, M., *The Moon: Resources, Future Development, and Colonization*. Wiley–Praxis, New York and Chichester, UK, 1999.

Thangavelu, M., MALEO: Modular Assembly in Low Earth Orbit. Alternative strategy for lunar base development. *Journal of the British Interplanetary Society*, January 1993.

Thangavelu, M., MALEO: Modular Assembly in Low Earth Orbit. Strategy for lunar base development. *Journal of the British Interplanetary Society*, January 1993.

Appendix V

Rapid prototyping: Layered metals fabrication technology development for support of lunar exploration at NASA/MSFC

Kenneth G. Cooper   *NASA Marshall Space Flight Center*
James E. Good   *MEI Technologies, Inc*
Scott D. Gilley   *Tec-Masters, Inc*

NASA's human exploration initiative poses great opportunity and risk for missions to the Moon and beyond. In support of these missions, engineers and scientists at the Marshall Space Flight Center are developing technologies for ground-based and in situ fabrication capabilities utilizing provisioned and locally refined materials. Development efforts are pushing state-of-the-art fabrication technologies to support habitat structure development, tools and mechanical part fabrication, as well as repair and replacement of ground support and space mission hardware such as life support items, launch vehicle components, and crew exercise equipment. This appendix addresses current fabrication technologies relative to meeting targeted capabilities, near-term advancement goals, and process certification of fabrication methods.

V.1 INTRODUCTION

A Fabrication Technologies research and hardware development program for support of space missions is underway at the NASA Marshall Space Flight Center (MSFC) in Huntsville, Alabama. This effort is a sub-element of the In Situ Fabrication and Repair (ISFR) element, which is working in conjunction with the In Situ Resources Utilization (ISRU) element. ISFR has the charter to provide fabrication and repair capabilities for space mission equipment with a long-term goal of operation in extraterrestrial environments while ISRU will extract useable materials from those local environments. ISFR activities include fabrication of mechanical components and assemblies on Earth as well as within partial gravity environments with optional consideration of micro-gravity operations capability for in-transit phases.

The effort includes development of supporting fabrication, repair, and habitat structure technologies to maximize use of in situ resources. However, provisioned fabrication feedstocks may be launched with the carrier vehicle and utilized either during in-transit phases or in situ until ISRU product feedstocks are available.

The current research effort has some past history in that layer-by-layer manufacturing has been investigated for spaceflight endeavors. The level of maturity of this technology as recent as a few years ago left a large gap between state-of-the-art additive manufacturing techniques before and what is achievable currently. Although this is true, MSFC was looking ahead (Cooper, 2001) in the 1990s and investigated the possibility of microgravity manufacturing using additive fabrication techniques. Additionally, applications in outer space (Cooper, 2002) were investigated and these efforts provided a foundation for the current development program. Significant rapid manufacturing techniques and material development advancements have been made in the last two years and require a serious investigation to determine the feasibility of in situ manufacturing for space applications and to determine technology gaps.

The ISFR element supports the entire life cycle of the Exploration program by: reducing downtime due to failed components; decreasing risk to crew by recovering quickly from degraded operation equipment; improving system functionality with advanced geometry capabilities; and enhancing mission safety by reducing assembly part counts of original designs where possible. Non-destructive Evaluation (NDE) capabilities will help reduce crew exposure to harsh space environments by providing autonomous technologies capable of identifying and confirming a failure or validating that a repair or fabrication operation was successful (Bodiford *et al.*, 2005). NDE will also be incorporated within fabrication systems where possible to verify integrity, material properties, and dimensional accuracy during the build cycle.

ISFR also provides habitat manufacturing and assembly technologies that incorporate in situ resources to produce autonomous, affordable, and pre-positioned habitat environments. These habitats will feature radiation shielding and protection from micrometeoroids and exhaust plumes. ISFR strives to reduce launched mass and volume resource requirements for supply of spares and materials from Earth by utilizing in situ resources. In addition, ISFR is investigating just-in-time repair capability using soldering, patching, and adhesives to return functionality to components that may be repairable rather than require complete replacement.

This appendix strives to provide the current technologies available for direct metal fabrication with a specific emphasis on the Arcam EBM process. Significant advances have allowed the gap to close between rapid prototypes and rapid manufacturing. These current processes can provide an important capability on the lunar surface while minimizing the upmass of traditional feedstock. These are crucial components for a successful long-duration mission to the Moon.

V.2   FABRICATION TECHNOLOGIES OVERVIEW

This appendix specifically focuses on fabrication of metal components intended to initially produce flight hardware on Earth prior to mission launch as well as on-demand replacement of failed or new components in situ. Earth-based fabrication is targeted at providing rapid turn around of prototypes production as well as future flight components with complex geometries that are difficult or impossible to fabricate with traditional machining, casting, or forming methods. Future in situ fabrication goals include creation of replacement components and unforeseen tools that may enhance mission success as well as correction of design deficiencies during the mission for which spares are of little use if failures occur.

The main goal of fabrication technologies will be to provide rapid manufacture of parts and tools via a quality-controlled approach that may be a single process or a hybrid mix of additive and subtractive processes. An important near-term benefit will be the production of advanced geometry components and assemblies on Earth prior to launch as part of the mission equipment design. Earth-based component production activities will serve to provide immediate benefits of the development investment that may lead to future in situ fabrication operations. Development demonstrations will include production of components for functional testing with certification of the fabrication processes to produce pedigreed components.

V.2.1 Fabrication processes discussion

An initial study of fabrication technologies performed at MSFC showed that there is no single comprehensive solution for fabrication systems in space (Hammond *et al.*, 2006). A majority of space flight hardware is fabricated using typical subtractive processes such as Computer Numerically Controlled (CNC) machining that provide excellent dimensional accuracy and surface finish. However, they require feedstock material that is larger than the outer part envelope and also usually require part-specific hold down tooling that is not very mass efficient for complex geometry components. Machining processes also typically require bulky and heavy equipment to provide a stable platform for significant material removal operations. An anticipated approach will involve a combination of additive and subtractive technologies to provide the necessary fabrication capabilities.

Additive techniques typically involve building parts layer-by-layer using filaments, powders, liquids, or stacked sheets of feedstock materials successively joined together to build up a part in three dimensions. These methods offer reduction of feedstock requirements since parts are built up to size from a common feedstock size rather than subtractively machined down from bulk stock that must be larger than the part envelope. Current additive limitations include reduced accuracy and surface finish compared to traditional machining. Incorporating additive methods with light CNC machining processes integral to the system design is currently envisioned to provide required accuracies and feedstock efficiency. This also supports machining of certified feedstocks in situations where additive techniques may not be able to produce acceptable material properties by a constrained resource fabrication system. These two techniques, functioning in tandem, would thus provide improved capability over either single process. Based on a materials utilization study (Hammond *et al.*, 2006), the initial focus of fabrication technologies development is focusing on

Table V.1. Primary metallic materials set targeted for fabrication technologies development.

Material	Alloy
Aluminum	6061
Aluminum	7075
Titanium	Ti6Al4V
Stainless steel	316
Stainless steel	17-4PH
Inconel	625
Inconel	718

metallic components. This appendix discusses additive manufacturing technologies in reference to future application for space flight missions.

V.2.2   Materials set discussion

Before evaluating specific additive techniques, the desired material set was first defined as discussed (Hammond *et al.*, 2006). Technical interchange meetings with multiple flight hardware development programs at NASA/MSFC were conducted to determine material needs. Also, a statistical analysis of the materials utilized in several previously flown space flight hardware projects including the Material Identification and Usage Lists (MIULs) of Space Station and Shuttle Mid-Deck Payloads. Another source of information used to determine the material set was a compilation of failure data representing selected space vehicles. This list contained failed components as well as the component material. The list provided insight into the type of components that may require replacement, as well as the type of components requiring fabrication during a long-duration stay on the Moon or Mars. The primary set of metallic materials was thereby identified as shown in Table V.1.

V.2.3   Additive fabrication processes assessment

A trade study was performed to investigate additive techniques that could manufacture metal parts in support of space flight missions. The systems selected for evaluation included the following technologies and assessed for their metal fabrication capabilities and the ability to process the desired materials set. Also included in the assessment were part accuracy, surface finish, feedstock type, feedstock usage efficiency, and part build volume among other factors. The systems selected for evaluation included the following technologiesare shown in Table V.2. These systems produce parts by progressively consolidating feedstocks one layer at a time formed to shape per the part cross section geometry at that particular build height.

The process trade study focused on two primary areas: the part quality and material set. A 50% weighting was placed on the quality of the part, including

Table V.2. Assessed additive fabrication processes.

System or method	Description
Selective Laser Sintering (SLS®)-EOS®	Sinters successively spread layers of powder in raster fashion using laser beam.
Arcam Electron Beam Melting (EBM)	Fuses successively spread layers of powder in raster fashion using electron beam.
Laser Engineered Net Shaping (LENS®)	Fuses flowing powder sprayed by gas nozzle into laser beam to build parts progressively in vertical direction.
Prometal 3DP	Consolidates successively spread layers of powder by adhesive spray will follow-up sintering and infiltration to densify parts.
POM Group – Direct Metal Deposition (DMD®)	Fuses flowing powder sprayed by gas nozzle into laser beam to build parts progressively in vertical direction.
SLS®-3D SYSTEMS	Sinters successively spread layers of powder in raster fashion using laser beam. Requires infiltration for densification.
Ultrasonic Consolidation (UC)	Consolidates successively applied metal tape using ultrasonic welding with integral CNC functions to machine to net shape.

capabilities of the system with respect to the part build volume, geometrical tolerances, and surface finish among others. Relative rankings were then assigned to each system for all part quality criteria. A 50% weighting was also placed on the system ability to process a set of 14 typical aerospace materials, including the primary set listed in Table V.1. The evaluation was performed by members of the Fabrication Technologies team along with the members of MSFC Rapid Prototyping Laboratory. Two processes that distinguished themselves based on current technological maturity and NASA requirements were Electron Beam Melting (EBM) and Selective Laser Sintering (SLS). These two processes scored very closely in the trade study on their overall weighted score total with individual strengths and weaknesses taken into account. The EBM score was slightly higher and was chosen for further in-house development. Deciding factors included the EBM metallic materials set better matches those desired, it produces fully dense parts and build operations are performed in a vacuum chamber, which fits well with the space environment. The trade off with EBM is that it has a rough surface finish with lower absolute accuracy but these can be corrected via post processing with CNC machining, which is envisioned for an integrated system anyway.

V.2.4   The Electron Beam Melting (EBM) technology

The EBM technology was chosen for further in-house development due to its ability to produce a fully-dense metal material that yields high strength properties (Lindhe and Harrysson, 2003). While surface finish is not at the level desired by many industries or customers, this can be addressed through secondary processes such as CNC machining. Many components used in the aerospace field do not require a polished finish as a requirement, but merely as a desire. As this technology becomes established as a viable solution to component manufacturing, the designers must change the way these components are designed.

There are similarities between the EBM process and the SLS process in that both use a power source to heat a bed of powder layer-by-layer to additively "grow" a part. The main difference being the source of the power required to change the metal powder to a finished part. The laser has been used in the rapid prototyping industry for years and is the backbone for the SLS process. The electron beam gun is used in the Arcam machine to fully melt the metal powder in the EBM process. The electron beam has shown several advantages over the laser and its popularity is increasing as the early adopters have had success in part-fabrication and material characterization, validating the claims of the vendor that the EBM machine can produce fully dense material. The efficiency of the electron beam is five to ten times greater than laser technology. This is important as it results in less power consumption and lower maintenance and manufacturing costs. Arcam is using a 4000 W electron beam gun in their current EBM machines. It is this power that results in the full melt of the powder that allows for good material properties at high build speeds.

The EBM process starts with the pre-heating of the powder bed. The 4 kW electron beam gun is used to perform the pre-heat by using a low beam current but a high scan speed. The layers are heated after the initial 0.1 mm layer is spread across the bed. The scan speed can be as high as 1000 meters/sec with a build rate up to 60.6 cm^3/hour. The preheat is performed for two reasons:

- The preheat lightly sinters the metal powder to hold it in place for the subsequent layers used to fabricate the part.
- The preheat provides heat to the fabricated part during the build which reduces the thermal gradient between the last melted layer and the previously melted layers. The consistent temperature of the build will reduce the chance of residual stresses.

How does the beam get to the powder bed? The beam is generated by the electron beam gun which is fixed to the top of the vacuum chamber. The beam is deflected to reach the entire build volume. The deflection is achieved by the set of 2 magnetic coils. The first magnetic coil acts as a lens and focuses the beam to the desired diameter. The second magnetic coil deflects the beam to the desired location on the powder bed. Take note that no moving parts are needed to deflect the beam. The beam is created by heating a tungsten filament to a high temperature where many electrons are accelerated to half the speed of light and stream through the gun. After proceeding through the magnetic coils, the beam begins melting each layer of the fabricated part.

After each layer is melted, the build platform lowers by one layer-thickness. A rake spreads the powder which is then melted. This pattern continues until the part has been completed. At this point a helium purge is initiated to minimize the cool-down time. Without the purge, the part may require in excess of 20 hours to cool while the addition of the helium purge reduces the wait time to between 3–8 hours depending upon the part size. A completed part that has a large dimension in the Z-direction will require a longer time to cool the vat of powder because of the depth. A good scenario is one in which the part, or parts, is completed at the end of the day and can cool down overnight. Prior to taking the part out of the chamber, air is purged into the chamber to equalize the pressure to that of the exterior environment. At this point, the door is opened and the loose powder is removed from the part, it is bead-blasted to remove any stubborn powder attached to the surface. The bead blast is performed with the same material as the finished part. The part is now ready for any post-processing of any critical interfaces, such as machining a good surface at a mated interface.

This is a good time to note that the finished part is fully dense and does not require a secondary process where it is infiltrated to reduce porosity. Again, this is the major difference between the EBM and SLS processes. A characteristic of both processes is the ability to produce unique geometries such as internal cavities that are used as conformal cooling channels. These cooling channels cannot be produced easily from conventional CNC machining. The EBM process can fabricate a part that contains a matrix core with a solid shell, effectively having a hollow part supported with an internal scaffold. This is a situation where CNC machining is not an option. A similar part is shown in Figure V.1 where a scaffold is integrated within the four chambers of the flanged component. The scaffold provides a flow path where surface area is a critical component. It provides high surface area and low pressure drop through the matrix. This is advantageous for the purpose of this component which involves a life support system performing in a space environment.

The propulsion group at NASA/MSFC has been investigating the EBM process for parts fabrication. Turbo-machinery components have been made (see Figure V.2) to be used in functional testing alongside similar parts made by traditional methods. Tests are on-going and results are not available at this time but with the ability to redesign the components and fabricate the parts quickly, the EBM process can speed up the design process.

As a demonstration part for the Crew Exploration Launch Vehicle (CEVCLV), the Rapid Prototyping Lab at MSFC worked with the CEV CLV design team to look for situations where the EBM process could aid the design. A titanium end fitting (see Figure V.3) was chosen and fabricated in 8 hours in order to demonstrate the quick turn-around time and the capability to fabricate such a labor-intensive part. Typically, these end fittings, that were part of the initial design of the Interstage, are machined out of a large chunk block of titanium; however, CNC machining could not access part of the geometry that needed to be eliminated. In the end, the machined part was approximately 0.23 kg heavier than the designed part, where the baseline weight was approximately 1.8 kg. The EBM part was able to build the geometry, as designed, and at the expected weight.

V.2.5   Material feasibility studies of selected materials

Marshall Space Flight Center has contracted Arcam to perform feasibility studies on selected materials that are of interest in the aerospace industry, as well as the military.

Figure V.1. ECLSS component.

Figure V.2. Turbopump component.

Figure V.3. CLV end fitting.

These materials include Inconel 625 and 718, aluminum 6061, stainless steel 316, and 17-4 steel. Research has shown that these materials have been used extensively in space systems found on the space shuttle, space station, the MIR, and other vehicles. A suite of materials such as these would increase the demand for the EBM process. Currently, the Inconel 625 feasibility study has been completed. The results show the material performs very well in the EBM machine.

Inconel 625 is a nickel-based superalloy well-known for its high temperature and wet corrosion resistance. The study that was performed focused on the melting properties and whether the process was stable over hundreds of melted layers. No actual parts were made and parameter optimization was not a goal of this study.

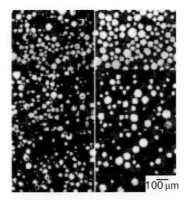

Figure V.4. Powder cross-section.

Two different materials were used in this study and supplied by two independent powder manufacturers. It was seen from the results that the pedigree of the powder can make a difference in the performance of the powder. Figure V.4, illustrates the cross-section of the powder particles. Impurities can be seen on the left-column samples as well as some spherical porosity.

The right column photos show a powder that is almost perfectly spherical and is virtually free from impurities and pores.

The impurities of the one powder source was observed in the melt pool throughout the builds and may be the result of silicon, magnesium, iron, and oxygen.

The tensile strength was consistent with the impurity levels of the two powders. The powder with the greater level of impurities and porosity reflected lower strengths in all samples tested.

The sample parts were fabricated using both powders and analyzed. Once again, the best surface quality was obtained from the "better" powder.

These parts were limited in height due to the availability of the powders. The base of the samples measured 15×80 mm with a height of 20–30 mm. Because of the limited powder, only horizontal tensile bars were fabricated for testing.

The images at the bottom of Figure V.4 show dark particles in the top surface of the samples and are believed to come from impurities in the powder.

Samples (Figure V.5) were tested in their original state, directly from the EBM machine. No post-build treatments were performed outside of sand blasting to remove excess powder. In Figure V.5, the samples labeled (a)–(d) have dimensions 15×80 mm $\times 25$ mm high.

In addition to the previously mentioned tensile tests, the samples underwent additional investigations including:

- X-Ray fluorescence (XRF) for metallic content.
- High-temperature combustion with IR detection for oxygen content.
- Portable Brinell hardness tester.
- Optical microscopy for microstructure examination.

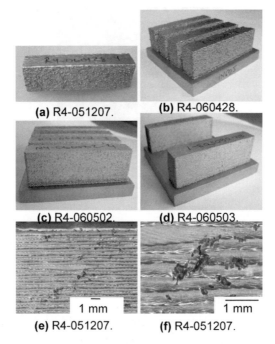

(a) R4-051207.   **(b)** R4-060428.

(c) R4-060502.   **(d)** R4-060503.

1 mm   1 mm
(e) R4-051207.   **(f)** R4-051207.

Figure V.5. EBM fabricated test samples.

Table V.3. Brinell hardness of built material.

Wrought annealed IN 625	R4-051207	R4-060428	R4-060502	R4-060503
190 ksi	179, 176, 178 ksi	199, 202, 200 ksi	199, 202, 200 ksi	193, 192, 192 ksi

The results of the Brinell hardness test are shown on the following page. The results include the wrought annealed Inconel 625 hardness as documented in the *ASM Handbook* (10th edition). The measurements were taken at various positions along the build direction (Z).

The tensile tests reflect material strengths that fall between cast and wrought Inconel 625. Optimization of the build process and the addition of post-treatments can provide improved strength characteristics. The test results are shown in Table V.4.

At the conclusion of the Inconel 625 feasibility study, it was determined that the material can be used in the machine based on the promising results. The powder melts very well and the process was shown stable after hundreds of melted layers. After initial smoking in the chamber, the material does not smoke easily. The initial smoking can be minimized but not totally eliminated by an extra preheat of the powder surface.

Table V.4. Tensile properties of built material.

Material properties	Wrought annealed IN625	Polycrystalline cast IN625	R4-051207	R4-060428	R4-060502	R4-060503
UTS, MPa (RT)	930	710	702, 673	733, 780	784, 789	746
YTS, MPa (RT)	517	350	349, 355	423, 424	426, 430	395
Elongation, % (RT)	42.5	48	34.5, 27	41.5, 44.5	38.5, 40	44.5
(RT)	Not specified	Not specified	37, 28	59, 664	1, 66	61

V.3   CONCLUSIONS

The Fabrication Technologies research team selected an additive metal-fabrication system after performing a trade study assessment of layered fabrication technologies. Targeted materials, representative of typical space flight hardware requirements, were identified using historical parts failure data and material-usage analyses. A trade study was then performed to assess part quality and material-set capabilities that are currently existing or anticipated to occur in the near future. Based on the trade study factors, it was concluded that the EBM process was the best suited to the capability requirements of the ISFR team with application to future space environments.

An EBM metal fabrication system at MSFC has been purchased and is in operation and will be used for material-set expansion development and fabrication of demonstration parts to support active programs within NASA, such as the CEV, CLV, and ECLSS, among others. Post-processing techniques will be utilized to improve the fabricated part quality, including CNC machining. The Fabrication Technology team will continue to develop the EBM technology and look for opportunities within NASA/MSFC and other agencies to fulfill the Vision for Space Exploration.

V.4   ACKNOWLEDGMENTS

The authors would like to express their sincere thanks to the colleagues and industry contributors who provided information that made this research possible. Much information was gathered from system vendors and users who are experts in their respective fields. We would also like to thank Arcam AB of Sweden for providing research data as well as photographs as input to this publication.

V.5   REFERENCES

Bodiford, M.P., Gilley, S.D., Howard, R.W., Kennedy, J.P., Ray, J.A., "Are We There Yet? . . . Developing In Situ Fabrication and Repair (ISFR) Technologies to Explore and Live on the Moon and Mars" in proceedings of *1st Space Exploration Conference: Continuing the Voyage of Discovery*, edited by NASA Marshall Space Flight Center, Huntsville, Alabama, 2005, AIAA-2005-2624.

Cooper, K.G., *Extending Rapid Prototyping Past the Horizon: Applications in Outer Space*, PEO1-235, Society of Manufacturing Engineers, Dearborn, Michigan, 2001.

Cooper, K.G., "Microgravity Manufacturing Via Rapid Prototyping Techniques", in proceedings of *Georgia Tech Rapid Prototyping and Manufacturing Symposium*, edited by NASA Marshall Space Flight Center, Huntsville, Alabama, 2001.

Hammond, M.S., Good, J.S., Gilley, S.D., and Howard, R.W., "Developing Fabrication Technologies to Provide On-Demand Manufacturing for Exploration of the Moon and Mars," in proceedings of *44th AIAA Aerospace Sciences Meeting and Exhibit*, edited by NASA Marshall Space Flight Center, Huntsville, Alabama, 2006, AIAA-2006-0526.2006.

Lindhe, U., Harrysson, O., "Rapid Manufacturing with Electron Beam Melting (EBM) – A Manufacturing Revolution?" in proceedings of *Rapid Manufacturing: Utilizing Direct Metals*, edited by Arcam AB, Molndal, Sweden, 2003, TP03PUB397.

Bibliography

INTRODUCTION (within the frontmatter)

Bush, G.W., *New Vision for NASA*, NASA Hq Speech, April 14th, 2004. Office of the President, White House, Washington, D.C., 2004.

Duke, M. (ed.), *Workshop on using in situ resources for construction of planetary outposts*. LPI Technical Report Number 98-01, Lunar and Planetary Institute, Houston, TX, 1998.

Durst, S., Bohannan, C., Thomason, C., Cerney, M., and Yuen, L., *Proceedings of the International Lunar Conference 2003/International Lunar Exploration Working Group 5*. American Astronautical Society/Univelt, San Diego, CA, 2003.

Galloway, R.G. and Lokaj, S. (eds.), *Space '98, Proceedings of the 6th International Conference and Exposition on Engineering, Construction and Operations in Space*. American Society of Civil Engineers, Reston, VA, 1998.

Garriott, O. and Griffin M., *Extending Human Presence into the Solar System*, a report. Planetary Society, Pasadena, CA, July 25th, 2004.

Hanley, Jeff (Program Manager), *NASA Constellation Program*, 2007. *http://www.nasa. gov/ mission_pages/constellation/orion/index.html*

Harris, P.R., *Living and working in space: Human behavior, culture and organisation* (2nd edn.). Wiley/Praxis, Chichester, U.K., 1996.

Horowitz, Scott (Associate Administrator, Exploration Systems Mission Directorate – ESMD – Constellation Program), Nations in Space. *Ejournal USA* (U.S. Dept. of State Publication), **11**, No. 3, Washington D.C., October, 2006.

Johnson, S.W. *et al.* (ed.), Seven Proceedings, *Space '88, '90, '92, '94, '96, 2000, and 2002, International Conference and Exposition on Engineering, Construction and Operations in Space*. American Society of Civil Engineers, New York, 1988–1990–1992–1994–1996–2000–2002.

Kalam, A.A.P.J., *Creative Leadership for Future Challenges of Space*, Address to the International Space University, Strasbourg, France, April 24th. Presidential Archives, Rashtrapathi Bhavan, New Delhi, India, 2007.

Marburger, J. (Director, Office of Science and Technology Policy, Executive Office of the President, White House, Washington, D.C.), Keynote Address, *44th Robert H. Goddard Memorial Symposium, Greenbelt, MD, March 15th, 2006.*

McKay, M.F., McKay, D.S., and Duke, M.B., *SPACE RESOURCES: Overview; Scenarios, Volume I; Energy, Power, and Transport, Volume II, Social Concerns, Volume III, and Materials, Volume IV*, NASA SP-509. U.S. Government Printing Office, Washington, D.C., 1992.

Mendell, W.W. (ed.), *Lunar bases and space activities of the 21st century*. Lunar and Planetary Institute, Houston, TX, 1985.

Mendell, W.W. (ed.), *2nd Conference on Lunar Bases and Space Activities of the 21st Century*, NASA Conference Publication 3166, Vol. 1. NASA, Washington, D.C., 1992.

Schrunk, D., THE PLANET MOON: Opportunities for Space Exploration, *Proceedings of the 4th International Conference on the Exploration and Utilization of the Moon*, ESA SP-462. ESA, Noordwijk, The Netherlands, July 10–14, 2000.

Schrunk, D., *The Rationale for Establishing an International Lunar Base*, presentation. American Astronomical Society, Pasadena, CA, October, 2000.

Schrunk, D., The Moon: Optimum location for the first industrial/scientific base in space, *Proceedings of the 8th International Space Conference 2002*, pp. 122–128. American Society of Civil Engineers, Reston, VA, 2002.

Schrunk, D., Cooper, B., Sharpe, B., and Thangavelu, M., A Coherent Vision for Space Exploration and Development in the 21st Century, *Proceedings of the 7th International Conference on Space 2000*. American Society of Civil Engineers, Reston, VA, 2000.

Schrunk, D., Cooper, B., Sharpe, B., and Thangavelu, M., The Planet Moon Project, *Proceedings of the 7th International Conference on Space 2000*. American Society of Civil Engineers, Reston, VA, 2000.

Spudis, P., *The Once and Future Moon*. Smithsonian Library of the Solar System, November, 1996.

A sampling of visionary writings on the exploration and development of space

Bova, B., *Welcome to Moon Base*. Ballantine Books, 1987.

von Braun, W. *et al.*, *Man on the Moon*. Colliers, 1952.

Clarke, A.C., *Interplanetary Flight*. Temple, London, 1950.

Clarke, A.C., *The Exploration of the Moon*, illustrated by R.A. Smith. Harper Row, New York, 1954.

Ehricke, K.A., *Space Flight*, Vols. 1–2. Van Nostrand, Princeton, 1960–1962.

Ehricke, K.A., Lunar industrialization and settlement – Birth of polyglobal civilization, *Lunar Bases and Space Activities of the 21st Century*, pp. 827–855. Lunar and Planetary Institute, Houston, TX, 1985.

Gagarin, Y., *Road to the Stars*. Foreign Languages Publishing House, Moscow, 1962.

Goddard, R.H., *The Ultimate Migration*. Goddard Library Manuscript, January, 1918.

Godwin, B.F., *The Man In the Moone; Or a Discourse Of A Voyage Thither, by Domingo Gonsales, The Speedy Messenger*. Printed by John Norton for Ioshua Kirton and Thomis Waren, London, 1638.

Heppenheimer, T.A., *Colonies in Space*. Stackpole Books, 1977.

Kepler, J., *Astronomia Novo*, 1609.

Ley, W., *Rockets, Missiles and Space Travel*. Viking, New York, 1957.

Ley, W. and Bonestell, C., *The Conquest of Space*. The Viking Press, New York, 1956.

O'Neill, G.R., *The High Frontier: Human Colonies in Space*. Space Studies Institute Press, Princeton, NJ, 1989.

Ryan, C. (ed.) and Bonestell, C. (illustrator), *Conquest of the Moon*. Viking, New York, 1953.

Sagan, C., *Cosmos*. Ballantine, New York, 1980.

Savage, M.T., *The Millennial Project: Colonizing the Galaxy in Eight Easy Steps*. Little, Brown & Company, Boston, 1992.

Tsiolkovsky, K.E., *Na Luna [On the Moon]*. Goncharov, Moscow, 1893.

Verne, J., *Round the Moon*. Jules Hetzel, Paris, 1870.

Wells, H.G., *The First Men on the Moon*. Newnes, London, 1901.

Wilkins, B.J., *The Discovery of A World on the Moone; Or a Discourse tending to Prove 'tis Probable There May Be Another Habitable World In That Planet*, A Founder and First Secretary of The Royal Society (the 3rd edn. includes an appendix: The Possibility Of A Passage Thither). Michael Sparke and Edward Forrest, London, 1638.

CHAPTER 1: LUNAR ORIGINS AND PHYSICAL FEATURES

Baldwin, R.B., *The Face of the Moon*. Chicago University Press, 1949.

Bates, J.R., Lauderdale, W.W., and Kernaghan, H., *ALSEP Termination Report*, NASA Reference Publication 1036. NASA, Washington, D.C., 1977.

Belbruno, E.A. and Gott, J.R., Where did the Moon come from? *Astronomy Journal*, **129**, 1724–1745, 2005.

Binder, A.B., Lunar Prospector: Overview. *Science Magazine*, **281**, September 4th, 1475–1476, 1998.

Burke, J.D., The hunt for lunar ice. *The Planetary Report*, **XVIII**, No, 4, pp. 6–11, July/August, 1998.

Canup, R.M., Dynamics of Lunar Formation. *Annual Review of Astronomy and Astrophysics*, **42**, 441–475, 2004.

Canup, R.M. and Asphaug, E., The Lunar-Forming Giant Impact. *Nature*, **412**, 708–712, 2001.

Canup, R.M. and Righter, K. (eds.), *Origin of the Earth and the Moon*. University of Arizona Press, 2000.

Compton, W.D., *Where no man has gone before: A history of Apollo lunar exploration missions*. NASA SP-4214. U.S. Government Printing Office, Washington, D.C., 1989.

Cooper, B.L., Sources and Subsurface Reservoirs of Lunar Volatiles, *Proceedings of the 20th Lunar and Planetary Science Conference*. Lunar and Planetary Institute, Houston, TX, 1990.

DMAAC, *Lunar Maps*. Defense Mapping Agency Aerospace Center, St. Louis, MO, November, 1971.

Fielder, G.H., *Structure of the Moon's Surface*. Pergamon Press, Oxford, 1961.

Galimov, E.M, Krivtsov, A.M, Legkostupov, M.S., Eneev, T.M., and Sidorov, Y.I., Dynamic Model of the Earth–Moon System. *Geochemistry International*, **43**, No. 11, 1045–1055, 2005.

Harland, D., *Exploring the Moon: The Apollo Expeditions*. Springer/Praxis, Chichester, U.K., 1999.

Hartmann, W.K., A Brief History of the Moon. *The Planetary Report*, **XVII**, No. 5, The Planetary Society, October, 1997.

Heiken, G., Vaniman, D., and French, B.M. (eds.), *The Lunar Source Book*. Cambridge University Press and Lunar and Planetary Institute, Houston, TX, 1991, 1998.

Hood, L.L. and Williams, C.R., Lunar swirls: Distribution and possible origins, *Proceedings of the 19th Lunar and Planetary Science Conference*. Cambridge University Press and Lunar and Planetary Institute, Houston, TX, 1991.

Hörz, F., Lava tubes: Potential shelters for habitats, *Lunar Basins and Space Activities of the 21st Century*, edited by W. Mendell. Lunar and Planetary Institute, Houston, TX, 1985.

Jolliff, B.L. *et al.*, *New Views of the Moon*. American Minerological Society, 2005.

Konopliv, A.S. *et al.*, Improved gravity field of the Moon from Lunar Prospector. *Science Magazine*, **281**, September 4th, 1476–1480, 1998.

Laskar, J., Joutel, F., and Roubutel, P., Stabilization of the Earth's obliquity by the Moon. *Nature*, **361**, 615–617, 1993.

Lin, R.P. *et al.*, Lunar surface magnetic fields and their interaction with the solar wind: Results from Lunar Prospector. *Science Magazine*, **281**, September 4th, 1480–1484, 1998.

Long, Kim, *The Moon Book: Fascinating Facts about the Magnificent, Mysterious Moon*. Johnson Books, Boulder, CO, ISBN 1555662307, 1988.

LSI, *Post-Apollo lunar science*, report of a study. Lunar Science Institute, Houston, TX, July, 1972.

NASA, *Apollo Traverse Maps*, prepared by USGS for NASA, NASA Lunar Planning Charts LOC 1–4. Aeronautical Chart and Information Center, USAF, St. Louis, MO, under the direction of Dept. Of Defense for NASA, May, 1971.

Nozettte, S. *et al.*, *Clementine: Back to the Moon and First Flyby of a Near Earth Asteroid*. NASA Clementine Science Team, NASA HQ/NRl., Washington, D.C., 1994.

Oberbeck, V.R., Greely, R., Morgan, R.B., and Lovas, M.J., On the origin of lunar sinuous rilles. *Modern Geology*, **1**, 75–80, 1969.

Schonfeld, E., High spatial resolution Mg/Al maps of the western Crisium and Sulpicius Gallus regions, *Proceedings of the 12th Lunar and Planetary Science Conference*. Lunar and Planetary Institute, Houston, TX, 1981.

Stacey, N.J.S. and Campbell, D.B., A search for ice at the lunar poles (Abstract), *Proceedings of the 26th Lunar and Planetary Science Conference*. Lunar and Planetary Institute, Houston, TX, 1995.

Taylor, S.R., *Solar System Evolution: A New Perspective* (2nd edn.). Cambridge University Press, 2001.

Urey, H.C., *The Planets*. Yale University Press, 1952.

USAF, *Lunar Reference Mosaic Maps*. Aeronautical Chart and Information Center, U.S. Air Force, St. Louis, MO, November, 1962.

USGS, *Atlas of the Moon*. U.S. Geological Survey, Washington, D.C., 1971.

Zuber, M.T., Smith, D.E., Lemoine, F.G., and Neumann, G.A., The shape and internal structure of the Moon from the Clementine mission. *Science*, **266**, 1839–1843, 1994.

CHAPTER 2: SCIENCE AND CHALLENGES

Adler, T., Remote surgery: Operating on patients from afar. *Science News*, **146**, 266–267, October 22nd, 1994.

Angel, J.R. and Woolf, N.J., Searching for life on other planets. *Scientific American*, April, 60–66, 1996.

Arnett, D. and Bazan, G., Nucleosynthesis in stars: Recent developments. *Science*, **276**, 1359–1362, May 30th, 1997.

Benaroya, H., Reliability of telescopes for the lunar surface. *Lunar Based Astronomy, Journal of the British Interplanetary Society*, **48**, 99–106, Dept. of Mechanical and Aerospace Engineering, Rutgers University, Piscataway, NJ, 1995.

Bhandari, N., Chandrayaan-1: Science Goals, *International Conference on the Exploration and Utilization of the Moon, November 22–26, 2004, Udaipur, Rajasthan, India. Journal of Earth System Science*, **114**, No. 6, 701–709, December, 2005.

Burke, B.F., Astrophysics from the Moon. *Science*, **250**, 1365–1370, December 7th, 1990.

Burnham, Jr., R., *Burnham's Celestial Handbook: An Observer's Guide to the Universe beyond the Solar System* (Vols. 1–3). Dover Publications, New York, 1978.

Chen, P.C., Scoping the Moon: Lunar-based telescopes put astronomy in a new light. *Ad Astra*, 42–45, May/June, 1994.

Chevalier, R.A., Type II supernovae SN 1987A and Sn1993J. *Science*, **276**, 1374–1378, May 30th, 1997.

Connors, M.M., Harrison, A.A., and Akins, F.R., *Living aloft: Human requirements for extended spaceflight*, NASA SP-483. U.S. Government Printing Office, Washington, D.C., 1985.

Cooper, B., *Applications of Electromagnetic Radiation for Exploration of Lunar Regolith for Potential Resources*, Dissertation. The University of Texas at Dallas, 530 pp., 1992.

Cronin, J.W., Gaisser, T.K., and Swordy, S.P., Cosmic rays at the energy frontier. *Scientific American*, 44–49, January 1997.

Crossett, C., *Lunar Base Observatories* (M. Thangavelu, Study Director). AE 599 Space Concepts Studio Project Report, University of Southern California, May, 1995.

Cruikshank, D.P., Stardust memories (Comet Hale–Bopp). *Science*, **275**, 1895–1896, March 28th, 1997.

Day, R.M., XTE: Astronomy with autonomy. *Aerospace America*, 34–39, June, 1996.

Dyson, F.J., Major observatories in space, *Observatories in Earth and Beyond*, edited by Yoji Kondo. Kluwer Academic Publishers, 1990.

ESA, *Global Exploration Strategy: The Framework for Coordination*. Head of the European Space Agency, Paris, France, May, 2007.

Enserink, M., Mining the deep field. *Science*, **274**, 2006–2007, December 20th, 1996.

Fielder, J. and Leggett, N., Lunar agricultural requirements definition, *Engineering and Construction, and Operations in Space*, pp. 344–351. Aerospace Division/American Society of Civil Engineers, Albuquerque, NM, August, 1988.

Fishman, G.J. and Hartmann, D.H., Gamma-ray bursts. *Scientific American*, 46–51, July, 1997.

Fogleman, G.C., Advanced human support technologies program: Providing support to humans for future space exploration missions, *26th International Conference on Environmental Systems, Monterey, CA, July 8–11*, SAE Technical Paper Series 961594, 1996.

Foing, B.H., Highlights, *COSPAR 1992 Symposium: Astronomy and Space Science from the Moon. Lunar Based Astronomy, Journal of the British Interplanetary Society*, **48**, 71–76, ESA/SSD ESTEC, 1995.

Foing, B.H., Astronomy and space science from Station Moon. *Lunar Based Astronomy, Journal of the British Interplanetary Society*, **48**, 67–70, ESA/SSD ESTEC, 1995.

Frank, A., STARMAKER: The new story of stellar birth. *Astronomy*, 53–57, July, 1996.

Glanz, J., "First light" for giant sky survey. *Science*, **280**, 1337–1338, May 29th, 1998.

Goodwin, S., *Hubble's Universe*. Penguin Books USA, New York, 1996.

Haymes, R.C., Lunar Based Gamma Ray Astronomy, *Lunar Bases and Space Activities of the 21st Century*, pp. 307–313. Lunar and Planetary Institute, Houston, TX, 1984.

Hilchey, J.D. and Nein, M.E., Lunar based optical pelescopes: Planning the astronomical tools of the twenty-first century. *Lunar Based Astronomy, Journal of the British Interplanetary Society*, **48**, 77–82, Program Development Directorate, G.C. Marshall Space Flight Center, Huntsville, AL, 1995.

Hisahiro, K. *et al.*, SELENE Project Status, *International Conference on the Exploration and Utilization of the Moon, November 22–26, 2004, Udaipur, Rajasthan, India. Journal of Earth System Science*, **114**, No. 6, 771–775, December, 2005.

Hood, L.L. and Sonett, C.P., The next generation geophysical investigation of the Moon, *Lunar Bases and Space Activities of the 21st Century*, pp. 253–263. Lunar and Planetary Institute, Houston, TX, 1984.

Johnson, S.W., Chua, M.C., and Burns, J.O., Luna Dust, Lunar Observatories, and other Operations on the Moon. *Lunar Based Astronomy, Journal of the British Interplanetary Society*, **48**, 87–92, Dept. of Civil Engineering, University of New Mexico, 1995.

Johnson, S., Chua, K.M., and Carrier, D., III, Lunar Soil Mechanics. *Journal of the British Interplanetary Society*, **48**, 43–48, 1995.

Kalam, A.A.P.J., International Partnership in Lunar Missions, Inaugural Address, *International Conference on the Exploration and Utilization of the Moon, November 22–26, 2004, Udaipur, Rajasthan, India. Journal of Earth System Science*, **114**, No. 6, 577–585, December, 2005.

Kalam, A.A.P.J., Address to Boston University, *Symposium of the Future of Space Exploration, April 12th, 2007.*

Kaufmann, W.J., III, *Discovering the Universe*. W. H. Freeman & Co., New York, 1995.

Kouveliotou, C., Gamma ray bursts. *Science*, **277**, 1257–1258, August 29th, 1997.

Leonard, P.J.T., The challenge of gamma ray burst observations. *Science*, **281**, 525–526, July 24th, 1998.

Lowman, Jr., P.D., Candidate site for a robotic lunar observatory: The central peak of Riccioli Crater. *Lunar Based Astronomy, Journal of the British Interplanetary Society*, **48**, 83–86, Goddard Space Flight Center, MD, 1995.

Macchetto, F.D. and Dickinson, M., Galaxies in the young universe. *Scientific American*, 92–99, May, 1997.

McFarland, D.E., *Managerial imperative: Age of macromanagement*. Ballinger/Harper & Row, Cambridge, MA, 1985.

Mendell, W. (ed.), *International Lunar Farside Observatory and Science Station*. International Space University Summer Session Program Report, 1993, University of Alabama, Huntsville.

Mizutani, H. *et al.*, Lunar A Mission: Outline and Current Status, *International Conference on the Exploration and Utilization of the Moon, November 22–26, 2004, Udaipur, Rajasthan, India. Journal of Earth System Science*, **114**, No. 6, 763–768, December, 2005.

Moore, G.T. and Heubner-Moth, J., Genesis II Advanced Lunar Outpost: Human Factors Design Response, NASA–USRA Grant Program Report, *1st International Design for Extreme Environments Assembly, Houston, TX, 1991.*

Morbidelli, A., New insights on the Kuiper Belt. *Science*, **280**, June 26th, 1998.

Murdin, P. and Allen, D., *Catalogue of the Universe*. Book Club Associates, London, 1980.

NASA, *Planetary Exploration through Year 2000*, a report. Solar System Exploration Committee, NASA Advisory Council, U.S. Government Printing Office, Washington, D.C., 1986.

NASA, *Exploration Studies Technical Report* (Vols. 1–3), NASA Technical Memorandum TM 4075. Office of Exploration, NASA Headquarters, Washington, D.C., 1988–1989.

NASA, *Exploration Systems Architecture Study Report*, NASA TM 2005 214062. Exploration Systems Mission Directorate, NASA Headquarters, Washington, D.C., November, 2005.

Nein, M.E. and Hilchey J.D., The Lunar Ultraviolet Telescope Experiment (LUTE): Enabling technology for an early lunar surface payload. *Lunar Based Astronomy, Journal of the British Interplanetary Society*, **48**, 93–98, G.C Marshall Space Flight Center, Huntsville, AL, 1995.

Nomoto, K., Iwamoto, K., and Kishimoto, N., Type Ia supernovae: Their origin and possible applications in cosmology. *Science*, **276**, 1378–1382, May 30th, 1997.

NRC, *The decade of discovery in astronomy and astrophysics*. National Research Council, National Academy Press, Washington, D.C., 1991.

O'Dell, C.R. and Beckwith, V.W., Young stars and their surroundings. *Science*, **276**, 1355–1359, May 30th, 1997.

Peebles, P.J.E., *Principles of Physical Cosmology*. Princeton University Press, Princeton, NJ, 1993.

Rechtin, E., *Systems Architecting: Creating and Building Complex Systems*. Prentice Hall, ISBN-10: 0138803455, 1990.

Rechtin, E., *System Architecting of Organizations: Why Eagles Can't Swim* (280 pp.). CRC Press, Boca Raton, FL, ISBN-10: 0849381401, 1999.

Reuhauser, R., Low-mass pre-main sequence stars and their X-ray emission. *Science*, **276**, 1363–1370, May 30th, 1997.

Robson, I., *Active Galactic Nuclei*. Wiley/Praxis, Chichester, U.K., 1996.

Sayles, L.R. and Chandler, M.K., *Managing Large Systems: Organizations for the Future*. Harper & Row, New York, 1961.

Shu, F.H., *The Physical Universe, an introduction to astronomy*. University Science Books, 1989.

Simmons, S.C. and Butler, D.J., *Critical medical technologies for human space exploration*, SAE Technical Paper Series 972288, *26th International Conference on Environmental Systems, Monterey, CA, July 8–11, 1996.*

Smoot, G. and Davidson, K., *Wrinkles in Time*. William Morrow & Co., New York, 1993.

Sorenson T.C. and Spudis P.D., The Clementine Mission: A 10-year Perspective, *International Conference on the Exploration and Utilization of the Moon, November 22–26, 2004, Udaipur, Rajasthan, India. Journal of Earth System Science*, **114**, No. 6, 645–668, December, 2005.

Spudis, P.D. and Taylor, G.J., The roles of humans and robots as field geologists on the Moon, *2nd Conference on Lunar Bases and Space Activities of the 21st Century*, NASA Conference Publication 3166, Vol. 2, pp. 307–314. NASA, Washington, D.C., 1992.

Spudis, P.D. *et al.*, *Status and future of lunar geoscience*, NASA SP-484. NASA, Washington, D.C., 1986.

Taylor, G.J., The need for a lunar base: Answering basic questions about planetary science, *Lunar Bases and Space Activities of the 21st Century*, pp. 711–716. Lunar and Planetary Institute, Houston, TX, 1985.

Thorne, K.S., *Black Holes and Time Warps*. W.W. Norton & Co., New York, 1994.

Villard, R., Hubble's new view of the cosmos. *Aerospace America*, 20–25, May, 1996.

Watson, A., Star-watchers team up telescopes for a sharper view. *Science*, **271**, February 16th, 1996.

Zuckerman, B.M. and Malkan, M.A., *The origin and evolution of the Universe*. Boston University Press, Boston, 1995.

CHAPTER 3: RESOURCES

Bustin, R. and Gibson, E.K., Jr., Availability of hydrogen for lunar base activities, *Proceedings of 2nd Conference on Lunar Bases and Space Activities of the 21st Century*, edited by W.W. Mendell, NASA Conference Publication 3166. NASA, Washington, D.C., 1988.

Cadogan, P., *The Moon: Our Sister Planet*. Cambridge University Press, Cambridge, U.K., 1981.

Carter, J.L., Lunar regolith fines: A source of hydrogen, *Lunar Bases and Space Activities of the 21st Century*, edited by W.W. Mendell, pp. 571–581. Lunar and Planetary Institute, Houston, TX, 1985.

Cooper, B.L., Sources and subsurface reservoirs of lunar volatiles, *Proceedings of the 20th Lunar and Planetary Science Conference*, pp. 259–269. Lunar and Planetary Institute, Houston, TX, 1990.

DoI/USGS, *Geologic Atlas of the Moon*. Department of the Interior/U.S. Geological Survey, Washington, D.C., 1971.

Elphic, R.C. *et al.*, Lunar Fe and Ti abundances: Comparison of Lunar Prospector and Clementine data. *Science Magazine*, **281**, September 4th, 1998, 1493–1496.

Feldman, W.C. *et al.*, Fluxes of fast and epithermal neutrons from Lunar Prospector: Evidence for water ice at the lunar poles. *Science Magazine*, **281**, September 4th, 1998, 1496–1500.

Feldman, W.C. *et al.*, Major compositional units of the Moon: Lunar Prospector thermal and fast neutrons. *Science Magazine*, **281**, September 4th, 1998, 1489–1493.

Gibson, E.K., Jr. and Moore, G.W., Variable carbon contents of lunar soil 74220. *Earth and Planet. Sci. Lett.*, **20**, No. 3, 404–408, 1973.

Gibson, E.K., Jr., Moore, G.W., and Johnson, S.M., *Summary of analytical data from gas release investigations, volatilization experiments, elemental abundance measurements of lunar samples, meteorites, minerals, volcanic ashes and basalts*. NASA Johnson Space Center, Houston, TX, 1974.

Glaser, P.E., Energy for Lunar Resource Exploitation, *Proceedings of Lunar Materials Technology Symposium*. NASA Space Engineering Research Center, Arthur D. Little Inc., 1992.

Haskin, L.A., Water and cheese from the lunar desert: Abundances and accessibility of H, C, and N on the Moon. *2nd Conference on Lunar Bases and Space Activities of the 21st Century*, NASA Conference Publication 3166, Vol. 2, pp. 393–396. NASA, Washington, D.C., 1992.

Heiken, G.H., Vaniman, D.T., and French, B.M. (eds.), *Lunar Source Book*. Cambridge University Press/Lunar and Planetary Institute, Houston, TX, 1991.

Johnson, S.W., Characterization of lunar surface materials for Use in construction, *Proceedings of Lunar Materials Technology Symposium*. NASA Space Engineering Research Center, BDM International Inc., 1992.

Lawrence, D.J. *et al.*, Global elemental maps of the Moon: The Lunar Prospector gamma-ray spectrometer. *Science Magazine*, **281**, September 4th, 1998, 1484–1489.

Lucas J.W. (ed.), Thermal Characteristics of the Moon. *Progress in Astronautics and Aeronautics*, **28**, AIAA, MIT Press, 1972.

Morris, R.V., Score, R., Dardano, C., and Heiken, G., *Handbook of Lunar Soils*, NASA Planetary Materials Branch Publication 67, JSC 19069. NASA Johnson Space Center, Houston, TX, 1983.

NRC, *New Frontiers in the Solar System: An Integrated Exploration Strategy*. Space Studies Board, National Research Council, Washington, D.C., 2003.

Taylor, L.A., Return to the Moon: Lunar Robotic Science Missions, *Proceedings of Lunar Materials Technology Symposium*. NASA Space Engineering Research Center, Dept. of Geological Sciences, University of Tennessee, 1992.

Taylor, L.A., Resources for a lunar base: Rocks, minerals, and soil of the Moon, *2nd Conference on Lunar Bases and Space Activities of the 21st Century*, NASA Conference Publication 3166, Vol. 2, pp. 361–378. NASA, Washington, D.C., 1992.

Turkevich, A.L., The average chemical composition of the lunar surface, *Proceedings of the 4th Lunar Science Conference*, Suppl. 4. *Geochim. et Cosmochim. Acta*, 1159–1168, 1973.

Weaver, L.B. and Laursen, E.F., Techniques for the Utilization of Extraterrestrial Resources, *39th Congress of the International Astronautical Federation, Bangalore*. Lockheed Missiles and Space Company, 1988.

Williams, R.J. and Jadwick, J.J., *Handbook of Lunar Materials*, NASA Reference Publication 1057. NASA, Washington, D.C., 1980.

CHAPTER 4: ROBOTICS/COMMUNICATIONS

Albus, J.A., Intelligent Robots for Planetary Exploration and Construction, *Proceedings of Lunar Materials Technology Symposium*. NASA Space Engineering Research Center, National Institute of Standards and Technology, 1992.

Budden, N.A., Virtual presence: One step beyond reality. *Ad Astra*, 30–35, January/February, 1997.

Chobotov, V.A. (ed.), *Orbital Mechanics*, AIAA Education Series. American Institute of Aeronautics and Astronautics, Washington, D.C., 1991.

Cohen, M., The Suitport's Progress, *Life Sciences and Space Medicine Conference, Houston, TX, April, 1995*, AIAA-95-1062. NASA Ames Research Center.

Cooper, B.L., Sharpe, B., Schrunk, D., and Thangavelu, M., Telerobotic Exploration and Development of the Moon, *International Conference on the Exploration and Utilization of the Moon, November 22–26, 2004, Udaipur, Rajasthan, India. Journal of Earth System Science*, **114**, No. 6, 815–822, December, 2005.

David, L., Robots for all reasons. *Aerospace America*, 30–35, September, 1995.

Demsetz, L.A., Byrne, R.H., and Wetzel, J.P. (eds.), *Robotics '98: Proceedings of the 3rd ASCE Specialty Conference on Robotics for Challenging Environments, Albuquerque, NM, April 26–30, 1998* (49 papers).

Gagliardi, R.M. and Karp, S., *Optical Communications* (2nd edn.). John Wiley Series in Telecommunications and Signal Processing, ISBN-0471542872, February 1995.

Hall, J.R. and Hastrup, R.C., Deep Space Telecommunications, Navigation, and Information Management: Support of the Space Exploration Initiative, *41st Congress of the International Astronautical Federation, Dresden, 1990*, AF 90-445. NASA Jet Propulsion Laboratory, Pasadena, CA/Caltech.

Huixian Sun *et al.*, Scientific Objectives and Payloads of Chang E-1 Lunar Satellite, *International Conference on the Exploration and Utilization of the Moon, November 22–26, 2004, Udaipur, Rajasthan, India. Journal of Earth System Science*, **114**, No. 6, 789–794, December, 2005.

Iannotta, B., Rockets take aim at booming market. *Aerospace America*, 34–41, February, 1998.

Isakowitz, S.J., *International Reference Guide to Space Launch Systems*. AIAA Space Transportation Systems Committee, American Institute of Aeronautics and Astronautics, Washington, D.C., 1991.

Jeffrey, J., Biesiadecki, P., Leger, C., and Maimone, M.W. (2007) Tradeoffs between directed and autonomous driving on the Mars Exploration Rovers. *International Journal of Robotics Research*, **26**, No. 1, January, 2007.

Katzman, M., *Laser Satellite Communications*. Prentice Hall, Englewood Cliffs, NJ, 1987.

Korn, P. (ed.), *Humans and Machines in Space: The Payoff*. American Astronautical Society, San Diego, CA, 1992.

Krotkov, K.E., Hebert, M., Henriksen, L., Levin, P., Maimone, M., Simmons, R., and Teza, J., Evolution of a prototype lunar rover: Addition of laser-based hazard detection, and results from field trials in lunar analog terrain, *Autonomous Robots*, **7**, No. 2, 1999.

Kubota, T. *et al.*, Japanese Lunar Robotics Exploration by Cooperation with Lander and Rover, *International Conference on the Exploration and Utilization of the Moon, November 22–26, 2004, Udaipur, Rajasthan, India. Journal of Earth System Science*, **114**, No. 6, 777–785, December, 2005.

Lesh, J.R., *Recent progress in deep space optical communications*, SPIE OE Laser 93 1866-05. NASA Jet Propulsion Laboratory, Pasadena, CA/Caltech, 1993.

Marohn, C. and Hanly, C., Twenty-first century surgery using twenty-first century technology: Surgical robotics. *Current Surgery*, **61**(5), 466–473.

Minsky, M. (ed.), *Robotics*. Omni Press, Garden City, 1985.

NASA, *Exploration Studies Technical Report* (Vols. 1–3), NASA Technical Memorandum TM 4075. Office of Exploration, NASA Headquarters, Washington, D.C., 1988–1989.

Regan, F.J., *Re-entry Vehicle Dynamics*, AIAA Education Series. American Institute of Aeronautics and Astronautics, Washington, D.C., 1984.

Russell S., DARPA Grand Challenge Winner: How It Works: 132 miles, 23 vehicles, 0 drivers – Stanley, a VW Touareg, wins the race of the century (so far). *Popular Mechanics*, January, 300 West 57 Street, New York, 2006.

Simmons, R. *et al.*, *Experience with Rover Navigation for Lunar-like Terrains*. Robotics Institute, Carnegie Mellon University, Pittsburgh, 1995.

Skaar, S.G. and Ruoff, C.F. (eds.), *Teleoperations and Robotics in Space*. American Institute of Aeronautics and Astronautics, Washington, D.C., 1994.

CHAPTER 5: FIRST LUNAR BASE

Allen, C.C., Graf, J.C., and McKay, D.S., Sintering Bricks on the Moon: Engineering, Construction and Operations in Space IV, *Space '94, Proceedings of the 4th International Conference*, pp. 1220–1229. American Society of Civil Engineers, New York, 1994.

Allen, C.C., Bricks and Ceramics, *Workshop on Using In Situ Resources for Construction of Planetary Outposts*, LPI Technical Report 98-01, edited by M.B. Duke. Lockheed Martin Space Mission Systems and Services, Houston, TX, 1998.

Altenberg, B.H., Oxygen Production on the Lunar Materials Processing Front. *Proceedings of Lunar Materials Technology Symposium*. NASA Space Engineering Research Center, Bechtel Group, 1992.

Anthony, D.L., Cochran, W.C, Haupin, W.E., Keller, W.E., and Larimer, K.T., Dry extraction of silicon and aluminum from lunar ores (Preprint), *2nd Symposium on Lunar Bases and Space Activities of the 21st Century*, edited by W.W. Mendell, paper no. LBS-88-066, 1988.

Beall, G.H., Glasses, Ceramics and Composites from Lunar Materials, *Proceedings of Lunar Materials Technology Symposium*. NASA Space Engineering Research Center, Corning Inc., 1992.

Beck, T.R., Metals Production, *Proceedings of Lunar Materials Technology Symposium*. NASA Space Engineering Research Center, Electrochemical Technologies Corp., 1992.

Benaroya, H., Lunar industrialization. *Journal of Practical Applications in Space*, **VI**, No. 1, 85–94, Fall, 1994.

Benaroya, H., *In Situ* Resources for Lunar Base Applications, *Workshop on Using In Situ Resources for Construction of Planetary Outposts*, LPI Technical Report 98-01, edited by M.B. Duke, Dept. of Mechanical and Aerospace Engineering, Lunar and Planetary Institute, Houston, TX, 1998.

Bhogeswara, R., Choudaray, U., Erstfield, T., Williams, R., and Chang, Y., Extraction processes for the production of aluminum, titanium, iron, magnesium and oxygen from nonterrestrial sources, *Space Resources and Space Settlements*, edited by J. Billingham and W. Gilbreath, NASA SP-428, pp. 257–274. NASA, Washington, D.C., 1979.

Binder, A.B., Culp, M.A., and Toups, L.D., Lunar derived construction materials: Cast basalt, *Engineering, Construction and Operations in Space: Proceedings of Space '90*, edited by S.W. Johnson and J.P. Wetzel. American Society of Civil Engineers, New York, 1990.

Burke, J.D., Where do we locate the Moon Base? *Spaceflight*, **19**, 363–366, 1977.

Burke, J.D., Ballistic Transport of Lunar Construction Materials, *Workshop on Using In Situ Resources for Construction of Planetary Outposts*, LPI Technical Report 98-01, edited by M.B. Duke. Lunar and Planetary Institute, Houston, TX, 1998.

Burke, J.D., Energy Conversion Evolution at Lunar Polar Sites, *International Conference on the Exploration and Utilization of the Moon, November 22–26, 2004, Udaipur, Rajasthan, India. Journal of Earth System Science*, **114**, No. 6, 633–635, December, 2005.

Bussey, D.B.J., Spudis, P.D., and Robinson, M.S., Illumination Conditions at the Lunar South Pole. *Geophysical Research Letters*, **26**, No. 9, 1187–1190, 1999.

Bussey, D.B.J. *et al.*, Permanent Shadow in Simple Craters Near the Lunar Poles. *Geophysical Research Letters*, **30**, No. 6, 1278, 2003.

Bussey, D.B.J. *et al.*, Constant Illumination at the Lunar North Pole. *Nature*, **434**, No. 7035, 842, 2005.

Capps, S. and Wise, T., Lunar basalt construction materials, *Engineering, Construction and Operations in Space: Proceedings of Space '90*, edited by S.W. Johnson and J.P. Wetzel. American Society of Civil Engineers, New York, 1990.

Carter, J.L., Lunar regolith fines: A source of hydrogen, *Lunar bases and space activities of the 21st century* (A86-30113 13-14). Lunar and Planetary Institute, Houston, TX, 1985.

Cooper, B.L, Lunar Ilmenite for solar power cells. *Space Resource News*, **4**, No. 6, League City, TX, June, 1995.

Cooper, B.L., Sharpe, B., Schrunk, D., and Thangavelu, M., Telerobotic Exploration and Development of the Moon (Presentation), *ILEWG Meeting-6, Udaipur, India, November, 2004.*

Criswell, D.R., Powder metallurgy in space manufacturing. *Space Manufacturing*, **4**, 389–398, American Institute of Aeronautics and Astronautics, New York, 1981.

Curreri, P.A., *Ore nonspecific process for differentiation and restructuring of minerals in a vacuum*. NASA Tech. Briefs No. 420, p. 70, October. NASA, Washington, D.C., 1993.

Cutler, A.H. and Waldron, R.D., A Reassessment of Criteria Development Methodology for Comparative Rating of Refining Processes for Non Terrestrial Resources, *ASCE SPACE '92 Conference*. American Society of Civil Engineers, Reston, VA, 1992.

Fahey, G.M. *et al.*, Regolith Excavation and Sintering Machine. *Journal of Aerospace Engineering*, ASCE, 1992.

Franklin, H.A., Materials Transportation, *Workshop on Using In Situ Resources for Construction of Planetary Outposts*, LPI Technical Report 98-01, edited by M.B. Duke. Bechtel Corp./Lunar and Planetary Institute, Houston, TX, 1998.

Geiss, J., Eberhardt, P., Signer, P., Buehler, F., and Meister, J., *The Solar-Wind Composition Experiment, Apollo 11 Preliminary Science Report*, NASA SP-214. NASA, Washington, D.C., 1970.

Haskin, L.A., The Moon as a Practical Source of Hydrogen and Other Volatile Elements (Abstract), *Lunar and Planetary Sciences Conference XX*. Lunar and Planetary Institute, Houston, TX.

Iwata, T., Evolutionary scenario of lunar manufacturing. *Journal of the British Interplanetary Society*, **47**, 539–542, 1994.

Jakes, P., Cast Basalt, Mineral Wool, and Oxygen Production: Early Industries for Planetary (Lunar) Outposts, *Workshop on Using In Situ Resources for Construction of Planetary Outposts*, LPI Technical Report 98-01, edited by M.B. Duke, Institute of Geochemistry, Charles University, Czech Republic/Lunar and Planetary Institute, Houston, TX, 1998.

Kanamori, H. *et al.*, *Considerations on the technologies for lunar resource utilization*, LPI Technical Report Number 98-01, pp. 10–11. Lunar and Planetary Institute, Houston, TX, 1998.

Keller, R., *Dry extraction of silicon and aluminum from lunar ores* (Final report), SBIR Contract NAS9-17575. EMEC Consultants, 1986.

Khalili, N.E., Regolith and Local Resources to Generate Lunar Structures and Shielding, *Lunar Bases and Space Activities in the 21st Century*, LBS 88-027, 1988.

Landis, G.A., Materials refining for structural elements from lunar resources, *Workshop on Using In Situ Resources for Construction of Planetary Outposts*, LPI Technical Report 98-01, edited by M.B. Duke. Ohio Aerospace Institute, NASA Lewis Research Center, Cleveland, OH/Lunar and Planetary Institute, Houston, TX, 1998.

Lewis, J.S., Processing Non Terrestrial Materials, *SME Annual Meeting, Phoenix, AZ*, SME 92-17, 1992.

Lin, T.D. *et al.*, Lunar and Martian Resource Utilization – Cement and Concrete, *Workshop on Using In Situ Resources for Construction of Planetary Outposts*, LPI Technical Report 98-01, edited by M.B. Duke, LinTek Inc./Lunar and Planetary Institute, Houston, TX, 1998.

Margot, J.L., Campbell, D.B., Jurgens, R.F., and Slade, M.A. (1999) Topography of the lunar poles from radar interferometry: A survey of cold trap locations. *Science*, **284**, No. 5420, 1658–1660.

Meegoda, J.N. *et al.*, Use of lunar type soil for concrete construction, *Proceedings of the 5th International Conference on Space '96, Albuquerque, NM*, pp. 614–620, 1996.

Meek, T.T., Interaction of microwave radiation with matter: A thermodynamic approach, *Space 90: Engineering, Construction and Operations in Space, Proceedings of Space '90*, edited by S.W. Johnson and J.P. Wetzel. American Society of Civil Engineers, New York, 1992.

Oder, R.R., Beneficiation of lunar soils: Case studies. *Magnetics*, **27**, No. 6, 567–570, 1991.

Phinney, W.C., Criswell, D., Drexler, E., and Garmirian, J., Lunar resources and their utilization, *Space Manufacturing Facilities II: Proceedings of the 3rd Princeton/AIAA Conference*, edited by J. Grey, pp. 171–182.

Prisbrey, K. and White, H., Plasma Based Steel Rod or Re-Bar Production from In Situ Materials, *Workshop on Using In Situ Resources for Construction of Planetary Outposts*, LPI Technical Report 98-01, edited by M.B. Duke. Lunar and Planetary Institute, Houston, TX, 1998.

Ramohalli, K., Propellant Production and Useful Materials: Hardware Data from Components and Systems, *Proceedings of Lunar Materials Technology Symposium*. NASA Space Engineering Research Center, University of Arizona, 1992.

Rosenberg, S.S., On Site Manufacture of Propellant Oxygen from Lunar Resources, *Proceedings of Lunar Materials Technology Symposium*. NASA Space Engineering Research Center, Aerojet Propulsion Division, 1992.

Rowley, J.C. and Neudecker, J.W., In situ rock melting applied to lunar base construction and for exploration drilling and coring on the Moon, *Lunar Bases and Space Activities of the 21st Century*, edited by W.W. Mendell. Lunar and Planetary Institute, Houston, TX, 1985.

Schrunk, D.G. *et al.*, Physical transportation on the Moon: The lunar railroad, *Proceedings of the 6th International Conference on Space '98*, pp. 347–353. American Society of Civil Engineers, Reston, VA, 1998.

Semkow, K.W. and Sammells, A.F., The indirect electrochemical refining of lunar ores. *J. Electrochem. Soc.*, **134**, No. 8, 2088–2089, 1987.

Sharpe, B. and Schrunk, D., An Operationally Ideal Location for the First Permanent Base on the Moon, *Proceedings of the 7th International Conference on Space 2000*. American Society of Civil Engineers, Reston, VA, 2000.

Sharpe, B. and Schrunk, D., Malapert Mountain: Gateway to the Moon, *Proceedings of the Cospar Session, World Space Congress, Houston, TX, October, 2002*.

Sharpe, B., Schrunk, D., and Thangavelu, M., Lunar Reference Mission: Malapert Station (Presentaation), *International Lunar Conference 2003, Kohala Coast, Hawaii, November, 2003*.

Sherwood, B., Lunar Materials Processing System Integration, *Proceedings of Lunar Materials Technology Symposium*. NASA Space Engineering Research Center, Boeing Defense and Space Group, 1992.

Steurer, W.H. and Nerad, B.A., Vapor phase reduction, *Research on the Use of Space Resources*, edited by W.F. Carroll. NASA Jet Propulsion Laboratory, Pasadena, CA, 1983.

Sullivan, T.A., Process engineering concerns in the lunar environment, *Proceedings of the AIAA Space Progress and Technologies Conference 1990*. American Institute of Aeronautics and Astronautics, Washington, D.C., 1990.

Taylor, L.A., Generation of native Fe in lunar soil, *Proceedings of the 1st International Conference on Engineering, Construction and Operation in Space '88*, p. 67. American Society of Civil Engineers, New York, 1988.

Taylor, L.A and Oder, R.R., Magnetic beneficiation and hi-Ti mare soils: Rock, mineral, and glassy components, *Engineering, Construction and Operations in Space: Proceedings of Space '90*, edited by S.W. Johnson and J.P. Wetzel. American Society of Civil Engineers, New York, 1990.

Teti, F., Whittaker, W., Kherat, S., Barfoot, T., and Sallaberger, C., Sun-Synchronous Lunar Polar Rover as a First Step to Return to the Moon, *Proceedings of the ISAIRAS 2005 Conference, Munich, Germany*, ESA SP-603. ESA, Noordwijk, The Netherlands.

Vaniman, D.T. and Heiken, G.H., Getting lunar ilmenite: From soils or rocks? *Engineering, Construction and Operations in Space: Proceedings of Space '90*, edited by S.W. Johnson and J.P. Wetzel. American Society of Civil Engineers, New York, 1990.

Vaniman, D., Pettit, D., and Heiken, G., Uses of lunar sulfur (Abstract/Preprint), *Proceedings of the 2nd Symposium on Lunar Bases and Space Activities of the 21st Century*. Lunar and Planetary Institute, Houston, TX, 1988.

Waldron, R.D., Lunar manufacturing: A survey of products and processes. *Acta Astronautica*, **17**, No. 7, 691–708, Pergamon Press, U.K., 1988.

Waldron, R.D., Recent Developments in Lunar Resource Process Evolution and Evaluation, *AIAA Space Programs and Technologies Conference, Huntsville, AL, 1990.*

Waldron, R.D. and Criswell, D.R., Overview of methods for extraterrestrial materials processing, *4th Princeton/AIAA Conference on Space Manufacturing Facilities.* American Institute of Aeronautics and Astronautics, New York, 1979.

Waldron, R.D. and Criswell, D.R., Materials processing in space, *Space Industrialization* (Vol. 1), edited by B. O'Leary, pp. 97–130. CRC Press, New York, 1982.

Wittenberg, L.J., Santarius, J.F., and Kulcinski, G.L., Helium-3 fusion fuel resources for space power, *Transactions of the 4th Symposium on Space Nuclear Power Systems*, pp. 327–330, SEE N88-24254 17-73.

CHAPTER 6: RETURN OF HUMANS

Alred, J. *et al.*, *Lunar Outpost.* Systems Definition Branch, Advanced Programs Office, Johnson Space Center, Houston, TX, 1989.

Angelo, J. and Easterwood, G.W., *Lunar Base Concepts, Technologies and Applications.* Orbit Books/Krieger Publishing, Melbourne, FL, 1989.

Balsiger, H., Huber, M.C.E., Lena, P., and Battrick, B., *International lunar workshop: Towards a world strategy for the exploration and utilisation of our natural satellite*, ESA SP-1170. ESA, Noordwijk, The Netherlands, 1994.

Bearden, D.A. and Ardalan, S.M., *Lunar Base Transportation and Logistics Architecture. The Lunar Polar Cycler*, AE 599 Space Concepts Studio Project Report (M. Thangavelu, Study Director). University of Southern California, May, 1995.

Bonnet. R.M., Taking the Next Step: The European Moon Program. *Planetary Report*, **15**, No. 1, 8–11, February, 1995.

Burke, J.D., Merits of a lunar polar base location, *Lunar Bases and Space Activities of the 21st Century*, pp. 77–84. Lunar and Planetary Institute, Houston, TX, 1985.

Cooke, D., *NASA's Exploration Architecture.* Deputy Associate Administrator, NASA Exploration Systems Mission Directorate, NASA Headquarters, Washington, D.C., October 4th, 2005.

Dowling, R., Staeble, R.L., and Svitek, T., A lunar polar expedition, *2nd Conference on Lunar Bases and Space Activities of the 21st Century*, NASA Conference Publication 3166, Vol. 2, pp. 175–182. NASA, Washington, D.C., 1992.

Duke, M.B., Mendell, W.W., and Roberts, B.B., Strategies for a permanent lunar base, *Lunar Bases and Space Activities of the 21st Century*, pp. 57–68. Lunar and Planetary Institute, Houston, TX, 1985.

Durst, S., *Proceedings of the International Lunar Conference 2003.* American Astronomical Society, Univelt, San Diego, CA, 2003.

Eckart, P., *Spaceflight Life Support and Biospherics.* Space Technology Library, Microcosm Press, Torrance, CA, 1994.

Eckart, P., *The Lunar Base Handbook.* McGraw-Hill Space Science Series, New York, 1999.

ESA, *A Moon Programme: The European View*, ESA BR-101. ESA, Noordwijk, The Netherlands, May, 1994.

Freeman, M., *Challenges of Human Space Exploration.* Springer/Praxis, Chichester, U.K., 2000.

Gibbons, J., *Exploring the Moon and Mars: Choices for the Nation*. Congress of the United States, Office of Technology Assessment. U.S. Government Printing Office, Washington, D.C., 1991.

Harris, P.R., Why not use the Moon as a space station? *Earth Space Review*, **4**, No. 4, 1995.

Harrison, A., *Spacefaring*. University of California Press, Berkeley, CA, 2001.

Harrison, A., Clearwater, Y., and McKay, C., *From Antarctica to Outer Space*. Springer-Verlag, New York, 1990.

Johnson, S.W. and Leonard, R.S., Evolution of concepts for lunar bases, *Lunar Bases and Space Activities of the 21st Century*, pp 47–56. Lunar and Planetary Institute, Houston, TX, 1985.

Koelle, H.H. (chairman), *Recommended lunar development scenario*. IAA Subcommittee on Lunar Development draft report, January, 1995.

Lockheed Martin Corp, *Orion Crew Vehicle, http://www.lockheedmartin.com/wms/find Page. do?dsp=fec&ci=17675&sc=400*, 2007.

Lowman, P.D., Lunar bases: A post-Apollo evaluation, *Lunar Bases and Space Activities of the 21st Century*, pp. 35–46, Lunar and Planetary Institute, Houston, TX, 1985.

Mendell, W., Alred, J., and Kuznetz, L. (Study Directors), International Lunar Initiative Organization, *International Space University Inaugural Summer Session at MIT* (Report), 1988.

NASA, *Orion Spacecraft, http://www.aerospaceguide.net/spaceexploration/orion.html*, 2007.

Seltzer, J.C., *A lunar base bibliography* (A collection of 915 government and industry references on lunar base activities). Solar System Exploration Division, Planetary Materials Laboratory Data Center, NASA Johnson Space Center, Houston, TX, March, 1988.

Siegfried, W.H., Return to the Moon: A commercial program to benefit the Earth, *2nd International Lunar Workshop, Kyoto, Japan, 1996*.

Thangavelu, M. (Faculty Adviser, USC School of Architecture), Architectural Elements for a Sustainable Colony in the South Polar Region of the Moon, *International Lunar Conference, Hawaii, 2003*.

Thangavelu, M. (Faculty Adviser, USC Dept. of Aerospace and Mechanical Engineering), HERCULES Project, *AIAA Space 2004 Conference, Long Beach, CA, September, 2004*.

Thangavelu, M. (Faculty Adviser, Division of Astronautics and Space Technology, USC Viterbi School of Engineering), LEWIS and CLARKE Project, *AIAA Space 2007 Conference, Long Beach, CA, September 2006*.

Thangavelu, M. (Faculty Adviser, Division of Astronautics and Space Technology, USC Viterbi School of Engineering), Jules Verne Project, *AIAA Space 2007 Conference, Long Beach, CA, September 2007*.

CHAPTER 7: CIRCUMFERENTIAL UTILITIES

Bents, D.J., *High-temperature solid oxide regenerative fuel cell for solar photovoltaic energy storage*, NASA TM-89872, AIAA 87-9203, 1987.

Brandhorst, Jr., H.W., Photovoltaic Technology, *Space Resources*, NASA SP-509, Vol. 2, pp. 12–20. U.S. Government Printing Office, Washington, D.C., 1992.

Brown, W.C., Receiving Antenna and Microwave Power Rectification. *Microwave Power*, **5**, 279–292, December, 1970.

Brown, W.C., Satellite Power Stations: A New Source of Energy? *IEEE Spectrum*, 1973.

Burden, D., Nuclear Energy Technology, *Space Resources*, NASA SP-509, Vol. 2, pp. 32–56. U.S. Government Printing Office, Washington, D.C., 1992.

Carter, N.E., *The challenge of macro-engineerings*, Batelle Today Reprint No. 24. Battelle Institute, Columbus, OH, 1985.

Criswell, D.R., The initial lunar supply base, *Space Resource and Space Settlements*, NASA SP-428, pp. 207–224. NASA, Washington, D.C., 1979.

Criswell, D.R., Solar power system based on the Moon, *Solar Power Satellites*, pp. 272–288. Ellis Horwood, New York, 1993.

Criswell, D., Lunar Solar Power System: System Options, Costs, and Benefits to Earth, *Intersociety Energy Conversion Engineering Conference, Orlando, FL, 1995*, Vol. 1, pp. 595–600.

Criswell, D.R. and Curreri, P.A., Photovoltaics using in situ resource utilization for HEDS, *Proceedings of the 6th International conference and Exposition on Engineering, Construction, and Operations in Space (Space '98)*, pp. 286–289, 1998.

Criswell, D.R. and Waldron, R.D., Lunar system to supply solar electric power to Earth, *Proceedings of 25th Intersociety Energy Conversion Engineering Conference, Reno, NV, August, 1990*, pp. 61–71.

Criswell, D.R. and Waldron, R.D., Results of a lunar-based power system to supply Earth with 20,000 GW of electric power, *A Global Warming Forum*, pp. 111–126, CRC Press, 1992.

Criswell, D, and Waldron, R., International lunar base and lunar-based power system to supply Earth with electric power. *Acta Astronautica*, **29**, No. 6, 469–480. Pergamon Press, 1993.

Criswell, D.R., Waldron, R.D., and Aldrin, B., Lunar power systems, *Conference Proceedings, Space: Our Next Frontier*, p. 262. National Center for Policy Analysis, Dallas, TX, 1984.

Curreri, P.A., *In Situ* Processing of Photovoltaic Devices, *Workshop on Using In Situ Resources for Construction of Planetary Outposts*, LPI Technical Report 98-01, edited by M.B. Duke. NASA Marshall Spaceflight Center, Huntsville, AL, 1998.

Davidson, F.P. and Meador, C.L. (eds.), *Macroengineering: Global Infrastructure Solutions*. Ellis Horwood, Chichester, U.K., 1992.

DeRonck, H.J., Fuel Cell Technology for Lunar Surface Operations, *Proceedings of Lunar Materials Technology Symposium*. NASA Space Engineering Research Center, International Fuel Cells Corp., 1992.

Ehricke, K.A., *The Power Relay Satellite, A Means of Global Energy Transmission through Space*, E74-3-1. Rockwell Intl. Corp., El Segundo, CA, 1974.

Ewert, M.K., Keller, J.R., and Hughes, B., Conceptual design of a solar powered heat pump for lunar base thermal control system, *26th International Conference on Environmental Systems, Monterey, CA, July 8–11, 1996*, SAE Technical Paper Series 961535.

Fang, P.H., A lunar solar cell production plant. *Solar Cells*, **25**, 31–37, 1988.

Fordyce, S.J., *Space Power (Advanced)*, International Space University Core Lecture at MIT, July 1988.

Fordyce, S.J. and Faymon, K.A., Space Power Technology into the 21st Century, *Space Power Systems Technology Conference, Costa Mesa, CA, June, 1984*, NASA Technical Memorandum 83690.

Freundlich, A., Ignatiev, A., Horton, C., Duke, M., Curreri, P., and Sibille, L., Manufacture of Solar Cells on the Moon, *Proceedings of the 31st IEEE Photovoltaic Specialist Conference*, IEEE Catalog Number 05CH37608C, pp. 794–797, 2005.

Galloway, G.S., Polar Lunar Power Ring: Propulsion Energy Resource. *Lunar Bases and Space Activities of the 21st Century*, April, 1988.

Greenwood, D.L., Criswell, D., and Peterson, M.A., Rationale and Plans for a Lunar Power System, *Proceedings of the Power Beaming Workshop*. Pacific Northwest Laboratories, Pasco, WA, 1991.

Griffin, B.N., *An Infrastructure for Early Lunar Development*. The Boeing Company, Huntsville, AL, 1992.

Hanford, A.J. and Ewert, M.K., An assessment of advanced thermal control system technologies for future human space flight, *26th International Conference on Environmental Systems, Monterey, CA, July 8–11, 1996*, SAE Technical Paper Series 961480.

Hertzberg, A., Thermal Management in Space, *Space Resources*, NASA SP-509, Vol. 2, pp. 57–69. U.S. Government Printing Office, Washington, D.C., 1992.

Hickman, M.J., Curtis, H.B., and Landis, G.A., Design Considerations for Lunar Base Photovoltaic Power Systems, *21st Photovoltaic Specialists Conference, Kissimmee, FL, 1990*, NASA Technical Memorandum 103642. NASA, Washington, D.C.

Horton, C., Gramajo, A., Alemu, L., Williams, A., Ignatiev, A., and Freundlich, A., First Demonstration of Photovoltaic Diodes on Lunar Regolith Based Substrate. *Acta Astronautica*, **56**, 537, 2005.

Houts, M.G., Poston, D.I., and Berte, M.V., Fission Power Systems for Surface Outposts, *Los Alamos National Laboratory Workshop on Using In Situ Resources for Construction of Planetary Outposts*, LPI Technical Report 98-01, edited by M.B. Duke. Lunar and Planetary Institute, Houston, TX, 1998.

Ignatiev, A. and Freundlich, A., The production of photovoltaic devices in space, *Proceedings of the 5th International Conference on Space '96, Albuquerque, NM, 1996*, Vol. 1, pp. 287–292.

Ignatiev, A. and Freundlich, A., Solar Cell Fabrication on the Moon from Lunar Resources, *Proceedings of the 57th International Astronautical Congress, Valencia, Spain, 2006*.

Ignatiev, A., Freundlich, A., and Kubricht, T., *Thin Film Solar Cell Growth on the Surface of the Moon by Vacuum Evaporation*, ISRU TIM. Lunar and Planetary Institute, Houston, TX, February, 1997.

Katzan, C.M. and Edwards, J.L., *Lunar Dust Transport and Potential Interactions with Power System Components*, NASA Contractor Report 4404. NASA, Washington, D.C., November 1991.

Landis, G.A., Solar Power for the Lunar Night, *9th Biennial SSI/Princeton Conference on Space Manufacturing, Princeton, NJ, May 10–13, 1989*, NASA Technical Memorandum 102127. NASA, Washington, D.C.

Landis, G.A., Moonbase night power by laser illumination. *Journal of Propulsion and Power*, **8**, No. 1, 251–254, January/February, 1992.

Landis, G.A. and Perino, M.A., Lunar production of solar cells: A near term product for lunar industry facility, *Proceedings of the 9th Princeton/AIAA/SII Conference, May 1989*, p. 148.

Lewis, N.S., Wet solar cells electrolyze water while providing power. *Space Resource News*, **4**, No. 11, November, 1995.

Lotker, M., Solar Thermal Electric Systems: A Cost Effective Utility Option, *Solar '90: The 19th ASES Annual Conference*. Luz International Limited, Los Angeles, CA, 1990.

Lunar Energy Park Study Group, A proposed plan for a lunar-based power plant: space technology utilized for global energy supply in the future. *Science and Technology in Japan*, **14**, No. 53, 54–59, April, 1995.

Margot, J.L., Campbell, D.B., Jurgens, R.F., and Slade, M.A., Topography of the lunar poles from radar interferometry: A survey of cold trap locations. *Science*, **284**, No. 5420, 1658–1660, June 4th, 1999.

Mason, L.S., *SP-100 Power Systems Conceptual Design for Lunar Base Applications*. NASA Johnson Space Center, Houston, TX, May, 1988.

Mendell, W., A lunar vision. *Ad Astra*, 26–31, National Space Society, Washington, D.C., 1996.

Peterson, M.N.A. *et al.*, Progress and plans for a lunar power system, *Proceedings of the 3rd Annual World Energy System Symposium, Uhzgorod, Ukraine, November, 1993.*

Potter, S.D., Applications of thin-film technology in space power systems, *Proceedings of the High Frontier Conference XII, Princeton, NJ, May 4–7, 1995.*

Repic, E. (Project Manager), SEI Projects at Rockwell: An Update (Presentation), *Planetary Surface Systems.* NASA Johnson Space Center, Houston, TX, April, 1991.

Sammells, A.F. and Semkow, K.W., Electrolytic cell for lunar ore refining and electric energy storage (Preprint), *Proceedings of the 2nd Symposium on Lunar Bases and Space Activities of the 21st Century, Houston, TX, 1988.*

Schrunk, D.G., Cooper, B.L., and Sharpe, B.L., Concept for a Permanent Lunar Utilities System, *Proceedings of the 5th International Conference on Space '96*, pp. 935–941, 1996.

Space Power Inc., *Topaz II Space Power System Design Study*, Air Force Philips Laboratory Quarterly Briefing, April 1991.

Stone, J.L., Photovoltaics: Unlimited electrical energy from the Sun. *Physics Today*, September 1993, 22–29, 1993.

Sunkara, S.S., *Growth and evaluation of ilmenite wide bandgap semiconductor for high temperature electronic applications*, Ph.D. dissertation. Texas A & M University, 1995.

Taylor, G.J., The Need for a Lunar Base: Answering Basic Questions about Planetary Science, *Lunar Bases and Space Activities of the 21st Century*, pp. 189–197. Lunar and Planetary Institute, Houston, TX, 1984.

Waldron, R.D. and Criswell, D.R., Concept of the lunar power system. *Space Solar Power Review*, **5**, 53–75, 1984.

Waldron, R.D. and Criswell, E.R., System requirements/constraints and design options for a global lunar power system, *Proceedings of the 26th International Energy Conversion Engineering Conference*, Vol. 4, p. 128, 1991.

Waldron, R.D. and Criswell, D.R., Overview of lunar industrial operations, *AIP Conference on Alternative Power from Space, Albuquerque, January, 1995*, p. 965.

Walker, S.T., Alexander, R.A., and Tucker, S.P., Thermal Control on the Lunar Surface. *Journal of the British Interplanetary Society*, **48**, 27–32, 1995.

Woodcock, G.R., Parametric analysis of lunar resources for space energy systems, *Space Manufacturing 7: Space Resources to Improve Life on Earth.* American Institute of Aeronautics and Astronautics, Washington, D.C., 1989.

Woodcock, G., Electrical Power Integration for Lunar Operations, *Proceedings of Lunar Materials Technology Symposium.* NASA Space Engineering Research Center, Boeing Defense and Space Group, 1992.

CHAPTER 8: PLANET MOON

Aldrin, B., Cyclic Trajectory Concepts (SAIC Presentation), *Interplanetary Transit Study Meeting.* NASA Jet Propulsion Laboratory, Pasadena, CA/Caltech, 1985.

Allen, J., *BIOSPERE 2. The Human Experiment*, edited by A. Blake. Penguin Books, 1991.

Allen, J. and Nelson, M., *Space Biospheres.* Orbit Books, Malabar, FL, 1987.

Austin, G. and Vinopal, T., Designing the Space Transfer Vehicle: The Lunar Piloted Mission, *41st Congress of the International Astronautical Federation, Dresden, 1990*, IAF 90-170.

Baracat, W.A. and Butner, C.L., *Tethers in Space Handbook.* NASA Office of Spaceflight Advanced Programs, 1986.

Bedini, D., Selfconstructing, deployable, ready-to-use habitat for the Moon: A cheap challenge for a lunar base, *26th International Conference on Environmental Systems, Monterey, CA, July 8–11, 1996*, SAE Technical Paper Series 961398.

Bekey, I. and Herman, D., Space Stations and Space Platforms: Concepts, Design, Infrastructure, and Uses, *Progress in Astronautics and Aeronautics* (Vol. 99). American Institute of Aeronautics and Astronautics, New York, 1985.

Bell, L. and Trotti G. (Study Directors), *The Manned Lunar Outpost*. Sasakawa International Center for Space Architecture, School of Architecture, University of Houston, TX, 1989.

Bernold, L.E., Cable-Based Lunar Transportation System. *ASCE Journal of Aerospace Engineering*, **7**, No. 1, Paper No. 3012, January, 1994, Dept. of Civil Engineering, North Carolina State University, Raleigh, NC.

Bernold, L.E., Compaction of Lunar Type Soil. *ASCE Journal of Aerospace Engineering*, **7**, No. 2, Paper No. 1660, April, 1994, Dept. of Civil Engineering, North Carolina State University, Raleigh, NC.

Booher, C. (Manager), *Manned Systems Integration Standards* (Vols. 1–4), NASA STD-3000. NASA Johnson Space Center, Houston, TX, 1992.

von Braun, W., Whipple, F.L., Ley, W., and Ryan, C., *Man on the Moon*. Colliers, 1952.

Burke, J.D. (Study Director), *The Lunar Polar Orbiter Mission*, International Space University Report, 1989.

Carroll, J.A., *Guidebook for Analysis of Tether Applications*, RH4-394049. Martin Marietta Corp., 1985.

Chamberland, D., Advanced Life Support Systems in Lunar and Martian Environments Utilizing a Higher Plant Base Engineering Paradigm, *22nd International Conference on Environmental Systems, 1992*, SAE-921286.

Clarke, A.C., *The Exploration of the Moon*, illustrated by R.A. Smith. Harper Row, New York, 1954.

Clarke, A.C., The Space Elevator: "Thought Experiment or Key to the Universe?". *Advanced Earth Oriented Applied Space Technology*, **1**, 39–48, 1981.

Colombo, G., *The use of tethers for payload orbital transfer*, NASA Contract 33691, Vol. 2. NASA, Washington, D.C., 1982.

Colombo, G., Applications of tethers in space, *Workshop Proceedings, Virginia, 1985*, Vols. 1–2, NASA CP 2364-5. NASA, Washington, D.C.

Criswell, M.E., Habitat Construction Requirements, *Workshop on Using In Situ Resources for Construction of Planetary Outposts*, LPI Technical Report 98-01, edited by M.B. Duke. Dept. of Civil Engineering, Colorado State University, Fort Collins/Lunar and Planetary Institute, Houston, TX, 1998.

Dalton, C. and Hohmann, E., *Conceptual Design of a Lunar Colony*, NGT 44-005-114. NASA/ASEE Systems Design Institute, 1972.

Dalton, C. and Hohmann, E., *Conceptual Design of a Lunar Colony*, NASA CR-129164. NASA/ASEE Systems Design Institute, University of Houston/MSC, Rice University, 1972.

Dempster, W.F., Biosphere 2: Overview of System Performance during the First Nine Months, *22nd International Conference on Environmental Systems, 1992*, SAE-921129.

Drake, B.G. and Joosten, B.K., Preparing for human exploration, *Proceedings of the 6th International Conference and Exposition on Engineering, Construction and Operations in Space '98*, pp. 541–554. American Society of Civil Engineers, Albuquerque, NM, April, 1998.

Eckart, P., Lunar base model, *Proceedings of the 6th International Conference and Exposition on Engineering, Construction and Operations in Space '98*, pp. 601–607. American Society of Civil Engineers, Albuquerque, NM, April, 1998.

Edeen, M.A. and Brown, M.F., Approaches to Lunar Base Life Support, *AIAA Programs and Technologies Conference, 1990*, AIAA 90-3740. American Institute of Aeronautics and Astronautics, Washington, D.C.

Ehricke, K.A., Lunar industrialization and settlement-birth of polyglobal civilization, *Lunar Bases and Space Activities of the 21st Centrury*, pp. 827–855. Lunar and Planetary Institute, Houston, TX, 1985.

Elrod, M., Considerations of a habitat design. *Journal of the British Interplanetary Society*, **48**, 39–42, 1995.

Folsome, C.E. and Hanson, J.A., The Emergence of Materially Closed System Ecology, *Ecosystem Theory and Applications*, edited by N. Polunin. Wiley, 1986.

Fradin, D.B., *Space Colonies*. Children's Press, Chicago, 1985.

Frye, R.J. and Mignon, G., Closed Ecological Systems: From Test Tubes to the Earth's Biosphere, *Proceedings of Lunar Materials Technology Symposium*. NASA Space Engineering Research Center, Environmental Research Laboratory, University of Arizona, 1992.

Fujita, J.S., *Back to the Future: The Lunar Humanity Center*, AE 599 Space Concepts Studio Project Report (M. Thangavelu, Study Director). University of Southern California, May, 1995.

Griffin, M.D and French, J.R., *Space Vehicle Design*, AIAA Education Series. American Institute of Aeronautics and Astronautics, Washington, D.C., 1991.

Gruener, J.E. and Ming, D.W., Requirements for Planetary Outpost Life Support Systems and the Possible Use of In Situ Resources, *Workshop on Using In Situ Resources for Construction of Planetary Outposts*, LPI Technical Report 98–01, edited by M.B. Duke. Hernandez Engineering Inc., Houston, TX/Lunar and Planetary Institute, Houston, TX, 1998.

Harrison, R.A., *Cylindrical Fabric-Confined Soil Structures*. TRW, Los Angeles, 1990.

Harwood, O.P., An Evolutionary Space Station Architecture. *Journal of the British Interplanetary Society*, **38**, 305–314, 1985.

Hoffman, S.J. and Hiehoff, J.C., Preliminary design of a permanently manned lunar surface research base, *Lunar Bases and Space Activities of the 21st Century*, pp. 69–75. Lunar and Planetary Institute, Houston, TX, 1985.

Koelle, H.H., Space Freighters for the 21st Century, *41st Congress of the International Astronautical Federation, Dresden, 1990*, IAF-90-195. International Astronautical Federation, Paris.

Koppeschaar, C., *Moon Handbook: A 21st-century Travel Guide*. Moon Publications, Chico, CA, 1995.

Kozlov, I.A. and Shevchenko, V.V., Mobile lunar base project. *Journal of the British Interplanetary Society*, **48**, 49–54, 1995.

Land, P., Lunar base design, *Lunar Bases and Space Activities of the 21st Century*, pp. 363–374. Lunar and Planetary Institute, Houston, TX, 1985.

Laursen, E.F., Jones, Jr., C.H., and Niehoff, J., "Common Base" Surface Facilities for the Space Exploration Initiative, *41st Congress of the International Astronautical Federation, Dresden, 1990*, IAF 90-441. International Astronautical Federation.

Matsumoto, S. *et al.*, Inflatable Lunar Structure with Reinforcing Rings: Self Shaping and Self Sinking System for Lunar Base, *43rd Congress of the International Astronautical Federation, Washington, D.C., 1992*, IAF 92-0768.

Matsumoto, S. *et al.*, Lunar Base System Design. *Journal of the British Interplanetary Society*, **48**, 11–14, 1995.

McCallum, Taber *et al.*, Biosphere 2 Test Module: A ground based sunlight driven prototype of a closed ecological life support system. *Advanced Space Research*, **12**, 151–156, 1992.

McElroy, J.F., SPE Water Electrolyzers in Support of the Lunar Outpost, *Proceedings of Lunar Materials Technology Symposium*. NASA Space Engineering Research Center, Hamilton Standard Division, United Technologies Corp., 1992.

Moore, G.T. and Rebholz, P.J., Aerospace Architecture: A Comparative Analysis of Five Lunar Habitats, *AIAA Aerospace Design Conference, Irvine, CA, 1992*, AIAA 92-1096. American Institute of Aeronautics and Astronautics, Washington, D.C.

Mount, F.E. *et al.*, *Human Factors Assessments of the STS-57 SpaceHab-1 Mission*, NASA Technical Memorandum 104802. NASA, Washington, D.C., August, 1994.

NASA ARC, *Space Station/Antarctic Analogs*, NASA Grants NAG 2-255 and NAGW-659. NASA Ames Research Center, Moffett Field, CA, February, 1981.

NCS, *Pioneering the Space Frontier*. National Commission on Space/Bantam Books, New York, 1986.

Paine, T.O., *The Next Forty Years in Space*, International Astronautical Federation Congress, Malaga, Spain, 1989.

Pearson, J., Anchored Lunar Satellites for Cislunar Transportation and Communication, *Journal of Aastronautical Sciences*, **27**, No. 1, 39–62, 1979.

Penzo, P.A., *Tethers in Space Studies Report*. NASA Jet Propulsion Laboratory, Pasadena, CA, 1986.

Richter, P.J. *et al.*, Concepts for lunar outpost development. *ASCE Journal of Aerospace Engineering*, **3**, No. 4, October, 1990, Fluor Daniel Corp.

Rinderle, E.A., *Galileo User's Guide: Mission Design Systems, Satellite Tour Analysis and Design Subsystem*, JPL D-263. NASA Jet Propulsion Laboratory, Pasadena, CA/Caltech, July, 1986.

Roberts, B., Options for a Lunar Base Surface Architecture, *Proceedings of Lunar Materials Technology Symposium*, pp. IV-1-15. NASA Space Engineering Research Center, NASA Johnson Space Center, Houston, TX, 1992.

Sadeh, W.Z. and Criswell, M.E., Inflatable structures for a lunar base. *Journal of the British Interplanetary Society*, **48**, 33–38, 1995.

Shevchenko, V.V., Manned lunar base site selection. *Journal of the British Interplanetary Society*, **48**, No. 1, 15–20, January 1995.

Simon, M.C. and Bialla, P.H., *Analysis of Alternative Infrastructures for Lunar and Mars Exploration*, 41st Congress of the International Astronautical Federation, Malaga, Spain, 1990, IAF-90-442.

Stroup, T.L., Lunar bases of the 20th Century: What might have been. *Journal of the British Interplanetary Society*, **48**, 3–10, 1995.

Stuhlinger, E. and Ordway, III, F.I., *Wernher Von Braun: Crusader for Space*, A biographical memoir. Krieger Publishing Company, 1994.

Switzer, K.L., *Lunar Base Agriculture and Environmental Control and Life Support System*, AE 599 Space Concepts Studio Project Report (M. Thangavelu, Study Director). University of Southern California, May, 1995.

Tascione, T.F., *Introduction to the Space Environment*. Orbit Books, Melbourne, FL, 1988.

Thangavelu, M., The Nomad Explorer Vehicle for Global Lunar Development, *Proceedings of the 1992 International Astronautical Federation Meeting, Washington, D.C., 1992*.

Thangavelu, M., Siting the millennial time capsule and presidential library, *Proceedings of the 6th International Conference and Exposition on Engineering, Construction and Operations in Space '98*, pp. 666–673. American Society of Civil Engineers, Reston, VA, 1998.

Thangavelu, M. *et al.*, Evolution of a Satellite Service Facility, *Proceedings of the 6th International Conference and Exposition on Engineering, Construction and Operations in Space*, pp. 710–723. American Society of Civil Engineers, Reston, VA, 1998.

Thangavelu, M., Khalili, E.N., and Girardey, C.C., *In Situ* Generation of A "To Scale" Extraterrestrial Habitat Shell and Related Physical Infrastructure Utilizing Minimally Processed Local Resources, *Workshop on Using In Situ Resources for Construction of Planetary Outposts*, LPI Technical Report 98-01, edited by M.B. Duke. Lunar and Planetary Institute, Houston, TX, 1998.

Vondrak, R.R., Creation of an artificial lunar atmosphere. *Nature*, **248**, No. 5450, 657–659, 1974.

Walden, B.E. *et al.*, Utility of Lava Tubes on Other Worlds, *Workshop on Using In Situ Resources for Construction of Planetary Outposts*, LPI Technical Report 98-01, edited by M.B. Duke. Oregon Moon Base L5 Society/Lunar and Planetary Institute, Houston, TX, 1998.

White, F., *The Overview Effect*. Houghton Mifflin Company, Boston, 1987.

CHAPTER 9: GOVERNANCE OF THE MOON

Anaejionu, P., Goldman, N.C., and Meeks, P.J. (eds.), *Space and Society: Challenges and Choices*. Univelt, San Diego, CA, 1984.

Bormanis, A. and Logsdon, J.M. (cds.), *Emerging policy issues for long-duration human exploration*, December Workshop Report. Space Policy Institute, The George Washington University, Washington, D.C., 1992.

Christol, C.Q., *The Modern International Law of Outer Space*. Pergamon Press, Los Angeles, 1982.

Cocca, A.A., The Common Heritage of Mankind: Doctrine and Principle of Space Law: An Overview, *Proceedings of the 29th Colloquium on the Law of Outer Space, IAF Congress, Innsbruck, Austria, 1986*.

Cooper, B.L., The Beatenberg Declaration. *Space Resource News*, **4**, No. 8, 3–4, September, 1995.

Gabrynowicz, J.I., The "Province" and "Heritage" of mankind reconsidered: A new beginning, *2nd Conference on Lunar Bases and Space Activities of the 21st Century*, NASA Conference Publication 3166, Vol. 2, pp. 691–698. NASA, Washington, D.C., 1992.

Goldman, N., *American Space Law: Domestic and International*. Iowa State University Press, IO, 1988.

Goldman, N.C., *American Space Law: International and Domestic*. Univelt, San Diego, CA, 1996.

Harris, P.R., Human dimensions in space policy. *Space Policy*, **5**, No. 2, 147–154, May, 1989.

Harris, P.R. and O'Donnell, D.J., Creating new social institutions to develop business from space: Moon/Mars economic development authorities, *Proceedings of the 48th International Astronautical Congress, October 6–10, Turin, Italy, 1997*.

Harris, P.R., Moran, R., and Moran, S., *Managing Cultural Differences: Global Leadership Strategies for the 21st Century* (6th edn.). Elsevier Publishing, San Diego, CA, 2004.

Jasentuliyana, N., Space Law and the United Nations: A Research Guide, *Proceedings of the 29th Colloquium on the Law of Outer Space, IAF Congress, Innsbruck, Austria, 1986*.

Jasentuliyana, N., *Space Law, Development and Scope*. International Institute of Space Law, Westport, CT, 1992.

Johnson-Freese, J., A Model for Multinational Space Cooperation: The Interagency Consultancy Group. *Space Policy*, 288–300, November, 1989.

Joyner, C.C. and Schmitt, H.H., Extraterrestrial law and lunar bases: General legal principles and a particular regime proposal (INTERLUNE), *Lunar Bases and Space Activities of the 21st Century*, pp. 741–750. Lunar and Planetary Institute, Houston, TX, 1985.

Koelle, H.H., A Methodology to Determine the Sociopolitical Forces Influencing Lunar Development Policies. *Space Policy*, Elsevier Science, U.K., May, 1996.

Kopal, V.M., The Law of Outer Space: Its Place in the System of International Law and some Related Problems of its Teaching, *Proceedings of the 29th Colloquium on the Law of Outer Space, IAF Congress, Innsbruck, Austria, 1986*.

Kopal, V., Vladimir Mandl: Founding Writer in Space Law, *First Steps toward Space*, Smithsonian Annals of Flight No. 10, pp. 87–90, edited by J. Durant, III.

Logsdon, J.M. (ed.), *Exploring the Unknown* (Vols. 1–6), NASA SP-4077. NASA History Office, Office of History and Plans, NASA Headquarters, Washington, D.C., 1995–2000.

Logsdon, J., A Sustainable Rationale for Human Spaceflight. *Issues in Science and Technology* (Winter 2004), **20**, No. 3, 10, March 22nd, 2004, Tomas Gale.

Logsdon, J., *Proposals for Space Development in the 21at Century*, JAXA interview. Japanese Space Agency, 2007.

Lubos, P., Management of outer space. *Space Policy*, **10**, No. 3, 189–198, August, 1994.

Matte, N.M., Legal implications of the exploration and uses of the Moon and other celestial bodies (Presentation), *43rd Congress of the International Astronautical Federation, Washington, D.C., August 28–September 5, 1992*.

Milde, M. (ed.), *Annals of Air and Space Law*. McGill University/Institute of Air and Space Law, Montreal, Canada, 1996.

Montaner, M.G., Establishment of an international space organization. *Space Governance*, **4**, No. 1, 78–81, January, 1997.

Moran, R.T., Harris, P.R., and Stripp, W.G., *Developing Global Organizations*. Gulf Publishing, Houston, TX, 1993.

O'Donnell, D.J., An archenemy revisited: The 1979 Moon Treaty, *Proceedings of the 6th International Conference and Exposition on Engineering, Construction and Operations in Space (Space '98), Albuquerque, NM, April, 1998*, pp. 681–687.

O'Donnell, D.J. and Harris, P.R., Legal Strategies for a Lunar Economic Development Authority. *Annals of Air and Space Law*, **XXI**, Part II, 121–130.

O'Donnell, D., Aldrin, B., Blair, B., and Schrunk, D., The Lunar Economic Development Authority: A Municipal Governance Tool, *International Astronomical Congress, Valencia, Spain, 2006*, Session IAC-06-A5.1.7.

Pedersen, K.S., Is it time to create a world space agency? *Space Policy*, **9**, No. 2, May, 1993.

Rechtin, E., *System Architecting of Organizations: Why Eagles Can't Swim* (280 pp.). CRC Press, Boca Raton, FL, July 27, 1999, ISBN-10: 0849381401.

Roberts, L.D., Needed: A private property standard for space. *Ad Astra*, 42–44, November/December, 1997.

Roberts, L.D., Pace, S., and Reynolds, G.H., Playing the commercial space game: Time for a new rule book? *Ad Astra*, 44–47, May/June, 1996.

Robinson, G.S. and White, Jr., H.M., *Envoys of Mankind*. Smithsonian Institution Press, Washington, D.C., 1986.

Schrunk, D.G., Lawmaking Standards for Space Governance. *Space Governance,* **4**, No. 1, January, 1997 (a publication of the World Space Bar Association).

Schrunk, D.G., *THE END OF CHAOS: Quality Laws and the Ascendancy of Democracy.* QL Press, Poway, CA, 2005.

Sterner, E.R. and Benaroya, H., Lunar industrialization and colonization: Towards a policy framework. *Journal of the British Interplanetary Society,* **47**, 516–520, 1994.

UN, *Treaty on Principles Governing the Activities of States in the Exploration and use of Outer Space, including the Moon and Other Celestial Bodies* (The Outer Space Treaty), 1967.

UN, *Treaty Governing the Activities of States on the Moon and Other Celestial Bodies* (The Moon Treaty), 1979.

Van Reeth, G. and Madders, K., Reflections on the Quest for International Cooperation. *Space Policy,* 221–231, August, 1993.

CHAPTER 10: ENDLESS FRONTIERS

Aldrin, B., Jones, R., Davis, H., Talay, T., Thangavelu, M., and Repic, E., *Evolutionary Space Transportation Plan for Mars Cycling Concepts,* NASA/JPL Contract No. 123098. NASA Jet Propulsion Laboratory, Pasadena, CA, 2002.

Asphaug, E., New views of asteroids. *Science,* **278**, 2070–2071, December 19th, 1997.

Bekey, I., Tethering a new technique for payload deployment. *Aerospace America,* 36–40, March, 1997.

Benningfield, D., Where do comets come from? *Astronomy,* 30–36, September, 1990.

Bishop, R.H., Byrnes, D.V., Newman, D.J., Carr, C.E., and Aldrin, B., Earth–Mars Transportation Opportunities: Promising Options for Interplanetary Transportation, *Richard H. Battin Astrodynamics Conference, College Station, Texas, March, 2000,* AAS 00-255. American Astronomical Society, Pasadena, CA.

Byrnes, D.V., Longuski, J.M., and Aldrin, B., Cycler Orbits between Earth and Mars. *AIAA Journal of Spacecraft and Rockets,* **30**, No. 3, 334–336, May/June, 1993.

Clark, S., *Life on Other Worlds and How to Find It.* Springer/Praxis, Chichester, U.K., 2000.

David, L., Assessing the threat from comets and asteroids. *Aerospace America,* 24–38, August, 1996.

David, L., Incredible shrinking spacecraft. *Aerospace America,* 20–24, January, 1996.

Davidson, W.L. and Stump, W.R., Lunar stepping stones to a manned Mars exploration scenario, *2nd Conference on Lunar Bases and Space Activities of the 21st Century,* NASA Conference Publication 3166, Vol. 2, pp. 677–682. NASA, Washington, D.C., 1992.

Drexler, K.E., *Nanosystems: Molecular Machinery, Manufacturing, and Computation.* Wiley, 1992.

Duke, M.B., Mendell, W.W., and Roberts B.B., *Lunar Base: A Stepping Stone to Mars,* pp. 84–162. American Astronomical Society, Pasadena, CA.

Dyson, F.J., 21st-century spacecraft. *Scientific American,* 114–116, September, 1995.

Erb, R.B., Power from Space for the Next Century. *International Astronautical Federation Proceedings, Montreal, Canada, October, 1991,* Paper No. IAF 91-231. International Astronautical Federation, Paris.

Finney, B., SETI and Interstellar Migration. *Journal of the British Interplanetary Society,* **38**, 1985.

Finney, B. and Jones, E.M., From Africa to the Stars: the Evolution of the Exploring Animal, *Proceedings of the 6th Princeton/SSI Conference on Space Manufacturing,* Space

Manufacturing, Vol. 53: Advances in the Astronautical Sciences. Univelt, San Diego, CA, 1983.

Flint, E.M., *Thin film disc shaped large space structures – sunsat or solar sail – a proposed construction/assembly method.* American Institute of Aeronautics and Astronautics, Inc. and the Space Studies Institute, 1995.

Forward, R.L., Roundtrip interstellar travel using laser-pushed lightsails. *Journal of Spacecraft,* **21**, No. 2, 1984.

Forward, R.L., Starwisp: An ultra-light interstellar probe. *Journal of Spacecraft,* **22**, No. 3, 1985.

Friedman, L., *Starsailing.* Wiley, New York, 1988.

Glanz, J., Engineers dream of Practical star flight: From "News of the week". *Science,* **281**, 765–767, August 7th, 1998.

Glaser, P.W., Davidson, F.P., and Csigi, K.I., *Solar Power Satellites – The Emerging Energy Option.* Ellis Horwood, Chichester, U.K., 1993.

Grant, P., Starr, C., and Overbye, T., A power grid for the hydrogen economy. *Scientific American,* 77–81, July 2006.

Hazen, R.M., The stuff of life: What was life's first energy source? *Planetary Report,* **XVIII**, No. 4, 16–17, July/August, 1998.

Jones, R.M., Electromagnetically launched microspacecraft for space science missions. *Journal of Spacecraft and Rockets,* **26**(5), 338–342, October, 1989.

Kerr, R.A., Life goes to extremes in the deep Earth – and elsewhere? *Science,* **276**, 703–704, May 2nd, 1997.

Kowal, C.T., *Asteroids: Their Nature and Utilization* (2nd edn.). Wiley/Praxis, Chichester, U.K., 1996.

Landis, G.A., An evolutionary path to satellite solar power systems, *Space Power,* **9**, No. 4, 365–370, 1990.

Lewis, J.S., Jones, T.D., and Farrand, W.H., Carbonyl extraction of lunar and asteroidal materials, *Engineering, Construction and Operations in Space: Proceedings of Space '88,* edited by S.W. Johnson and J.P. Wetzel. American Society of Civil Engineers, New York, 1988.

Lewis, J.S., *Mining the Sky: Untold Riches from the Asteroids, Comets, and Planets.* Helix Books/Addison-Wesley, Reading, MA, 1996.

Mackenzie, B.A., Bootstrapping space resource utilization with tethers, regolith rockets and micro rovers, *Proceedings of the 5th International Conference on Space '96, Albuquerque, NM, 1996,* pp. 321–327.

Mallove, E. and Matloff, G., *Star Flight Handbook.* Wiley, New York, 1989.

Mankins, J.C., The space solar power option. *Aerospace America,* 30–36, May, 1997.

Marburger, John (Director, Office of Science and Technology Policy Executive Office of the President, White House Washington D.C.), Keynote Address, *44th Robert H. Goddard Memorial Symposium, Greenbelt, MD, March 15, 2006.*

Maryniak, G.E. and O'Neill, G.K., Nonterrestrial resources for solar power satellite construction, *Solar Power Satellites,* pp. 253–271. Ellis Horwood, New York, 1993.

McInnes, C.R., *Solar Sailing.* Springer/Praxis, Chichester, U.K., 1999.

Milstein, M.C., Europa's ocean. *Astronomy,* 39–43, October, 1997.

Nelson, R.M., Mercury: The forgotten planet. *Scientific American,* 56–67, November, 1997.

Nock, K., *Cyclical Visits to Mars via Astronaut Hotels.* National Institute of Advanced Concepts, June, 2001.

Okushi, J., First Mars outpost architectural study I, *26th International Conference on Environmental Systems, Monterey, CA, July 8–11, 1996,* SAE technical paper series 961445.

O'Leary, B., Mining the Apollo and Amor Asteroids. *Science*, **197**, 363–366, July 22nd, 1977.

O'Neill, G., *The High Frontier*. Space Studies Institute Press, Princeton, NJ, 1989.

Potter, S.D., *Low mass solar power satellites built from lunar or terrestrial materials: Final Report*. Space Studies Institute, Princeton, NJ, 1994.

Repic, E., Richter, P., and Roy, C., The Lunar Resource Base: Stepping Stone to Mars, *43rd Congress of the International Astronautical Federation, Washington, D.C., 1992*, IAF 92-0538.

Savage, M.T., *The millennial project: Colonizing the galaxy in eight easy steps*. Little, Brown & Co., Boston, 1992.

Snow, W. *et al.*, *Preliminary design of a superconducting quenchgun for launching one ton payloads from the lunar surface* (Final report). Large Scale Programs Institute/EML Research, Inc., Hudson, MA, May, 1988.

Space Research Associates, Inc. (Seattle, WA), *Near-term non-terrestrial materials usage in solar power satellites: Final Report*. Space Studies Institute, Princeton, NJ, August 22nd, 1989.

Van Pelt, M., *Space Tourism*. Springer/Praxis, Chichester, U.K., 2005.

Veverka, J. *et al.*, NEAR's flyby of 253 Mathilde: Images of a "C" asteroid. *Science*, **278**, 2109–2114, December 19th, 1997.

Webb, S., *Where Is Everybody?* Copernicus/Praxis, New York, 2002.

Wright, J.L., *Space Sailing*. Gordon & Breach Science Publishers, 1992.

Yeomans, D.K. *et al.*, Estimating the mass of Asteroid 253 Mathilde from tracking data during the NEAR flyby. *Science*, **278**, 2106–2109, December 19th, 1997.

Zubrin, R., *The case for Mars*. Free Press, New York, 1996.

Index

Printing: Mercedes-Druck, Berlin
Binding: Stein+Lehmann, Berlin